AF Handbook 10-644

Survival Evasion Resistance Escape (SERE) Operations

27 March 2017
DEPARTMENT OF THE AIR FORCE

BY ORDER OF THE
SECRETARY OF THE AIR FORCE

AIR FORCE HANDBOOK 10-644
27 March 2017

Operations

SURVIVAL EVASION RESISTANCE ESCAPE
(SERE) OPERATIONS

COMPLIANCE WITH THIS PUBLICATION IS MANDATORY

ACCESSIBILITY: Publications and forms are available on e-Publishing website at www.e-publishing.af.mil for downloading or ordering.
RELEASABILITY: There are no releasability restrictions on this publication.

OPR: HAF/A3XX (CMSgt Filby) Certified by: HAF/A3 (Lt Gen MARK C. NOWLAND)
Supersedes: AFH10-644, 21 March 2017 Pages: 644

This handbook describes the various environmental conditions affecting human survival, and describes isolated personnel (IP) activities necessary to survive during successful evasion or isolating events leading to successful recovery. It is the fundamental reference document providing guidance for any USAF service member who has the potential to become isolated; deviations require sound judgment and careful consideration. This publication provides considerations to be used in planning and execution for effective mission accomplishment of formal USAF Survival, Evasion, Resistance, and Escape (SERE) training, environmentally specific SERE training, and combat survival continuation training programs. The tactics, techniques, and procedures in this publication are recognized best practices presenting a solid foundation to assist USAF service members to maintain life and return with honor from isolating events. This handbook also applies to US Air Force Reserve and Air National Guard units and members. Refer recommended changes and questions about this publication to the office of primary responsibility (OPR) using the AF Form 847, *Recommendation for Change of Publication*; route the AF Form 847 from the field through major command (MAJCOM) publications/forms managers. Ensure that all records created as a result of processes prescribed in this publication are maintained IAW Air Force Manual (AFMAN) 33-363, Management of Records, and disposed of IAW the Air Force Records Disposition Schedule (RDS) in the Air Force Records Information Management System (AFRIMS). The use of the name or mark of any specific manufacturer, commercial product, commodity, or service in this publication does not imply endorsement by the Air Force. See **Attachment 1** and **Attachment 2** for glossary and acronyms.

Table of Contents

1. Mission 1
1.1. IP Mission 2
1.2. Goals 2
1.3. Survival 3
1.4. Decisions 3
1.5. Elements 3
2. Conditions Affecting Survival 4
2.1. Environmental Conditions 5
2.2. The IP's Condition 7
2.3. Duration-The Time Condition 10
2.4. Sociopolitical Condition 11
2.5. Chemical, Biological, Radiological, Nuclear and Explosive (CBRNE) Conditions ... 11
3. The IP's Needs 12
3.1. Maintaining Life 12
3.1.1. Personal Protection 12
3.1.2. Sustenance 15
3.1.3. Health (Physical and Psychological) 15
3.1.4. Travel 16
3.2. Maintaining Honor 17
3.3. Returning 17
4. The Psychological Aspects of SERE 18
4.1. Captivity 18
4.2. The Will to Survive 19
4.3. Overcoming Stress 21
4.4. Crisis Period 21
4.5. The Coping Period 22
4.6. Situation Awareness 22
4.7. Attitude 23
4.8. Contributing Factors 24
4.9. Survival Stresses 25
4.10. Hunger 28
4.11. Frustration 29
4.12. Fatigue 30
4.13. Maintaining Group Morale and Efficiency 31
4.14. Sleep Deprivation 31
4.15. Isolation 32
4.16. Insecurity 32
4.17. Depression 33
4.18. Emotional Reactions 33
4.19. Anxiety 36
4.20. Panic 36
4.21. Hate 36
4.22. Resentment 36
4.23. Anger 37
4.24. Impatience 37
4.25. Dependence 37
4.26. Loneliness 38
4.27. Boredom 38
4.28. Hopelessness 38

- 4.29. Summary 39
- 5. SERE Medicine 40
- 5.1. Health Preservation 40
- 5.2. Basic Health Care 42
- 5.3. Clothing and Bedding 45
- 5.4. Illness 46
- 5.5. Injury 49
- 5.6. Environmental Injury 62
- 5.7. Marine Animals that Bite 71
- 5.8. Marine Animals that Sting 73
- 5.9. Marine Animals that Puncture 75
- 5.10. Medical Plants for SERE Scenarios 79
- 5.11. Conclusions 86
- 6. Weather 87
- 6.1. Knowledge of Weather 87
- 6.2. Atmosphere 87
- 6.3. Elements Affecting Weather 87
- 6.4. Storms 89
- 6.5. Weather Forecasting 92
- 6.6. Summary 97
- 7. Environment 98
- 7.1. Terrain 98
- 7.2. Life Forms 100
- 7.3. Climate 101
- 7.4. Effects of Climate on Terrain 102
- 7.5. Effects of Terrain on Climate 103
- 7.6. Effects of Climate and Terrain on Life Forms 104
- 8. Global Climate Characteristics 106
- 8.1. Tropical Climates 107
- 8.2. Dry Climates 115
- 8.3. Warm Temperate Climates 120
- 8.4. Snow and Ice Climates 129
- 9. Open Seas 135
- 9.1. Seas are a Large Percentage of the Earth's Surface 135
- 9.2. Seas are Geographically Aligned along Distinct Boundaries 135
- 9.3. Ocean Climatic Conditions 135
- 9.4. Procuring Drinking Water on the Open Seas 137
- 9.5. Shelter for Open Seas 138
- 9.6. Life Forms 138
- 9.7. Traveling on Open Seas 139
- 9.8. Physical Considerations 144
- 9.9. Life Preserver Use 146
- 9.10. Raft Procedures 147
- 9.11. Making Landfall 155
- 9.12. Methods of Getting Ashore 157
- 10. Local People 159
- 10.1. Contact with People 159
- 10.2. IP Behavior 160
- 10.3. Political Allegiance 161
- 10.4. Population in Built-Up Areas 161
- 10.5. Summary 162

11. Proper Body Temperature	163
11.1. Optimum Core Temperature	163
11.2. Water as a an Effective Way to Transfer Body Heat	164
11.3. Factors Which Affect Survival Time	164
11.4. Factors that can Transfer Body Heat	165
12. Clothing	168
12.1. Protection	168
12.2. Clothing Materials	168
12.3. Insulation Measurement	170
12.4. Clothing Wear in Snow and Ice Areas	174
12.5. Clothing in the Summer Arctic	177
12.6. Clothing at Sea	179
12.7. Anti-exposure Garments	179
12.8. Warm Oceans	181
12.9. Tropical Climates	181
12.10. Dry Climates	181
12.11. Care of the Feet	182
13. Shelter	186
13.1. Shelter Considerations	186
13.2. Principles of Shelter Locations and Types	187
13.3. Types of Shelters	190
13.4. Shelter for Tropical Areas	202
13.5. Shelters for Hot and Dry Climates	203
13.6. Shelters for Snow and Ice Areas	206
13.7. Tree-Line Areas	208
13.8. General Construction Techniques	211
13.9. Arctic or Cold Weather Shelter Living Considerations	212
13.10. Summer Considerations for Arctic and Arctic-Like Areas	212
13.11. Maintenance and Improvements	213
14. Firecraft	215
14.1. Considerations	215
14.2. Elements of Fire	215
14.3. Fire Site Preparation	218
14.4. Firecraft Tips	219
14.5. Fire Making with Matches (or Lighter)	219
14.6. Heat Sources	221
14.7. Other Methods of Fire Starting	228
14.8. Burning Aircraft/Vehicle Fuel	228
14.9. Fire Lays	229
15. Equipment	231
15.1. Issued Equipment	231
15.2. Parachute	241
15.3. Other Improvised Equipment	247
15.4. Ropes and Knots	250
15.5. Personal Survival Kit	270
16. Water	271
16.1. Water Sources	271
16.2. Locating and Procuring Water	273
16.3. Water in Snow and Ice Areas	274
16.4. Water in Tropical Areas	276
16.5. Water in Dry Areas	280

16.6. Preparation of Water for Consumption	283
17. Food	286
17.1. Nutrition	286
17.2. Food	289
17.3. Animal Food	290
17.4. Plant Food	309
17.5. Food in Tropical Climates	320
17.6. Food in Dry Climates	330
17.7. Food in Snow and Ice Climates	334
17.8. Food on the Open Seas	343
17.9. Preparing Game Food	346
17.10. Preparing Plant Food	353
17.11. Cooking	356
17.12. Preserving Food	359
18. Land Navigation	362
18.1. Maps	362
18.2. Aeronautical Charts	365
18.3. Marginal Information	369
18.4. Topographic Map Symbols and Colors	374
18.5. Coordinate Systems	375
18.6. Elevation and Relief	382
18.7. Representative Fraction (RF)	388
18.8. Graphic (Bar) Scales	388
18.9. Protractors	390
18.10. Map Orientation	392
18.11. Determining Cardinal Directions Using Field Expedients	394
18.12. Determining Specific Position	398
18.13. Determining Specific Location without a Compass	398
18.14. Dead Reckoning	399
18.15. Position Determination	401
18.16. The Compass and Its Uses	411
18.17. Using a Map and Compass, and Expressing Direction	414
19. Land Travel	418
19.1. Decision to Stay or Travel	418
19.2. Travel	420
19.3. Land Travel Techniques	421
19.4. Glaciers and Glacial Travel	427
19.5. Snow and Ice Areas	432
19.6. Dry Climates	434
19.7. Tropical Climates	437
19.8. Forested Areas	438
19.9. Mountain Walking Techniques	438
19.10. Burden Carrying	442
19.11. Rough Land Travel and Evacuation Techniques	446
19.12. Specialized Knots for Climbing and Evacuation	446
19.13. Seat Harness	450
19.14. Route Selection	451
19.15. Dangers to Avoid	451
19.16. Climbing	452
19.17. Overland Snow Travel	459
19.18. Snow and Ice Climbing Procedures and Techniques	461

19.19. Evacuation Principles and Techniques	464
20. River Travel	468
20.1. River Travel	468
20.2. Using Safe Judgment and Rules for River Travel	468
20.3. River Hydraulics	478
20.4. Emergency Situations	488
20.5. Improvised Rafts	492
20.6. Fording Streams	495
21. Signaling and Communication	496
21.1. Communications Equipment	496
21.2. Emergency Equipment	496
21.3. Visualizing Emergency Development	497
21.4. Furnishing Information in Permissive Environments	497
21.5. Furnishing Information in Non-Permissive Environments	497
22. Recovery	511
22.1. IP's Role	511
22.2. National Search and Rescue (SAR) Plan	511
22.3. IP Responsibilities	512
22.4. Recovery Site	513
22.5. Recovery Procedures	513
22.6. Preparations for Open Seas Recovery	520
22.7. Reintegration	520
23. Evasion	522
23.1. Convert Survival	522
23.2. Definitions	522
23.3. Five Phases of Evasion	522
23.4. Value of Evasion	524
23.5. Preparation	527
23.6. Evasion Principles	529
23.7. Camouflage	532
23.8. Types of Observation	532
23.9. Preventing Detection	534
23.10. Factors of Recognition	534
23.11. Principles and Methods of Camouflage	538
23.12. Concealment in Various Geographic Areas	546
23.13. Concealment Factors for Areas Other Than Temperate	549
23.14. Camouflage and Concealment Techniques for Shelters	551
23.15. Firecraft under Evasion Conditions	554
23.16. Sustenance for Evasion	556
23.17. Security	562
23.18. Evasion Movement	564
23.19. Observing Terrain	565
23.20. Evasion Considerations	566
23.21. Movement Techniques Which Limit the Potential for Detection of an Evader (Single)	567
23.22. Movement Techniques Which Limit the Potential for Detection of Evaders (Group)	571
23.23. Rally Points	576
23.24. Barriers to Evasion Movement	577
23.25. Evasion Aids	586
23.26. Non-Permissive Recovery	591
23.27. Ground Rescue	593
23.28. Water Recovery	594

24. Urban	595
24.1. Key Components	595
24.2. Five Phases of Urban Evasion	601
24.2.1. Immediate Action and Initial Movement	601
24.2.2. Movement	603
24.2.3. Recovery	606
24.2.4. BLISS	607
24.2.5. Disguise	609
24.2.6. Urban Navigation	614
24.2.7. Urban Barriers	618
24.2.8. Structures	620
24.2.9. Subsurface	621
24.2.10. Ground Vehicle	624
24.2.11. Urban Foraging	631
Attachment -1 –Glossary of References and Supporting Information	634
Attachment -2 – Acronyms	639

Air Force Handbook 10-644
Survival, Evasion, Resistance, and Escape Operations

THE ELEMENTS OF SURVIVING

Chapter 1

MISSION

1. Mission

An isolating event ends one mission-for the individual, but starts another-to successfully return from the event. Are they prepared? Can they handle the new mission, not knowing what it entails? Unfortunately, many Isolated Personnel (IP) are not fully aware of this new mission or are not fully prepared to carry it out. All instructors teaching survival, evasion, resistance, and escape (SERE) must prepare the IP to face and successfully complete this new mission (Figure 1-1).

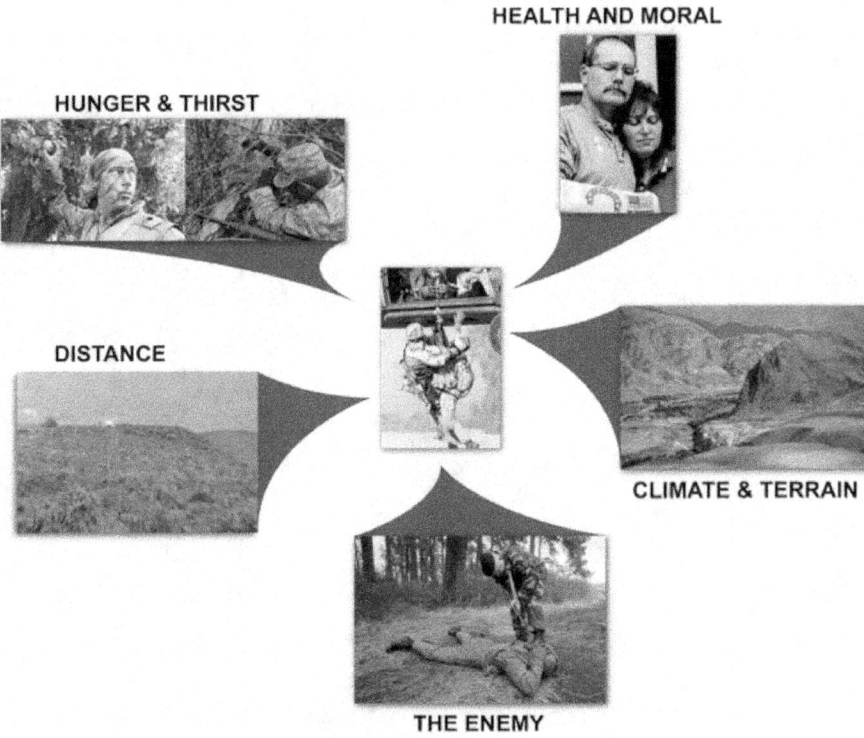

Figure 1-1 Elements of Surviving

1.1. IP Mission

The moment an individual or team encounters an isolating event, the assigned mission is to: "return to friendly control without giving aid or comfort to the enemy, to return early and in good physical and mental condition."

1.1.1. Friendly Doesn't Always Mean Friendly

On first impressions, "Friendly Control" seems to relate to a combat situation. However, even in peacetime, the environment may be quite hostile. Imagine a pilot parachuting into the arctic when it is minus 40°F or an explosive ordnance disposal crew separated by the detonation of a roadside bomb in the desert where temperatures may soar above 120°F; neither of these situations is particularly agreeable. The possibilities for encountering hostile conditions affecting human survival are endless. IP who egress an aircraft or are isolated from team mates on the ground may confront situations difficult to endure.

1.1.2. Without Giving Aid or Comfort
The second segment of the mission, "without giving aid or comfort to the enemy," is directly related to a combat environment. This part of the mission may be most effectively fulfilled by following the moral guide - the Code of Conduct. Remember, however, that the Code of Conduct is useful to an IP at all times and in all situations. Moral obligations apply to the peacetime situation as well as in wartime.

1.1.3. Return Early and in Good Physical and Mental Condition

The final phase of the mission is "to return early and in good physical and mental condition." A key factor in successful completion of this part of the mission may be the will to survive. This will is present, in varying degrees, in all human beings. Although successful survival is based on many factors, those who maintain this important attribute will increase their chances of success.

1.2. Goals

Categorizing this mission into organizational components, the three goals of an IP are to maintain life, maintain honor, and return. SERE training courses provide training in the skills, knowledge, and attitudes necessary for an IP to successfully perform the fundamental survival goals shown in Figure 1-2.

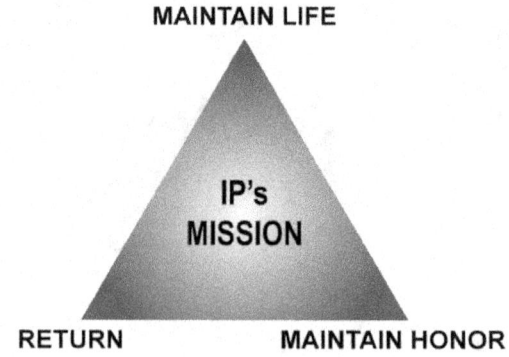

Figure 1-2 IP Goals

1.3. Survival

Surviving is extremely stressful and difficult. The IP may be constantly faced with hazardous and difficult situations. The stresses, hardships, and hazards (typical of an isolating event) are caused by the cumulative effects of existing conditions (see chapter 2 pertaining to conditions affecting survival.) Maintaining life and honor and returning, regardless of the conditions, may make surviving difficult or unpleasant. The IP mission forms the basis for identifying and organizing the major needs of an IP, so the IP can reach their goals. (See IP needs in chapter 3).

1.4. Decisions

The decisions an IP makes and the actions taken in order to survive determine their prognosis for surviving.

1.5. Elements

The three primary elements of the IP's mission are: the conditions affecting survival, IP needs, and the means for surviving.

Chapter 2

CONDITIONS AFFECTING SURVIVAL

2. Conditions Affecting Survival

Five basic conditions affect every isolating event (Figure 2-1). These conditions may vary in importance or degree of influence from one event to another and from individual to individual. At the onset, these conditions can be considered to be neutral-being neither an advantage nor a disadvantage for the IP. The IP may succumb to their effects-or use them to their best advantage. These conditions exist in each isolating event, and they will have great bearing on the IP's every need, decision and action.

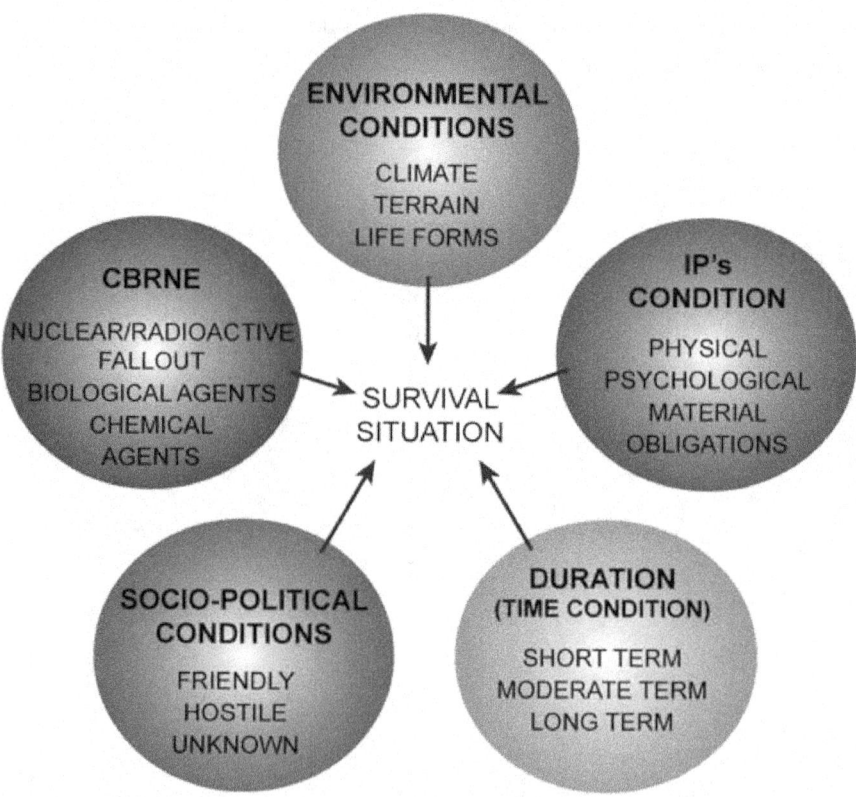

Figure 2-1 Five Basic Conditions

2.1. Environmental Conditions

Climate, terrain, and life forms are the basic components of all environments. These components can present special problems for the IP. Each component can be used to the IP's advantage. Knowledge of these conditions will contribute to the success of the IP's mission.

2.1.1. Climate

Temperature, moisture, and wind are the basic climatic elements. Extreme cold or hot temperatures, complicated by moisture (rain, humidity, dew, snow, etc.) or lack of moisture, and the possibility of wind, may have a life threatening impact on the IP's needs, decisions, and actions. The primary concern, resulting from the effects of climate, is the need for personal protection (consisting of clothing, shelter and fire). Climatic conditions also have a significant impact on other aspects of survival (for example, the availability of water and food, the need and ability to travel, recovery capabilities, physical and psychological problems, etc.) (Figure 2-2).

U.S. Army photo by Cpl. Rich Mattingly

Figure 2-2 Harsh Climates

2.1.2. Terrain

Mountains, prairies, hills, and lowlands, are only a few examples of the variety of land forms which describe "terrain." Each of the land forms has a different effect on an IP's needs, decisions, and actions. An IP may find a combination of several terrain forms in a given situation. The existing terrain will affect the IP's needs and activities in such areas as travel, recovery, sustenance and to a lesser extent, personal protection. Depending on its form, terrain may provide security and concealment for an evader, cause travel to be easy or difficult, provide protection from cold, heat, moisture, wind, or chemical, biological, radiological, nuclear and explosive (CBRNE) conditions, or make surviving a seemingly impossible task (Figure 2-3).

Figure 2-3 Terrain

2.1.3. Life Forms

When considering an isolating event, there are two basic life forms other than human/plant life and animal life. (NOTE: The special relationship and effects of people on the isolating event are covered separately). Geographic areas are often identified in terms of the abundance of life (or lack thereof). For example, the barren arctic or desert, primary (or secondary) forests, the tropical rain forest, the polar ice cap, etc., all produce images regarding the quantities of life forms. These examples can have special meaning not only in terms of the hazards or needs they create, but also in how an IP can use available life forms (Figure 2-4).

Figure 2-4 Life Forms

2.1.3.1. Plant Life

There are hundreds of thousands of different types and species of plant life. In some instances, geographic areas are identified by the dominant types of plant life within that area. Examples of this are savannas, tundra and deciduous forests. Some species of plant life can be used advantageously by an IP - if not for the food or the water, then for improvising camouflage, shelter or providing for other needs.

2.1.3.2. Animal Life

Reptiles, amphibians, birds, fish, insects, and mammals are life forms which directly affect an IP. These creatures affect the IP by posing hazards (which must be taken into consideration) or by satisfying needs.

2.2. The IP's Condition

The IP's condition and the influence it has in each isolating event are often overlooked. The primary factors which constitute the IP's condition can best be described by the four categories shown in Figure 2-5. IP must prepare themselves in each of these areas before each mission, and be in a state of "constant readiness" for the possibility of an isolating event. IP must be aware of what role their condition plays both before and during the isolating event.

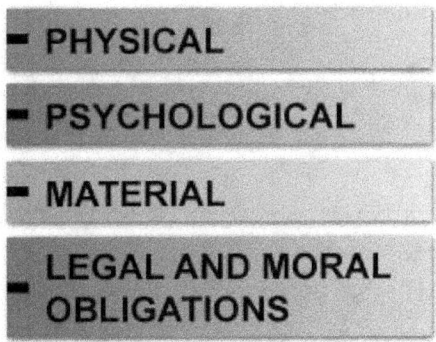

Figure 2-5 IP's Condition

2.2.1. Physical

The physical condition and fitness level of the IP are major factors affecting survivability. IP who are physically fit will be better prepared to face isolating events than those who are not. Further, an IP's physical condition (injured or uninjured) during the initial phase of an isolating event will be a direct result of circumstances surrounding the event. In short, high levels of physical fitness and good post-event physical condition will enhance an IP's ability to cope with such diverse variables as:

- Temperature extremes.
- Rest or lack of it.
- Water availability.
- Food availability.

- Long term isolating events.

2.2.1.1. Physical Weakness

During long term isolating event physical weakness may increase as a result of nutritional deficiencies, disease, lack of adequate rest, etc.

2.2.2. Psychological

The IPs' psychological state greatly influences their ability to successfully return from an isolating event.

2.2.2.1. Psychological Effectiveness

Psychological effectiveness in an isolating event (including captivity) results from effectively coping with the following factors which may occur naturally or be induced by coercive manipulation:

- Initial shock.
- Pain.
- Hunger.
- Thirst.
- Cold or Heat.
- Frustration.
- Fatigue (including Sleep Deprivation).
- Isolation - Includes forced (captivity) and the extended duration of any episode.
- Insecurity - Induced by anxiety and -self-doubts.
- Loss self-esteem - Most often induced by coercive manipulation.
- Loss of self-determination - Most often induced by coercive manipulation.

2.2.2.2. Depression - Mental "Lows"

An IP may experience emotional reactions during an isolating event due to the previously stated factors, previous life experiences (including training) and the IP's psychological tendencies. Emotional reactions commonly occurring in isolating (including captivity) events include:

- Boredom
- Loneliness
- Impatience
- Dependency
- Humiliation
- Resentment
- Anger

- Hate
- Anxiety
- Fear
- Panic

2.2.2.3. Crisis and Coping Phases

Psychologically isolating events may be divided into "crisis" phases and "coping" phases. The initial crisis period will occur at the onset of the isolating event. During this initial period, "thinking" as well as "emotional control" may be disorganized. Judgment is impaired and behavior may be irrational (possibly to the point of panic). Once the initial crisis is under control, the coping phase begins and the IP is able to respond positively to the event. Crisis periods may well recur, especially during extended events (evasion or captivity). An IP must strive to control if avoidance is impossible.

2.2.2.4. Will to Survive - The Most Important Tool

The most important psychological tool that will affect the outcome of an isolating event is the will to survive. Without it, the IP is surely doomed to failure-a strong will is the best assurance of survival.

2.2.3. Material

At the beginning of a isolating event, the IP's clothing and the contents of available equipment, survival kits, salvageable resources from the vehicle, parachute, – aircraft, or environment are the sum total of the IP's material assets. Adequate equipment familiarity and pre-mission preparations are required (and must be stressed during training). Although important all the time, once the isolating event has started, special attention must be given to the care, use, and storage of all materials to ensure they continue to be serviceable and available. Items of clothing and equipment should be selectively augmented with improvised items.

2.2.3.1. Appropriate Clothing

Clothing appropriate to anticipated environmental conditions (on the ground) should be worn or carried as space and mission permit.

2.2.3.2. Appropriate Equipment

The equipment available to an IP affects all decisions, needs, and actions. The IP's ability to improvise may provide ways to meet some needs.

2.2.4. Legal and Moral Obligations

An IP has both legal and moral obligations and responsibilities. Members of the military service will find their legal obligations expressly identified in the Geneva Conventions, Uniform Code of Military Justice (UCMJ), Department of Defense (DOD) and Air Force directives and policies. Before deploying to their operational environment, the potential IP should understand the rules of engagement and their legal status in all likely types of isolating events. Moral obligations are expressed in the Code of Conduct (Figure 2.6). Legal and moral obligations are addressed in greater detail in Chapter 23.

THE CODE OF CONDUCT

I

I am an American, fighting in the forces which guard my country and our way of life. I am prepared to give my life in their defense.

II

I will never surrender of my own free will. If in command, I will never surrender the members of my command while they still have the means to resist.

III

If I am captured, I will continue to resist by all means available. I will make every effort to escape and aid others to escape. I will accept neither parole nor special favors from the enemy.

IV

If I become a prisoner of war, I will keep faith with my fellow prisoners. I will give no information or take part in any action which might be harmful to my comrades. If I am senior, I will take command. If not, I will obey the lawful orders of those appointed over me and will back them up in every way.

V

When questioned, should I become a prisoner of war, I am required to give name, rank, service number, and date of birth. I will evade answering further questions to the utmost of my ability. I will make no oral or written statements disloyal to my country and its allies or harmful to their cause.

VI

I will never forget that I am an American, fighting for freedom, responsible for my actions, and dedicated to the principles which made my country free. I will trust in my God and in the United States of America.

Figure 2-6 Code of Conduct

2.2.5. Other Responsibilities that Influence Behavior

Other responsibilities influence behavior during isolating events and influence the will to survive. Examples include feelings of obligation or responsibilities to family, self, and spiritual beliefs.

2.2.5.1. Perception of Influence

An IP's individual perception of responsibilities influences survival needs and affects the psychological state of the individual both during and after the isolating event. These perceptions will be reconciled either consciously through rational thought or subconsciously through attitude changes. Training specifically structured to foster and maintain positive attitudes provides a key asset to survival.

2.3. Duration-The Time Condition

The duration of the isolating event has a major effect upon the IP's needs. Every decision and action will be driven in part by an assessment of when recovery or return is probable. Air superiority, rescue capabilities, the distances involved, climatic conditions, the ability to locate the IP, or potential for or actual captivity are major factors which directly influence the duration (time condition) of the isolating event. An IP can never be certain that rescue is imminent and must prepare as though it is not.

2.4. Sociopolitical Condition

Sociopolitical conditions are the social, cultural, and behavioral factors characterizing the relationships and activities of the population of a specific region or operational environment. The local people, their social customs, cultural heritage, politics, and attitudes will affect the IP's status within the spectrum of conflict. Due to these sociopolitical differences, the interpersonal relationship between the IP and any people with whom contact is established is crucial to surviving. To an IP, the attitude of the people contacted will be friendly, hostile, or unknown.

2.4.1. Friendly People

The IP who comes into contact with friendly people, or at least those willing (to some degree) to provide aid, is indeed fortunate. Immediate return to home, family, or home station, however, may be delayed. When in direct association with even the friendliest of people, it is essential to maintain their friendship. These people may be of a completely different culture in which a commonplace American habit may be a gross and serious insult. In other instances, the friendly people may be active insurgents in their country and constantly in fear of discovery. Every IP action, in these instances, must be appropriate and acceptable to ensure continued assistance.

2.4.2. Hostile People

A state of war need not exist for an IP to encounter hostility in people. Contact with hostile people must be avoided. If captured, regardless of the political or social reasons and the IP's legal status, the IP must make all efforts to adhere to the Code of Conduct and the legal obligations of the UCMJ, the Geneva Conventions, and DOD and USAF policy.

2.4.3. Unknown People

The IP should consider all factors before contacting unknown people. Some primitive cultures and closed societies still exist in which outsiders are considered a threat. In other areas of the world, differing political and social attitudes can place an IP "at risk" in contacting unknown people.

2.5. Chemical, Biological, Radiological, Nuclear and Explosive (CBRNE) Conditions

CBRNE conditions may occur during combat operations. CBRNE events create life-threatening conditions from which an IP needs immediate protection. The longevity of CBRNE conditions further complicates an IP's other needs, decisions, and actions.

Chapter 3

THE IP'S NEEDS

3. The IP's Needs

The three fundamental goals of an IP are to 1) maintain life, 2) maintain honor, and 3) return. This may be further divided into nine basic needs which include: 1) personal protection, 2) sustenance, 3) health, 4) travel, 5) communications, 6) recovery, 7) evasion, 8) resistance, and 9) escape. Meeting the individual's needs during the isolating event is essential to achieving the IP's fundamental goals (Figure 3-1).

Figure 3-1 IP Needs

3.1. Maintaining Life

Four elementary needs of an IP in any situation which are categorized as the integral components of maintaining life are: 1) personal protection, 2) sustenance, 3) health and 4) travel.

3.1.1. Personal Protection

The human body is comparatively fragile. Without protection, the effects of environmental conditions (climate, terrain and life forms) and CBRNE conditions may be fatal. The IP's primary defenses against the effects of the environment are protective clothing, equipment, shelter, and fire (Figure 3-2).

Figure 3-2 Personal Protection

3.1.1.1. Clothing

The need for adequate clothing and its proper care and use cannot be overemphasized. The human body's tolerance for temperature extremes is very limited with clothing being the first line of defense. However, its ability to regulate heating and cooling is extraordinary. The availability of clothing and its proper use is extremely important to an IP, enhancing their body's ability to regulate temperature. Clothing also provides excellent protection against the external effects of alpha and beta radiation, and may serve as a shield against the external effects of some chemical or biological agents.

3.1.1.2. Equipment

Survival equipment is designed to aid IP throughout their event. It must be cared for to maintain its effectiveness. Items found in a survival kit, ground/water vehicle, or aircraft can be used to help satisfy the nine basic needs. Quite often, however, an IP must improvise to overcome an equipment shortage, failure, or deficiency.

3.1.1.3. Shelter

The IP's need for shelter is threefold-as a place for protection from the conditions (environmental and socio-political), safety from hazards, and is big enough for the IP and their equipment. The duration and location of the isolating event will have some effect on shelter choice. In areas that are warm and dry, the IP's need is easily satisfied using natural resting places. In cold climates, the criticality of shelter can be measured in minutes, and rest is of little immediate concern. Similarly, in areas of residual radiation, the criticality of shelter may also be measured in minutes (Figure 3-3).

Figure 3-3 Shelter

3.1.1.4. Fire

Fire serves many IP needs such as providing a source of heat to warm the body, dry clothing, purifying water, preparing food, and signaling (Figure 3-4).

Figure 3-4 Fire

3.1.2. Sustenance

Sustenance is the IP's need for water and food to maintain normal body functions and to provide strength, energy, and endurance to overcome the physical stresses of survival (Figure 3-5).

3.1.2.1. Water

The IP must be constantly aware of the body's primary and continuing need for water.

3.1.2.2. Food

Less important than water initially, the need for food receives little attention during the first hours of an isolating event. Additionally, the human body must have adequate amounts of water to process any food appropriately. The successful IP will need to overcome aversions to food before physical or psychological deterioration sets in.

Figure 3-5 Evasion Sustenance

3.1.3. Health (Physical and Psychological)

The IP must be the doctor, nurse, medic, and psychologist. Self-aid and buddy care may be the IP's sole recourse.

3.1.3.1. Physical Health

Attention to personal protection, sanitation, and personal hygiene are major factors in preventing physical, morale, and attitudinal problems. Safety must be foremost in the mind of the IP. One miscalculation with a knife or tool can result in self-inflicted injury or death. In the event of injury or illness, the IP's existence may depend on the ability to perform self-aid or buddy care. In many instances, common first aid procedures will suffice; in others, more primitive techniques will be required (Figure 3-6). The need for cleanliness in the treatment of injuries and illness is self-evident. Proper use of clothing and equipment can prevent serious medical or physical problems.

Figure 3-6 Self Aid

3.1.3.2. Psychological Health

Perhaps the IP's greatest strength is a positive and optimistic attitude. An individual's ability to cope with psychological stresses will enhance successful survival. Optimism, determination, dedication, and humor, as well as many other psychological attributes, are all helpful for an IP to overcome psychological stresses. Refer to chapter 4 for more information on psychological health.

3.1.4. Travel

An IP may need to travel on land or water to meet any or all of their goals. In any isolating event, the IP must weigh the need to travel against capabilities and/or safety (Figure 3-7).

Figure 3-7 Travel

3.2. Maintaining Honor

The three needs associated with maintain honor are evasion, resistance, and escape.

3.2.1. Captivity- Worse than Evasion

Evasion will be one of the most difficult and hazardous situations an IP will face. However difficult and hazardous evasion may be necessary. Captivity is always worse. Evasion needs are specifically addressed in the Evasion chapter.

3.2.2. The Need to Resist Exploitation

The captive's need to resist exploitation is self-evident. This need is both a moral and a legal obligation. Resistance is much more than refusing to divulge information. Resistance is made up of two distinctly separate behaviors expected of the captive:

- Complying with legal and authorized requirements.
- Preventing exploitation. This includes the captive doing such things as disrupting enemy activity through resisting, refusing to divulge information, subtle harassment, and distracting enemy guards who could be used on the front lines. For a captive, escape is the ultimate form of resistance.

3.3. Returning The need to return involves communication, recovery, and/or traveling (on land or water). Communication requires a hands-on knowledge of signaling procedures and devices to report and locate the IP (Figure 3-8). The IP must be familiar with the various platforms, devices, and methods that may be used during recovery. The IP must also be familiar with the various methods and techniques for traveling overland and on the water to safely and accurately reach the recovery location, if necessary.

Figure 3-8 Return

Chapter 4

PSYCHOLOGICAL ASPECTS OF SERE

4. The Psychological Aspects of SERE The psychological aspects of SERE involve the will to survive, emotional reactions, and many other contributing factors. An IP's attitude and mental strength have a considerable influence on their ability to survive an isolating event. The will to survive is defined as the desire to live despite seemingly insurmountable mental and/or physical obstacles. The tools for survival are furnished by the military, the individual, and the environment. The training for SERE comes from training publications, instruction, and the individual's own efforts. But tools and training are not enough without a will to survive. In fact, the records prove that will alone have been the deciding factor in many survival cases. While these accounts are not classic examples of how to survive, they illustrate that a single-minded IP with a powerful will to survive can overcome most hardships. There are cases where people have eaten their belts for nourishment, boiled water in their boots to drink as broth, or have eaten human flesh - though this certainly wasn't their cultural instinct.

4.1. Captivity

Captive environments may vary, but none are good. The will to survive has historically been a very powerful motivator to overcome the worse captivity scenarios.

- Such was the case of Greg Williams, who was held hostage by an Al Qaeda linked terrorist in the Philippines (Figure 4-1). Mr. Williams' situation was unique. He had divested himself of all worldly goods, going to the Philippines for missionary work, but had changed who he was going to work with at the last minute, so no one in country really knew who he was or what he was doing there. He was captured by terrorist who thought of him as a source of income. Unfortunately, no one at "home" or at the mission knew where he was, as a result, there was no one to pay any kind of ransom. The terrorist had killed other hostages. His treatment by the terrorist just confirmed what he knew; his life expectancy was very short once they found out he was of no monetary value to them. Chained by the ankle inside a coral rock cave, he had tried to dig out the bolted chain from the cave's wall by using his fingers and chicken bones from his food, but had no luck. The use of the bones had led him to the only thing available to him, his own teeth! Mr. Williams figured he could pull out a tooth with a crown to dig into the wall; which is exactly what he did. His captors had beaten him knocking out and loosening several teeth; he found a loose tooth in his mouth with a crown and pulled it out. He used the yanked out tooth with a crown for approximately two days to dig around the bolt securing him to the coral wall. Unfortunately, a few days later while using the tooth to dig, his plan and "escape tool" were discovered by one of his captors. Just a little after his escape attempt was discovered, one of the captors aided Mr. Williams to escape. Mr. Williams' will to survive and determination led him to take extraordinary means to escape.

Figure 4-1 Al Qaeda Linked Terrorists

4.2. The Will to Survive

The will to survive is determination. It is a refusal to surrender.

- On April 18, 1995, Capt. Brian Udell, an F-15E fighter pilot and Capt. Dennis White, an F-15E weapons systems officer, assigned to Seymour-Johnson AFB, NC, were flying a routine training mission over the Atlantic (Figure 4-2). They lost control of their F-15E which plummeted straight for the earth. Just above 10,000 feet and exceeding 600 knots (nearly 700 mph), Udell gave the order: 'Bail out! Bail out! Bail out!' By the time the canopy blew off, White ejected at 4,500 feet. With the aircraft still picking up speed (more than 780 mph) Udell ejected at 3,000 feet.
  - Because of the high speed, Udell's parachute opening shock felt like he had been hit by a train. Udell felt some pain, but had no clue to the extent of his injuries. He began going through his post-ejection checklist. During his checklist he attempted to inflate his life preserver, but found it shredded. He figured he'd better reel in the life raft that automatically deploys during ejection to ensure he had some kind of flotation device when entering the water. That's when he discovered his left arm was injured. He hauled in the raft with his right arm and his teeth. Just about the time he got his functional hand on the raft, he hit the water, which is when his struggle both physically and mentally began. Stranded in the middle of the Atlantic Ocean with 5-foot waves and 17 mph winds making the 60 degree water feel like a giant ice chest, he knew he was in trouble. Udell had no life preserver, an injury rendered his left arm useless, and it was a pitch-black night. It would have been easy to surrender to the environment, but actually it was just the start of his struggle to survive. With his one good arm, he desperately clung to the side of a partially inflated life raft trying to climb in. Salt water made him painfully aware of open wounds - gashes, cuts, and scrapes - scattered over his broken body. The thought of his blood pouring into the sea and inviting sharks to

surround him sent another kind of chill up his spine. Newly motivated, he kicked his legs to assist his one good arm.

- Udell's lower limbs, from his left ankle and right knee down, felt as though they were barely attached. Seemingly held loosely together only by the skin, they proved useless as paddles - trailing lifelessly behind the upper portion of his legs. Kicking them through the water was like trying to stir a glass of ice tea with a wet noodle. Thoughts of death started to infiltrate his mind. Once in the life raft, Udell had immobilized his legs and his left arm, Udell searched his 6-foot-1 frame for other injuries. Finding nothing that appeared life-threatening, he went into prevent-shock mode. He drank some water out of an emergency pack that automatically releases during ejection, then tried to get warm. Udell spent four hours in the water before a Coast Guard helicopter found him. Using an emergency radio, he directed them to his location where he was rescued. Capt Udell's determination and faith provided him the strength and presence of mind to overcome his isolating event, use his training and equipment, and survive to contact rescue forces.

Figure 4-2 F-15E

4.2.1. Desert Environment Tests the Will to Survive

The desert environment has long been one to test the will and determination of any IP. An event showing a great deal of the will to survive was that of an Australian sergeant who bailed out of his aircraft, and landed in the desert with both legs injured. He used his parachute to bind up his wounded legs, but found he could not stand on his feet. Knowing his need for medical attention, the unlikelihood of rescue forces finding him, and his lack of supplies, he crawled dragging his legs for four days and nights, trying to affect his own recovery. The Sergeant found that at night he couldn't sleep, due to a combination of anxiety and pain. The Sergeant kept crawling, using the stars to aid him in navigation. During the day, he found himself partially shaded while crawling around the large rocks which blocked his path. He reached the point of total exhaustion on the fifth day and found that his mind was beginning to wander, blacking in and out, and hallucinating. On the fifth day he was found by a desert patrol. He had been without food or water for the entire period (Figure 4-3).

Figure 4-3 Will to Survive

4.2.2. Loss of Hope

The exact opposite of the will to survive is the lack of will or loss of hope. For example, in the Canadian wilderness, a pilot ran into engine trouble and chose to dead stick his plane onto a frozen lake rather than punch out. The pilot did a beautiful job and slid to a stop in the middle of the lake. He left the aircraft and examined it for damage. After surveying the area, he noticed a wooded shoreline only 200 yards away where food and shelter could be provided. Approximately halfway there, the pilot changed his mind and returned to the cockpit of his aircraft where he smoked a cigar, took out his pistol, and blew his brains out. Less than 24 hours later, a rescue team found him. Why did the pilot give up? Why was he unable to survive? Why did he take his own life?

4.3. Overcoming Stress

The ability of the mind to overcome stress and hardship becomes most apparent when there appears to be little chance of a person surviving. When there seems to be no escape from a situation, the will to survive enables a person to begin to win the battle of the mind. This mental attitude can bridge the gap between the crisis period and the coping period.

4.4. Crisis Period

The crisis period is the point at which the person realizes the gravity of the situation and understands that the problem will not go away. At this stage, action is needed. Most people will experience shock in this stage as a result of not being ready to face this new challenge. Most will recover control of their faculties, especially if they have been prepared through knowledge and training. Shock during a crisis is normally a response to being overcome with anxiety. Thinking will be disorganized. At this stage, direction will be required because the individual is being controlled by the environment. The person's center of control is external. In a group isolating event, a natural leader may appear who will direct and reassure the others. The greater the group's cohesiveness and discipline, the easier the group will deal with the situation. But if the situation continues to control the individual or the group, the response may be panic, behavior may be irrational, and judgment is impaired. In a solitary IP episode, the individual must gain control of the situation and respond constructively. In either case, IP must evaluate the situation and develop a plan of action. During the evaluation, the IP must determine the most critical needs to improve the chance of living and being rescued.

4.5. The Coping Period

The coping period begins after the IP recognizes the gravity of the situation and resolves to endure it rather than succumb. The IP must tolerate the effects of physical and emotional stresses. These stresses can cause anxiety which becomes the greatest obstacle to self-control and solving problems. Coping with the situation requires considerable internal control. For example, the IP must often subdue urgent desires to travel when that would be counterproductive and dangerous. A person must have patience to sit in a protective shelter while confronted with an empty stomach, aching muscles, numb toes, and suppressed feelings of depression and hopelessness. Those who fail to think constructively may panic. This could begin a series of mistakes which result in further exhaustion, injury, and sometimes death. Death comes not from hunger pains but from the inability to manage or control emotions and thought processes.

4.6. Situation Awareness

When a potential IP is required to make critical choices – usually at a fast pace – the vast majority of errors that occur are a direct result of failures in situation awareness (SA). SA can allow the IP to see likely answers to complex situations. The loss of SA, in some cases, causes the isolating event or at the very least, can complicate it.

4.6.1. Impact of Self Perception

SA is an IP's continuous perception of self (and their team/crew) in relation to the dynamic environment including their mission, vehicle (aircraft or ground vehicle), threats, resources, and the ability to forecast future events, and then execute tasks based on that perception and understanding.

4.6.2. Adaptation

SA, as we define it, is a specific brand of adaptation. This adaptation is the process by which an IP channels their knowledge and behavior to attain goals, tempered by the conditions and constraints imposed by the task environment. The continuous process of SA involves four actions:

1. Observe/Perceive: Observe, perceive, and recognize the available facts that comprise the individuals, situation, event, or environment. Examine and note activities, people, objects, and states. It includes using all of one's senses to gather data. Observing and perceiving can also include listening to one's inner intuition. SA is all about perceiving the dynamic state of the environment and making predictions from that perception. These predictions form the basis for decisions to perform tasks that support your goals.

2. Analyze/Interpret: Assess the observed facts as they relate to the situation. Include what was learned in past training, pre-deployment briefings, and study about the area or situation. Answer the question, "What does this mean to me?" Keep an open mind and

remain flexible. Pre-conceived expectations can reduce the ability to analyze accurately. Another skill is being able to choose which piece of information is important in any given situation. The IP must focus on the right information from their observations. As a guide, ask "What matters most for me now and in the future?"

3. Decide: Formulate a course of action based on sound judgment and the available information. Judgment is a process where choices are weighed by considering their risks verses their benefits and evaluation the possible consequences of acting on a given choice over the others. Sound judgment equals a good decision which equals effective action.

4. Execute & Modify: As stated before, the individual should attempt to anticipate and predict how the situation is likely to develop, understand their options and course(s) of action, and be prepared to modify those actions if necessary. However, while evading or during captivity, an IP often has to execute decisions under intense time pressure. Critical incidents often force a decision, especially those that must be made under extreme time pressure. If something has not gone according to plan, take corrective action following the same steps outlined above.

4.6.3. Thinking Ahead

This understanding enables the IP to think ahead, project the future state of the isolating event, and determine a course of action to meet, use, or mitigate the conditions affecting them. The more the IP anticipates accurately, the more efficient they become at meeting their needs and goals.

4.6.4. Situational Awareness

SA represents the cumulative effects of everything an IP is and does as applied to successful mission accomplishment to maintain life, maintain honor, and return.

4.7. Attitude

The IP's attitude is the most important element of the will to survive. With a positive mental attitude, almost anything is possible. The desire to live is sometimes based upon the feelings toward another person and/or thing. Love and hatred are two emotional extremes which have moved people to do exceptional things physically and mentally. The lack of a will to survive can sometimes be identified by the individual's lack of motivation to meet essential survival needs, emotional lack of control resulting in reckless, panic-like behavior, and a lack of self-esteem.

4.7.1. Controlling Emotions

Controlling emotions is essential to strengthen the will to survive during an emergency. The first step is to avoid a tendency to panic or lose self control. Sit down, relax, and analyze the situation rationally. Once thoughts are collected and thinking is clear, the next step is to make decisions. In normal living, people can avoid decisions and let others do their planning. But in an isolating situation, this will seldom work. Failure to decide on a course of action is actually a decision for inaction. This lack of decision making may result in capture or even death. However, decisiveness must be tempered with flexibility and planning for unforeseen circumstances. As an example, an aircrew member down in an arctic noncombat situation decides to construct a shelter for protection from the elements. The planning and actions must allow sufficient flexibility so the aircrew can monitor the area for indications of rescuers and be prepared to make contact - visually, electronically, or by any other means - with potential rescuers.

4.7.2. Tolerance

Tolerance will typically be a factor in isolating situations. An IP will have to deal with many physical and psychological discomforts, such as unfamiliar animals, insects, loneliness, and depression. IP are trained to tolerate uncomfortable situations. That training must be applied to deal with the stress of environments.

4.7.3. Overcoming Fear

IP in both combat and noncombat situations must face and overcome fears to strengthen the will to survive. These fears may be founded or unfounded, be generated by the IP's uncertainty or lack of confidence or based on the proximity of enemy forces. Indeed, fear may be caused by a wide variety of real and imagined dangers. Despite the source of the fear, IP must recognize fear and make a conscious effort to overcome it.

4.7.4. Optimism

Optimism is one of an IP's assets during an isolating event. IP must maintain a positive, optimistic outlook on their circumstance and how well they are doing. Setting achievable goals (simple to complex) and working toward them can help maintain optimism. Faith, hope, prayer, and meditation can all be helpful.

4.8. Contributing Factors

IP must recognize that coping with the psychological aspects of SERE are at least as important as handling the environmental factors. In virtually any isolating event, the IP will be in an environment that can support human life. The IP's problems will be compounded because they never really expected to become isolated in the desert, over the ocean, urban, or anywhere else. No matter how well prepared, the IP probably will never completely convince themselves that it can happen to them. However, the USAF is filled with historic records that show it can and does happen. Before an IP learns about the physical aspects of SERE, they must first understand that psychological problems may occur and that solutions to those problems must be found if the isolating event is to reach a successful conclusion (Figure 4-4).

Figure 4-4 Psychological Aspects

4.9. Survival Stresses

The emotional aspects associated with isolation must be completely understood just as survival conditions and equipment are understood. An important factor bearing on success or failure during an isolating event is the individual's psychological state. Maintaining an even, positive psychological state or outlook depends on the individual's ability to cope with or manage many factors. Some include:

- Understanding how various physiological and emotional signs, feelings, and expressions affect one's bodily needs and mental attitude.
- Managing physical and emotional reactions to stressful situations.
- Knowing individual tolerance limits, both psychological and physical.
- Exerting a positive influence on companions.

4.9.1. Adapting to Stress

Nature has endowed everyone with biological mechanisms which aid in adapting to stress. The bodily changes as a result of fear and anger, for example, tend to increase alertness and provide extra energy to either run away or fight. These and other mechanisms can hinder a person during an isolating event. For instance, an IP in a raft could cast aside reason and drink sea water to quench a thirst or IP in enemy territory, driven by hunger pangs, could expose themselves to capture when searching for food. These examples illustrate how normal reactions to stress could create problems for an IP.

4.9.2. Threats to Maintaining Life

Two of the gravest threats to maintaining life are concessions to comfort and apathy. Both threats represent attitudes which must be avoided. To survive, an IP must focus planning and effort on fundamental needs.

4.9.3. Comfort

Many people consider comfort their greatest need. Yet, comfort is not essential to human survival. IP must value life more than comfort, and be willing to tolerate heat, hunger, dirt, itching, pain, and any other discomfort. Recognizing discomfort as temporary will help IP concentrate on effective action.

4.9.4. Apathy

As the will to keep trying to survive lessens, drowsiness, mental numbness, and indifference will result in apathy. This apathy usually builds on slowly, but ultimately takes over and leaves an IP helpless. Physical factors can contribute to apathy. Exhaustion due to prolonged exposure to the elements, loss of body fluids (dehydration), fatigue, weakness, or injury are all conditions which can contribute to apathy. Proper planning and sound decisions can help an IP avoid these conditions. Finally, the IP must watch for signs of apathy in companions and help prevent it. The first signs are resignation, quietness, lack of communication, loss of appetite, and withdrawal from the group. Preventive measures could include maintaining group morale by planning, activity, and getting the organized participation of all members.

4.9.5. Common Survival Stressors

Many common stresses cause reactions which can be recognized and dealt with appropriately during isolating events. An IP must understand that stresses and reactions often occur at the same time. Although IP will face many stresses, the following common stresses will occur in virtually all survival episodes: pain, thirst, cold, and heat, hunger, frustration, fatigue, sleep deprivation, isolation, insecurity, loss of self-esteem, loss of self-determination, and depression.

4.9.6. Pain

Pain, like fever, is a warning signal calling attention to an injury or damage to some part of the body. Pain is discomforting but is not in itself harmful or dangerous. Pain can be controlled, and in an extremely grave situation, survival must take priority over giving in to pain.

- The biological function of pain is to protect an injured part by warning the individual to rest it or avoid using it. During an isolating event, the normal pain warnings may have to be ignored in order to meet more critical needs. People have been known to complete a fight with a fractured hand, to run on a fractured or sprained ankle, to land an aircraft despite severely burned hands, and to ignore pain during periods of intense concentration and determined effort. Concentration and intense effort can actually stop or reduce feelings of pain. Sometimes this concentration may be all that is needed to survive.

- Despite pain, an IP can move in order to live. Pain can be reduced by understanding its source and nature. Recognizing pain as a discomfort to be tolerated, concentrating on necessities such as thinking, planning, and keeping busy. When personal goals are maintaining life, honor, and returning, and these goals are valued highly enough, an IP can tolerate almost anything.

4.9.7. Thirst

The lack of water and its accompanying problems of thirst and dehydration are among the most critical problems facing IP. Although thirst indicates the body's need for water, it does not indicate how much water is needed. If a person drinks only enough to satisfy thirst, it is still possible to slowly dehydrate. Prevention of thirst and the more debilitating dehydration is possible if IP drink plenty of water any time it is available, and especially when eating (Figure 4-5). While prevention is the best way to avoid dehydration, virtually any degree of dehydration is reversible simply by drinking water. Thirst, like fear and pain, can be tolerated if the will to carry on, supported by calm, purposeful activity, is strong.

Figure 4-5 Thirst

4.9.8. Cold and Heat

The average normal body temperature for a person is 98.6°F. Individuals have survived body temperatures well below and well above normal, but consciousness is clouded and thinking numbed at a much smaller change. An increase/decrease just a few degrees above/below normal for any prolonged period may prove fatal. Any deviation from normal temperature, even as little as one or two degrees, reduces efficiency.

4.9.8.1. Cold

Cold is a serious stress since even in mild degrees it lowers efficiency. Extreme cold numbs the mind and dulls the will to do anything except get warm again. Cold numbs the body by lowering the flow of blood to the extremities, and results in sleepiness. IP have endured prolonged cold and dampness through exercise, proper hygiene procedures, shelter, and food (Figure 4-6). Wearing proper clothing and having the proper climatic survival equipment when in cold weather areas are essential to enhance survivability.

Figure 4-6 Cold

4.9.8.2. Heat

Just as numbness is the principal symptom of cold, weakness is the principle symptom of heat. Most people can adjust to high temperatures, whether in the heat of the desert during evasion or in an isolation box during captivity. It may take from two days to a week before circulation, breathing, heart action, and sweat glands are all adjusted to a hot climate. Heat stress also accentuates dehydration, which was discussed earlier. In addition to the problem of water, there are many other sources of discomfort and impaired efficiency, which are directly attributable to heat or to the environmental conditions in hot climates. Extreme temperature changes, from extremely hot days to very cold nights, occur in desert and plains areas. Proper use of clothing and shelters can decrease the adverse effects of such extremes.

4.9.9. Effects of Wind

Blowing wind, in hot summer, has been reported to irritate some IP. Wind can constitute an additional source of discomfort and difficulty in desert areas when it carries particles of sand and dirt. Protection against sand and dirt can be provided by tying a cloth around the head after cutting slits for vision.

4.9.9.1. Fear as a Result of Sandstorms Snowstorms

Acute fear has been experienced among IP in sandstorms and snowstorms. This fear results from both the terrific impact of the storm itself and its obliteration of landmarks showing direction of travel. Finding or improving shelter for protection from the storm itself is important.

4.9.10. Mirages and Illusions

Mirages and illusions distort visual perception but sometimes account for serious incidents. These illusions are common in desert areas. In the desert, distances are usually greater than they appear and, under certain conditions, mirages obstruct accurate vision. Inverted reflections are a common occurrence.

4.10. Hunger

A considerable amount of edible material which IP may not initially regard as food may be available during an isolating event. Early symptoms include impulsivity, irritability, hyperactivity, and possibly passiveness or compliance. Research has revealed no evidence of permanent damage or any decrease in mental efficiency from short periods of total fasting. Long term starvation recovery depends on physical re-nourishment as well as possible psychological treatment.

4.10.1. Starvation

The prolonged and rigorous Minnesota semi-starvation studies during World War II observed in detail the physical and psychological effects of prolonged, famine-like semi-starvation on healthy men and their subsequent rehabilitation from this condition. However, even today, knowledge of the biology of starvation in humans is still imperfect due to the great difficulty of obtaining reliable data from subjects undergoing severe food deprivation. Famines do not generally lend themselves to scientific investigation. A recent study done by the British Nutrition Foundation in 2006 has examined the literature on starvation, using Body Mass Index (BMI) to define the limits of human survival to starvation. Data included is based on normal weight subjects who died from starvation, famine or anorexia nervosa. The following behavioral changes were revealed:

- Dominance of the hunger drive over other drives.
- Lack of spontaneous activity.
- Tired and weak feeling.
- Inability to do physical tasks.
- Dislike of being touched or caressed in any way.
- Quick susceptibility to cold.
- Dullness of all emotional responses (fear, shame, love, etc.).

- Lack of interest in others - apathy.
- Dullness and boredom.
- Limited patience and self-control.
- Lack of a sense of humor.
- Moodiness - reaction of resignation.

4.10.2. Food Procurement

Frequently, in the tension or distress of some SERE episodes, hunger is forgotten. IP have gone for considerable lengths of time without food or awareness of hunger pains. An early effort should be made to procure and consume food to reduce the stresses brought on by food deprivation. Both the physical and psychological effects described are reversed when food and a protective environment are restored. Return to normal is slow and the time necessary for the return increases with the severity of starvation. If food deprivation is complete and only water is ingested, the pangs of hunger disappear in a few days, but even then the mood changes of depression and irritability occur. When the food supply is limited, even strong friendships are threatened. The individual tendency is still to search for food to prevent starvation and such efforts might continue as long as strength and self-control permit.

4.10.3. Aversions to Strange Foods

Food aversion may result in hunger. Adverse group opinion may discourage those who might try foods unfamiliar to them. In some groups, the barrier would be broken by someone eating the particular food rather than starving. The solitary individual has only personal predispositions to overcome and will often try strange foods.

4.10.4. Controlling Food Aversions

Controlling food aversions during isolation or captivity can help mitigate hunger for the IP. Food aversion is reflected in the excerpt from a survival account that follows: *"It had consisted of a small bowl of "soup" – really lukewarm water with a few pieces of leafy matter, some bitter-tasting kohlrabi, and a moldy chunk of bread full of weevils and hair. Lt Rodney Knutson had not been hungry, and the appearance and odor of the food had done nothing to awaken his appetite. Even so, it was his duty now to keep up his strength..."*

4.11. Frustration

Frustration occurs when one's efforts are stopped, either by obstacles blocking progress toward a goal or by not having a realistic goal. It can also occur if the feeling of self-worth or self-respect is lost.

4.11.1. Obstacles

A wide range of obstacles, both environmental and internal, can lead to frustration. Frustrating conditions often create anger, accompanied by a tendency to attack and remove the obstacles to goals.

4.11.2. Controlling Frustration

Frustration must be controlled by channeling energies into a positive and worthwhile obtainable goal. The IP should complete the easier tasks before attempting more challenging ones. This will not only instill self-confidence, but also relieve frustration.

4.12. Fatigue

In an isolating event, an IP must continually cope with fatigue and avoid the accompanying strain and loss of efficiency. An IP must be aware of the dangers of over-exertion. In many cases, an IP may already be experiencing strain and reduced efficiency as a result of other stresses such as heat or cold, dehydration, hunger, or fear. An IP must judge capacity to walk, carry, lift, or do necessary work, and plan and act accordingly. During an emergency, considerable exertion may be necessary to cope with the situation to ensure survival or escape. If an individual understands fatigue and the attitudes and feelings generated by various kinds of effort, that individual should be able to call on available reserves of energy when they are needed (Figure 4-7).

Figure 4-7 Fatigue

4.12.1. Avoiding Complete Exhaustion

An IP must avoid complete exhaustion which may lead to physical and psychological changes. Although a person should avoid working to complete exhaustion, in emergencies certain tasks must be done in spite of fatigue.

4.12.2. The Need for Rest

Rest is a basic factor for recovery from fatigue and is also important in resisting further fatigue. It is essential that the rest following fatiguing effort be sufficient to permit complete recovery; otherwise, the residual fatigue will accumulate and require longer periods of rest to recover from subsequent effort. During the early stages of fatigue proper rest provides a rapid recovery. This is true of muscular fatigue as well as mental fatigue. Sleep is the most complete form of rest available and is basic to recovery from fatigue.

4.12.3. The Need for Short Rest Breaks to Improve Morale

Short rest breaks during extended stress periods can improve total output and increase morale and motivation. There are four ways in which rest breaks are beneficial: 1) by providing opportunities for partial recovery from fatigue, 2) by helping to reduce energy expenditure, 3) by increasing efficiency by enabling a person to take maximum advantage of planned rest, and 4) by relieving boredom by breaking up the uniformity and monotony of the task.

4.12.4. Rest During Strenuous Activity

IP should rest before they show a definite decline in what they can do physically. When efforts are highly strenuous or monotonous, rest breaks should be more frequent. Rest breaks providing relaxation are the most effective. In mental work, mild exercise may be more relaxing. When work is monotonous, changes of activity, conversation, and humor are effective relaxants. In deciding on the amount and frequency of rest periods, the loss of efficiency resulting from longer hours of effort must be weighed against the absolute requirements of the isolating event.

4.12.5. Reducing Fatigue

Fatigue can be reduced by working smarter. An IP can do this in two practical ways:

1. Adjusting the pace of the effort. Balance the load, the rate, and the time period. For example, walking at a normal rate is a more economical effort than fast walking.
2. Adjusting the technique of work. The way in which work is done has a great bearing on reducing fatigue. Economy of effort is most important. Rhythmic movements suited to the task are best.

4.13. Maintaining Group Morale and Efficiency

A person should form an opinion of individual capacity based on actual experience. Likewise, a group leader must form an opinion of the capacities of fellow IP. Group support, cooperation, team work, and competent leadership are important factors in maintaining group morale and efficiency, thereby reducing stress and fatigue. An IP usually feels tired and weary before the physiological limit is reached. In addition, other stresses experienced at the same time - such as cold, hunger, fear, or despair - can intensify fatigue. The feeling of fatigue involves not only the physical reaction to effort, but also subtle changes in attitudes and motivation. Remember, a person has reserves of energy to cope with an important emergency even when feeling very tired.

4.14. Sleep Deprivation

The effects of sleep loss are closely related to those of fatigue. Sleeping at unaccustomed times, sleeping under strange circumstances (in a strange place, in noise, in light, or in other distractions), or missing part or all of the accustomed amount of sleep will cause a person to react with feelings of weariness, irritability, emotion tension, and some loss of efficiency. The extent of an individual's reaction depends on the amount of disturbance and on other stress factors which may be present at the same time.

4.14.1. Compensating for Sleep Loss

Strong motivation is one of the principal factors in helping to compensate for the impairing effects of sleep loss. Superior physical and mental conditioning, opportunities to rest, consumption of food and water, and companions has been shown to aid in the fight against the effects of sleep deprivation. If a person is in reasonably good physical and mental condition, sleep deprivation can be endured for five days or more without damage, although efficiency during the latter stages may be poor. A person must learn to get as much sleep and rest as possible. Restorative effects of sleep are felt even after short naps. In some instances, IP may need to stay awake. Activity, movement, conversation, eating, and drinking are some of the ways a person can stimulate the body to stay awake.

4.14.2. Sleepiness Comes in Waves during Sleep Depravity

When one is deprived of sleep, sleepiness usually comes in waves. A person may suddenly be sleepy immediately after a period of feeling wide awake. If this can be controlled, the feeling will soon pass and the person will be wide awake again until the next wave appears. As the duration of sleep deprivation increases, these periods between waves of sleepiness become shorter. The need to sleep may be so strong in some people after a long period of deprivation that they become desperate and do careless or dangerous things in order to escape this stress.

4.15. Isolation

Loneliness, helplessness, and despair which are experienced by IP when they are isolated are among the most severe stresses. People often take their associations with family, friends, military colleagues, and others for granted, but IP quickly begin to miss the daily interaction with other people. However, these, like the other stresses already discussed, can be conquered. Isolation can be controlled and overcome by knowledge, understanding, deliberate countermeasures, and a determined will to resist it.

4.16. Insecurity

Insecurity is the feeling of helplessness or inadequacy resulting from varied stresses and anxieties. These anxieties may be caused by uncertainty regarding individual goals, abilities, and the future during an isolating event. Feelings of insecurity may have widely different effects on the IP's behavior. An IP should establish challenging but attainable goals. The better an IP feels about individual abilities to achieve goals and adequately meet personal needs, the less insecure the IP will feel.

4.16.1. Loss of Self-Esteem

Self-esteem is the state or quality of having personal self-respect and pride. Lack or loss of self-esteem in an IP may bring on depression and a change in perspective and goals. A loss of self-esteem may occur in individuals in captivity. Humiliation and other factors brought on by the captor may cause them to doubt their own worth. Humiliation comes from the feeling of losing pride or self-respect by being disgraced or dishonored, and is associated with the loss of self-esteem. Prisoners must maintain their pride and not become ashamed either because they are Prisoners of War (POWs) or because of the things that happen to them as a result of being a POW. The IP who loses respect (both personally and of the enemy) becomes more vulnerable to captor exploitation attempts. To solve this problem, IP should try to maintain proper perspective about both the situation and themselves. Their feelings of self-worth may be bolstered if they recall the implied commitment in the Code of Conduct - POWs will not be forgotten.

4.16.2. Loss of Self-Determination

A self-determined person is relatively free from external controls or influences over his or her actions. In everyday society, these controls and influences are the laws and customs of our society and of the self-imposed elements of our personalities. During an isolating event, the controls and influences can be very different. IP may feel as if events, circumstances, and (in some cases) other people, are in control of the situation. Some factors which may cause IP to feel they have lost the power of self-determination are a harsh captor, captivity, bad weather, or rescue forces that make time or movement demands. IP must decide how unpleasant factors will be allowed to affect their mental state. They must have the self-confidence, fostered by experience and training, to live with their feelings and decisions, and to accept responsibility for both the way they feel and how they let those feelings affect them.

4.17. Depression

As an IP, depression is the biggest psychological problem that has to be conquered. Everyone experiences mental highs as well as mental lows. Feeling depressed is a normal reaction to many situations including loss, disappointment, and loss of self-esteem. While most of these emotional changes in mood are temporary and do not become chronic, these feelings may become overwhelming and lead to a deep depression. Depressed IP may feel fearful, guilty, or helpless. They may lose interest in the basic needs of life. Many cases of depression also involve pain, fatigue, loss of appetite, or other physical ailments. Some depressed IP have tried to injure or kill themselves.

4.17.1. Factors which Can Create A Continuous Cycle of Depression

The main reason depression is a difficult problem is that it can affect a wide range of psychological responses. Factors can become mutually reinforcing, creating a continual cycle. For example, fatigue may lead to a feeling of depression, and that depression may increase the loss of self-esteem, and in turn, lead to deeper depression, and so on.

4.17.2. Onset of Depression in an Survival Situation

Depression usually begins after an IP has met the basic needs for sustaining life, such as water, shelter, and food. Once the IP's basic needs are met, there is often too much time for that person to dwell on the past, the present predicament, and on future problems. The IP must be aware of the necessity to keep the mind and body active to eliminate the feeling of depression. One way to keep busy is by checking and improving shelters, signals, and food supply on a daily basis.

4.18. Emotional Reactions

IP may depend more upon their emotional reactions to a situation than upon calm, careful analysis of potential danger - the enemy, the weather, the terrain, the nature of the isolating event, etc. Whether they will succumb to the fear, or use it as a stimulant for greater sharpness, is more dependent on the IP's reactions to the situation than on the situation itself. Although there are many reactions to stress, the following are the most common: fear, anxiety, panic, hate, resentment, anger, impatience, dependence, loneliness, boredom, and hopelessness.

4.18.1. Fear

Fear can either save or cost one their life. Some people are at their best when they are scared. Many IP faced with survival emergencies have been surprised at how well they remembered their training, how quickly they could think and react, and what strength they had. The experience gave them a new confidence in themselves. On the other hand, some people become physiologically paralyzed when faced with a simple survival situation. Some of them have been able to recover their senses before it was too late. In other cases, a fellow IP was on hand to assist them. However, others have not been so fortunate and have either been captured or killed.

4.18.1.1. Different Reactions to Fear

How a person will react to fear depends more upon the individual than it does upon the situation. This has been demonstrated both in actual survival situations and in laboratory experiments. It isn't always the physically strong or the happy-go-lucky people who handle fear most effectively. Timid and anxious people have met emergencies with remarkable coolness and strength.

4.18.1.2. Fear is Common in Life Threatening Emergencies

Anyone who faces life-threatening emergencies experiences fear. Fear is conscious when it results from a recognized situation (such as an immediate prospect of bailout) or when experienced as apprehension of impending disaster. Fear also occurs at a subconscious level and creates feelings of uneasiness, general discomfort, worry, or depression. Fear may vary widely in intensity, duration, and frequency of occurrence, and affect behavior across the spectrum from mild uneasiness to complete disorganization and panic. People have many fears; some are learned through personal experiences, and others are deliberately taught to them. These fears may control behavior, and an IP may react to feelings and imagination rather than to the problem causing fear.

4.18.1.3. Controlling the Fear

When a person's imagination distorts a moderate danger into a major catastrophe, or vice versa, behavior can become abnormal. There is a general tendency to underestimate and this leads to reckless, foolhardy behavior in individuals and may affect the group. The principal means of fighting fear (in an isolating situation) is to pretend that it does not exist. There are no sharp lines between recklessness and bravery. It is necessary to check behavior constantly to maintain proper control.

4.18.1.4. Physical Signs and Symptoms of Fear

One or more of the following signs or symptoms may occur in those who are afraid. However, they may also appear in circumstances other than fear.

- Quickening of pulse; trembling.
- Dilation of pupils.
- Increased muscular tension and fatigue.
- Perspiration of the palms of hands soles of feet, and armpits.
- Dryness of mouth and throat; higher pitch of voice; stammering.
- Feeling of "butterflies in the stomach," emptiness of the stomach, faintness, and nausea.

4.18.1.5. Psychological Signs and Symptoms of Fear

The following common psychological symptoms typically accompany the physical symptoms listed above:

- Irritability and/or increased hostility.
- Talkativeness in early stages, leading finally to speechlessness.
- Confusion, forgetfulness, and inability to concentrate.
- Feelings of unreality, flight, panic, or stupor.

4.18.1.6. Understanding, Admitting, and Accepting Fear

IP cannot run away from fear and must take action to control it. Appropriate actions include:
- Understanding fear.
- Admitting that it exists.
- Accepting fear as reality.

4.18.2. Thinking Logically in the Face of Fear

Training can help IP recognize what individual reactions may be. Using prior training, IP should learn to think, plan, and act logically, even when afraid.

4.18.3. Coping Effectively with Fear

To effectively cope with fear, an IP must:

- Develop confidence. Use training opportunities; increase capabilities by keeping physically and mentally fit; know what equipment is available and how to use it; learn as much as possible about all aspects of survival.
- Be prepared. Accept the possibility that an isolating event can happen. Be properly equipped and clothed at all times, and have a plan ready. Hope for the best, but be prepared to cope with the worst.
- Keep informed. Listen carefully and pay attention to all briefings. Know when danger threatens and be prepared if it comes; increase knowledge of survival environments to reduce the unknown.
- Keep constructively busy at all times. Prevent hunger, thirst, fatigue, idleness, and ignorance about the situation, since these increase fear.
- Learn how fellow IP respond to fear. Work together in emergencies - to live, work, plan, and help each other as a team.
- Practice religion. Don't be ashamed of having spiritual faith.
- Cultivate a survival attitude. Keep the mind focused on a goal and keep everything else in perspective. Learn to tolerate discomfort. Don't exert energy to satisfy minor desires which may conflict with the overall goals - to maintain life, maintain honor, and return.
- Cultivate mutual support. The greatest support under severe stress may come from a tightly knit group. Teamwork reduces fear while making the efforts of every person more effective.
- Exercise leadership. The most important test of leadership and perhaps its greatest value lies in the stress situation.
- Practice discipline. Attitudes and habits of discipline developed in training carry over into other situations. A disciplined group has a better chance of survival than an undisciplined group.
- Lead by example. Calm behavior and demonstration of control are contagious. Both reduce fear and inspire courage.

4.18.4. Overcoming Fear

Every person has goals and desires. The greatest values exercise the greatest influence. Because of strong religious, moral, or patriotic values, people have been known to face torture and death calmly rather than reveal information or compromise a principle. Fear can kill or it can save lives. It is a normal reaction to danger. By understanding and controlling fear through training, knowledge, and effective group action, fear can be overcome.

4.19. Anxiety

Anxiety is a universal human reaction. Its presence can be felt when changes occur which affect an individual's safety, plans, or methods of living. It is generally felt when individuals perceive something bad is about to happen. A common description of anxiety is "butterflies in the stomach." Anxiety creates feelings of uneasiness, general discomfort, worry, or depression. Anxiety and fear differ mainly in intensity. Anxiety is a milder reaction and the specific cause(s) may not be readily apparent, whereas fear is a strong reaction to a specific, known cause. Common characteristics of anxiety are: fear of the future, indecision, feelings of helplessness, resentment.

4.19.1. Overcoming Anxiety

To overcome anxiety, the IP must take positive action by adopting a simple plan. It is essential to keep your mind off of your injuries, or the other stressors associated with the isolating event, and do something constructive. For example, one hostage attempted to learn his captors' language.

4.20. Panic

In the face of danger, a person may panic or freeze and cease to function in an organized manner. A person experiencing panic may have no conscious control over individual actions. Uncontrollable, irrational behavior is common in emergency situations. Panic is brought on by a sudden overwhelming fear, and can often spread quickly through a group of people. Every effort must be made to bolster morale and calm the panic with leadership and discipline. Panic has the same signs as fear and should be controlled in the same manner as fear.

4.21. Hate

Feelings of intense dislike, extreme aversion, or hostility - is a powerful emotion which can have both positive and negative effects on an IP. An understanding of the emotion and its causes is the key to learning to control it. Hate is an acquired emotion rooted in a person's knowledge or perceptions. The accuracy or inaccuracy of the information is irrelevant to learning to hate.

4.21.1. Emotional Feelings of Hate as a Motivator

Any person, any object, or anything that may be understood intellectually, such as political concepts or religious dogma, can promote feelings of hate. Feelings of hate (usually accompanied with a desire for vengeance, revenge, or retribution) have sustained former POWs through harsh ordeals. If an individual loses perspective while under the influence of hate and reacts emotionally, rational solutions to problems may be overlooked, and the IP may be put in jeopardy.

4.21.2. Examining Why Hate is Present

To effectively deal with this emotional reaction, the IP must first examine the reasons why the feeling of hate is present. Once that has been determined, IP should then decide what to do about those feelings. Whatever approach is selected, it should be as constructive as possible. IP must not allow hate to control them.

4.22. Resentment

Resentment is the experiencing of an emotional state of displeasure or indignation toward some act, remark, or person that has been regarded as causing personal insult or injury. An IP must always remember that factors beyond the IP's control may play a role in any isolating event. A hapless IP may feel jealous resentment toward a fellow IP, travel partner, etc., if that other person is perceived to be enjoying a success or advantage not presently experienced by the observer. The IP must understand that events cannot always go as expected. It is detrimental to morale and could affect survival chances if feelings of resentment over another's attainments become too strong. Imagined slights or insults are common. The IP should try to maintain a sense of humor and perspective about ongoing events and realize that stress and lack of self-confidence play roles in bringing on feelings of resentment.

4.23. Anger

Anger is a strong feeling of displeasure and hostility produced by a real or supposed wrong. People become angry when they cannot fulfill a basic need or desire which seems important to them. When anger is not relieved, it may turn into a more enduring attitude of hostility, characterized by a desire to hurt or destroy the person or thing causing the frustration. When anger is intense, the IP loses control over the situation, resulting in impulsive behavior which may be destructive in nature. Anger is a normal response which can serve a useful purpose when carefully controlled. If the situation warrants and there is no threat to survival, one could yell or scream, take a walk, do some vigorous exercise, or just get away from the source of the anger, even if only for a few minutes.

4.24. Impatience

The psychological stresses brought about by feelings of impatience can quickly manifest themselves in physical ways and affect mental well-being. IP who allow impatience to control their behavior may find that their efforts prove to be counterproductive and possibly dangerous. For example, IP who do not have the ability or willingness to suppress annoyance when confronted with delay may expose themselves to capture or injury.

4.24.1. Overcoming Impatience

Potential IP must understand they have to bear pain, misfortune, and annoyance without complaint. In the past, many IP have displayed tremendous endurance, both mental and physical, in times of distress or misfortune. Each potential IP should learn to recognize the things which may make them impatient to avoid acting unwisely. One IP who couldn't wait stated: "I became very impatient. I had planned to wait until night to travel but I just couldn't wait. I left the ditch about noon and walked for about two hours until I was caught."

4.25. Dependence

An IP can become dependent on their equipment creating a lack of balance, interfering with decisions, loss of situational awareness, and hindering the development of alternate methods/solutions to problems. An IP may experience feelings of dependency in a captivity environment. The captor will try to develop in POWs/hostages feelings of need, support, and trust for the captor. By regulating the availability of basic needs like food, water, clothing, social contact, and medical care, captors show their power and control over the prisoners' fate. Through emphasis on the prisoners' inability to meet their own basic needs, captors seek to establish strong feelings of prisoner dependency, making prisoners extremely vulnerable to captor exploitation - a major captor objective. IP's recognition of this captor tactic is key to countering it.

POWs/hostages must understand that, despite captor controls, they do control their own lives. Meeting even one physical or mental need can provide an IP with a small victory and provide the foundation for continued resistance against exploitation.

4.26. Loneliness

Loneliness can be very debilitating during a survival episode. Some people learn to control and manipulate their environment and become more self-sufficient while adapting to changes. Others rely on routines and familiarity of surroundings to function and obtain satisfaction.

4.26.1. Combating Feelings of Loneliness

A potential IP must develop the ability to combat feelings of loneliness during an isolating event long before the event occurs. Self-confidence and self-sufficiency are key factors in coping with loneliness. People develop these attributes by developing and demonstrating competence in performing tasks. As the degree of competence increases, so does self-confidence and self-sufficiency. Military training, more specifically SERE training, is designed to provide individuals with the competence and self-sufficiency to cope with and adapt to survival living.

4.26.2. Being Active to Counter Loneliness

In an isolating event, the countermeasures to conquer loneliness are to be active, plan, and think purposely. Development of self-sufficiency is the primary protection since all countermeasures in isolation require the IP to have the ability to practice self-control.

4.27. Boredom

Boredom is accompanied by a lack of interest and may include feelings of strain, anxiety, or depression, particularly when no relief is in sight and the person is frustrated. Relief from boredom must be based on correction of the two basic sources: repetitiveness and monotony. Boredom can be relieved by a variation of methods – solving problems, rotating duties, broadening the scope of a particular task or job, taking rest breaks, or other diversification techniques. The under gratifying nature of a task can be counteracted by clearing up its meaning, objectives, and, in some cases, relation to the total plan. An IP should not be bored as there are always problems to be solved. An IP should own the situation and not the other way.

4.27.1. Alleviating Boredom

Ray Rising was kidnapped by Colombian guerrillas on March 31, 1994. He spent 810 days as a hostage in multiple jungle camps, in some cases spending weeks shackled to small tarp covered shelter, only being allowed to move to relieve himself. Ray felt "like a dog on the end of a leash". During Ray Rising's captivity to develop rapport with his captures and to alleviate his own boredom he would offer to repair his guards electronic devises like their CD-players and hand-held radios.

4.28. Hopelessness

Hopelessness is a state of feeling that success is impossible, regardless of actions taken, or the certainty that future events will turn out for the worst no matter what a person tries to do. Feelings of hopelessness can occur at virtually any time during an isolating event. IP have experienced hopelessness when unable to maintain health due to an inability to care for sickness, broken bones, or injuries; considering their chances of returning home alive, seeing their loved ones again, or believing in their physical or mental ability to deal with the situation, for example, evading long distances or withholding information from an interrogator.

4.28.1. Situations Where an IP Loses the Will to Survive

During situations where physical exhaustion or exposure to the elements affects the mind, a person may begin to lose hope and give up. During some instances of captivity, deaths have occurred for no apparent cause. These individuals actually appeared to will themselves to die. The premise in the minds of these people is that they are going to die and the situation seems totally futile. People who died in this manner withdrew themselves from the group, became despondent, then lay down and gave up. In some cases, death followed rapidly.

4.28.2. Eliminating Stress to Treat Hopelessness

One way to treat hopelessness is to eliminate the cause of the stress. Rest, comfort, and morale building activities can help eliminate this psychological problem.

4.28.3. Purposeful Attitude to Treat Hopelessness

Another method is to make the person so angry that they wanted to get up and attack the tormentors. A purposeful attitude has a powerful influence on morale and combating the feeling of hopelessness.

4.28.4. Compromise

Since many situations cannot be dealt with successfully by either withdrawal or direct attack, it may be necessary to work out a compromise solution. The action may entail changing an IP's method of operation or accepting substitute goals.

4.29. Summary

Psychological factors can be overcome by IP if they can recognize the problem, work out alternative solutions, decide on an appropriate course of action, take action, and evaluate the results. Perhaps the most difficult step in this sequence is deciding on an appropriate course of action. IP may face either one or several psychological problems. These problems are quite dangerous and must be effectively controlled or countered for survival to continue. IP do not choose their situation and would escape it if they could. IP decide how they will deal with the hand they are dealt and can determine their fate. The isolating event is not an easy one, but it is one in which success can be achieved.

Chapter 5

SERE Medicine

5. SERE Medicine

Among the many things which can compromise an IP's ability to return are medical problems encountered during the isolating event. Although military members are provided with Self Aid-Buddy Care (SABC) instruction, SERE medicine addresses the use of expedient techniques which can aid survival in a field or captivity environment when more traditional methods are unavailable. An IP's situational awareness will help determine the appropriate treatment based on their situation. The procedures in this chapter must be viewed in the reality of a true isolating event. The results of treatment may be substandard when compared with present medical standards. However, these procedures will not compromise professional medical care which becomes available following recovery. Moreover, in the context of an isolating event, they may represent the best available treatment to extend the individual's life expectancy.

Figure 5-1 POW Medicine

5.1. Health PreservationSERE medicine encompasses procedures and expedients that are required and available for the preservation of health and the prevention, improvement, or treatment of injuries and illnesses encountered during an isolating event. It must also be suitable for application by nonmedical personnel to themselves or others in the circumstances of the isolating event. SERE medicine is more than first aid in the conventional sense. It approaches final definitive treatment in that it is not dependent upon the availability of technical medical assistance within a reasonable period of time.

5.1.1. Triage in Group Survival

In a group survival situation, the methods of treating illnesses and injuries do not change but the decisions and outcomes involved may change. It may become necessary to conduct a rudimentary triage to determine if individuals have injuries that are incompatible with life. Treating these individuals may waste critical supplies and time. This waste could endanger the remaining personnel and jeopardize their chances for successful survival and evasion. Consideration should be given to the potential of having to leave critically injured, unable to move, or deceased individuals behind to insure the safety and freedom of the remaining IP (Figure 5-2). These considerations should be thought of as part of IPs pre-mission preparation and discussion.

- During Operation Gothic Serpent, Warrant Officer Michael Durant was the pilot of *Super Six Four*, the second MH-60L Black Hawk helicopter to crash during the Battle of Mogadishu on October 3, 1993. The helicopter was hit by a rocket-propelled grenade in the tail, which led to its crash about a mile southwest of the operation's target.

- Durant and his crew of three (Bill Cleveland, Ray Frank, and Tommy Field), survived the crash, though they were badly injured. Durant suffered a broken leg and a badly injured back. Two snipers, Master Sergeant Gary Gordon and Sergeant First Class Randy Shughart, had been providing suppressive fire from the air at hostile Somalis who were converging on the area. Both volunteered for insertion and fought off the advancing Somalis, killing an undetermined number, until they ran out of ammunition and were overwhelmed and killed, along with Cleveland, Frank, and Field. Both Gordon and Shughart received the Medal of Honor posthumously for this action.

- The Somalis captured Durant and held him for several days. Durant was the only one of his crew to survive. During part of Durant's time in captivity, he was cared for by Somali General Mohamed Farrah Aidid's propaganda minister Abdullahi "Firimbi" Hassan. After eleven days in captivity, Durant was released, along with a captured Nigerian soldier, to the custody of the International Committee of the Red Cross.

Figure 5-2 Survival Medicine

5.1.2. Factors that Reduce Survival Expectancy

Injuries and illnesses peculiar to certain environments can reduce survival expectancy. In cold climates, and often in an open sea survival situation, exposure to extreme cold can produce serious tissue trauma, such as frostbite, or death from hypothermia. Exposure to heat in warm climates, and in certain areas on the open seas, can produce heat cramps, heat exhaustion, or life-threatening heatstroke.

5.1.3. Loss of Gender Boundaries in a Survival Situation

Gender boundaries and social customs may need to be crossed during an isolating event. This may involve a female treating a male or male treating a female. The relationship will be of a clinical nature and could entail wound tending, feeding, and personal hygiene.

- During Desert Storm, then Major Rhonda Cornum was on a Black Hawk helicopter which was shot down on February 27, 1991. She suffered two broken arms, a broken finger, a gunshot wound in the back, and other injuries. She and two other survivors of the crash were captured by the Iraqis. After three days of being roughly shuttled from bunkers to primitive prisons, Major Cornum received medical attention. During their captivity and until their repatriation, Sergeant Troy Dunlap, a US Army Pathfinder, assisted Major Cornum, acting as her nurse and tending to her personal needs due to her inability to use her arms.

5.1.4. How Illnesses Can Interfere with a Successful Survival Situation

Illnesses contracted during evasion or in a captivity environment can interfere with successful survival. Among these are gastrointestinal disorders, respiratory diseases, skin infections and infestations, malaria, typhus, cholera, etc. Captives are, in the physical sense at least, under the control of their captors. Thus, the application of SERE medicine principles will depend on the amount of medical service and supplies the captors can and are willing to give to their prisoners. An enemy may both withhold supplies and confiscate IP's supplies. Some potential enemies (even if they wanted to provide captive medical support) have such low standards of medical practice that their best efforts could jeopardize the recovery of the patient.

5.1.5. Probable Medical Injury in an IP Situation

Looking at combat recovery missions for the past five years, all IP recovered had some type of medical injury ranging from minor to major including fractures, lacerations, burns, and combat trauma. Potential IP must understand, that even in the unlikely event of ending up un-injured during the initial part of their isolating event, 10 percent of all IP have some type of injury by the time they are rescued. It is important to emphasize that even minor injuries or ailments, when ignored, become major problems in a survival situation. Thus, prompt attention to the most minor medical problem is essential in a survival episode. Applying principles of survival medicine should enable military members to maintain health and well-being until rescued and returned to friendly control.

5.2. Basic Health Care

In a survival situation, cleanliness is essential to prevent infection. Adequate personal cleanliness will not only protect against disease germs that are present in the individual's surroundings, but will also protect the group by reducing the spread of these germs.

5.2.1.1. Washing

Washing, particularly the face, hands, and feet, reduces the chances of infection from small scratches and abrasions. A daily bath or shower with hot water and soap is ideal. If a tub or shower is not available, the body should be cleaned with a cloth and soapy water, paying particular attention to the body creases (armpits, groin, etc.), face, ears, hands, and feet. After this type of "bath", the body should be rinsed thoroughly with clear water to remove all traces of soap, which could cause irritation.

5.2.1.2. Substitute when Soap is Unavailable

Soap, although an aid, is not essential to keeping clean. Ashes, sand, loamy soil, and other expedients may be used to clean the body as well as clean cooking utensils.

5.2.1.3. Air Baths When Water is in Short Supply

When water is in short supply, the IP should take an "air bath." All clothing should be removed and the body simply exposed to the air. Exposure to sunshine is ideal, but even on an overcast day or indoors, a two-hour exposure of the naked body to the air will refresh the body. Care should be taken to avoid sunburn when bathing in this manner. Exposure in the shade, shelter, sleeping bag, etc., will help if the weather conditions do not permit direct exposure.

5.2.1.4. Keeping Hair Trimmed

Hair should be kept trimmed; preferably two inches or less in length, and the face should be clean-shaven. Hair provides a surface for the attachment of parasites and the growth of bacteria. Keeping the hair short and the face clean-shaven will provide fewer habitats for these organisms. At least once a week, the hair should be washed with soap and water. When water is in short supply, the hair should be combed or brushed thoroughly and covered to keep it clean. It should be inspected weekly for fleas, lice, and other parasites. When parasites are discovered, they should be removed.

5.2.1.5. Infection and Open Wounds

The principal means of infecting food and open wounds is when they come in contact with unclean hands. Hands should be washed with soap and water, if available, after handling any material which is likely to carry germs. This is especially important after each visit to the latrine, when caring for the sick and injured, and before handling food, food utensils, or drinking water. The fingers should be kept out of the mouth and the fingernails kept closely trimmed and clean. A scratch from a long fingernail could develop into a serious infection.

5.2.2. Care of the Mouth and Teeth

Application of the following fundamentals of oral hygiene will prevent tooth decay and gum disease. The mouth and teeth should be cleansed thoroughly with a toothbrush and toothpaste at least once each day. When a toothbrush is not available, a "chewing stick" can be fashioned from a twig. The twig should be washed, and then chewed on one end until it is frayed and brush-like. The teeth can then be brushed very thoroughly with the stick, taking care to clean all tooth surfaces. If necessary, a clean strip of cloth can be wrapped around the finger and rubbed on the teeth to wipe away food particles which may have collected. When neither toothpaste nor toothpowders are available, salt, soap, or baking soda can be used as a substitute. Gargling with Willow bark tea will help protect the teeth and relieve pain. Gum tissues should be stimulated by rubbing them vigorously with a clean finger each day.

5.2.2.1. Removing Food Debris

Food debris which has accumulated between the teeth should be removed by using dental floss or toothpicks. The latter can be fashioned from small twigs. Parachute inner core can be used by separating the filaments of the inner core and using this as a dental floss. Dental floss can also be improvised from fishing line.

5.2.2.2. Cleaning Dentures and Removable Bridges

Use as much care cleaning dentures and other dental appliances, removable or fixed, as when cleaning natural teeth. Dentures and removable bridges should be removed and cleaned with a denture brush or "chew stick" at least once each day. The tissue under the dentures should be brushed or rubbed regularly for proper stimulation. Removable dental appliances should be removed at night or for a two- to three-hour period during the day.

5.2.2.3. Dental Problems

Dental Problems were a common issue among captives, not only during confinement, but also before capture. Periodontitis (inflammation of tissue surrounding the tooth), pyorrhea (discharge of pus), and damage to teeth consistent with poor hygiene and "wear and tear" were also present.

5.2.2.4. Toothaches

Specific complaints such as pain associated with the common toothache represented one of the most distressing problems faced by past POWs. Along with toothaches, captives may have to deal with facial injury during egress, or caused by physical abuse during interrogation

5.2.2.5. Maintaining a Well Maintained Cleaning Program

The basic principle of maintaining a well-planned cleaning program using fiber, brushes, or branches certainly contributed to the relatively low incidence of cavities and infection among POW's. The lancing of an abscess using bamboo sticks, although not a professional maneuver has merit insofar as the pressure is relieved and the tendency to develop into cellulitis (widespread infection) decreased. The application of aspirin directly into the cavity should be discouraged. The most effective tool against dental complications in captivity is proper preventive dentistry. The present program, if adhered to, is adequate to ensure a high state of dental hygiene while captive. Dental pain/problems were a major concern mentioned by POWs after their return from captivity during the Vietnam War. Dental pain made eating what little food they received difficult, disrupted times when they were left alone by their captors, and could undermine all other activities.

5.2.3. Care of the Feet

Proper care of the feet is of utmost importance in a survival situation, especially if the IP has to travel. Serious foot trouble can be prevented by observing the following simple rules.

5.2.3.1. Washing or Air Cleaning Feet

The feet should be washed, dried thoroughly, and massaged each day. If water is in short supply, the feet should be "air cleaned" along with the rest of the body.

5.2.3.2. Trimming Toenails

Toenails should be trimmed straight across to prevent the development of ingrown toenails.

5.2.3.3. Proper Fitting and Breaking in Boots

Boots should be broken in before wearing them on any mission. They should fit properly, neither so tight that they bind and cause pressure spots nor so loose that they permit the foot to slide forward and backward when walking. Insoles can be improvised to reduce any friction spots inside the shoes.

5.2.3.4. Socks

Socks should be large enough to allow the toes to move freely but not so loose that they wrinkle. Wool socks should be at least one size larger than cotton socks to allow for shrinkage. Socks with holes should be properly darned before they are worn. Wearing socks with holes or socks that are poorly repaired may cause blisters. Clots of wool on the inside and outside should be removed from wool socks because they may cause blisters. Socks should be changed and washed thoroughly with soap and water each day. Woolen socks should be washed in cool water to lessen shrinkage and turned inside out when cleaned. In camp, freshly laundered socks should be stretched to facilitate drying by hanging in the sun or in an air current. While traveling, a damp pair of socks can be dried by placing them inside layers of clothing or securing them on the outside of the pack. If socks become damp, they should be exchanged for dry ones at the first opportunity.

5.2.3.5. Examining Feet Regularly When Traveling

When traveling, the feet should be examined regularly to see if there are any red spots or blisters. If detected in the early stages of development, tender areas should be covered with adhesive tape to prevent blister formation.

5.3. Clothing and Bedding

Clothing and bedding can become contaminated with any disease germs which may be present on the skin, in the stool, in the urine, or in secretions of the nose and throat. Therefore, keeping clothing and bedding as clean as possible will decrease the chances of skin infection and decrease the possibility of parasite infestation. Outer clothing should be washed with soap and water when it becomes soiled. Socks and under clothes should be changed daily. If water is in short supply, clothing should be "air cleaned. "For air cleaning, the clothing is shaken outside, then aired and sunned for two hours. Clothing cleaned in this manner should be worn in rotation. Sleeping bags should be turned inside out, fluffed, and aired after each use. Bed linen should be changed at least once a week, and the blankets, pillows, and mattresses should be aired and sunned (Figure 5-3).

Figure 5-3 Bedding

5.3.1. Rest

Rest is necessary for the IP because it not only restores physical and mental vigor, but also promotes healing during an illness or after an injury.

5.3.1.1. Planning Session

In the initial stage of the survival episode, rest is particularly important. After those tasks requiring immediate attention are done, the IP should inventory available resources, decide upon a plan of action, and even have a meal. This "planning session" will provide a rest period without the IP having a feeling of "doing nothing."

5.3.1.2. Planned Rest Periods

If possible, regular rest periods should be planned in each day's activities. The amount of time allotted for rest will depend on a number of factors, including the IP's physical condition, the number of IP, the presence of hostile forces, etc. Usually, 10 minutes each hour should be sufficient. During these rest periods, the IP should change either from physical activity to complete rest or from mental activity to physical activity as the case may be. The IP must learn to become comfortable and to rest under less than ideal conditions.

5.4. Illness

Many illnesses which are minor in a normal medical environment become major in a survival situation when the individual is alone without medications or medical care. IP should use standard methods (treat symptoms) to prevent expected diseases since treatment in a survival situation is so difficult. In an isolating event, whether short-term or long-term, the dangers of disease are multiplied. Key preventive methods are to maintain a current immunization record, maintain a proper diet, and exercise.

5.4.1. Personal Hygiene

Application of the following simple guidelines regarding personal hygiene will enable the IP to safeguard personal health and the health of others:

- All water obtained from natural sources should be disinfected before consumption if possible.

- The ground in the camp area should not be soiled with urine or feces. Latrines should be used. When no latrines are available, individuals should dig "cat holes" and cover their waste.

- Fingers and other contaminated objects should never be put into the mouth. Hands should be washed before handling any food or drinking water, before using the fingers in the care of the mouth and teeth, before and after caring for the sick and injured, and after handling any material likely to carry disease or germs.

- After each meal, all eating utensils should be cleaned and disinfected in boiling water.

- The mouth and teeth should be cleansed thoroughly at least once each day.

- Bites and insects can be avoided by keeping the body clean, by wearing proper protective clothing, and by using a head net, improvised bed nets, and insect repellants.

- Wet clothing should be exchanged for dry clothing as soon as possible to avoid unnecessary body heat loss. Refer to chapter 15 for more information on the care and use of clothing and equipment.

- Personal items such as canteens, pipes, towels, toothbrushes, handkerchiefs, and shaving items should not be shared with others.
- All food scraps, cans, and refuse should be removed from the camp area and buried.
- If possible, an IP should get seven or eight hours of sleep each night.

5.4.2. Food Poisoning

Food poisoning is a significant threat to IP. Due to sporadic food availability, excess foods must be preserved and saved for future consumption. Methods for food preservation vary with the global area and situation. Bacterial contamination of food sources has historically caused much more difficulty in survival situations than the ingestion of poisonous plants and animals. Similarly, dysentery or water-borne diseases can be controlled by proper sanitation and personal hygiene. If the food poisoning is due to toxins such as staphylococcus or botulism (acute symptoms of nausea, vomiting, and diarrhea soon after ingestion of the contaminated food), supportive treatment is best. Keep the patient quiet, lying down, and ensure the patient drinks substantial quantities of water. If the poisoning is due to ingestion of bacteria which grow within the body, (delayed gradual onset of same symptoms), take antibiotics (if available). In both cases, symptoms may be alleviated by frequently eating small amounts of fine, clean charcoal. In POW situations, if chalk is available, reduce it to powder, and eat to coat and soothe the intestines. Proper sanitation and personal hygiene will help prevent spreading infection to others in the party or continuing re-infection of the patient.

5.4.3. Gastrointestinal Problems

Gastrointestinal problems often occur while in captivity.

5.4.3.1. Diarrhea and Dysentery

Diarrhea and Dysentery are common ailments in captivity. Symptoms of diarrhea and dysentery include violent explosive uncontrollable discharge with possible passage of mucous and blood. They are commonly caused by ingesting contaminated food and water, poor sanitation and hygiene, viral infections, and nutritional disturbances. Treatment can involve diet restrictions (solid food denial and increased liquid intake). Restrict intake to nonirritating foods (avoid vegetables and fruits), establish hygienic standards, increase fluid intake, and, when available, use anti-diarrheal agents. Captives can use a more exotic therapeutic regimen consisting of eating cooked green bananas, tea from specific tree bark, and drinking a quarter teaspoon of ground charcoal or chalk mixed with a quart of disinfected water.

5.4.3.2. Worms and Intestinal Parasites

Worms were extremely common in captivity. Twenty-eight percent of the released prisoners in the Vietnam War indicated worms as a significant medical problem during captivity.

- Danya Curry and Heather Mercer were held as prisoners by the Taliban, ultimately being rescued by Special Forces after 9/11. After three months of captivity, Heather had been given prescription medication to get rid of worms. Half way through the six pill treatment, one night she felt a "string" resting on her tongue, when she pulled it out; it was identified as a 10" worm. Later that night more worms attempted to crawl out, later diagnoses indicated that her infestation was so severe that the worms crawled out not because of the medicine, but they had "nowhere else to go".

5.4.3.2.1. Symptoms and Care of Worm Infestation

Seldom fatal, worms are significant, as they can lower the general resistance other illnesses of the individual. Common identifiers of worms include severe rectal itching, insomnia, and restlessness. Worms often cause gastrointestinal problems similar to those resulting from a variety of other causes. Prevention can be a simple and readily obtainable goal. Worms and intestinal parasites can be caused by undercooked or contaminated food, poor sanitation, and infected water sources. Treatments can include the ingestion of peppers which contain certain substances chemically similar to morphine. They are effective as a counter-irritant for decreasing bowel activity. Also, pumpkin seeds can be ground up and mixed with water into a "medicinal" porridge or a cup of tea made from one to two grams of thyme multiple times per day can be ingested. Lastly, shoes should be worn when possible, hands should be washed after defecation, and fingernails should be trimmed close and frequently.

5.4.4. Nutritional Deficiencies

Nutritional deficiencies are caused by a lack of a specific nutrient that the body requires. Prevention for all nutritional disease include keeping the IP in good physical condition prior to the isolating event, educate themselves on local food sources, and eating everything and anything which is not harmful to themselves (balanced diet).

5.4.4.1. Vitamin B Deficiency (Beriberi)

Beriberi is a nutritional disease resulting from a deficiency of vitamin B (Thiamine). Early signs and symptoms of the disease include muscle weakness and atrophy, loss of vibratory sensation over parts of the extremities, numbness, and tingling in the feet. According to captives, Beriberi's primary manifestation was pain in the feet described as "a minor frostbite that turned to shooting pains". It is difficult to formulate a diagnosis for Beriberi. The treatment is to increase caloric intake by eating anything of value, especially items high in vitamin B such as green peas, cereal grains, and unpolished rice.

5.4.4.2. Vitamin A Deficiency

There were several reported cases of decreased vision (primarily at night) attributed to vitamin A deficiency. This problem usually occurred during periods of punishment or politically provoked action when food was withheld as part of the discipline. The condition responded well to increased caloric intake and deserves little special mention. An understanding of the transient nature of this problem and its remedial response to therapy is important.

5.4.5. Skin Diseases

Boils, fungi, heat rash, and insect bites appeared frequently and typically remain a problem throughout the captivity experience. Preventing different types of skin diseases can be simple. Common preventions include practicing sterility when handling skin conditions, cleanliness, exposure to sunlight, keeping skin dry, and getting adequate nutrition.

5.4.5.1. Lesions

Dermatological lesions are common to various prison experiences. Their importance lies not in their lethality, but for their irritant quality and the debilitating and grating effect on morale and mental health.

5.4.5.2. Fungal Infections

Fungal infections were a common skin problem for those in tropical captivity settings. As with other skin lesions, they are significant for their noxious characteristics and weakening effect on morale and mental health. Superficial fungal infections of the skin are widespread throughout the world. Their frequency among POWs reflects the favorable circumstances of captivity for cultivating fungal infections. Treatment may consist of the removal of body hair (to prevent or improve symptoms in the case of heat rash), exposure to sunlight to dry out fungal lesions, and development of effective techniques to foster body cooling and to decrease heat generation. Considerable effort should be directed at keeping the body clean.

5.4.5.3. Boils and Blisters

Boils and blisters are typically deep-seated infection that usually involves the hair follicles and adjacent subcutaneous tissue, especially parts exposed to constant irritation. Boils and blisters are typically caused by warm humid dark places, poor sanitation and hygiene, and will increase with malnutrition and exhaustion. Boils seldom appear singularly. Once present, they are disseminated by fingers, clothing, and discharges from the nose, throat, and groin. Throughout the history of POW captivity, self-treatment was practiced primarily when there was distrust of captor techniques. Prisoners would attempt to lance the boil with any sharp instrument such as a needle, wire, splinters, etc., and exude their contents by applying pressure. The area was then covered with toothpaste and when available, iodine. Modern therapy consists of hot compresses to hasten localization, and then conservative incision and drainage. Topical antibiotics and systemic antibiotics are then used. Boils increase in frequency with a decrease in resistance as seen in malnutrition and exhaustion states in a tropical environment. The application of any material or medication with a cleansing effect may be used (soaks in saline, soap, iodine, and topical antibiotics).

5.5. Injury

Injuries are possible during any isolating event. In isolating events involving captivity, it is likely that an IP will begin their captivity experience with pre-capture injuries such as burns, wounds, fractures, and lacerations as well as the possibility of other injuries occurring from the captor's physical abuse. The practice of a few simple rules will generally lead to acceptable results in wound treatment. Also, if necessary, much can be done after reintegration to correct cosmetic and functional defects that may occur during an isolating event.

5.5.1. General Management of Injuries

5.5.1.1. Bleeding

Bleeding is the first and most important concern and must be controlled as soon as possible. Controlling bleeding is important because in survival situations, replacement transfusions are not possible. Immediate Self-Aid-Buddy-Care (SABC) steps should be taken to stop the flow of blood, regardless of its source. The method used should be commensurate with the type and degree of bleeding. Most bleeding can be controlled by direct pressure on the wound and that should be the first treatment used. To control bleeding by direct pressure on the wound, sufficient pressure must be exerted to stop the bleeding, and that pressure must be maintained long enough to "seal off" the bleeding sources. It is best to apply the pressure and keep it in place for up to 20 minutes. Alternate pressing and then releasing to see if the wound is still bleeding is not desirable. Oozing blood from a wound of an extremity can be slowed or stopped by elevating the wound above the level of the heart.

5.5.1.1.1. Pressure Points to Stop Hemorrhaging

If direct pressure fails, the next line of defense would be the use of classic pressure points to stop hemorrhaging.

5.5.1.1.2. Tourniquet

The last method for controlling hemorrhage would be the tourniquet. The tourniquet should only be used as a last resort. Even in more favorable circumstances where the tourniquet can be applied as a first aid measure and left in place until trained medical personnel remove it, the tourniquet may result in the loss of a limb. The tourniquet should be used only when all other measures have failed, and it is a life and death matter. A tourniquet, when required and properly used, will save life. The basic characteristics of a tourniquet and the methods of its use are well covered in standard first aid texts; however, certain points merit emphasis in survival medicine. A tourniquet should be used only after every alternate method has been attempted. If unable to get to medical aid, after 15 to 20 minutes, the IP should gradually loosen the tourniquet. If bleeding has stopped, remove the tourniquet; if bleeding continues, reapply and leave in place. The tourniquet should be applied as near the site of the bleeding as possible, between the wound and the heart, to reduce the amount of tissue lost. NOTE: Use of a constricting band that reduces rather than stops blood flow is NOT recommended. It does not allow for venous blood return.

5.5.1.2. Pain

The control of pain accompanying disease or injury under survival situations is both difficult and essential. In addition to its morale-breaking discomfort, pain contributes to shock and makes the IP more vulnerable to enemy influences. Ideally, pain should be eliminated by the removal of the cause. However, this is not always immediately possible. Therefore, measures for the control of pain are beneficial. Position, heat, and cold can be applied alone or in conjunction with each other.

5.5.1.2.1. Position

The part of the body that is hurting should be put at rest, or at least its activity restricted as much as possible. The position selected should be the one giving the most comfort, and be the easiest to maintain. Splints and bandages may be necessary to maintain immobilization of the affected area. Elevation of the injured part, with immobilization, is particularly beneficial for throbbing type pain such as is typical of the "mashed" finger. Open wounds should be cleansed, foreign bodies removed, and a clean dressing applied to protect the wound from the air and chance contacts with environmental objects to help reduce pain.

5.5.1.2.2. Heat and Cold

Generally, the application of warmth reduces pain such as toothache, bursitis, etc. However, in some conditions, application of cold on injuries such as strains and sprains has the same effect. Warmth or cold is best applied by using water soaked rags, materials, or water filled storage bags. This works well due to water's high specific heat. The IP can try both to determine which is most beneficial. Caution should be used when placing hot or cold materials in direct contact with skin to avoid additional injuries.

5.5.1.2.3. Pain Killers

Drugs are very effective in reducing pain. However, they are not likely to be available in the survival situation unless the IP adds them to a personal survival kit. "Natural" procedures can be an alternative to pain killers in a survival situation. If drugs are not available, there are some parts of vegetation which can be used. For example, parts of the Willow tree have been used for their pain-relieving and fever-lowering properties for hundreds of years. The fresh bark contains salicin, a substance that chemically resembles aspirin. It temporarily relieves headache, stomachache, and other body pain. Salicin is metabolized in to salicylic acid in the human body, which is a precursor of aspirin. Boiling eight to ten six-inch stalks in two cups of water until water turns into tea will yield a pain relieving drink. Wintergreen was also traditionally used for body aches and pains as well. Boil a half a cup of leaves into a tea and ingest the tea to reduce pains. Also, the boiled bark of the magnolia tree helps relieve internal pains and fever, and has been known to stop dysentery. To be effective controlling pain, stronger narcotic drugs such as codeine and morphine are required.

5.5.1.3. Shock

In essence, shock is a circulatory reaction of the body (as a whole) to an injury (mechanical or emotional). All IP should be familiar with the signs and symptoms of shock so that the condition may be anticipated, recognized, and dealt with effectively. However, the best survival approach is to treat all moderate and severe injuries for shock. No harm will be done, and such treatment will speed recovery. Normally, fluids administered by mouth are generally prohibited in the treatment of shock following severe injury. Such fluids are poorly absorbed when given by mouth, and they may interfere with later administration of anesthesia for surgery. In survival medicine, however, the situation is different in that the treatment being given is the final treatment. IP cannot be deprived of water for long periods just because they have been injured; in fact, their recovery depends upon adequate hydration. Small amounts of warm water, warm tea, or warm coffee given frequently early in shock are beneficial if the patient is conscious, can swallow, and has no internal injuries. In later shock, fluids by mouth are less effective as they are not absorbed from the intestines. Burns, in particular, require large amounts of water to replace fluid lost from injured areas. *Alcohol should never be given to a person in shock or who may go into shock.*

5.5.2. Infections

Infection is a serious threat to the IP. In survival medicine, one must place more emphasis on the prevention and control of infection by applying techniques identified throughout this handbook. In survival or captivity, consider all breaks in the skin due to mechanical trauma contaminated, and treat appropriately. Even superficial scratches should be cleaned with soap and water and treated with antiseptics, if available.

5.5.2.1. Basic Treatment

Water is the most universally available cleaning agent, and should be (preferably) sterile. At sea level, sterilize water by placing it in a covered container and boiling it for 10 minutes. Above 3,000 feet, water should be boiled for one hour (in a covered container) to ensure adequate sterilization. The water will remain sterile and can be stored indefinitely as long as it is covered. When water is not available for cleaning wounds, the IP should consider the use of urine. Urine may well be the most nearly sterile of all fluids available and, in some cultures, is preferred for cleaning wounds. IP should use urine from the midstream of the urine flow. Antiseptics should generally not be used in wounds which go beneath the skin's surface since they may produce tissue damage which will delay healing. Open wounds must be thoroughly cleansed with disinfected water. Irrigate wounds rather than scrubbing to minimize additional damage to the tissue. Bits of debris such as clothing, plant materials, etc., should be rinsed out of wounds by pouring large amounts of water into the wounds and ensuring that even the deepest parts are clean. In a fresh wound where bleeding has been a problem, care must be taken not to irrigate so vigorously that clots are washed away and the bleeding resumes. Allow a period of about an hour after the bleeding has been stopped before beginning irrigation with the disinfected water. Begin gently at first, removing unhealthy tissue, increasing the vigor of the irrigation over a period of time. If the wound must be cleaned, use great care to avoid doing additional damage to the wound. The wound should be left open to promote cleansing and drainage of infection.

5.5.2.2. The "Open Treatment" Method

This is the only safe way to manage survival wounds. Efforts to close open wounds by suturing or by other procedure should not be made. In fact, it may be necessary to open the wound even more to avoid entrapment or infection and to promote drainage. The term "open" does not mean that dressings should not be used. Good surgery requires that although wounds are not "closed," nerves, bone, and blood vessels should be covered with tissue. Such judgment may be beyond the capability of the IP, but protection of vital structures will aid in the recovery and ultimate function. A notable exception to "open treatment" is the early closure of facial wounds which interfere with breathing, eating, or drinking. Wounds, left open, heal by formation of infection resistant granulation tissue (proud flesh). This tissue is easily recognized by its moist red granular appearance, a good sign in any wound. Frequently, immobilization will hasten the healing of major lacerations. While in captivity, deep open wounds will become infested with maggots frequently. The natural tendency is to remove these maggots, but actually, they do a good job of cleansing a wound by removing dead tissue. Maggots may, however, damage healthy tissue when the dead tissue is removed. So the maggots should be removed if they start to affect healthy tissue.

5.5.2.3. Maggot Therapy

The scavenging activity of the maggots actually removes dead tissue. The maggots excrete a digestive fluid which liquefies the tissue proteins, which the maggots then ingest. The maggots exude calcium carbonate which alkalizes the wound and increases the destruction of bacteria and dead tissue by the body's white blood cells. The maggot's excretion also contains two chemicals which stimulate growth of healthy tissue and hasten wound healing. Excretions contain both allantoin and urea which stimulate growth of healthy tissue and hasten wound healing.

5.5.2.4. Physiological "Logistics"

Despite all precautions, some degree of infection is almost universal in survival wounds. This is the primary reason for the "open" treatment advocated above. The human body has a tremendous capacity for combating infections if it is permitted to do so. The importance of proper rest and nutrition to wound healing and infection control has been mentioned earlier. In addition, the "logistics" of the injured body part should be improved. The injury should be immobilized in a position to favor adequate circulation, both to and from the wound. Avoid constrictive clothing or bandages. Applying heat to an infected wound will further aid in mobilizing local body defense measures. Soaking the wound in lukewarm saltwater will help draw out infection and promote oozing of fluids from the wound, thereby removing toxic products. Poultices, made of clean clay, shredded tree bark with antiseptic qualities, ground grass seed, etc., will also draw out the infection.

- For bee stings and other common skin problems, use clay with disinfected water. Mix together, using only enough water to form a thick paste, and apply it on the problem area. Let the clay dry for about 20 minutes, and then rinse it off with cool disinfected water.

- For a ground seed poultice, insert the ground seed between the third and fourth layers of gauze or bandage material and close. Pour hot water over the gauze to release the flaxseed's essential oils. Set the poultice aside to cool. Apply the poultice to the wound using medical tape to secure it in place.

5.5.2.5. Drainage

Adequate natural drainage of infected areas promotes healing. Generally, wicks or drains are unnecessary. On occasion, it may be better to remove an accumulation of pus (abscess) and insert light, loose packing to ensure continuous drainage. The knife or other instrument used in making the incision for drainage must be sterilized to avoid introducing other types of organisms. The best way to sterilize in the field is with dry or moist heat.

5.5.2.6. Debridement

Debridement is the surgical removal of lacerated, devitalized, or contaminated tissue. The debridement of severe wounds may be necessary to minimize infection (particularly of the gas gangrene type) and to reduce septic (toxic) shock. In essence, debridement is the removal of foreign material and dead or dying tissue. The procedure requires skill and should only be done by nonmedical personnel in case of dire emergency. If required, follow these general rules. Dead skin must be cut away. Muscle may be trimmed back to a point where bleeding starts and gross discoloration ceases. Fat which is damaged tends to die and should be cut away. Bone and nerves should be conserved where possible and protected from further damage. Provide ample natural drainage for the potentially infected wound and delay final closure of the wound.

5.5.2.7. Sutures

On occasion, it may be necessary to suture the wound, despite the danger of infection, in order to control bleeding or increase the mobility of the patient. If a needle is available, thread may be procured from parachute lines, fabric, or clothing, and the wound closed by "suturing." If suturing is required, place the stitches individually and far enough apart to permit drainage of underlying parts. Do not worry about the cosmetic effect; just approximate the tissue. For scalp wounds, hair may be used to close after the wound is cleansed. Infection is less a danger in this area due to the rich blood supply. Remember that in most situations, it is imperative that the wound be left open and allowed to drain.

5.5.2.8. Dressings and Bandages

After cleansing, some wounds should be covered with a clean dressing. The dressing should be sterile. However, in the survival situation, any clean cloth will help to protect the wound from further infection. A proper bandage will anchor the dressing to the wound and afford further protection. Bandages should be snug enough to prevent slippage, yet not constrictive. Slight pressure will reduce discomfort in most wounds and help stop bleeding. Once in place, dressings should not be changed too frequently unless required. External soiling does not reduce the effectiveness of a dressing, and pain and some tissue damage will accompany any removal. In addition, changing dressings increases the danger of infection.

5.5.3. Lacerations

When soft tissue is split, torn, or cut, there are three primary concerns: 1) bleeding, 2) infection, and 3) healing of the wound. See bleeding section under General Management of Injuries 5.5.1.1.

5.5.3.1. Infection

Infection is the second concern when dealing with lacerations. It is extremely important that an IP take proper precaution to prevent wounds from becoming infected.

5.5.3.2. Healing

An open wound will heal by a process known as granulation. During the healing phase, the wound should be kept as clean and dry as possible. For protection, the wound may be covered with clean dressings to absorb the drainage and to prevent additional trauma to the wound. These dressings may be loosely held in place with bandages (clean material may be used for dressings and bandages). The bandages should not be tight enough to close the wound or to impair circulation. At the time of dressing change, disinfected water may be used to gently rinse the wound. The wound may then be air dried and a clean dressing applied (the old dressing may be boiled, dried, and reused). Nutritional status is interrelated with the healing process, and it is important to consume all foods available to provide the best possible opportunity for healing.

5.5.4. Fractures and Sprains

Throughout history, captive therapy was primarily that of helping each other to exercise or immobilize the injured area, and in severe cases, to provide nursing care. In survival medicine, the initial immobilization is part of the ultimate treatment. In addition, immobilization in proper position hastens healing of fractures and improves the ultimate functional result. In an isolating event, the immobilization must suffice for a relatively long period of time and permit the patient to maintain a fairly high degree of mobility. Materials for splinting and bandaging are available in most survival situations, and proper techniques are detailed in SABC.

5.5.4.1. Fractures

Bone fractures are of two general types, open and closed. The open fracture is associated with a break in the skin over the fracture site which may range all the way from a broken bone protruding through the skin to a simple puncture from a bone splinter. The general goals of fracture management are: restore the fracture to a functional alignment; immobilize the fracture to permit healing of the bone and rehabilitation.

5.5.4.1.1. Reducing Fractures

The reduction of fractures is normally beyond the scope of first aid; however, in the prolonged survival situation, the correction of bone deformities is necessary to hasten healing and obtain the greatest functional result. The best time for manipulation of a fracture is in the period immediately following the injury, before painful muscle spasms ensue (Figure 5-4). Restoring or reducing the fracture simply means realigning the pieces of bones, putting the broken ends together as close to the original position as possible. The natural ability of the body to heal a broken bone is remarkable and it is not necessary that an extremity fracture be completely straight for satisfactory healing to occur. However, in general, it is better if the broken bone ends are approximated so that they do not override. Fractures are almost always associated with muscle spasms which become stronger with time. The force of these muscle spasms tends to cause the ends of the broken bones to override one another, so the fracture should be reduced as soon as possible. To overcome the muscle spasm, force must be exerted to reestablish the length of the extremity. Traction is applied by firmly and steadily pulling on the extremity until overriding fragments of bone are brought into line (check by the other limb). If the IP is alone, the problem is complicated but not impossible. Traction can still be applied by using gravity. The distal portion of the extremity is tied to (or wedged) into the fork of a tree or similar point of fixation. The weight of the body is then allowed to exert the necessary traction, with the joint being manipulated until the dislocation is reduced. Frequently, it is advantageous to continue traction after reduction to ensure the proper alignment of the bones. Once the ends of the bone are realigned, the force of the muscle spasm tends to hold the bones together.

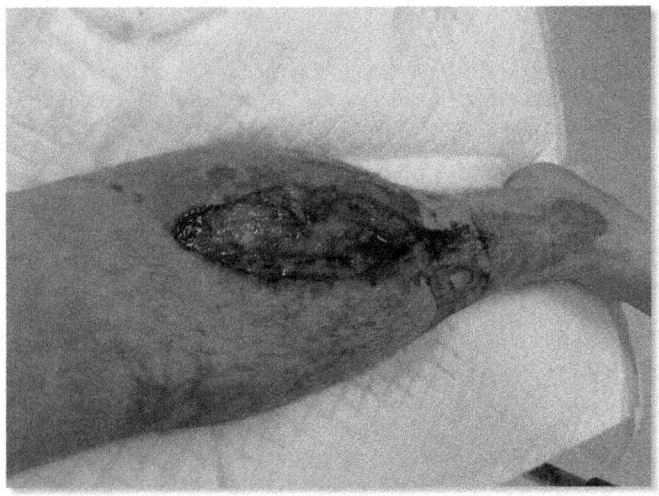

Figure 5-4 Fracture

5.5.4.1.2. Immobilizing Fractures

At this point, closed fractures are ready to be immobilized, but open fractures require treatment of the soft tissue injury in the manner outlined earlier. In other words, the wound must be cleansed and dressed, and then the extremity should be immobilized. The immobilization preserves the alignment of the fracture and prevents movement of the fractured parts which would delay healing. Once the extremity is immobilized the pulse should be repeatedly checked at the farthest part of the immobilized limb to ensure circulation. For fractures of long bones of the body, it becomes important to immobilize the joints above and below the fracture site to prevent movement of the bone ends. For example, in a fracture of the mid forearm, both the wrist and the elbow should be immobilized. In immobilizing a joint, it should be fixed in a "neutral" or functional position. That is, neither completely straight nor completely flexed or bent, but in a position approximately midway between. In splinting a finger, for example, the finger should be curved to about the same position the finger would naturally assume at rest.

5.5.4.2. Sprains

An acute non-penetrating injury to a muscle or joint is best managed by applying cold as soon as possible after the injury. Icepacks or cold compresses should be used intermittently for up to 48 hours following the injury. This will minimize hemorrhage and disability. Be careful not to use snow or ice to the point where frostbite or cold injury occurs. As the injured part begins to become numb, the ice should be removed to permit re-warming of the tissues. Then the ice can be reapplied. Following a period of 48 to 72 hours, the cold treatment can be replaced by warm packs to the affected area. A "sprain or strain" may involve a wide variety of damage ranging from a simple bruise to deep hemorrhage or actual tearing of muscle fibers, ligaments, or tendons. While it is difficult to establish specific guidelines for treatment in the absence of a specific diagnosis, in general, injuries of this type require some period of rest (immobilization) to allow healing. The period of rest is followed by a period of rehabilitation (massage and exercise) to restore function. For what appears to be a simple superficial muscle problem, a period of five to 10 days rest followed by a gradual progressive increase in exercise is desirable. Pain should be a limiting factor. If exercise produces significant pain, the exercise program should be reduced or discontinued. In captivity, it is probably safest to treat severe injuries to a major joint like a fracture with immobilization (splint, cast) for a period of four to six weeks before beginning movement of the joint.

5.5.4.3. Splints

A splint of any rigid material such as boards, branches, bamboo, metal boot insoles, or even tightly rolled newspaper may be almost as effective as plaster or mud casts. In conditions such as continuous exposure to wetness, the splint can be cared for more effectively than the plaster or mud cast. Also, in cases where there is a soft tissue wound in close proximity to the fracture, the splint method of immobilization is more desirable than a closed cast because it permits change of dressing, cleaning, and monitoring of the soft tissue injury. Improvising a splint using available materials may be done using the materials listed above or using several parallel, pliable branches, woven together with vines, cords, or parachute lines. Use care so that the extremity is not constricted when swelling follows the injury. The fracture site should be loosely wrapped with parachute, other type of cloth, or soft plant fibers. The splints can then be tied in place extending at least the entire length of the broken bone and preferably fashioned in such a way as to immobilize the joint above and below the fracture site. Splinting or immobilizing materials should be insulated to avoid heat/cold injuries as well as improve comfort (Figure 5-5). It may be necessary to preserve the mobility of the IP after reduction of the fracture. This is difficult in fractures of the lower extremities, although tree limbs may be improvised as crutches. With companions, the use of improvised litters may be possible.

Figure 5-5 Splinting

5.5.4.4. Immobilization Duration

The time required for immobilization to ensure complete healing is very difficult to estimate. It must be assumed that healing time will, in general, be prolonged. This means that for a fracture of the upper extremity of a "non-weight bearing bone," immobilization might have to be maintained for eight weeks or more to ensure complete healing. For a fracture of the lower extremity or a "weight bearing bone," it might require 10 or more weeks of immobilization.

5.5.4.5. Swelling

While a limb is immobilized swelling is a common side effect that will need to be dealt with. Keeping the limb elevated as much as possible throughout the day to aid in circulation will help.

5.5.4.6. Rehabilitation

Following the period of immobilization and fracture healing, a program of rehabilitation is required to restore normal functioning. Muscle tone must be re-established and the range of motion of immobilized joints must be restored. In cases where joints have been immobilized, the rehabilitation program should be started with "passive range of motion exercises". This means moving the joint through a range of motion without using the muscles which are normally used to move that joint. For example, if the left wrist has been immobilized, a person would begin the rehabilitation program by using the right hand to passively move the left wrist through a range of motion which can be tolerated without pain. When some freedom of motion of the joint has been achieved, the individual should begin actively increasing that range of motion using the muscles of the joint involved. Do not be overly forceful in the exercise program. Using pain as a guideline, the exercise should not produce more than minimal discomfort. Over a period of time, the joint movement should get progressively greater until the full range of motion is restored. Also, exercises should be started to restore the tone and strength of muscles which have been immobilized. Again, pain should be the limiting point of the program and progression should not be so rapid as to produce more than a minimal amount of discomfort.

5.5.5. Burns

Burns, encountered in isolating events, pose serious problems (Figure 5-6). Burns cause severe pain, increase the probability of shock and infection, and offer an avenue for the loss of considerable body heat, fluids, and salts. Initially, cooling a small area burn (10 percent of the body or less than any single extremity) by immersing in cool water or using a cold, moist, compress will provide relief from pain. Treating larger burns with cool water is risky due to the danger of hypothermia. Covering the wound with a clean, dry, dressing of any type reduces the chance for infection. Such protection enhances the mobility of the patient and the capability for performing other vital survival functions. In burns about the face and neck, ensure the victim has an open airway. Burns of the face and hands are particularly serious in a survival situation as they interfere with the capability of IP to meet their own needs. Breaking an aloe vera leaf and spreading the internal clear gel on burns will aid in the healing process. Additionally, soaking certain barks (Willow, Oak, and Maple) in water soothes and protects burns by astringent action. This is a function of the acid content of these barks.

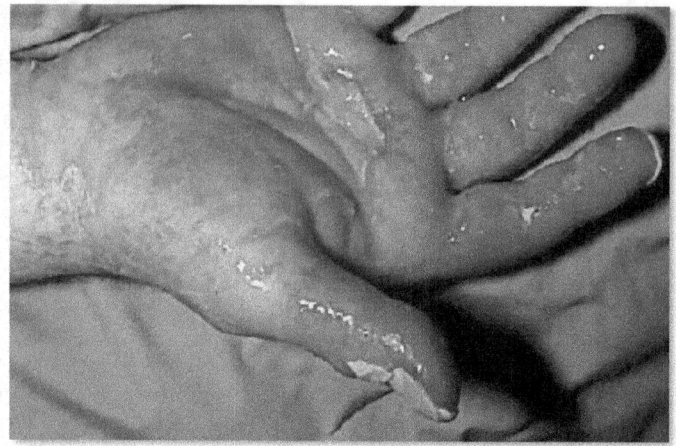

Figure 5-6 Burns

5.5.5.1. Maintenance of Fluids to Recover From Burns

Maintenance of body fluids and salts is essential to recover from burns. Usually the only way to administer fluids in a survival situation is by mouth; hence, the casualty should ingest sufficient water early before the nausea and vomiting of toxicity intervenes. Consuming the eyes and blood (both cooked) of animals can help restore electrolyte levels if salt tablets are not available. NOTE: The IP may also pack salt in personal survival kits to replace electrolytes (one quarter (1/4) teaspoon per quart of water).

5.5.6. Head Injuries

Injuries to the head pose additional problems related to brain damage as well as interfering with breathing and eating. Bleeding is more profuse in the face and head area, but infections have more difficulty in taking hold. This makes it somewhat safer to close such wounds earlier to maintain function. Cricothyroidotomy may be necessary if breathing becomes difficult due to obstruction of the upper airways. In the event of unconsciousness, watch the patient closely and keep him or her still. Even in the face of mild or impending shock, keep the head level or even slightly elevated if there is reason to expect brain damage. Do not give fluids or morphine to unconscious persons.

5.5.6.1. Performing a Cricothyroidotomy

Often, the safest and fastest method to establish a surgical airway is to perform a cricothyroidotomy. In this procedure, the cricothyroid membrane is opened to allow air to pass through into the trachea. The cricothyroid membrane is the soft spot just below the Adam's apple (thyroid cartilage) and just above the cricoid cartilage. In men, this is easily identified by running your finger down the center of the neck. Just beyond the Adam's apple the cricothyroid membrane is a small, soft indentation about the width of your finger.

5.5.6.2. Locating the Cricothyroid Membrane

The thyroid cartilage (Adam's apple) in women is not usually as prominent as it is in men. It is easier to find their cricothyroid membrane by sliding your finger up the midline of the neck to the first hard bump. That is the cricoid cartilage. Above the cartilage is the cricothyroid membrane (your target), and above the membrane is the thyroid cartilage. Practice identifying the cricothyroid membrane in your own neck, and in other personnel. Then, in an emergency, you will have no difficulty finding it.

5.5.6.3. Tools Needed to Perform a Cricothyroidotomy

To perform an emergency cricothyroidotomy, a sharp instrument, is needed. The following provides examples of instruments that can be used:

- Scalpel
- Pocket knife
- Scissors
- Razor blade
- Sharp edge of a tin can
- Broken glass

5.5.6.4. Speed is Necessary for Performing a Cricothyroidotomy

Ideally, this procedure is performed by well-trained and experienced medical personnel, using sterile technique and instruments designed for this purpose. In other than ideal conditions, speed is more important than sterile technique, special instruments or experience. In an operational setting, if the casualty needs a surgical airway, go ahead and do it, and do it quickly. Some of the most successful airway rescues have been performed by inexperienced, minimally-trained personnel, who have never done this before. There typically won't be time for any anesthetic, but that is not normally an issue since the people on whom you would be doing this procedure will be unconscious. Bleeding usually is not a problem because there are no large blood vessels either in the skin or beneath the skin in this area. Whatever device you use to keep the airway open, try to find some adhesive tape to hold it in place, without obstructing air flow. An emergency cricothyroidotomy can be left in place for up to 72 hours, but after that, it should be replaced by a tracheotomy, placed lower in the trachea by trained surgeons.

5.5.6.4.1. Cricothyroidotomy Procedure

Here is the process for performing this procedure:

1. Identify the cricothyroid membrane with your index finger.
2. Make a transverse incision using any available sharp object, directly over the cricothyroid membrane. The incision should be about an inch long.
3. Once through the skin, feel with your index finger for the soft, compressible cricothyroid membrane.
4. Take your sharp object and push it straight down through the cricothyroid membrane. There will be a distinct "pop" as you open the trachea. Don't worry about going too deep. The far side of the trachea at this point is made of very tough cartilage and it's not too easy to go all the way through it.
5. Once through the membrane, withdraw your sharp instrument and replace it with a hollow tube to keep the airway open.

6. Ideally, this hollow tube would be an endotracheal tube, but any tube-like structure will work fine. Examples include:
   - Ball point pen barrel
   - Nail clipper
   - Two keys
   - One key turned sideways
   - Bent paper clip

5.5.7. Abdominal Wounds

Wounds of the abdomen are particularly serious in a survival situation and the IP must seek immediate medical attention. Such wounds, without immediate and adequate surgery, have an extremely high mortality rate and render patients totally unable to care for themselves. If intestines are not extruded through the wound, a secure bandage should be applied to keep this from occurring. If the intestine is extruded, do not replace it due to the almost certain threat of fatal peritonitis. Cover the extruded bowel with a large dressing and keep the dressing wet with any drinkable fluid or if drinkable fluid is unavailable, then urine may be used. The IP should lie on their back and avoid any motions that increase intra-abdominal pressure which might extrude the bowels even more. Keep the IP in an immobile state or move on a litter. "Nature" will eventually take care of the problem either through death, or walling-off of the damaged area.

5.5.8. Chest Injuries

Injuries of the chest are common, painful, and disabling. In the case of a flail chest, ribs broken away from the sternum, apply a bulky dressing and wrap to immobilize. Severe bruises of the chest or fractures of the ribs may require that the chest be immobilized to prevent large painful movements of the chest wall. If a bandage is required, apply it while the patient deeply exhales. In an isolating event, it may be necessary for IP to wrap their own chest. This is more difficult but can be done by attaching one end of the long bandage (parachute material) to a tree or other fixed object, holding the other end in the hand, and slowly rolling body toward the tree, keeping enough counter pressure on the bandage to ensure a tight fit.

5.5.8.1. Sucking Chest Wounds

These wounds are easily recognized by the sucking noise and appearance of foam or bubbles in the wound. These wounds must be closed immediately before serious respiratory and circulatory complications occur. Ideally, the patient should attempt to exhale while holding the mouth and nose closed (Valsalva) as the wound is closed. This inflates the lungs and reduces the air trapped in the pleural cavity. This procedure may need to be repeated if symptoms return or worsen. Frequently, a taped, airtight dressing is all that is needed, but sometimes it is necessary to suture the wound to make sure the wound is closed.

5.5.9. Eye Injuries

Eye injuries are quite serious in a survival situation due to pain and interference with other survival functions. The techniques for removing foreign bodies and for treating snow blindness are covered in SABC. With more serious eye injuries involving disruption of the contents of the orbit or impaled objects additional steps may be required (note: Do not remove an object that is impaled in the eye unless required for IP survival). A protective covering could be made from a bandage, cup, or similar item to protect the eye from further injury and movement. This will require the unaffected eye be covered and a vertical slit placed in the covering to reduce sympathetic eye movement. As a last resort IP may consider removing the impaled object and risk losing the eye. The IP may need to tape the lid of the affected eye closed to prevent infection (Figure 5-7).

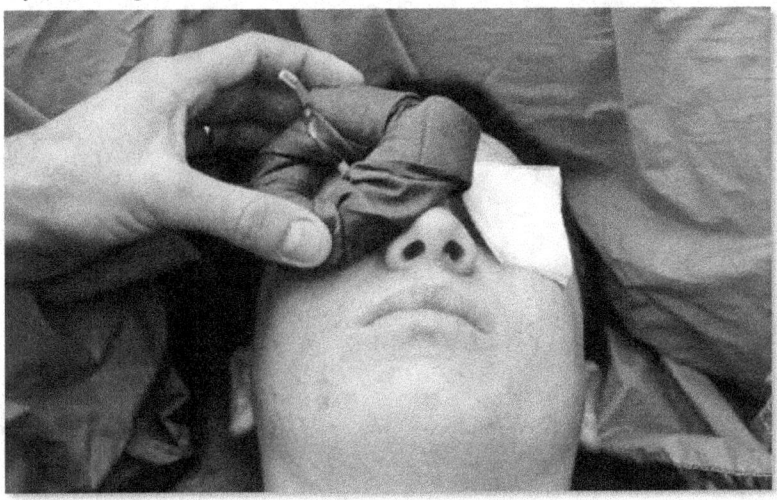

Figure 5-7 Eye Injuries

5.5.10. Thorns and Splinters

Thorns and splinters are frequently encountered in survival situations. Reduce their danger by wearing gloves and proper footgear. Their prompt removal is quite important to prevent infection. Wounds made by these agents can be quite deep compared to their width which increases chances of infection by those organisms (such as tetanus) which grow best in the absence of oxygen. Removal of splinters is aided by the availability of a sharp instrument (needle or knife), needle nose pliers, or tweezers. Take care to get all of the foreign body out; sometimes it is best to open the wound sufficiently to properly cleanse it and to allow air to enter the wound. When cleaned, treat as any other wound.

5.5.11. Blisters and Abrasions

Care for blisters and abrasions promptly. If redness or pain is noted, the IP should stop what they are currently doing (if at all possible) to find and correct the cause. Frequently, a protective dressing or bandage and/or adhesive will be sufficient to prevent a blister. If a blister occurs, do not remove the top. Apply a sterile (or clean) dressing. Small abrasions should receive attention to prevent infection. Using soap with a mild antiseptic will minimize the infection of small abrasions which may not come to the attention of the IP.

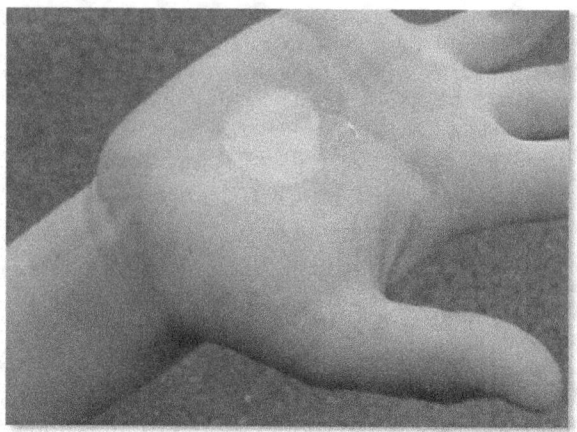

Figure 5-8 Blisters and Abrasions

5.5.12. Insect Bites

Bites from insects, leeches, ticks, chiggers, etc., pose several hazards. Many of these organisms transmit diseases, and the bite itself is likely to become infected, especially if it itches and the IP scratches it. The body should be inspected frequently for ticks, leeches, etc., and these should be removed immediately. If appropriate and possible, the IP should avoid infested areas. These parasites can best be removed by applying heat, irritants, or petroleum jelly to them or their entrance hole to encourage a relaxation of their hold on the host. Then the entire organism may be gently detached from the skin, without leaving parts of the head imbedded. Treat such wounds as any other wound. Applying cold wet dressings will reduce itching, scratching, and swelling.

5.6. Environmental Injury

5.6.1. Dehydration

Dehydration is the lack of adequate body fluids for the body to carry on normal functions at an optimal level (brought on by loss, inadequate intake, or a combination of both). Fluid losses up to five percent are considered mild, up to ten percent are considered moderate, and up to fifteen percent are considered severe. Severe dehydration can result in seizures, permanent brain damage, cardiovascular collapse and death if not treated quickly. To prevent dehydration, drink water and ration your sweat, not your water. Common causes and symptoms of dehydration include, but are not limited to excessive loss of fluid through vomiting or excessive urine, stools or sweating, poor intake of fluids, sunken eyes, dry or sticky mucus membrane in the mouth, lack of skins normal elasticity, decreased or absent urine output, and decreased tears. If an IP shows signs of dehydration, treatment includes oral rehydration which may be sufficient for mild dehydration. Intravenous fluids and hospitalization may be necessary for moderate to severe dehydration.

5.6.2. Heat Related Injuries

5.6.2.1. Preventing Heat Related Injuries

By sweating, breathing, shivering, and shifting the flow of blood between the skin and internal organs, the body can usually keep its temperature within a narrow range in hot or cold weather.

However, overexposure to high temperatures can result in heat disorders such as heat exhaustion, heatstroke, and heat cramps. The risk of heat disorders is increased by high humidity, which decreases the cooling effect of sweating, and by prolonged strenuous exertion, which increases the amount of heat produced by the muscles. Sweat is the body's main system for getting rid of extra heat. When an IP sweats, and the water evaporates from the skin, the heat that evaporates the sweat comes mainly from the skin. As long as blood is flowing properly to the skin, extra heat from the body's core is "pumped" to the skin and removed by sweat evaporation. If a person does not sweat enough, they cannot get rid of extra heat well, and you also can't get rid of heat as well if blood is not flowing to the skin. Dehydration will make it harder for you to cool in two ways: if you are dehydrated you won't sweat as much, and the body will try to keep blood away from the skin in order to keep blood pressure at the right level in the core of your body. Since people lose water when they sweat, they must make up that water to keep from becoming dehydrated. If the air is humid, it is harder for sweat to evaporate -- this means that the body cannot get rid of extra heat when it is muggy as it can when it is relatively dry. The best fluid to drink when a person is sweating is water. Using common sense is the best way to prevent heat-related illnesses. Stay well hydrated; to make sure that the body can get rid of extra heat, and be sensible about exertion. It's also important to be sensible about how much a person exerts themselves in hot weather. The hotter and more humid it is, the harder it will be for a person (IP) to get rid of excess heat. Clothing also makes a difference. Use clothing to protect from the sun/heat and aid in keeping the body cool. Many layers of clothing may hide heavy sweating, masking symptoms of heat disorders.

5.6.2.2. Heat Cramps

Heat cramps are severe muscle spasms resulting from heavy sweating during exertion in extreme heat. Heat cramps are caused by the excessive loss of fluids and salts (electrolytes)--including sodium, potassium, and magnesium--resulting from heavy sweating. Heat cramps often begin suddenly in the hands, calves, or feet. They are often painful and disabling. The muscles become hard, tense, and difficult to relax. Heat cramps can be prevented or treated by drinking beverages or eating foods that contain sodium (salt), potassium, and magnesium such as fiddlehead ferns, crickets nuts (acorn and pine work best), and fruits.

5.6.2.3. Heat Exhaustion

Heat exhaustion is a condition resulting from exposure to heat for many hours, in which excessive loss of fluids from heavy sweating leads to fatigue, low blood pressure, and sometimes collapse. Exposure to high temperatures can cause the loss of too much fluid through sweating, particularly during hard physical labor or exercise. Salts (electrolytes) are lost with the fluids, disturbing the circulation and the brain's functioning. As a result, heat exhaustion may develop. The major symptoms of heat exhaustion are increased fatigue, weakness, anxiety, and drenching sweats. A person may feel faint when standing still because blood collects (pools) in blood vessels of the legs, which are dilated by the heat. Also, the heartbeat becomes slow and weak, the skin becomes cold, pale, and clammy, and the IP may become confused. The loss of fluids reduces the volume of blood, lowers blood pressure, and may cause the person to collapse or faint. Usually, heat exhaustion can be diagnosed on the basis of the symptoms. The main treatment is replacing fluids (rehydration) and salt. Lying flat or with the head lower than the rest of the body and sipping cool, slightly salty beverages every few minutes are generally all that is needed. Moving to a cool environment helps. After rehydration, a person often recovers rapidly and fully. If blood pressure remains low and the pulse remains slow for more than an hour despite this treatment, another condition should be suspected, diagnosed and treated.

5.6.2.4. Heat Stroke

Heatstroke is a life-threatening condition resulting from long, extreme exposure to heat, in which a person cannot sweat enough to lower body temperature. This condition often develops rapidly and requires immediate intensive treatment. If a person reaches this state, body temperature may rise to dangerously high levels, causing heatstroke. Heatstroke may develop rapidly and is not always preceded by warning signs such as a headache, vertigo (a whirling sensation), or fatigue. If a person has suffered a heat stroke, sweating usually but not always decreases. The skin is hot, flushed, and unusually dry. The heart rate increases and may quickly reach 160 to 180 beats per minute, in contrast to the normal rate of 60 to 100 beats per minute. The breathing rate usually increases, but blood pressure rarely changes. Body temperature (which should be measured rectally), rises rapidly to 104° F. to 106° F., causing a feeling of burning up. A person may become disoriented and confused and can quickly lose consciousness or have convulsions. Heatstroke can cause permanent damage or death if not treated immediately. A temperature of 106° F is very serious; a temperature just one degree higher is often fatal. Permanent damage to internal organs, such as the brain, can occur quickly, often resulting in death. Heatstroke is an emergency, and lifesaving measures should be started immediately. A person who can't be taken to a hospital quickly should be wrapped in wet bedding or clothing, immersed in a lake, stream, pond, or river. Pay attention to their body temperature (through their signs and symptoms) to avoid overcooling. After severe heatstroke, rest is advised for a few days to avoid other possible complications to include heart failure. Body temperature may fluctuate abnormally for weeks causing the potential for other environmental injuries.

5.6.3. Cold Related Injuries

5.6.3.1. Preventing Cold Related Injuries

The four main factors that make an IP more susceptible to localized cold injury are inadequate insulation from cold or wind, restricted circulation, fatigue, and poor nutrition.

5.6.3.1.1. Inadequate Insulation from Cold or Wind

The easiest and most effective way to prevent the body from losing heat is the proper wear of clothing. Keep the head and ears covered. IP will lose as much as 50 percent of their total body heat from an unprotected head at 50°F. When exerting the body, prevent perspiration by clothing at the neck and wrists and loosening it at the waist. If the body is still warm, comfort can be obtained by removing the inner layers of clothing, one layer at a time. The outer protective layer should be worn unless it is too bulky and a factor in causing the body to overheat. When work stops, the individual should put the clothing on again to prevent chilling. Change and dry wet clothing throughout the day.

5.6.3.1.2. Restricting Circulation

Avoid restricting the circulation. Clothing should not be worn so tight that it restricts the flow of blood that distributes the body heat and helps prevent cold injuries.

5.6.3.1.3. Fatigue

When a person is severely fatigued his body generates less heat. This is why it is extremely important that when in cold environment to take more rest breaks and drink plenty of water during these breaks. Once the body is overworked it is very difficult to gain the strength back. IP should take their time accomplishing tasks (like building shelters) and stay hydrated.

5.6.3.1.4. Poor Nutrition

The amount and type of food IPs eat in cold environment directly affects the amount of heat generated by the body. A large portion of an IPs diet should consist of fats. The energy contained in fats is more slowly released that the energy in carbohydrates. Because of this, it is a longer lasting form of energy.

5.6.3.2. Hypothermia

Literally, hypothermia means "low temperature" and is diagnosed when the inner (core) temperature of the body falls below 95°F (35°C). Generalized, progressive cooling of the body results in hypothermia. It can develop quickly. The body can usually tolerate a drop of a few degrees of internal body temperature. Below 90°F, the body loses its ability to regulate its temperature and to generate body heat. Progressive loss of body heat then begins. Hypothermia occurs when the core temperature is between 90 and 95°F. Hypothermia is a silent killer of those who are not prepared to face a cool, wet day. It happens when your body is not able to make enough heat to replace the warmth you lose to the environment around you.

5.6.3.2.1. Mild Hypothermia

Even mild degrees of hypothermia can have serious consequences and complications. Involuntary shivering begins in response to a drop in the body core temperature. The patient is usually alert and shivering. Shivering is an attempt to generate more heat through muscular activity. Shivering usually stops once the core temperature drops below 90° - 92°F (30° - 31°C). That is an indication of extreme danger. Warming the skin of someone with hypothermia may stop the shivering, even though the core temperature has not changed.

5.6.3.2.2. Additional Symptoms of Hypothermia

As the IP begins to sink farther into hypothermia they will have the additional symptoms of clumsiness and mental confusion (i.e., slowing of thought and action). If left untreated they become unconscious. Eventually all cardio-respiratory activity may cease and the patient may appear dead. Never assume a patient that is cold and without a pulse is dead.

5.6.3.2.3. Proper Emergency Measures

Patients can survive even after severe hypothermia if you carry out proper emergency measures. First, prevent further body heat loss, and then try to warm the victim. If outdoors, place the victim out of the wind in the best shelter possible. Change the victim into dry clothes, if possible, and place on top of as much insulation as possible to keep the victim off the ground. Placing a hypothermic person in a cold sleeping bag is not sufficient. The bag should be pre-warmed by another person, if possible, to transfer a maximum amount of body heat to the bag. If the bag is large enough, it can be a lifesaving step to put both people in together (two non-hypothermic people to one hypothermic person is the best ratio). If unconscious, place the victim on his/her side to ensure an open air passage. If the victim is conscious, administer warm fluids such as tea or broth. Sweets should be given if the victim is able to eat; they are quickly transformed into heat and energy. Never use alcoholic beverages as a way to re-warm a hypothermic individual.

5.6.3.3. Frostbite

Frostbite is literally the freezing of body tissue; usually skin (Figure 5-9). Fingers, toes, ears, and the nose are the area's most vulnerable to frostbite. There are two degrees of frostbite which an

IP should be concerned about. The first is Frostnip, which usually affects skin on the face, ears, or fingertips. Frostnip may cause tingling, numbness, and/or blue-white skin color for a short time (with normal feeling and color returning quickly when re-warmed). No permanent tissue damage occurs. The second is deep frostbite, in which the skin and underlying tissue freezes, looking pale or blue and feels cold, numb, and stiff or rubbery to the touch. Permanent damage is possible, depending on how long and how deeply the tissue is frozen.

5.6.3.3.1. Symptoms of Frostbite

Frostbite is caused by either prolonged exposure to cold temperatures or shorter exposure to very cold temperatures. Many people with frostnip or frostbite experience numbness. A "pins and needles" sensation, severe pain, itching, and burning are all common when the affected area is warmed and blood starts flowing again. Skin may look white, grayish-yellow, or even black with severe frostbite, and it may feel hard, waxy, and numb. Blistering is also common.

5.6.3.3.2. Treating Frostbite

Keep the affected part elevated in order to reduce swelling. Move to a warm area to prevent further heat loss. NOTE: Many people with frostbite may be experiencing hypothermia. Saving their lives is more important than preserving a finger or foot. Remove all constrictive jewelry and clothes because they may further block blood flow. Give the person warm, nonalcoholic, non-caffeinated fluids to drink.

5.6.3.3.3. Frostnip

In cases of frostnip re-warm the IP by putting the injured area against warm areas of the body to re-warm (i.e., frostnip fingers in the armpit). With frostbite, never re-warm an affected area if there is any chance it may freeze again. This thaw-refreeze cycle is very harmful and leads to disastrous results. Also, avoid a gradual thaw. The most effective method is to re-warm the area quickly. Therefore, keep the injured part away from sources of heat until a proper re-warming can take place.

5.6.3.3.4. What to do Once the Area is Re-Warmed

Once the area is re-warmed, apply a dry, sterile bandage, place cotton between any involved fingers or toes (to prevent rubbing). Do not rub the frozen area with snow (or anything else, for that matter). The friction created by this technique will only cause further tissue damage. Above all, keep in mind that the final amount of tissue destruction is proportional to the time it remains frozen, not to the absolute temperature to which it was exposed. Therefore, the act of re-warming and avoiding refreezing are critical. After re-warming, treatment of the affected parts includes leaving blisters intact, elevation of the affected parts, and using any Ibuprofen as indicated.

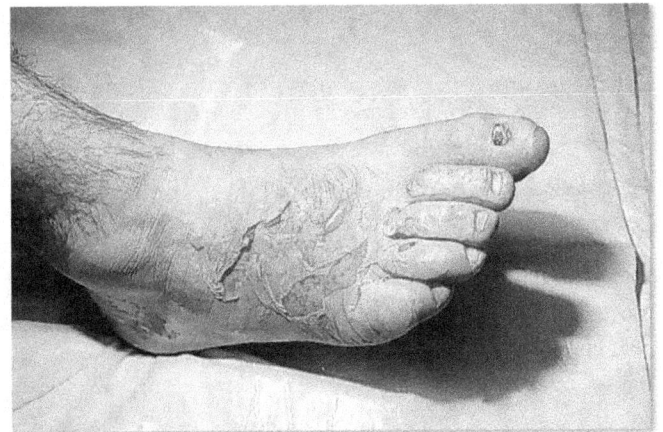

Figure 5-9 Frostbite

5.6.3.4. Trench Foot or Immersion Foot

Trench foot (immersion foot) is a cold injury that occurs gradually over several days of exposure to cold, but not freezing, temperatures. The name comes from World War I troops who developed symptoms after standing in cold, wet trenches. Symptoms of trench foot include red wrinkling skin that turns pale, swelling, numbness or burning pain, leg cramps, slow or absent pulse in the foot, and the development of blisters or ulcers after two to seven days, but no actual freezing of the skin. For immersion foot, remove wet shoes, boots, and socks. Re-warm the affected areas (do not use heat from a fire to dry feet), relieving pain, protecting it from further cold exposure, and preventing problems such as infection or dead skin (gangrene).

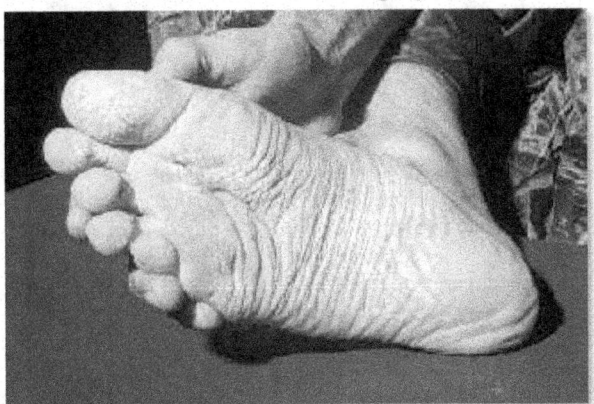

Figure 5-10 Trench Foot or Immersion Foot

5.6.4. Bites and Stings

5.6.4.1. Prevention

Prevention is the key to treatment. Follow these simple rules to reduce the chance of accidental bites and stings:

- Don't put your hand into dark places such as rock crevices, heavy brush, or hollow logs without first investigating. When crossing over debris, look to see what is on the other side.
- Look where you are stepping (especially in heavy brush or tall grass).
- Use extra caution when walking at night.
- Don't pick up any insect if you are not sure if it is venomous.
- Don't pick up freshly killed snakes without severing the head (dead snakes can still bite).
- Check all clothing and bedding before using.

5.6.4.2. Snakebites

No single physical characteristic distinguishes a poisonous snake from a harmless one except the presence of poison fangs and glands. A determination of the presence of fangs and glands safely can only be made when the snake is dead. When in doubt, treat a snake bite as if it poisonous.

5.6.4.2.1. Hemotoxic Venom Types

Some examples of these types of snakes are the copperhead, common adder, puff adder, sand viper, bushmaster, cottonmouth, and eastern/western diamondback rattlesnake. This type of venom affects the circulatory system, destroying blood cells, damaging skin tissue and causing internal hemorrhaging. Local effects include strong pain, swelling and necrosis at the site. Pain is much like that of a severe burn. Systemic effects include hemorrhaging, internal organ breakdown and destroying of blood cells. Signs and symptoms of a hemotoxic snakebite are: fang marks (can be one or more), purplish discoloration around site (developing within two to three hours after bite), numbness and possible blistering around the bite, nausea/vomiting, rapid heartbeat, low blood pressure, weakness and fainting, headache, dimmed vision, convulsions, excessive sweating, fever with chills, muscular twitching, and the entire extremity generally swells within eight to thirty-six hours.

5.6.4.2.2. Neurotoxic Venom Types

Some examples of these types of snakes are the coral snake, death adder, common cobra, Egyptian cobra, king cobra, green mamba, sea snakes, and krait. This type of venom affects the nervous system making the victim unable to breath. Local effects include little to no pain or swelling and possible necrosis. Systemic effects include respiratory collapse. Signs and symptoms of a neurotoxic snakebite are: fang marks or small scratches such (as in the case of a coral snake), little to no pain or swelling (if pain does occur it is confined to the bite area), blurred vision, drooping eyelids, slurred speech, increased salivation/sweating, drowsiness, difficulty breathing, paralysis, convulsions, nausea, and coma. Neurotoxic venom effects might not develop for one to eight hours after the bite.

5.6.4.2.3. Neurotoxic and Hemotoxic Venom Types

The venom of the Gabon viper, rhinoceros viper, tropical rattlesnake, and Mojave rattlesnake are both strongly neurotoxic and hemotoxic.

5.6.4.2.4. Snake Bite Treatment

- **What to do if bitten by a venomous snake.** If bitten by a venomous snake, it is important to do the following:
    - Remain calm (this is critical to reduce the spread of venom throughout the body).
    - Keep the extremity at heart level.
    - Disinfect the puncture wound, immobilize the extremity (be aware of swelling that can complicate splinting).
    - Limit movement.
    - Remove any restrictive items such as watches, rings, and boots.
    - Treat for shock.
    - Use an extractor kit if available.
- **What not to do if bitten by a venomous snake.** If bitten by a venomous snake, it is important **not** to:
    - Make incisions across each fang bite.
    - Apply mouth suction.
    - Use ice on the bite.
    - Place a tourniquet on the bitten extremity, or
    - Kill the snake for identification.

5.6.4.3. Spider Bites and Scorpion Stings

Signs and symptoms of a black widow (Figure 5-11) and funnel web spider bites are a pinprick sensation at the site of the bite, becoming a dull ache within thirty to forty minutes. Expect pain and spasms in the shoulder, back, chest, and abdominal muscles within thirty minutes to three hours after the bite. The IP may develop a rigid, board-like abdomen. They will usually become nauseous (with vomiting), restlessness, anxious, and start running a fever and may also develop a rash.

5.6.4.3.1. Brown Recluse Spider

Signs and symptoms of a brown recluse spider (Figure 5-11) bite are mild transitory stinging at the time of the bite, but there is little associated early pain. After two to eight hours, pain will occur varying from mild to severe. Several days later, an ulcer may form at the site of the bite. The bite may also produce serious systemic symptoms including fever, chills, weakness vomiting, joint pain, and a spotty skin eruption, all occurring within 24-48 hours after the venom injection. Scorpion bites will usually involve a sharp pain at the sting site. There will be swelling and discoloration at the sting site, which spreads gradually. Other signs and symptoms may include increased salivation, restlessness, poor coordination, seizures, incontinence, and nausea/ vomiting.

- General treatment for bites and stings is simple. The individual should remain calm. Disinfect the puncture site and in the case of a black widow, funnel web spider, and scorpions bites apply a cold compress to the puncture site if possible. Guard against infection and treat symptomatically and for shock.

Figure 5-11 Black Widow and Brown Recluse

5.6.5. Contact Skin (Dermatitis) Irritation

Common plants that cause skin irritation are poison ivy, poison oak, and poison sumac, but other plants can cause stinging, irritation, or an allergic reaction. Irritant contact, the most common type, involves inflammation resulting from contact with acids or alkaline materials. The reaction may resemble a burn. Allergic contact is caused by exposure to a substance or material to which you have become extra sensitive or allergic. The allergic reaction is often delayed, with the rash appearing 24 - 48 hours after exposure. The skin inflammation varies from mild irritation and redness to open sores, depending on the type of irritant, the body part affected, and your sensitivity.

5.6.5.1. Preventing Skin Irritation

To prevent skin irritation, avoid contact. Use protective gloves and keep sleeves rolled down with other barriers if contact with substances is likely or unavoidable. Wash skin surfaces thoroughly after contact with substances. Avoid over treating skin disorders. Symptoms of skin irritation include itching (pruritus) of the skin in exposed areas, skin redness or inflammation in the exposed area, tenderness of the skin in the exposed area, localized swelling of the skin, warmth of the exposed area, or skin lesions or rash at the site of exposure. Skin lesions which may consist of redness, rash, papules (pimple-like), vesicles, or bullae (blisters) may involve oozing, draining, or crusting and may become scaly, raw or thickened.

5.6.5.2. Initial Treatment

Initial treatment includes thorough washing with lots of water to remove any trace of the irritant that may remain on the skin. You should avoid further exposure to known irritants or allergens. In some cases, the best treatment is to do nothing to the area. Use of tannic acid "tea" poured on to the affected area may reduce some of the symptoms. Jewelweed is known for its ability to counter poison ivy and skin rashes. Jewelweed has been used as a poultice and salve. It is also used for bruises, burns, cuts, eczema, insect bites, sores, sprains, warts, and ringworm. Jewelweed is a smooth annual plant that can grow three to five feet. Trumpet-like shaped flowers hang from the plant in yellow or orange flowers with dark red dots. The Spotted Jewelweed variety is most commonly used for rashes although the Pale Jewelweed may also have medicinal properties.

5.6.5.3. Jewelweed

Jewelweed blooms May through October in the eastern part of North America from Southern Canada to the northern part of Florida. It is found most often in moist woods growing on the edges of creek beds. Slice the stem of the jewelweed plant (Figure 5-12), and then rub its juicy inside on exposed parts. This will promptly ease irritation and usually prevents breakout for most people.

Figure 5-12 Jewelweed Plant

5.7. Marine Animals that Bite

5.7.1. Sharks

Sharks live in almost all oceans, seas, and in river mouths. Sharks vary greatly in size, however, there is no correlation between the size of a shark and the risk of attack (Figure 5-13).

5.7.1.1. Behavior of Hungry Sharks

Hungry sharks sometimes follow fish up to the surface and into shallow waters along the shore. When sharks explore such waters, they are more likely to come in contact with people. Sharks seem to feed most actively during the night and particularly at dusk and dawn. After dark, they show an increased tendency to move toward the surface and into shore waters. Evidence indicates that a shark first locates food by smell or sound. Such things as garbage, body wastes, and blood probably stimulate the desire for food. A shark is also attracted by weak fluttery movements similar to those of a wounded fish. While a shark will investigate any large floating object as a possible food source, it probably will not attack a human unless it is hungry. Often the shark will swim away after investigating. At other times, it may approach and circle the object once or twice, or it may swim close and nudge the object with its snout. People on rafts are relatively safe unless they dangle their hands, arms, feet or legs in the water.

5.7.1.2. Preventing Shark Attack

Individuals on or in the water should keep a sharp lookout for sharks. Clothing and shoes should be worn. If sharks have been noticed, IP must be especially careful of the methods in which body wastes are eliminated and must avoid dumping blood and garbage. Vomiting, when it cannot be prevented should be done into a container or hand and thrown down current as far away as possible. The sea anchor can also be pulled from the water for a short period of time to move away.

5.7.1.2.1. What to do if Threatened by a Shark

If a group in the water is threatened or attacked by a shark, they should bunch together, form a tight circle, and face outward so an approaching shark can be seen. Ward off an attack by kicking or stiff-arming the shark. Striking with the bare hand should be used only as a last resort. Instead, survivors should use a hard and heavy object. During World War II the USS Indianapolis with 1,196 personnel on board was torpedoed by a Japanese submarine. Approximately 900 officers and sailors made it into the water alive before the ship sank twelve minutes later. Shark attacks began with sunrise on the first day and continued until the men were physically removed from the water. The survivors were not rescued until five days later by US Navy Sea Planes. When the survivors were recovered, only 317 remained alive.

5.7.1.3. Staying Quiet when Faced with Threatened by a Shark

Individuals should stay as quiet as possible and float to save energy and swim only if required to get to raft, land, or recovery.

- Do not swim directly away from the shark, but face the shark and swim to one side, with strong rhythmic movements.
- If a shark threatens to attack or damage a raft, jabbing the snout or gills with an oar or other hard object may discourage it. Check for sharks around and under the raft before going into the water.

Figure 5-13 Sharks

5.7.2. Barracuda

There are approximately 20 species of barracuda. Some are more feared in certain parts of the world than are sharks. If IP come down in any tropical or subtropical sea, they may encounter this fish. Barracuda are attracted by anything which enters the water and they seem to be particularly curious about bright objects. Accordingly, IP should avoid dangling dog tags or other shiny pieces of equipment in the water. Dark colored clothing is also best to wear in the water if no raft is available.

5.7.3. Moray Eels

If attacked by some species of moray eel, the IP may have to cut off their heads since some eels will retain their sharp crushing grip until dead. The knife used to do this should be very sharp since their skin is tough and difficult to cut. Their bodies are very slippery and hard to hold. An IP is most likely to come into contact with a moray eel when poking into holes and crevices around or under coral reefs. Use caution in these areas.

5.8. Marine Animals that Sting

Poisonous and Venomous Marine Animals (invertebrates). There are many marine animals that have no backbone and can inflict injuries by stinging.

5.8.1. Coelenterates

This group includes jellyfish, hydroids, sea anemones, and corals. Coelenterates are all simple, many-celled organisms. They all possess tentacles equipped with stinging cells or nematocysts in addition to other technical characteristics.

5.8.1.1. Corals

IP should treat coral cuts by thoroughly cleaning the wound and removing any coral particles. Some coral cuts have been helped by painting them with an antiseptic solution of tincture of iodine.

5.8.1.2. Sea Anemones

The sea anemone is one of the most plentiful marine creatures, with well over 1,000 species. Most of the stinging cells of the sea anemones are located on the outer ring of the tentacles.

5.8.1.3. Venom Apparatus of Coelenterates

All of the coelenterates have stinging cells or nematocysts located on the tentacles. Each of these cells is like a capsule. If the IP comes into contact with the capsule, part of it springs open and a very sharp, extremely small "thread"-type tube appears. The sharp tip of the tube penetrates the skin and the venom is injected. When coming in contact with the tentacles of any coelenterate, the IP brushes up against literally thousands of these small-stinging organs.

5.8.1.3.1. Symptoms of Stings

The symptoms produced by coelenterate stings will vary according to species, where the sting is located, and the physical condition of the IP. In general, though, the sting caused by hydroids and hydroid corals is primarily skin irritations of a local nature. Stings of the Portuguese Man-of-War may be very painful. True corals and sea anemones produce a similar reaction. Some of the sting of these organisms may be hardly noticeable, while others may cause death in three to eight minutes. Symptoms common to all of these may vary from an immediate mild prickly or stinging sensation, like that of touching a nettle, to a burning, throbbing, shooting-type pain which may cause the survivor to become unconscious. In some cases, the pain may be localized, while in others, it may spread to the groin, armpits, or abdomen. The area in which contact was made will usually become red, followed by severe inflammation, rash, swelling, blistering, skin hemorrhages, and sometimes ulceration. In severe cases of reaction, in addition to shock, the person may experience one or more of the following: muscular cramps, lack of touch and temperature sensations, nausea, vomiting, backache, loss of speech, constriction of the throat, frothing at the mouth, delirium, paralysis, convulsions, and death. Since some of these traits appear quickly, the victim should try to get out of the water if at all possible to avoid drowning.

5.8.1.3.2. Jellyfish

One of the most deadly jellyfish is the sea wasp (an uncommon creature which is found in tropical southern Pacific waters) (Figure 5-14). This animal can cause death anywhere from 30 seconds to three hours after contact. Most deaths take place within 15 minutes. The pain is said to be excruciating. The sea wasp can be recognized by the long tentacles that hang down from the four corners of its square shaped body.

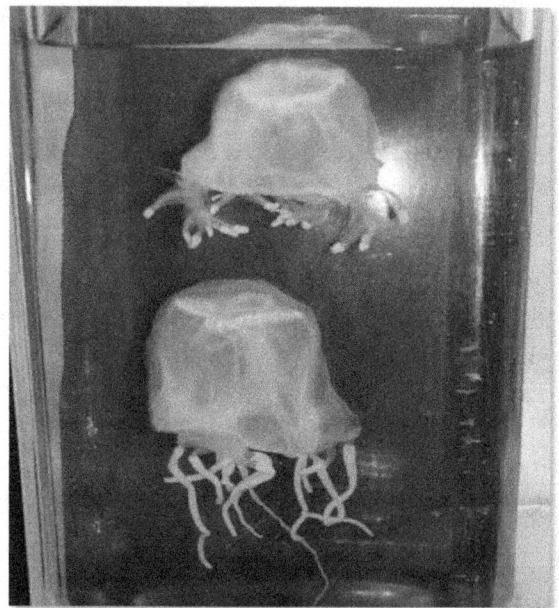

Figure 5-14 Sea Wasp Jellyfish

5.8.1.3.3. Relieve Pain

Tentacles or other matter on the skin should be removed immediately. This is important because as long as this matter is on the skin, additional stinging cells may be discharged. Use clothing, seaweed, or any other available material to remove the matter. DO NOT rub the wound with anything, especially sand, as this may cause the stinging cells to be activated. DO NOT suck the wound.

5.8.1.3.4. Alleviating the Effects of Poison

Alleviate poison effects. The following local remedies have been used in various parts of the world with varying degrees of success: papain (protein destroying enzyme) from the Papaya plant, soap, diluted ammonia solution or urine, (Urine - with its ammonia content - may be the only source of relief available to an IP).

5.8.1.3.5. The Potential need to Perform CPR

Artificial respiration and cardiopulmonary resuscitation may be required. There are no known specific antidotes for most coelenterate stings. However, there is one anti venom for the sea wasp which is papain, a proteolysis enzyme in the juice of the green fruit of the papaya. Even if the IP is in an area where the anti venom is available, it may be too late to obtain and use it. The venom acts so quickly that medical help is often too late.

5.9. Marine Animals that Puncture

5.9.1. Mollusks Octopus, squid, and univalve shellfish are in this category. Mollusks make up the largest single group of biotoxic marine invertebrates of direct importance to the IP. Stinging or venomous mollusks which concern the IP fall mainly into two categories:

5.9.1.1. Gastropoda (Stomach Footers)

These in general are un-segmented invertebrates. Sometimes their soft bodies will secrete a calcareous shell. They have a muscular foot which serves a variety of functions. Some breathe by means of a type of siphon while others use gills. Some types have jaws. In those that don't have jaws, food is obtained by a rasp-like device called a radula. In the cone shells, the radula is a barb or tooth (a hollow, needle-like structure).

- These univalves include marine snails, slugs, as well as land and freshwater snails. It is estimated that there are over 33,000 living species of gastropods. However, only members of the genus conus are of concern to the IP. Of these cone shells, there are over 400 species, but they will only be discussed in general terms with the emphasis placed on the more dangerous species. With few exceptions, these attractive shellfish are located in tropical or subtropical areas. All of these shells have a very highly developed venom apparatus designed for vertebrate or invertebrate creatures and are found from shallow tidal areas to depths of many hundreds of feet. The area in which the IP may come into contact with these shellfish is in coral reefs and sandy or rubble habitat. All cone-shaped shells in these areas should be avoided. Cone shells are usually nocturnal. During the daytime, they burrow and hide in the sand, rocks, or coral; they feed at night on worms, octopus, other gastropods, and small fish. Several of these shells have caused death in humans. The venom apparatus lies within a body cavity of the animal and the animal is capable of thrusting and injecting the poison via the barb into the flesh of the victim. The cone shell is able to inflict its wound only when the head of the animal is out of the shell.

- The sting made by a cone shell is a puncture-type wound. The area around the wound may exhibit one or more of the following: turning blue, swelling, numbness, stinging, or burning sensation. The amount of pain will vary from person to person. Some say the pain is like a bee sting, while others find it unbearable. The numbness and tingling sensations around the site of the wound may spread rapidly, involving the whole body, especially around the lips and mouth. Complete general muscle paralysis may occur. Coma may ensue and death is usually the result of cardiac failure.

- The pain comes from the injection of venom, slime, and other irritating foreign matter into the wound site. The treatment is primarily symptomatic because there is no specific treatment. Applying hot towels or soaking the affected area in hot water may relieve some of the pain. Artificial respiration may be needed.
- This group includes the nautili, squid, cuttlefish, and octopus. Since the octopus is the marine animal most likely to be encountered by an IP, it is the only one that will be discussed. The head of this animal is large and contains well-developed eyes. The mouth is surrounded by eight legs equipped with many suckers. It can move rapidly by expelling water from its body cavity, though it usually glides or creeps over the bottom. Most octopuses live in water ranging from very shallow to depths of over 100 fathoms. All are carnivorous and feed on crabs, and other mollusks. Octopuses like to hide in holes or underwater caves-avoid these areas. The small blue ring octopus is usually only three or four inches across although some may be slightly larger. Found throughout the Indo-Pacific area, this octopus is not aggressive toward humans. Because its venomous bite is so dangerous, it should not be handled at any time. When this animal is disturbed the intensity of its blue rings varies rapidly on a light yellow or cream to brown background (Figure 5-15).
- The sharp parrot-like beak of the octopus makes two small puncture wounds into which a toxic solution or venom is injected. Pain is usually felt immediately in the form of a burning, itching, or stinging sensation. Bleeding from the wound is usually very profuse which may indicate the venom contains an anticoagulant. In some cases, the area around the wound and the entire appendage, may swell, turn red, and feel hot.

Figure 5-15 Blue Ringed Octopus

5.9.1.1.1. Treatment

Treat for shock, stop bleeding, clean the wound area since more venomous saliva could be in the area, and treat symptoms as they arise. There is no known cure for the venom of the Blue Ringed octopus.

5.9.2. Echinoderms

Sea cucumbers, starfish, and sea urchins are members of this group. Sea urchins comprise the most dangerous type of echinoderms. Sea urchins have rounded, egg-shaped, or flattened bodies. They have hard shells that carry spines. In some species, the spines are venomous and present a hazard if stepped on or handled. Some urchins are nocturnal. They all tend to be omnivorous, eating algae, mollusks, and other small organisms. They can be found in tidal pools or in areas of great depth in many parts of the world. Sea urchins are not good food sources. At certain times of the year, certain species can be poisonous.

5.9.2.1. Complications

The needle-sharp points of sea urchin spines are able to penetrate the flesh easily. These spines are also very brittle and tend to break off while still attached to the wound and are very difficult to withdraw. Stepping on one of these spines produces an immediate and very intense burning sensation. The area of pain will also swell, turn red, and ache. Numbness and muscular paralysis, swelling of the face, and a change in the pulse have also been reported. Secondary infection usually sets in. While some deaths have been reported, other victims have experienced loss of speech, respiratory distress, and paralysis. The paralysis will last from 15 minutes to six hours.

5.9.2.2. Treatment

Spines that are detached from the animal will continue to secrete venom into the wound. The spines of some species will be easily dislodged whereas others must be surgically removed. There will also be some discoloration due to a dye the animal secretes -- do not be disturbed by this. Some experts say to apply grease to allow the spines to be scraped off. Others advise leaving them alone since some of the spines will dissolve in the wound within 24 to 48 hours. Still other experts say to apply citrus juice, if available, or soak the area in vinegar several times a day to dissolve them.

5.9.2.3. Prevention

No sea urchin should be handled. The spines can penetrate leather and canvas with ease.

5.9.3. Venomous Spine Fish (Fish That Sting)

Types of fish in this group are Stingrays (includes whip rays, bat rays, butterfly rays, cow-nosed rays, and round stingrays), Rat fish, Catfish, Scorpion fish, Surgeon fish and Star Gazers (Figure 5-16 and Figure 5-17). NOTE: For all wounds from these types of fish, aid should be directed to three areas: alleviating the pain of the sting, trying to halt the effects of the venom, and preventing infection.

5.9.3.1. Pain Inflicted by Fish That Sting

Certain types of these fish have up to 18 spines. The pain caused by the sting of one of these spines is so great in some species that the victim may scream and thrash about wildly. In one case, a man stung in the face by a weever fish begged for bystanders to shoot him, even after two shots of morphine sulfate. Many of these fish are bottom dwellers who will not move out of the way when being approached by humans. Instead, they will lie quietly camouflaged, put up their spines, and simply wait for the unlucky individual to step on them. Other people have been injured by them while trying to remove them from fishing nets and fishing lines.

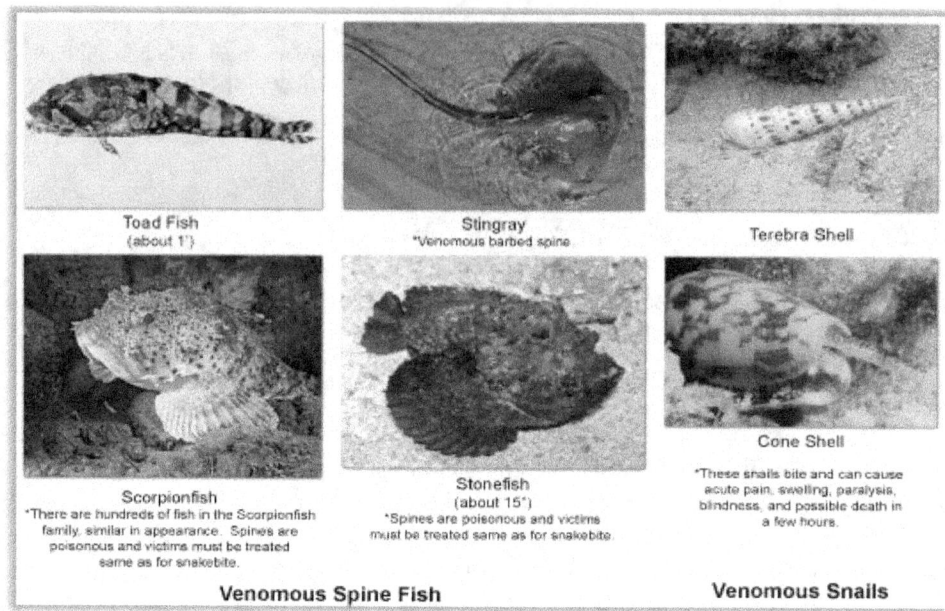

Figure 5-16 Venomous Spine Fish and Snails

5.9.3.2. How to Remove the Barb of a Stingray

In cases where humans are stung by stingrays, the barbs on the sharp spines may cause severe lacerations as well as introduce poison. These wounds should be irrigated without delay. Puncture wounds from the fish are small and make removal of the poison a difficult process. It may be necessary to remove the barb. A procedure which is fairly successful is to make a small cut across the wound (debride) and then apply suction. Even if no incision is made, suction should be tried since it is important to remove as much of the venom as possible. The more poison removed, the better. Morphine does not relieve the pain of some of these venoms.

5.9.3.3. Treating the Sting Injury

Most doctors agree that the injured part should be soaked in hot water from 30 minutes to one hour. The temperature of the water should be as hot as the patient can stand without injury. If the wound is on the face or body, hot moist cloth compresses can be used. The use of heat in this manner may weaken the effect of the poison in some cases. After soaking the wound, clean it again, if necessary. Cover the area of the wound with antiseptic and a clean sterile dressing. If antibiotics are available, it may be advisable to use them to help prevent infection. Treatment for shock is wise. Artificial respiration may be needed since different types of venom may cause cardiac failure, convulsions, or respiratory distress.

Figure 5-17 Fish With Poisonous Flesh

5.9.3.4. Treatment

As soon as any symptoms arise, vomiting should be induced by administering warm saltwater or the whites of eggs. If these procedures don't work, try sticking a finger down the person's throat. A laxative should also be given to the victim if one is available. The victim may have to be protected from injury during convulsions. If the victim complains of severe itching, cool showers may give some relief. Treat any other symptoms as they arise.

5.10. Medical Plants for SERE Scenarios

The Encyclopedia of Herbal Medicine (Chevallier, 2000) suggests that over the course of time, some 70,000 plant species have been used in the practice of medicine; the Chinese Pharmacopeia lists over 5,700 remedies that are mostly of plant origins. With the sheer volume of plants, the topic of herbal medicine can easily be overwhelming. Never the less, by applying a few guidelines, the information can benefit the potential IP. Plants to be used in SERE must be common, easily identified, readily obtainable and they should not have poisonous mimics in nature. They must be safe to use, have minimal side effects, and a high therapeutic index. Finally, the potential IP must know how to prepare and use medicinal plants or they have no value. The potential IP can learn medicinal plants that are available in most regions of the world (global in nature) and/or those medicinal plants native to the deployment location. Plants a potential IP learns should also treat symptoms of conditions (illnesses, wounds, and/or injuries) that are common for the most expected types of isolating event. The following list obtained from *The Encyclopedia of Herbal Medicine* (Chevalier, 2000), contains examples of plants that meet the criteria noted above.

5.10.1. Aloe Vera

Aloe Vera gel directly extracted from the leaf can be applied to treat burns, eczema, and other skin conditions (Figure 5-18).

Figure 5-18 Aloe Vera

5.10.2. Peppers

Cayenne Chili and other pepper species contain capsaicin. Oil infused with peppers can be massaged into the skin to relieve aching muscles and joints. Peppers have expectorant properties that help with congestion; eating them can also help eliminate intestinal parasites (Figure 5-19).

Figure 5-19 Cayenne Chili

5.10.3. Common Plantain

Common Plantain has several medicinal uses. Leaves can be macerated and applied to the skin to stop bleeding or promote wound healing. As an infusion (tea) it can aid in the treatment of diarrhea and gastritis. Plantain is also and expectorant/decongestant similar to Mullein (Figure 5-20).

Figure 5-20 Common Plantain

5.10.4. Dandelion

Dandelion is a great source of vitamins and all parts are edible. Dandelion has diuretic and detoxifying properties. Decoctions are a liquid medicine made from an extract of water-soluble substances, usually with the aid of boiling water. Herbal remedies or decoctions can be made from the root of the Dandelion to treat several skin conditions, arthritis, and constipation (Figure 5-21).

Figure 5-21 Dandelion

5.10.5. Dog Rose

Dog Rose, besides being vitamin rich, tea made from rose hips is used to treat diarrhea (Figure 5-22).

Figure 5-22 Dog Rose

5.10.6. Garlic

Garlic also treats multiple ailments. The juice from the bulbs can be mixed with water and used as an antiseptic to treat skin infections. Garlic cloves can be eaten to treat coughs and bronchitis (Figure 5-23).

Figure 5-23 Garlic

5.10.7. Mullein Leaves

Mullein leaves and flowers make a good expectorant for treating coughs and congestion when infused into a tea. The Aloe-like emollient in the leaves can be applied directly to skin wounds (Figure 5-24).

Figure 5-24 Mullein

5.10.8. Onions

Onions have diuretic, expectorant, anti-inflammatory, analgesic, antibiotic, and anti-rheumatic properties. Onions can be eaten or the juice applied topically; they also prevent oral infections and fight tooth decay (Figure 5-25).

Figure 5-25 Onions

5.10.9. White Willow

White Willow and Black Willow barks contain salicylic acid, a substance that chemically resembles aspirin. It temporarily relieves headache, stomachache, and other body pain. Salicin is metabolized into salicylic acid in the human body, which is a precursor of aspirin. White Willow has anti-inflammatory, analgesic, anti-pyretic, anti-rheumatic, and astringent properties; it can be prepared into a decoction (Figure 5-26).

Figure 5-26 White Willow

5.10.10. Yarrow

Yarrow made into a poultice can help treat wounds. In tea form, Yarrow helps with upper-respiratory infections, indigestion, and to regulate menses. All parts of the plant can be used (Figure 5-27).

Figure 5-27 Yarrow

5.10.11. Sweet Gum

Sweet Gum was traditionally used as incense and medicine well before settlers came to the Americas. The tree produces a light brown sap from the inner bark and wood, which is similar in color to honey from the inner bark and wood. Water from boiled Sweet Gum leaves can also be used as antiseptic for wounds (Figure 5-28).

Figure 5-28 Sweet Gum

5.10.12. Balsam

Balsam tree's boiled down sap is often used to cure dysentery, diarrhea diphtheria, and pulmonic catarrhs. When the balsam is combined with tallow or lard it becomes highly effective at curing skin problems. This mixture has been used to cure skin problems like ringworm. The medicinal properties of the balsam come from extractives found inside it. Cinnamyl cinnamate, ethyl, benzyl, and other forms of cinnamic acid are some of the extractives found in the balsam.

5.10.13. Preparing Plants for Medical Use

As previously stated, identifying plants with medicinal properties is only part of the equation. The potential IP must have some basic instruction on preparation and use. The following is a list of general guidelines for making herbal remedies (Chevallier, 2000).

5.10.13.1. Herbal Teas

Infusions (herbal teas) are made by steeping dried or fresh herbs (about one to two teaspoons for each cup [8 oz.] of water) for 10 minutes in boiling water then straining.

5.10.13.2. Macerations

Macerations are essentially infusions that are made by soaking the herbs in cold water versus boiling. Mixtures are allowed to sit overnight in a cool place and then strained.

5.10.13.3. Decoctions

Decoctions are also similar to infusions, but the process is used for tougher plant material such as roots, barks, seeds, berries, and stems. The plant matter is chopped into thin pieces. Approximately one to two teaspoons of herbs (dried or fresh) is added to 20 ounces of water and brought to a boil. The mixture is allowed to simmer until the volume has reduced by about one-third and then strained.

5.10.13.4. Infused Oils

Infused oils can be prepared by hot or cold methods. Any kind of vegetable oil (olive, canola, sesame, and almond) will work. With cold infusion, the herbs and oil are placed in a jar and allowed to sit for three to four weeks. The hot infusion method involves cooking the mixture over low heat for three to four hours. The mixtures are strained at the end of both processes.

5.10.13.5. Ointments

Ointments are made by adding three to four ounces of infused oil to a small amount (approximately a one inch cube) of melted beeswax. The mixture will thicken to the correct consistency once it cools.

5.10.13.6. Poultices

Poultices are ground herbs that are made into a paste with a small amount of water. The mixture is placed on a piece of gauze and then on the affected area of skin. The poultice is secured with a bandage and can be left in place for up to 24 hours.

5.11. Conclusions

When managing an IP's trauma during an isolating event, remember that the body will do the healing or repairing, and the purpose of the "treater" is to provide the body with the best possible atmosphere to conduct that self-repair. Some general principles include:

- Be in the best possible physical, emotional, and nutritional status before being exposed to a potential survival or captivity setting.
- Minimize the risk of injury at the time of survival or captivity by following appropriate safety procedures and properly using protective equipment.
- Maintain the best possible nutritional status while in captivity or the survival setting.
- Don't over treat!!! Overly vigorous treatment can do more harm than good.
- Use cold applications for relief of pain and to minimize disability from burns and soft tissue strains or sprains.
- Clean all wounds by gentle irrigation with large amounts of the cleanest water available.
- Leave wounds open.
- Splint fractures in a functional position.
- After the bone has healed, begin an exercise program to restore function.
- Remember that even improperly healed wounds or fractures may be improved by cosmetic or rehabilitative surgery and treatment upon rescue or repatriation.
- Common sense and basic understanding of the type of injuries are most helpful in avoiding complication and debilitation. Adequate nourishment and maintenance of physical condition will materially assist healing of burns, fractures, lacerations, and other injuries-the body will repair itself.

Chapter 6
WEATHER

6. Weather

History shows that the prepared IP will be more successful. This is especially true when an IP understands the hardships caused by the environment. Understanding weather patterns and being prepared for the climatic conditions likely to be encountered are extremely important to the IP.

6.1. Knowledge of Weather

Weather is not the same as climate, but rather the current state of the atmosphere with respect to wind, temperature, moisture, and air pressure. Climate, on the other hand, is the type of weather conditions generally prevalent over a region year after year.

6.2. Atmosphere

The atmosphere extends upward from the surface of the Earth for several miles, gradually thinning as it approaches its upper limit. Near the Earth's surface, the air is relatively warm due to contact with the Earth. As altitude increases, the temperature decreases by about 3.0°F for every 1,000 feet until air temperature reaches about 67°F below zero at seven miles above the Earth, this is known as the adiabatic lapse rate. An aircrew member ejecting at altitude or a ground team operating in mountainous terrain comes to understand this relationship quickly.

6.3. Elements Affecting Weather

Weather conditions in the lower layer of the atmosphere (troposphere) and on the Earth are affected by four elements: 1) temperature, 2) air pressure, 3) wind, and 4) moisture. An IP that understands and observes these elements and their corresponding effects on weather can utilize this knowledge to adjust their priorities and actions to meet their survival needs.

6.3.1. Temperature

Temperature is the measure of the warmth or coldness of an object or substance and, for this discussion, the various parts of the atmosphere. The sunlight entering the atmosphere reaches the Earth's surface and warms both the ground and the seas. Heat from the ground and the seas then warms the atmosphere. The atmosphere absorbs the heat and prevents it from escaping into space. This process is called the greenhouse effect because it resembles the way a greenhouse works. Once the Sun sets, the ground cools more rapidly than the air because it is a better conductor of heat. At night, the ground is cooler than the air, especially under a clear dry sky. Temperature also changes near the ground for other reasons. Dark surfaces are warmer than light-colored surfaces. Evening air settles in low areas and valleys creating spots colder than higher elevations. Seas, lakes, and ponds retain heat and create warmer temperatures at night near shore. The opposite is true during the day, especially in the spring when lakes are cold. On the beach, daytime temperatures will be cooler than the temperatures on land further from shore. Knowing this will help determine where an IP should build a shelter.

6.3.2. Air Pressure

Air pressure is the force of the atmosphere pushing on the Earth. The air pressure is greatly affected by temperature. Cool air is denser than warm air. As a result, warm air puts less pressure (less dense) on the Earth than cool air. A low-pressure area is formed by warm air whereas cool air forms a high-pressure area.

6.3.3. Wind

Wind is the movement of air from a high-pressure area to a low-pressure area. The larger the difference in pressure, the stronger the wind. On a global scale, the air around the Equator is replaced by the colder air around the poles. This same convection of air on a smaller scale causes valley winds to blow upslope during the day and down slope along the mountainside at night (diurnal effects). Relatively cool air blows in from the ocean during the day due to the heating and rising of the air above the land and reverses at night (Figure 6-1). This movement of air creates winds throughout the world. When cool air moves into a low-pressure area, it forces the air that was already there to move upward. The rising air expands and cools.

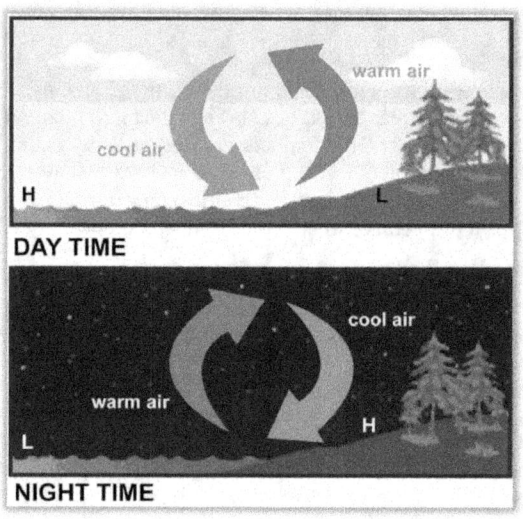

Figure 6-1 Air Transfer (Daytime/Nighttime)

6.3.4. Moisture

Moisture enters the atmosphere in the form of water vapor. Great quantities of water evaporate each day from the land and oceans causing vapor in the air called humidity. The higher the humidity, the higher the moisture content in the air relative to air temperature. Air holding as much moisture as possible is saturated. The temperature at which the air becomes saturated is called the dew point. When the temperature decreases to below the dew point, moisture in the air condenses into drops of water. Low clouds called fog may develop when warm, moist air near the ground is cooled to its dew point with light winds only. A cooling of the air may also cause moisture to fall to the Earth as precipitation (rain, snow, sleet, or hail).

6.3.4.1. Warm Fronts

In warm air over relatively cool air conditions (over-running), when a warm front moves forward, the warm air slides up over the wedge of the colder air lying ahead of it. This warm air has relatively higher humidity. As this warm air is lifted, its temperature is lowered. As the lifting process continues, condensation may occur, low nimbostratus and stratus clouds can form from which rain, mist, or fog may develop. As rain falls through the relatively cooler air below, increasing its moisture content while decreasing visibility. Any reduction of temperature in the colder air and increase in moisture might be caused by upslope motion or cooling of the ground after sunset, may result in extensive fog. As the warm air progresses up the slope, with constantly decreasing temperatures, clouds appear at increasing heights and thickness in the form of altostratus and cirrostratus, if the warm air is stable. If the warm air is unstable, cumulonimbus clouds and altocumulus clouds will form and may produce thunderstorms not visible to the naked eye. Finally, the air is forced up near the stratosphere and in the freezing temperatures at that level; the condensation appears as thin wisps of cirrus clouds. The upslope movement is usually very gradual, rising about 1,000 feet every 20 miles. Thus, the cirrus clouds, forming at approximately 25,000 feet altitude, may appear as far as 300-500 miles in advance of the point on the ground which marks the position of the front. Warm fronts produce more gradual changes in the weather than do cold fronts. The changes depend chiefly on the humidity of the advancing warm air mass. If the air is dry, cirrus clouds may form and there will be little or no precipitation. If the air is humid, light, steady rain or snow may fall for several days. Warm fronts usually have light winds. The passing of a warm front may bring a sharp rise in temperature, clearing skies, and an increase in humidity usually resulting in Southerly wind directions in the Northern hemisphere.

6.4. Storms

The main types of violent weather a person should be familiar with are thunderstorms, winter storms, tornadoes, and hurricanes/cyclones/typhoons (creating thunderstorms which may result in tornadoes). The IP's situational awareness regarding these phenomena will enable them to survive and depending on the situation, use them to their advantage, i.e., evasion and water procurement.

6.4.1. Thunderstorms and Lightning

Thunderstorms are the most frequent kinds of storms (Figure 6-2). As many as 50,000 thunderstorms occur throughout the world each day. Thunderstorms usually occur on warm, humid days. The temperatures inside the clouds are well below freezing which causes water vapor to condense very rapidly forming heavy rain. Lightning causes more fatalities than any other type of weather phenomenon. In the United States alone more than 200 lightning deaths occur each year. Hail which sometimes accompanies the thunderstorm can grow as large as baseballs and bring injuries, even fatalities, to IP if protective actions are not taken.

6.4.2. Lighting and Thunder

Lightning and thunder occur during the life of a thunderstorm. When the IP sees the flash of lightning, they should count the number of seconds until they hear the thunder. For every five seconds, it can be estimated the lightning is one mile away. A general rule to follow is if it is 30 seconds or less, the IP seek shelter or protection until the danger passes. IP should avoid being in, or near, high places, open fields, isolated trees, communications towers, metal fences, open vehicles, and water. The IP **SHOULD NOT** be the highest point on the surface. If an IP is caught in a lightning storm and feel their hair stands on end, their skin tingles, or they hear crackling noises, they should crouch on the ground with their weight on the balls of the feet, feet together, head lowered and their ears covered. Some experts recommend placing the hands on the forehead and elbows on the knees to create a path for lightning to travel to the ground through the extremities rather than through the core (heart).

Figure 6-2 Thunderstorms

6.4.3. Tornadoes

Tornadoes are the most violent entity within thunderstorms. Under certain conditions, violent thunderstorms will generate winds swirling in a funnel shape with rotational speeds of up to 400 miles per hour which extends out of the bottom of the thunderstorm. When this funnel-shaped cloud touches the surface, it can cause major destruction. The path of a tornado is narrow, usually not more than a couple of hundred yards wide. Tornadoes form in advance of a cold front and are usually accompanied by heavy rain and thunder. Indications of an approaching tornado include:

- Strong, persistent rotation in the cloud base.

- Whirling dust or debris on the ground under a cloud base -- tornadoes sometimes have no funnel!

- Hail or heavy rain followed by either dead calm or a fast, intense wind shift. Many tornadoes are wrapped in heavy precipitation and can't be seen.

- Day or night - Loud, continuous roar or rumble, that doesn't fade in a few seconds like thunder.

- Night - Small, bright, blue-green to white flashes at ground level near a thunderstorm (as opposed to silvery lightning up in the clouds). These mean power lines are being snapped by very strong wind, maybe a tornado.

- Night - Persistent lowering from the cloud base, illuminated or silhouetted by lightning -- especially if it is on the ground or there is a blue-green-white power flash underneath.

6.4.3.1. What to do if Caught in Open Wind

If caught in the open during a tornado, lie flat and face-down on low ground (a ditch is best), protecting the back of your head with your arms. Get as far away from trees and cars as you can; they may be blown onto you in a tornado (Figure 6-3).

Figure 6-3 Tornado

6.4.4. Winter Storms

Winter storms include ice storms and blizzards. An ice storm may occur when the temperature is just below freezing. During this storm, precipitation falls as rain but freezes on contact with the ground. A coating of ice forms on the ground and makes it very hazardous to the traveler. Snowstorms with high winds and low temperatures are called blizzards. The wind blows at 35 miles per hour or more during a blizzard, and the temperature may be 10°F or less increasing the risk of exposure injuries. Blowing snow makes it difficult to travel because of low visibility and drifting. An aware IP will be prepared to weather the storm, meet their needs, and use the conditions to cover their movements during evasion. Often these storms paralyze local citizens and municipal services creating opportunities for the IP (Figure 6-4).

Figure 6-4 Winter Storm

6.4.5. Hurricanes or Typhoons

A hurricane or typhoons (which are the same type of weather system with the name changed due to location on the planet) has a far more widespread pattern than a tornado. The storms form near the Equator over the oceans and are a large low-pressure area, about 500 miles in diameter. Winds swirl around the center (eye) of the storm at speeds over 75 miles per hour and can reach 190 miles per hour. Hurricanes break up over land and often bring destructive winds and floods. Thunderstorms often form within hurricanes and can produce tornadoes. Most hurricanes occurring in the United States sweep over the West Indies and strike the southeastern coast of the country. An early indication of a hurricane is a wind from an unusual direction. The arrival of high waves and swells at sea coming from an unusual direction may also give some warning. The high waves and swells are moving faster than the storm and may give several days warning. These swells can create a storm surge that can inundate low lying areas upon land fall and IP should move to higher ground early for protection.

6.4.5.1. Using Severe Winds, Rain, and Surf to the IP's Advantage

The severe winds, rain, and surf associated with hurricanes and typhoons can be both a danger and benefit to the IP. Movement is easily concealed within the storm but precautions must be taken to preserve life. Storm surge near a shore line is by far the most hazardous aspect of this type of event and can be dangerous for days (Figure 6-5).

Figure 6-5 Hurricanes

6.5. Weather Forecasting

Weather forecasting enables IP to make plans based on probable changes in the weather. Forecasts help IP decide what clothes to wear and type of shelter to build. During an evasion situation, it may help IP determine when to travel. While accurate weather prediction or forecasting normally requires special instruments, an awareness of changing weather patterns and attention to existing conditions can help an IP or evader prepare for and, when appropriate, use changing weather conditions to enhance their survivability. The following are some elementary weather indicators which could help predict the weather and help save lives.

6.5.1. Cloud Formations

The types of clouds that form above 20,000 feet above earth include:

- Cirrus - wispy, delicate looking clouds that are the sign of an approaching warm front.
- Cirrocumulus - a layer of tiny individual clouds, which look like scales on a fish, (hence the term mackerel sky) indicating unsettled weather.
- Cirrostratus - an opaque (almost see-through) sheet or layer of clouds, usually indicating the approach of rain within a day or so.

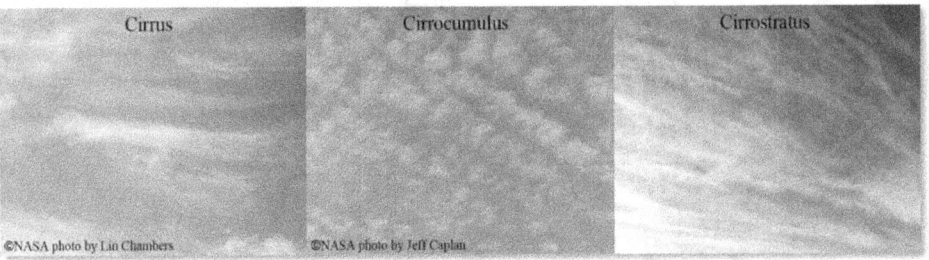

Figure 6-6 Clouds that form above 20,000 feet

6.5.2. Higher Elevation Cloud Types

The types of clouds that form between 7,000 and 20,000 feet above earth include:

- Cumulus - puffy cotton ball or cauliflower shaped clouds, indicating fair weather
- Altocumulus – composed of parallel bands or rounded masses, with a distinct shaded area, which may indicate thundershowers on a warm, humid day.
- Altostratus - a uniformly light gray sheet of clouds, indicating continuous rain or snow.

Figure 6-7 Clouds that form between 7,000 and 20,000 feet

6.5.3. Mid Elevation Cloud Types

The types of clouds that form closest to the Earth's surface, below 7,000 feet include:

- Stratus - a uniformly flat, horizontal, layered cloud, most often associated with fog.
- Nimbostratus - a formless, uniformly dark gray layer of clouds that produce light to moderate precipitation.
- Stratocumulus - a lumpy layer of clouds varying from light to dark gray, that typically produces drizzle or intermittent rain either when bad weather is on the way or when the weather is just about to clear.

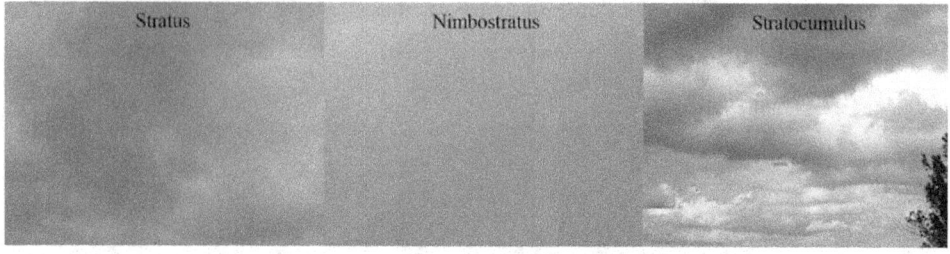

Figure 6-8 Clouds that form below 7,000 feet

6.5.4. Lower Elevation Cloud Types

One type of cloud formation that grows vertically and may extend beyond the tropopause is the Cumulonimbus (Figure 6-7) - these are thunder clouds, which can form as individual clouds or as a line of towers (called a squall line), and are associated with severe weather, including hail, lightning, tornadoes, as well as rain and snow. Notice the "anvil" to the top right of the cloud that helps define the Cumulonimbus formation.

Figure 6-9 Cumulonimbus

6.5.5. Other Weather Indicators

6.5.5.1. Low Hanging Clouds

"Low-hanging" clouds over mountains can mean a weather change especially if they are on the upwind side. If they get larger during the daytime, bad weather will arrive shortly. Diminishing clouds mean dry weather is on its way. Storms are often preceded by high thin cirrus clouds arriving from the west. When these thicken and are obscured by lower clouds, the chances increase for the arrival of rain or snow.

6.5.5.2. Moon, Sun, and Stars

The Sun, and stars are all weather indicators. A ring around the Moon or Sun means rain (Figure 6-10). The ring is created when tiny ice particles in fine cirrus clouds scatter the light of the Moon and the Sun in different directions. When stars appear to twinkle, it indicates that strong winds are not far off, and will become strong surface winds within a few hours. Also, a large number of stars in the heavens at night show clear visibility with a good chance of frost or dew in the morning.

Figure 6-10 Rings around Moon and Sun

6.5.5.3. Red Skies

The old saying "red skies at night, sailor's delight; red skies at morning, sailors take warning," has validity (Figure 6-11). The morning Sun turning the eastern sky crimson often signals the arrival of stormy weather. As the storm moves east, clouds may turn red as a clearing western sky opens for the setting Sun.

Figure 6-11 Red Sunset

6.5.6. What to Look for When Bad Weather is Near

"The farther the sight, the nearer the rain," is a seaman's chant. When bad weather is near, the air pressure decreases. High atmospheric pressure with stable and dusty air means fair weather.

6.5.7. Implications of a Cold Front in the Summer

A cold front arriving in the mountains during the summer usually means several hours of rain and thunderstorms. However, the passing of a cold front associated with stronger winds could mean several days of clear, dry weather. During this type of weather, sound will also travel shorter distances.

6.5.8. Morning Rainbow

A morning rainbow is often followed by rain. An afternoon rainbow often means unsettled weather, while an evening rainbow can mark a passing storm. A faint rainbow around the Sun may precede colder weather.

6.5.9. Increased Flower Fragrance is a Sign of Stormy Weather

Stormy weather may follow within hours when flowers seem to have increased their fragrance.

6.5.10. Looking to Plant Behavior for Signs of Rain

The flowers of many plants, like the dandelion, will close as humidity increases and rain is approaching due to the change in temperature with an updraft associated with a thunderstorm.

6.5.11. Sounds as a Predictor of Rain

People say "when sounds are clear, rain is near," because sound travels farther before storms.

6.5.12. Birds as a Predictor of Rain

Even birds can help predict the weather. Water birds may fly lower than normal across the water when a storm is approaching. Birds will huddle close together before a storm.

6.5.13. Rocks as a Predictor of Rain

As humidity increases, the rocks in high mountain areas will "sweat" and provide an indication of forthcoming rain.

6.5.14. Smoke as a Predictor of a Storm

Smoke, rising from a fire then sinking low to the ground, can indicate that a storm is approaching.

Figure 6-12 Smoke Inversion

6.6. Summary

Even with the modern equipment available, forecasting tomorrow's weather is often difficult. By understanding the basic characteristics that effect development of weather phenomena, the IP can take action to better prepare for and take advantage of its effects and how to meet the IP mission.

Chapter 7
ENVIRONMENT

7. Environment

The more IP know about environmental conditions (i.e., terrain, life forms, and climate), the better they can help themselves in a survival situation. Pre-mission planning and research of these components will better prepare the potential IP for land navigation, evasion, and meeting their other SERE needs. This chapter provides a brief introduction to these topics.

7.1. Terrain

Terrain is defined as a geographic area consisting of land and its features. The landmass of the Earth is covered with a variety of topography, including mountains, valleys, plateaus, and plains.

7.1.1. Mountains

The mountains vary greatly in size, structure, and steepness of slopes. For example, there is as much contrast between the large volcanic Cascade Mountains in the Northwest United States and those of the Hindu Kush Mountains, Afghanistan. Most major mountain systems will have corresponding foothills (Figure 7-1).

Figure 7-1 Composite of Mountains

7.1.2. Valleys

With two exceptions, valleys are formed as mountains are pushed up. The exceptions are massive gorges formed by glacial action and valleys carved out by wind and water erosion (Figure 7-2).

Figure 7-2 Gorges and Valleys

7.1.3. Plateaus

Plateaus are elevated and comparatively large, level expanses of land. Throughout the southwest, examples of the typical plateau can be seen (Figure 7-3). These plateaus were formed when a volcano deposited either lava or ash over a softer sedimentary area. Through years of erosion, the volcanic "cap" broke loose in places and allowed the softer ground to be carried away. This type of plateau is the least common; however, it is the largest. The Columbia Plateau of Washington State is one example which covers 200,000 square miles.

Figure 7-3 Plateau

7.1.4. Water

The water forms of the Earth include oceans, seas, lakes, rivers, streams, ponds, and ice.

7.1.4.1. Oceans

Oceans comprise approximately 70 percent of the Earth's surface. The major oceans include the Pacific, Atlantic, Indian, and Arctic. Oceans have an enormous effect on land, not only in their physical contact but in their effect on weather. In most cases, lakes today are descended from much larger lakes or seas.

7.1.4.2. Ice

Ice covers 10 percent of the Earth's surface. This permanent ice is found in two forms-pack ice and glaciers. Pack ice (normally seven to 15 feet thick) is frozen sea water and may be as much as 150-feet thick. Those pieces which break off form ice islands. The two permanent icepacks on Earth are found near the North (Arctic) and South (Antarctic) Poles. The Polar Icecaps, which are thousands of feet thick, partially, but never completely, thaw. An icecap is a combination of pack ice and ice sheets. The term is usually applied to an ice plate limited to high mountain and plateau areas. During glacial periods, an icecap will spread over the surrounding lowlands (Figure 7-4).

Figure 7-4 Pack Ice

7.2. Life Forms

Life forms can best be described in terms of vegetation and animal life, with special emphasis on humans (which are covered later).

7.2.1. Plants and Trees

There are as many as 400,000 plant species on Earth. An in-depth study is obviously impossible. To understand the plant kingdom better, it is important to understand basic plant functions and adaptations they have made to exist in diverse environments. Vegetation will be categorized into either trees or plants.

7.2.1.1. Trees

Of all the variety in species and types, trees can be divided simply into two types: 1) coniferous or 2) deciduous. Conifers are generally considered to be cone-bearing, evergreen trees. Some examples of conifers are pine, fir, and spruce. Deciduous trees are those which lose their leaves in winter and are generally considered as "hardwood". Some examples are maple, aspen, oak, and alder.

7.2.1.2. Plants

For discussion, we will divide plants into two categories: annuals and perennials. Annuals complete their life cycle in one year. They produce many seeds and regenerate from seed. Climatic conditions may not be conducive for growth the following year, so seeds may remain dormant for many years. A classic example is the 1977 desert bloom in Death Valley. Plants bloomed for the first time in 80 years. Perennials are plants which last year after year without regeneration from seed.

7.2.2. Animal Life

As with plants, the discussion of animal life has to be limited. Animals will be classified as either warm-blooded or cold-blooded. Using this division as a basic, it will be easier to describe animal adaptations to extreme climatic conditions. Warm-blooded animals are generally recognized as cold-adapted animals and include all birds and mammals. Obviously, humans are a part of this classification because they are cold-adapted. Cold-blooded animals gain heat from the environment. These are animals adapted for life in warm or moderate climates (lizards, snakes, etc.).

7.3. Climate

Climate can be described as an average condition of the weather at any given place. However, this description must be expanded to include the seasonal variations and extremes as well as the averages in terms of the climatic elements. In some areas, the climate is so domineering that the corresponding biome is named either in part or as a whole by the climate. Examples are the deserts and rain forests. The climate can be described in terms of its various elements - temperature, moisture, and wind (see chapter 6 for more information on weather).

7.3.1. Atmosphere

The atmosphere gains only about 20 percent of its temperature from the direct rays of the sun. Most of the atmospheric heat gain comes from the Earth radiating that heat (energy) back into the atmosphere and being trapped. This is the greenhouse effect.

7.3.1.1. Water and the Atmosphere

Thinking of the atmosphere as a greenhouse, it is easier to understand the relationship water has in this "closed system." As water evaporates, the amount of water vapor the atmosphere can absorb depends solely upon temperature. The dew point is achieved when the amount of water vapor in the air equals the maximum volume the air will hold at a given air temperature. Lowering of air temperature creates condensation. Condensation appears in the form of clouds, fog, and dew. Any additional temperature reduction results in precipitation, such as rain. If the temperature of the dew point is below freezing, precipitation may appear in the form of hail or snow.

7.3.2. Winds

Variation in air pressure is the primary cause of wind. When air is heated, it creates an area of lower pressure. As air cools, the pressure increases. Air movement occurs as the pressure tries to equalize, thus creating wind. Because wind is also a control of climate, people need to know why and how it affects climate. Let's look at wind in two aspects: localized wind (low altitude) and upper-air wind (high altitude). Localized wind is formed at low altitude, occurring due to dynamic topographical features and fluctuating air temperature and pressure. High-altitude winds surrounding the Earth are bands of stable high- and low-pressure areas (cells). Predictable winds move off these cells which are referred to as jet streams. These high-altitude winds control weather.

7.4. Effects of Climate on Terrain

The major effect climate has on terrain is erosion. Erosion can occur directly from heavy precipitation or indirectly by the accumulation of snow on snowpack and glaciers. Wind and temperature both have erosion potential.

7.4.1. Heavy Precipitation

Heavy precipitation or melting water from icepacks and glaciers can create deep ravines by cutting into mountainous areas. Broad flood basins along major rivers can also aid in the development of river deltas in lakes, oceans, and deep fjords. The action of glaciers throughout the years has carved out deep, broad valleys with steep valley walls (Figure 7-5).

Figure 7-5 Valleys

7.4.2. Effects of Wind Erosion

The effects of wind erosion are greatest in barren, dry areas. The Great Arches National Park has some of the most dramatic examples of the effect of wind erosion. This type of erosion is caused by the wind driving sand and dust particles against an exposed rock or soil surface, causing it to be worn away by the impact of the particles in an abrasive action (Figure 7-6). Another form of wind erosion involves the movement of loose particles lying upon the ground surface which may be lifted into the air or rolled along the ground. Dry river beds, beaches, areas of recently formed glacial deposits, and dry areas of sandy or rocky ground are highly susceptible to this type of erosion. Sand dunes are attributed to this phenomenon (Figure 7-7).

Figure 7-6 Erosion

Figure 7-7 Sand Dunes

7.4.3. Effects of Frost Action

Frost action will have a weathering or eroding effect on rock land formations and ground surfaces. The frost action is the repeated growth and melting of ice crystals in the pore space or fractures of soil and rock. The tremendous force of growing ice crystals can exert a pressure great enough to pry apart rock. Many scree and talus slopes are caused by this action. Scree is mixed gravel and loose dirt debris at the base of a slope, incline, or cliff. Talus is mixed rock (larger than scree) at the base of a slope, incline, or cliff. Where soil water freezes, it tends to form ice layers parallel with the ground surface, heaving the soil upward unevenly. The peat moss mounds of the tundra are an example of this action. The net effect of frost action will be dependent on the amount of surface moisture.

7.5. Effects of Terrain on Climate

The effect of terrain on climate is not nearly as subtle as the effect of climate on terrain. Three major factors exist which must be considered when studying the effects of terrain on weather.

7.5.1. Weather Systems as a Result of Evaporation

Moisture for most major weather systems comes from the evaporation of the oceans of the world. The temperature, location, and flow of ocean currents, combined with the prevailing winds will affect how much water will evaporate into the atmosphere. The warmer the ocean and associated current, the greater the rate of evaporation. Since the currents are deflected by landmasses, many warm currents flow parallel to major continents. When this moisture is blown inland by the prevailing winds, the net effect is the creation of a wet maritime climate, such as that found along the west coasts of Canada, Washington, Oregon, and Central Europe. If the temperature of the ocean and currents is cold, very little moisture will be yielded to the atmosphere. Examples of this occur along the Pacific coastline of Peru and Chile and along the Atlantic coastline of Angola and Southwestern Africa.

7.5.2. Interior of Large Continent Masses

The interior of large continent masses are dry because of the distance which isolates them from the effects of maritime climates. The large continents of the Northern Hemisphere create dry, high-pressure cells which isolate the interiors from the lower pressure moist air cells and keep them from having much effect.

7.5.3. Mountains can Serve as Moisture Barriers

Mountains serve as moisture barriers, separating the maritime influenced climates from the continental influenced climates. The barrier effect of mountains on weather will be dependent on the height, length, and width of the range and the severity of the weather fronts. In many cases, a lack of precipitation will extend for several hundred miles beyond the mountains. An example of this phenomenon occurs in the western states. The Cascade and Sierra Mountains block a great deal of Pacific Ocean moisture from the inland deserts of Washington, Oregon, and Nevada. The Rocky Mountains further block most of the moisture which is left in the atmosphere. Only the high cirrus clouds escape the barrier effect of these mountains. Another example can be seen in Asia. The Himalayan Mountains serve as a very effective barrier, blocking the Asiatic monsoon from central interior Asia, which helps create the Gobi Desert.

7.6. Effects of Climate and Terrain on Life Forms

Since plants require water and light, climate will greatly affect the type and number of plants in an area.

7.6.1. Rainfall and Plentiful Plants

In areas with a great deal of rainfall, plants will be plentiful. In these areas, plants must compete for available sunlight. In areas where the primary vegetation has been knocked down (by clear cutting, landslides, or along flood basins of rivers), a thick secondary growth will occur. In time, the secondary growth, if undisturbed, will become a climax forest. Some of the trees in these areas may grow to 300 feet. Because of the shade, vines and shade-tolerant perennials may sparsely cover the ground.

7.6.2. Plants Compete for Limited Rainfall

In areas where the amount of rainfall is limited, the plants must compete for the available water. The number of plants will also be sparse. Due to the harsh climatic and soil conditions, plant life is typically hardy. Plants have developed the following survival characteristics:

- Production of many seeds which germinate when water does come.
- Shallow root systems gather water quickly when they can.
- Ability to store water (cacti and other succulents).
- Rough, textured leaves (transpiration).
- Production of toxins (kill off competing plants).

7.6.3. How Vegetation is Affected by the Terrain

Vegetation is also affected by the terrain. In mountainous regions, the clouds begin to lose moisture as they pass over the tops of the mountains. The result is more water is available for the growth of vegetation. However, with any increase in elevation, the temperature becomes colder. This exposure to colder temperature has a drastic effect on plant life.

7.6.4. Animal Life in Terrain with Greater Rainfall

Generally, animal life is mostly dependent on two factors: water and vegetation. The greater the rainfall, the greater the corresponding numbers of animals. Conversely, the drier an area, the less vegetation there will be to support animals. The location of small animals is determined by the secondary growth and ground cover, used for protection.

7.6.5. The Affect of Temperature on the Behavior of Animals

Temperature also affects the behavior of animals. For example, animals may burrow to protect themselves from extreme heat or extreme cold or will be more active at night in hot, dry regions. Animals also respond physiologically to temperature extremes. During extreme cold, some species of mammals enter into a state of winter dormancy (hibernation). It is a special case of temperature regulation in which animals lower the setting of their "thermostats" to maintain lower than normal body temperatures in order to save energy while maintaining minimum body functions essential for survival. This is important since their normal food supply is not always available during the winter. During periods of excessive heat, some species of fish, reptiles, mammals, and amphibians will enter into a "summer sleep" called estivation. Estivation is a state in which the animal's body functions and activities are greatly reduced. Estivation and hibernation are not merely a result of temperature regulation but rather are methods by which the organisms survive unfavorable periods.

Chapter 8

GLOBAL CLIMATE CHARACTERISTICS

8. Global Climate Characteristics

Most IP will not have a choice of survival location. The ease or difficulty in maintaining life and honor and returning are dependent on the types and extremes of the climate, terrain, and life forms in the immediate area. The four major climate groups are designated as Tropical, Dry, Warm Temperate, and Snow and Ice Climates. Additionally, from a SERE perspective, IP may need to deal with specific circumstances within each climate such as open sea, costal, urban, evasion, and captivity.

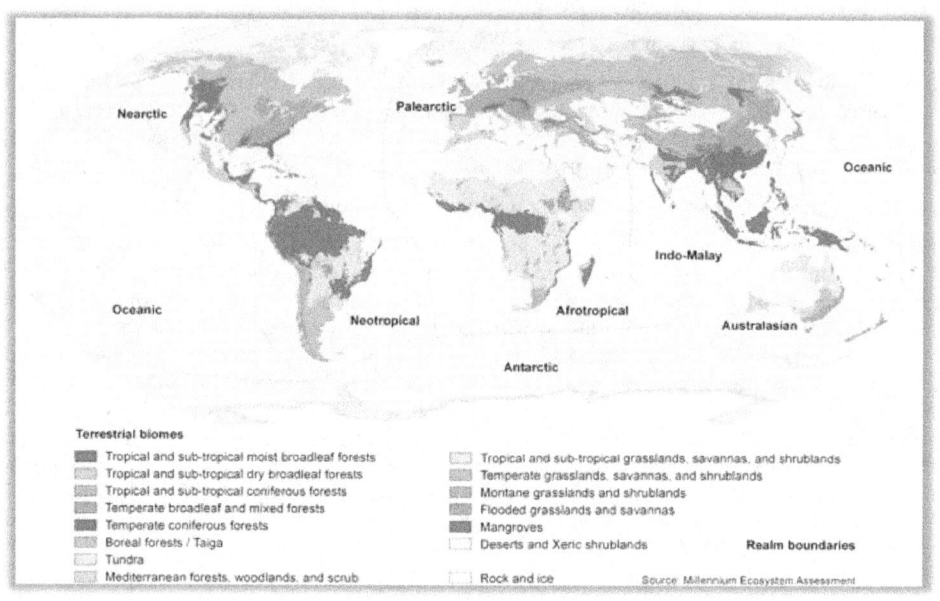

Figure 8-1 Global Climate Bands

- Tropical Climates. Average temperature of each month is above 64.4°F. These climates have no winter season. Annual rainfall is large and exceeds annual evaporation.

- Dry Climates. Potential evaporation exceeds precipitation on the average throughout the year. No water surplus; hence, no permanent streams originate in dry climate zones.

- Warm Temperate Climates. Coldest month has an average temperature under 64.4°F. but above 26.6°F. The warm temperate climates thus have both a summer and winter season.

- Snow and Ice Climates. Snow climates coldest month has an average temperature under 26.6°F. Average temperature of warmest month is above 50°F. Ice Climates average temperature of warmest month is below 50°F. These climates have no true summer.

8.1. Tropical Climates

Some people think of the tropics as an enormous and forbidding tropical rain forest through which every step taken must be hacked out and where every inch of the way is crawling with danger. Actually, much of the tropics are not rain forest. The tropical area may be rain forest, mangrove or other swamps, open grassy plains, or semi-dry brush land. The tropical area may also have deserts or cold mountainous districts. There is in fact, a variety of tropical climates. Each region, while subject to the general climatic condition of its own zone, may show special modifications locally. Each general climate is a whole range of basic minor climates. In all their diversity, the climates of the tropics have the following in common.

8.1.1. Constant Length of Day and Night

An almost constant length of day and night, a length that varies by no more than half an hour at the Equator to one hour at the limits of the tropics. As a result, the plant life thus has an evenly distributed period of daylight throughout the year.

8.1.2. Temperature Variation

Temperature variation throughout the tropics is minimal - 9°F to 18°F. Average temperature of each month is above 64.4°F. These climates have no winter season. Annual rainfall is large and exceeds annual evaporation.

8.1.3. Lack of Systematic Pattern of Major Tropical Landforms

There is no systematic pattern of major tropical landforms. There are high rugged mountains; such as the Andes of South America, karsts formations as in Southeast Asia, plateaus like the Deccan of India, hilly lands like those which back the Republic of Guinea in Africa, and both large and small plains like the extensive one of the upper Amazon River or the restricted plain of the Irrawaddy River in Burma. The arrangement of all these landforms is part of the pattern of the larger land masses, not of the tropics alone.

8.1.4. Tropical Rain Forests

The jungles in South America, Asia, and Africa are more correctly called tropical rain forests. These forests form a belt around the entire globe, bisected somewhat equally by the Equator. However, the tropical rain forest belt is not a continuous one, even in any of the various regions in which it occurs. Usually it is broken by mountain ranges, plateaus, and even by small semi-desert areas, according to the irregular pattern of climate which regulates the actual distribution of rain forest.

8.1.5. Characteristics of Tropical Rain Forests

Some of the leading characteristics of the tropical rain forest common to those areas in South America, in Asia, and in Africa, are:

- Temperatures average close to 80°F for every month.
- Vegetation in the rain forest consist of separate three levels of foliage
- High rainfall (80 inches or more) distributed fairly evenly throughout the year.
- Areas of occurrence lie between 23.5 North and 23.5 South Latitudes.
- Evergreen trees predominate; many have large girths up to 10 feet in diameter, with thick

leathery leaves.
- Vines (lianas) and air plants (epiphytes) are abundant.
- Herbs, grasses, and bushes are rare in the understory.
- Uniformity.
- Tree bark thin, green, smooth and usually lacking fissures.

8.1.6. Plant Life in Tropical Rain Forests

The majority of plants that grow in the forest of the rainy tropics are woody and are similar in size and dimensions of trees. Trees form the principal elements of the vegetation. The vines and air plants that grow on the trunks and branches of trees are woody. Grasses and herbs, which are common in the temperate woods of the United States, are rare in the tropical rain forest. The undergrowth consists of woody plants-seedling and sapling trees, shrubs, and young woody climbers. The bamboos, which are really grasses, grow to giant proportions, 20 to 80 feet high in some cases. Bamboo thickets in parts of some rain forests are very difficult to penetrate. The plants that produce edible parts in the jungle are often scattered, and require searching to find several of the same kind. A tropical rain forest (Figure 8-2) has a wider variety of trees than any other area in the world. Scientists have counted 179 species in one 8.5-acre area in South America. An area this size in a forest in the United States would have fewer than seven species of trees.

8.1.7. Tree Height in Tropical Rain Forests

The average height of the taller trees in the rain forest is rarely more than 150 to 180 feet. Some trees in the tropical rain forest attain 300 feet in height, but this is extremely rare. Trees more than 10 feet in diameter are also rare in the jungle. The trunks are, as a rule, straight and slender and do not branch until near the top. The bases of many of the trees in the rain forest have plank buttresses, (flag-like outgrowths), which are common in all tropical forests. The majority of mature tropical trees have large, leathery, dark-green leaves which resemble laurel leaves in size, shape, and texture. The general appearance is monotonous and large. Strikingly colored flowers are uncommon. Most of the trees and shrubs have inconspicuous flowers, often greenish or whitish in color.

8.1.8. Distribution of Tropical Rain ForestsRain Forest in the Americas

In the Americas, the largest continuous mass of rain forest is found in the basin of the Amazon River. This extends west to the lower slopes of the Andes and east to the Atlantic coast of the continent; it is broken only by relatively small areas of savanna and deciduous forest. This great South American rain forest extends south into the region of the Gran Chaco (south-central South America) and north along the eastern side of Central America into southern Mexico and into the Antilles chain of the West Indies. In the extreme northwest of South America (Ecuador, Colombia), there is a narrow belt of rain forest, separated from the Amazonian forest by a wide expanse of deciduous forest, extending from about latitude 6 degrees South to a little beyond the Tropic of Capricorn. The distribution of rain forest in Central America is perhaps less well known than any other major tropical region. The main areas are below the 500-foot elevation).

8.1.8.2. The Largest Rain Forest Area in the African Continent

In Africa, the largest area of rain forest lies in the Congo basin and extends westward into the Republic of Cameroon. As a narrow strip, the forest continues still farther west, parallel to the

Gulf of Guinea, through Nigeria and the Gold Coast of Liberia and Guinea. Southward from the Congo basin, the forest extends towards Rhodesia.

8.1.8.3. Rain Forest in the Eastern Tropics

In the eastern tropics, the rain forest extends from Ceylon and Western India to Southeast Asia and the Philippines, as well as through the Malay Archipelago to New Guinea. The largest continuous areas are in New Guinea, the Malay Peninsula, and the adjoining islands of Sumatra and Borneo, where the Indo-Malayan rain forest reaches its greatest luxuriance and floral wealth).

8.1.8.3.1. Rain Forest in India

In India, the area of rain forest is not large, but it is found locally in the western and eastern Ghats (coastal ranges) and, more extensively, in the lower part of the eastern Himalayas, the Khasia Hills, and Assam. In Burma and Southeast Asia, the rain forest is developed only locally; the principal vegetation being the monsoon forest. The monsoon forest is a tropical type which is partly leafless at certain seasons. In the eastern Sunda Islands from western Java to New Guinea, the seasonal drought (due to the dry east monsoon from Australia) is too severe for the development of a rain forest, except in locally favorable situations). In Australia, the tropical rain forest of Indo-Malaya is continued south as a narrow strip along the eastern coast of Queensland. Rain forest also extends into the islands of the western Pacific (Solomon, New Hebrides, Fiji, Samoa, etc.).

Figure 8-2 Rain Forest

8.1.9. Semi-Evergreen Seasonal Forest

In character, the semi-evergreen seasonal forest in Central and South America and Africa corresponds essentially to the monsoon forest of Asia. Characteristics of the semi-evergreen forest are:

- Two stories of tree strata-upper story 60 to 80 feet high; lower story 20 to 45 feet high.
- Large trees are rare; average diameter about two feet.
- Seasonal drought causes leaf fall; more in dry years.

The peculiar distribution of the rainy season and the dry season which occurs in the countries bordering the Bay of Bengal in southeastern Asia brings on the monsoon climate. The monsoons of India, Burma, and Southeast Asia are of two types. The dry monsoon occurs from November to April, when the dry northern winds from central Asia bring long periods of clear weather with only intermittent rain. The wet monsoon occurs from May to October, when the southern winds from the Bay of Bengal bring rain, usually in torrents, that lasts for days and often weeks at a time. During the dry season, most leaves drop completely off, giving the landscape a wintry appearance, but as soon as the monsoon rains begin, the foliage reappears immediately.

8.1.10. Tropical Scrub and Thorn Forest

Chief characteristics of the tropical scrub and thorn forest are:

- Definite dry season, with wet season varying in length from year to year. Rains appear mainly as downpours from thunderstorms.
- Trees are leafless during dry season; average height is 20 to 30 feet with tangled undergrowth in places (Figure 8-3).
- Ground is bare except for a few tufted plants in bunches; grasses are not common.
- Plants with thorns are most prominent.
- Fires occur at intervals.

Figure 8-3 Shrub and Thorn Forest

8.1.11. Savannas

General Characteristics of the Savanna include the following:

- Savannas lie wholly within the tropical zone in South America and Africa.
- The savanna looks like a broad, grassy meadow with trees spaced at wide intervals.
- The grasses of the tropical savanna often exceed the height of a man. However, none of the savanna grasses are sod-forming in the manner of lawn grasses, but are bunch grasses with a definite space between each grass plant.

- The soil in the savanna is frequently red.
- The scattered trees usually appear stunted and gnarled like old apple trees.
- Palms may be found on savannas.

8.1.11.1. Savanna of South America

For the most part, the vegetation is of the bunch-grass type. A long, dry season alternates with a rainy season. In these areas, both high and low grasses are present. Bright colored flowers appear between the grass bunches during the rainy season. The grains from the numerous grasses are useful as survival food, as well as the underground parts of the many seasonal plants that appear with and following the rains.

8.1.11.2. Savanna in Africa

The high grass tropical savanna of Africa is dominated by very tall, coarse grasses which grow from five to 15 feet high. Unless the local populace burns the grass during the dry season, the savanna becomes almost impenetrable. This type of savanna occurs in a broad belt surrounding the tropical rain forest and extends from western Africa eastward beyond the Nile River. From the Nile, it extends southward and westward. The tropical bunch grass savanna comprises the greatest part of the African savanna consisting of grasses about three feet tall. The African savanna has both dwarf and large trees. The most renowned of these large trees is the monkey bread or baobab (Figure 8-4).

Figure 8-4 Savanna

8.1.12. Vegetation The abundance of climbing plants is one of the characteristic features of rain forest vegetation. The great majority of these climbers are woody and many have stems of great length and thickness. Stems as thick as a man's thigh are not uncommon. Some lianas cling closely to the trees that support them, but most ascend to the forest canopy like cables or hang down in loops or festoons. In the rain forest, there is no winter or spring, only perpetual midsummer. The appearance of the vegetation is much the same at any time of year. There are seasons of maximum flowering during which more species bloom than at any other time, and also seasons of maximum production of young leaves, plant growth, and reproduction is continuous and some flowers can be found at any time. The margins of a tropical rain forest clearing and areas around abandoned dwellings abound in edible plants. However, in the center of the virgin rain forest, trying to find food is more difficult due to accessibility. The lofty trees are so tall that fruits and nuts are generally out of reach.

8.1.12.1. Food Plants

Some of the available food plants in the rain forest include the Indian or Tropical Almond, Wild Fig, Breadfruit, Rice, Rattan Palm, and Taro. There are many other edible plants in this environment that should be researched through environmental study of your area of operation (AO).

8.1.12.1.1. Plant Foods of the Semi-Evergreen Seasonal Forest

Plant Foods of the Semi-Evergreen Seasonal Forest include the banana, bamboo, chestnut, mango, water lily, and rose apple but there are many others available. Potential IP should conduct study of their AO to identify additional plants available for sustenance. Within the tropical scrub and thorn forest areas, the IP will find it difficult to get food plants in the dry season. During the height of the drought period, the primary kinds of foods come from the Tubers, Bulbs, nuts, gums, seeds and grains. During the rainy season in the tropical scrub and thorn forest, plant food is considerably more abundant. During this time, IP should look for the plants such as the St. John's Bread, Wild Caper, and Air Potato. Many of the food plants found on the savanna are also found in other vegetation areas. Some examples include the Wild Apple, Wild Chicory and Water Plantain.

8.1.13. Animal Life

The tropics abound in animal life. The tremendous varieties of animal species found in tropical areas throughout the world preclude discussions of each animal. It is essential that the survivor realize that just as people have an inherent fear of some animals, most animals also fear people. With some exceptions, animals of the tropics will withdraw from any encounter with humans. Being primarily nocturnal animals, most will never be seen by the survivor. By becoming familiar with the wild inhabitants of the tropics, the survivor will better understand this type of environment and will respect, not fear, the surroundings in which survival takes place.

8.1.13.1. The Pig in Tropical Areas

All tropic areas have members of the pig family. By habit, pigs are gregarious and are omnivorous in diet. They will eat any small animals they can kill, although they feed mainly on roots, tubers, and other vegetable substances. The most common species found in the Old World tropics are the peccary, the Indian wild boar, the Babirussa of Celebes, and the Central African Giant Forest Hog. In Central and South American tropics, peccaries are common. These pigs are represented by two species, the "white-lipped" peccary and the "collared" peccary. Both are grizzled black color, distinguishable by markings from which they derive their names. The white-lipped peccary, the larger of the two (height of approximately 18 inches), is black with white under the snout, and has the reputation of being the more ferocious. The collared peccary, reaching a height of 14 inches, is identified by the white or gray band around the body where the neck joins the shoulder. The collared peccary often travels in groups of five to 15. While alone, they are not particularly dangerous, but a pack can effectively repel any enemy and can make short work of a jaguar, cougar, or human. Both types of peccaries have musk glands which are located four inches up from the tail on the spine. This gland must be removed soon after the animal is killed, otherwise the flesh will become tainted and unfit for consumption.

8.1.13.2. Reptiles and Amphibians in the Tropics

Tropical areas harbor many species of reptiles and amphibians. Most of them are edible when skinned and cooked. Hazards from these animals are mostly imagined; however, some are venomous or dangerous if encountered. Individual species of crocodylidae family (alligators, crocodiles, caiman, and gavials) are usually only abundant in remote areas away from humans. Most dangerous are the saltwater crocodiles of the Far East and the Nile crocodiles in Africa. Poisonous snakes, while numerous in the tropics, are rarely seen and pose little danger to the wary survivor. There are no known poisonous lizards in the tropics. Several species of frogs and toads contain poisonous skin secretions. The large, pan-tropic toad, Bufo Marinus, exudes a particularly irritating secretion if handled roughly. Aside from skin irritations, these amphibians pose little danger to humans, unless the secretion gets into the eyes, where it may cause blurred vision, intense burning, and possible blindness. Tropic areas also have a large number of the various species of mice, rats, squirrels, and rabbits.

8.1.13.3. Deer

Deer are found in most jungle areas; however, their population is normally small. In the Asian jungles, several species of deer frequent the low, marshy areas adjacent to rivers. In the Central and South American jungles, two species of deer are most common. The jungle species is found in thick upland forests. It is much smaller than the North American species and seldom attains a weight of more than 80 pounds. Another deer found in the Central and South American jungles is the "brocket" or "jungle deer." This small reddish-brown deer, which attains a height of about 23 inches, is extremely shy and is found mostly in dense cover since it has no defense against other animals.

8.1.13.4. Insects in the Jungle

The real dangers lie in the insects located in the jungle, which can pass on diseases or parasites.

8.1.13.4.1. Malaria

Malaria is transmitted by mosquitoes, which are normally encountered from late afternoon until early morning. Guard against bites by camping away from swamps on high land and sleep under mosquito netting, if available; otherwise, use mud on the face as a protection against insects. Wear full clothing, especially at night, and tuck pants into the tops of socks or shoes. Wear mosquito head net and gloves. Take anti-malaria tablets (if available) according to directions.

8.1.13.4.2. Ant Species

The greatest number of ant species is found in the jungle regions of the world. Nesting sites may be in the ground or in the trees. Ants can be a considerable nuisance especially if near a campsite. They inflict pain by biting, stinging, or squirting a spray of formic acid. Before selecting a campsite, a close check of the area should be made for any nests or trails of ants.

8.1.13.4.3. Ticks

Ticks may be numerous in many vegetated areas. Use a protected area and undress often, inspecting all parts of the body for ticks, leeches, bed bugs, and other pests. If there are several people in the group, examine each other.

8.1.13.4.4. Fleas

Fleas are common in dry, dusty buildings. The females will burrow under the toenails or into the skin to lay their eggs. Remove them as soon as possible. In India and southern China, bubonic plague is a constant threat. Rat fleas carry this disease and discovery of dead rats usually means a plague epidemic in the rat population. Fleas may also transmit typhus fever and in many parts of the tropics, rats also carry parasites which cause jaundice and other fevers. Keep food in rat-proof containers or in rodent-proof caches. Do not sleep with any food in the shelter!

8.1.13.4.5. Typhus Fever

In many parts of the Far East, a type of typhus fever is carried by tiny red mites. These mites resemble the chiggers of southern and southwestern United States. They live in the soil and are common in tall grass, cut-over jungle, or stream banks. When a person lies or sits on the ground, the mites emerge from the soil, crawl through clothes, and bite. Usually people don't know they have been bitten, as the bite is painless and does not itch. Mite typhus is a serious disease and the survivor should take preventive measures to avoid this pest. The survivor should clear the camping ground and burn it off, sleep above the ground, and treat clothing with insect repellant.

8.1.13.4.6. Leeches

Leeches are primarily aquatic and their dependence on moisture largely determines their distribution. The aquatic leeches are normally found in still, freshwater lakes, ponds, and waterholes. They are attracted by disturbances in the water and by a chemical sense. Land leeches are quite bloodthirsty and easily aroused by a combination of odor, light, temperature, and mechanical sense. These leeches are the most feared of all since they may enter air passages from which they cannot escape once they have fed and become distended. Normally, there is little pain when leeches attach themselves, and after they fill with blood, they drop off unnoticed. Some leeches, living in springs and wells, may enter the mouth or nostrils when drinking and may cause bleeding and obstruction.

8.1.13.4.7. Spiders

Spiders, scorpions, hairy caterpillars, and centipedes are often abundant. The survivor should shake out shoes, socks, and clothing and inspect bedding morning and evening. A few spiders have poisonous bites which may cause severe pain. The black widow and the brown recluse spiders are venomous and should be considered very dangerous. The large spiders called tarantulas rarely bite, but if touched, the short, hard hairs which cover them may come off and irritate the skin. Centipedes bite if touched and their bite is like that of a wasp's sting. Avoid all types of many-legged insects. Scorpions are real pests as they like to hide in clothing, bedding, or shoes and strike without being touched. Their sting can cause illness or death.

8.1.14. Human Population

Density of human population varies with the climate and sub-climates. Cultivation is difficult in areas of tropical rain forests along the Equator. The torrential rains leach out the soil and weeds grow rapidly. Consequently, cultivated food sources must be supplemented by game and other products of the forest. Villages are usually scattered along rivers since movement is easier by water than through the dense forest. Numerous people are also located along coastal areas where farming takes place and people can obtain food from the sea.

8.2. Dry Climates

Dry climates are generally thought of as hot, barren areas that receive scanty rainfall. Rainfall is limited but dry climates are not barren wastelands and many kinds of plants and animals thrive (Figure 8-5).

Figure 8-5 Desert Scrub and Waste Areas

8.2.1. Deserts

Most deserts are located between the latitudes of 15 and 35 degrees on each side of the Equator and are dry regions where the annual evaporation rate exceeds the annual precipitation rate (generally less than 10 inches of rain annually). Extremes of temperature are as characteristic of deserts as is lack of rain and great distances. Hot days and cool nights are usual. A daily low-high spread of 45°F in the Sahara Desert and a 25°F to 35°F difference in the Gobi Desert is the rule. The difference between summer and winter temperatures is also extreme.

- Deserts occupy nearly 20 percent of the Earth's land surface, but only about four percent of the world's population lives there. The term "desert" is applied to a variety of areas. There are alkali deserts, rock deserts, and sand deserts. Some are barren gravel plains without a spear of grass, a bush, or cactus for a hundred miles. In other deserts, such as the Sonora Desert in Southern Arizona, there are grasses and thorny bushes where camels, goats, or even sheep find a subsistent diet. Anywhere they are found; deserts are places of extremes. They can be extremely dry, hot, cold, and often devoid of plants, trees, lakes, or rivers. Most important to the survivor is the extremely long time (distance) between water sources. There are many desert areas and climatic characteristics and seasonal variations of world deserts.

8.2.1.1. Sahara Desert

The Sahara is the Earth's largest desert. It stretches across North Africa from the Atlantic Ocean to the Red Sea and from the Mediterranean and the Sahara Atlas Mountains in the north to the Niger River in tropical Africa. It consists of three million square miles of level plains and jagged mountains, rocky plateaus, and graceful sand dunes. There are thousands of square miles where there is not a spear of grass, not a bush or tree, nor a sign of any vegetation. But Sahara oases-low spots in the desert where water can be reached for irrigation-are among the most densely populated areas in the world. Date groves and garden patches supporting 1,000 people per square mile are surrounded by barren plains devoid of life. Only 10 percent of the Sahara is sandy. The greater part of the desert is flat gravel plain from which the sand has been blown away and accumulated in limited areas forming dunes. There are rocky mountains rising 11,000 feet above sea level and there are a few depressions 50 to 100 feet below sea level. The change from plain to mountain is abrupt in the Sahara. Mountains generally go straight up from the plain like jagged skyscrapers from a city street. Sharp rising mountains on a level plain are especially noticeable in many desert landscapes because there is no vegetation to modify that abruptness. The lack of trees or bushes makes even occasional foothills appear more abrupt than in temperate climates (Figure 8-6).

Figure 8-6 The Sahara

8.2.1.2. Arabian Desert

Some geographers consider the Arabian Desert as a continuation of the Sahara. Half a million square miles in area, the Arabian Desert covers most of the Arabian Peninsula except for fertile fringes along the Mediterranean Sea, Red Sea, Arabian Sea, and the valleys of the Tigris-Euphrates Rivers. Along much of the Arabian coastline, the desert meets the sea. There is more sand in the Arabian Desert than in the Sahara and there are fewer date grove oases. These are on the east side of the desert at Gatif, Hofuf, and Medina. Also, there is some rain in Arabia each year, in contrast to the decades in the Sahara that pass without a drop. Arabia has more widespread vegetation, but nomads find scanty pasture for their flocks of sheep and goats and must depend on wells for water. Oil is carried across the desert in pipelines which are regularly patrolled by aircraft. Pumping stations are located at intervals. All these evidences of modern civilization have increased the well-being of the desert people and, as a result, chances for a safe journey afoot. However, the desert of Arabia is rugged, and native Arabs still get lost and die from dehydration.

8.2.1.3. Gobi Desert - "Waterless Place"

As used here, "Gobi" means only the 125,000 square mile basin or saucer-like plateau north of China which includes Inner and Outer Mongolia. On all sides of the Gobi, mountains form the rim of the basin. The basin itself slopes so gently that much of it appears to be a level plain. The Gobi has rocks, buttes, and numerous badlands, or deeply gullied areas (Figure 8.12.). For a hundred miles or so around the rim of the Gobi, there is a band of grassland. In average years, the Chinese find this to be a productive farmland. In drought years, agriculture retreats. Moving toward the center of the Gobi, there is less and less rainfall; soil becomes thinner, and grass grows in scattered bunches. This is the home of the Mongol herdsman. Their wealth is chiefly horses, but they also raise sheep, goats, camels, and a few cattle. Beyond the rich grassland, the Gobi floor is a mosaic of tiny pebbles which often glisten in sunlight. These pebbles were once mixed with the sand and soil of the area, but in the course of centuries, the soil has been washed or blown away and the pebbles have been left behind as loose pavement. What rain there is in the Gobi drains toward the basin; almost none of it cuts through the mountain rim to the ocean. There are some distinct and well-channeled watercourses, but these are usually dry. Many are remnants of prehistoric drainage systems. In the east, numerous shallow salt lakes are scattered over the plain. They vary in size and number with the changes of rainfall in the area. Sand dunes are found in the eastern and western Gobi, but these features are not as pronounced as they are in certain sections of the Sahara Desert. The Gobi is not a starkly barren wasteland like the Sahara Desert. Grass grows everywhere, although it is often scanty. Mongols live in collective type farm systems and habitations instead of being concentrated in oases (Figure 8-7).

Figure 8-7 The Gobi

8.2.1.4. Australian Desert

More than one-third of Australia's total area is desert. Rainfall in the area is unpredictable with an average of less than 10 inches per year. There are three connecting deserts which occupy Western Australia and one desert located in the center of the continent. They are the Great Sandy Desert, Great Victorian Desert, Gibson Desert, and Simpson or Arunta Desert. The three largest deserts, Sandy, Victoria, and Arunta are of the sandy type, held in place by vegetation. The Gibson in the western portion is a stone-type desert. Most of the deserts of Australia have elevations of 1,000 to 2,000 feet.

8.2.1.5. Atacama-Peruvian Desert of South America

Generally, there are two regions of desert in South America.

8.2.1.5.1. The First Desert Region

The first, and by far the largest, is along the west coast, beginning in the southern part of Ecuador, extending the entire coastline of Peru, and reaching nearly as far south as Valparaiso in Chile. This region, of about 2,000 miles in length and approximately 100 miles in width, is classified as true desert. Even so, along the shoreline of Peru as far south as Africa and inland a few miles, there is often a low-cloud or misty-fog layer. The layer is approximately 1,000 feet thick and produces a fine drizzle. Because of this frequent cloud cover and other phenomena, the temperature along the coastal desert is remarkably cool, averaging about 72°F in the summer daytime and about 55°F in the winter daytime. From about 30 degrees south, the cloud cover does not exist and this region may truly be called rainless. The rare and uncertain showers are valueless for cultivated vegetation. Behind the coastal ranges in the higher elevations, the dryness is at a maximum. In the nitrate fields of the Great Atacama Desert, the air is very dry and the slightest shower is very rare. Here the summer daytime temperatures are from 85°F to 90°F.

8.2.1.5.2. The Second Desert Region

The second desert region is entirely in Argentina (east of the Andes) extending in a finger-like strip from about 30 degrees south, southwest to about 50 degrees south. This region is approximately 1,200 miles long and 100 miles wide. In this highly dissected plateau region, the temperature ranges from a yearly average of 63°F in the north to 47°F in the south. The average annual rainfall pattern is from about 4 inches in the north to about 6 inches in the south.

8.2.1.6. United States and Mexican Deserts

Southwest Deserts of the United States and Mexico have four major subdivisions:

- Great Basin-the basin between the Rocky Mountains and the Sierra Nevada-Cascade Ranges of southern Nevada and western Utah.
- Mojave Desert-Southwestern California.
- Sonora Desert-Southeastern California across southern Arizona into the southwest corner of New Mexico and from Sonora and Baja, California, into Mexico.
- Chihuahua Desert-Lies to the east of the Great Sierra Madre Occidental, spreading north into southwest Texas, southern New Mexico, and southeast corner of Arizona (Figure 8-8).

Figure 8-8 Composite of American Deserts

8.2.1.6.1. Characteristics

The flat plains with scanty vegetation and abruptly rising buttes of our Southwest are reminders of both the Gobi and Sahara. But the spectacular rock-walled canyons found in the Southwest have few counterparts in the deserts of Africa and Asia. The gullied badlands of the Gobi resemble similar formations in both the Southwest and the Dakotas, but our desert rivers - the lower Colorado, lower Rio Grande, and tributaries such as the Gila and the Pecos - have a more regular supply of water than is found in Old World deserts. The Nile and Niger are, in part, desert rivers but get their water from tropical Africa. They are desert immigrant rivers (like the Colorado, which collects the melting snows of the southern Rockies) and gain sufficient volume to carry them through the desert country. In general, the southwest deserts have more varied vegetation, greater variety of scenery, and more rugged landscape than either the Gobi or the Sahara. In all three areas, it is often a long time (distance) between water sources. Death Valley, a part of the Mojave lying in southern California, probably has more waterholes and more vegetation than exist in vast stretches of the Sahara. The evil reputation of the Valley appears to have been started by unwise travelers who were too terrified to make intelligent searches for food and water. The dryness of the Death Valley atmosphere is unquestioned, but it lacks the vast barren plains stretching from horizon to horizon in the Sahara. Compared to the Sahara, the desert country of southwestern United States and Mexico sometimes looks like a luxuriant garden. There are many kinds of cactus plants in the desert, but these are not found in either the Sahara or Gobi.

8.2.1.7. Kalahari Desert

This desert is located in the southern part of Africa. The wasteland covers about 200,000 square miles and lies about 3,000 feet above sea level. Some parts are largely covered with grassland and scrubby trees. The climate is similar to that of the Atacama-Peruvian Desert of South America.

8.2.2. Vegetation

There are some common xerophytes plants (those plants that can live with a limited water supply) which are found in the major deserts of the world. Some examples would be the cactus family, wild onions, tulip, tubers, shrubs, dates, and other succulent plants. More specific information on procurement, preparation, and edibility can be found in the sustenance chapter.

8.2.3. Animal Life

There are over 5,000 species of birds, reptiles, mammals, and insects found in desert areas. The raven, dove, woodpecker, owl, and hawk are common bird species. Reptiles such as lizards and snakes are numerous due to their adaptation to desert areas and ability to conserve body fluids. Many types of mammals live in desert areas and are primarily found near water sources. Using many of the same survival techniques as local animal life can increase the chances of successful survival.

8.2.4. Human Population

Humans are greatly influenced by the presence of water and they live close to rivers, wells, cisterns, or oases. For example, in the Sahara Desert, the people inhabiting this area are located near about 50 desert oases and small coastal cities. In the Gobi Desert, the Mongols live in scattered camps and move from one well to the next as they travel. In the southwest deserts of America, the population is greater along the Colorado and Rio Grande rivers.

8.3. Warm Temperate Climates

The temperate zone is the area or region between the Tropic of Cancer and the Arctic Circle and between the Tropic of Capricorn and the Antarctic Circle. The latitudes which comprise the temperate zone are 23 ½ north and south latitude to 66 ½ north and south latitude.

8.3.1. Types of Climate in a Temperate Zone

There are two main types of climate which comprise the temperate group-a mild type, dominated by oceanic or marine climate; and a more severe one called continental climate.

8.3.1.1. Temperate Oceanic Climate

The temperate oceanic climate is the result of warm ocean currents where the westerly winds carrying moisture have a warming effect on the landmass. This oceanic type climate cannot develop over an extensive area on the eastern or leeward side of large continents in the middle latitudes. The extended effect of the ocean climate can be limited by mountain ranges. Such is the case with the Olympic, Cascade, and Rocky Mountains. As the oceanic weather system moves across the Olympic Mountains, it drops nearly 300 inches of precipitation annually. On the windward side of the Cascades, the annual precipitation ranges from 80 to 120 inches annually. In contrast, the region from the leeward side of the Cascades is a relatively dry area, receiving between 10 to 20 inches precipitation annually. As the system moves across the Rocky Mountains, most of the remaining moisture is lost.

8.3.1.2. Temperate Continental Climate

The temperate continental climate is a land-controlled climate which is a product of broad middle latitude continents. Because of this, the continental climate is not found in the Southern Hemisphere. This type of climate is very characteristic of the leeward side of mountain barriers and eastern North America and Asia. These areas are associated with dry interiors since there are few major warm-water sources available for formation of water systems. The average temperature in the winter and summer are not only extreme but also variable from one year to the next. The severe winter temperature is caused by the polar airflow toward the Equator, and neither winter nor summer temperatures are moderated by the effects of large water masses (oceans).

8.3.2. Variations in Climate in the Temperate Zone

The climate within the temperate zone varies greatly in temperature, precipitation, and wind. The temperate (mid-latitude) zone is divided into four major climate zones which are controlled by both tropic and polar air masses.

8.3.2.1. The Humid Subtropical Zone

The humid subtropical zone is located generally between 20 and 30 degrees north and south latitude. This climate also tends to occur on the east coast of the continents which are at these latitudes. An example of this zone in the United States is the area between Missouri to lower New York and east Texas to Florida. The temperature ranges from 75°F to 80°F in the summer months to 27°F to 50°F in the winter months. The total average precipitation is 30 to 60 inches or more. During the summer months, convectional rainfall is common and thunderstorms frequent. In the winter, the rain is more widespread and is usually associated with passing mid-latitude cyclones. The wind has a great influence in this area. The area is affected by both the prevailing westerly's and eastern trade winds. During the summer, the winds are influenced by eastern moist maritime air mass flows. Winters are influenced by westerly continental polar airflows. The weather is also influenced by low latitudes. The equatorial current which turns pole ward forms warm currents (Gulf Stream and Japanese and Brazilian) that parallel the coasts.

8.3.2.2. The Marine West Coast Climate

The marine west coast climate (Figure 8-9) is sometimes referred to as the temperate oceanic climate. This climate is generally between 40 and 60 degrees north and south latitudes, on the west side of the continent. Examples are the west coasts of Washington to Alaska, Chile, nearly all of Europe, and New Zealand. The summer months are cool with average temperatures of 60°F to 70°F. The winter months are mild with temperatures averaging 27°F to 50°F. The total average rainfall ranges from 20 to 200 inches. Since the maritime climates are under the influence of the westerly winds all year, rainfall is nearly uniform from season to season. These climates are probably cloudier than any other. They are characterized by widespread stratus and nimbostratus clouds and frequent fog. One of the main reasons for the tremendous rainfall in these climatic areas is the warm ocean currents. These currents yield moisture to the air which is blown inland by the westerly winds.

©National Park Service
Figure 8-9 Marine West Coast Climate

8.3.2.3. Middle Latitude Desert and Steppe Climates

Middle latitude desert and steppe climates of complex origins are found generally between latitudes 35 to 50 degrees and in the interior of Asia and North America. Mountain ranges serve as barriers to the moist maritime air masses, thus resulting in low levels of precipitation. In the summer, these interiors generate tropical air masses, while in winter they are overrun by polar air masses originating in Canada and Siberia. Deserts are also characterized by considerable differences between the average summer and winter temperatures.

- Of greater importance are the vast semi-dry steppes. Their annual precipitation of 10 to 20 inches supports short-grass vegetation. They comprise the great sheep and cattle ranges of the world; for example, the veldt of South Africa and the American Great Plains support vast numbers of animals.

- The Mediterranean climate is sometimes referred to as subtropical dry summer climate. It is generally located from 30 to 45 degrees north and south latitudes. Examples of this climate occur in the Mediterranean region, most notably Spain, Italy, and Greece. Summer temperatures usually average 75°F to 80°F; but in coastal locations near cool currents, the average is 5°F to 10°F lower. Typical temperature averages for the coldest months are 45°F to 55°F. Coastal locations are usually somewhat warmer in the winter than inland locations. Total annual rainfall is normally 15 to 30 inches along the equatorial margins and increases pole ward. This climate is a transitional zone between the dry west coast desert and the wet west coast climate. The westerly winds and cold ocean currents are the controlling influences of the Mediterranean climate. An example of a cold current which affects climates is the Humboldt Current (Peru Current) along the coast of Chile, Peru, and California.

8.3.3. Major Topographical Characteristics Major topographical characteristics found in temperate regions include:

- Mountains. Areas of steep slopes with local relief of more than 2,000 feet. Examples of this land form are the Rocky Mountains of North America, the Andes Mountains of South America, and the Himalayan Mountains of Asia.

- High Tablelands.
- Upland surfaces over 5,000 feet in elevation and having local relief of less than 1,000 feet, except where cut by widely separated canyons such as the High Tableland of the Wyoming Basin.
- Hills and Low Tablelands. Hill areas having local relief of more than 325 feet, but less than 2,000 feet. At the ocean shore land, however, local relief may be as low as 200 feet. A low tableland is an area less than 5,000 feet in elevation with local relief less than 325 feet, but which (unlike plains) either does not reach the sea or where it does, terminates in a bluff overlooking a low coastal plain. Examples of this terrain can be found in the Appalachian Mountains, Quebec, and Southern Argentina.
- Plains. Surfaces with local relief of less than 325 feet. On the marine side, the surface slopes gently to the sea. Plains rising continuously inland may attain elevations of high plains-over 2,000 feet. The greatest expanses of plains occur in the center of the North American Continent, Eastern Europe, and Western Asia (Figure 8-10).
- Depressions. Basins surrounded by mountains, hills, or tablelands which abruptly outline the basins. Examples of depressions can be found in the southwestern United States.

Figure 8-10 Plains

8.3.4. Temperate Zone Biomes

There are several biomes of plants and animal life within the temperature zone, and the characteristic life forms are dependent upon climatic characteristics within a specific area. The biomes are named for the plants most plentiful in the area.

8.3.4.1. Coniferous Forests

Coniferous forests (Figure 8-11) occur in a broad band across the northern portions of the continents of North America, Europe, and Asia. The northern boundary is the tundra and the southern limits are generally around 50 degrees north latitude. However, this zone extends down to 35 degrees north latitude in the mountainous regions of the western North American continent. This biome corresponds with the humid continental climate, except in the mountainous portions of North America below 50 degrees north latitude. The main life forms in this zone are the conifers or needle leaf, cone-producing trees, such as pines, firs, spruces, and hemlock. In these areas, the trees may grow closer together, not being severely limited by a need for sunlight, and are subject to frequent fires caused by lightning. When this occurs, the ecological succession is reversed, allowing low shrubs to spring up in the burned-over areas. Although the conifer is the predominate tree, there is more sub-climax or secondary growth in these biomes than in climax forests (mature or primary forests). In these areas, the pines, alders, aspen, and poplars are the dominant trees. The dominant shrubs are heather, small maples, and yews.

Figure 8-11 Coniferous Forest

8.3.4.2. Deciduous Forests

Deciduous (broad leaved) forests are found extensively in the eastern portion of the United States, in Europe, between 40 to 50 degrees north latitude; and also in eastern portions of Russia, China, Korea, and Japan from 35 to 50 degrees north latitude. This biome corresponds with the sub tropic and humid continental climatic zones; the area in which any deciduous forest group determine the predominant trees or climax vegetation found there. For example, in north central United States, Beech and Maple trees assume the dominant role; in Wisconsin and Minnesota, it is Basswood and Maple; in the eastern and southern regions, the dominant trees are Oak and Hickory. There are also spots in this biome where pines and broadleaf evergreens grow.

8.3.4.3. Deciduous and Mixed Deciduous-Coniferous Forest

Deciduous and mixed deciduous-coniferous forests manifest the following characteristics:

- Warm summer with rain; winters cold and drier; short drought periods.
- Only three stories of vegetation (trees, scrubs, herbs).
- Broadleaf trees without leaves in winter.
- Mature trees, uniform in height.
- Unimpeded view into interior of forest.
- Few herbs, ferns, mosses in summer, and abundance of edible fungi in spring and autumn.
- Trunks of trees covered with thick-fissured, dark-colored bark.
- Resting buds enclosed in hard scaly protecting leaves frequently covered with gum or resin.
- For the most part, leaves are thin and delicate, rarely thick and leathery like those of tropical rain forest trees.

8.3.4.3.1. Deciduous and Mixed Deciduous Coniferous Forests

The deciduous and mixed deciduous-coniferous forests that predominate over much of eastern United States are typical of this vegetation type. The deciduous forest is wholly temperate in character. By contrast with the tropical evergreen forest with its richly shaded but chiefly dark glossy green canopy, the broadleaved temperate forest extends in a uniformly bright green expanse. The temperate deciduous and mixed deciduous-coniferous forest vegetation type occupies extensive areas in several parts of the world (Figure 8-12):

- North America. Eastern United States.
- South America. Southern Chile, southeastern Brazil.
- Europe. Western and northern Europe, southern Scandinavia, southeastern Europe (Balkans).
- Asia. South central Siberia, southeastern Siberia and part of Manchuria, Korea (throughout), Japan (throughout), China (throughout except the extreme south and extreme north).
- Oceania. New Zealand.

Figure 8-12 Hardwood (Seasonal)

8.3.5. Steppes and Prairies

The prairie and steppe areas (Figure 8-13) are very closely related. However, the true prairie supports a somewhat different flora than the steppe areas due to the difference in precipitation found in each. In both Steppes and Prairies, the precipitation comes during the short growing season (spring). Summers are hot with intermittent showers.

- Prarie: 30-40 inches of rainfall per year with permanently moist subsoil
- Steppe: 15-30 inches of rainfall per year with permanently dry subsoil

Figure 8-13 Steppes

8.3.5.1. Steppes in Russia

The part of Russia extending from the Volga River through central Asia to the Gobi Desert has been referred to as the steppes. However, as a vegetation type, the steppe grasslands occur in many other parts of the world. The rainfall in steppe areas averages 15 to 30 inches per year, as compared to prairie areas which average 30 to 40 inches per year. The general aspect of a steppe area, like the prairie, is a broad treeless expanse of open countryside which may be quite rolling in places.

8.3.5.2. Main Plants

The main plants in these biomes are grasses. Due to different conditions, various characteristic grasses grow in specific areas on the prairies. The tall grasses are found near the edges of deciduous forests where larger amounts of water are available. The mid-grasses grow farther west, close to the Great Basin within the United States with short grasses growing in the rain shadows of the mountains. Wild flowers and other annuals are found throughout these regions.

8.3.6. Evergreen Scrub Forests

These biomes occur in southern California, in countries around the Mediterranean Sea, and in southern portions of Australia and correspond with the Mediterranean climate (Figure 8-14). The major life form in this area is vegetation composed of broad-leaved evergreen shrubs, bushes, and trees usually less than eight feet tall. This vegetation generally forms thickets. Sage and evergreen oaks are the dominant plants in North America in areas with rainfall between 20 and 30 inches. Areas with less rainfall or poorer soil have fewer, more drought-resistant shrubs such as Manzanita. Scrub forest vegetation becomes extremely dry by late summer. The hot, quick fires that commonly occur during this period are necessary for germination of many shrub seeds and also serve to clear away dense ground cover. This ground cover is difficult for the survivor to penetrate. The branches are tough, wiry, and difficult to bend. Trees are usually widely scattered, except where they occur in groves near a stream. Usually, both trees and shrubs have undivided leaves. Grasses and brightly colored spring-flowering bulbs and other flowers may also be found. The IP will find relatively few kinds of edible plant food within the scrub forest.

Figure 8-14 Evergreen Scrub Forest

8.3.7. Vegetation

If forced to survive in these areas for long periods, especially in winter, the IP will find that edible food plants are scarce. In the temperate environment edible plants are widely varied with bulbs, seeds, resins, cattail, chicory, fiddlehead fern, sorrel and others available. A general characteristic of a climax forest is the stratification of layers of plant growth similar to the canopy systems in the tropical rain forest. In a climax forest, there are usually a limited number of flowering plants, ferns, and shrubs for ground cover. More information is available in the sustenance chapter and through self study.

8.3.7.1. Desert Plants on the Fringe of the Deserts

On the fringes of the desert, desert plants may have moved into the grasslands. The following are some examples of food plants sound in the steppes: Sweet Acacia, Baobab, Cattail, Wild Onion, and Wild Dock. Other examples and specific information can be found in the sustenance chapter and through self study.

8.3.8. Animal Life

Animal life associated with deciduous forests is more varied and plentiful than in evergreen forests, though some animals such as certain species of deer, squirrels, martins, lynx, and wildcats are common in both areas. Wolves, foxes, and other small carnivores (flesh-eating animals) feed mainly on small rodents. Some forest dwellers, such as rodents, dig their dens below the ground while other dens are dug near streams where food and shelter are found. In the aquatic environment, the beaver builds dams for food and shelter. Muskrat, otter, and mink also seek the water's edge, while snakes, turtles, and frogs are found in the streams or lakes.

8.3.8.1. Common Herbivores in Steppes and Prairies

In the Steppes, common herbivores of the prairies are ground squirrels, prairie dogs, rabbits, gophers, and a great many species of mice that can be used for sustenance. A number of birds nest among the grasses. These include the meadowlark, prairie chicken, and grouse. During the dry season, some of these birds migrate to places better suited to raising their young.

8.3.8.2. Insects

Insects like grasshoppers are well adapted to a grassland environment. The natural enemies of such insects are birds and reptiles which in turn become the prey of owls and hawks.

8.3.8.3. Deer and Birds

Deer and birds usually inhabit these forests only during the wet season, which is the growth period for most scrub forest plants. Small dull colored animals such as lizards, rabbits, chipmunks, and quail are year-round residents.

8.3.9. Human Population

The majority of the world's population is found in the temperate zones of the world and may be a benefit or hazard to the IP. As with other environments high density of populations are often found near coastal areas, rivers, and lakes.

8.4. Snow and Ice Climates

8.4.1. Snow Climates

Snow Climates are defined as the interior continental areas of the two great landmasses of North America and Eurasia that lie between 35 and 70 degrees north latitude. The tree line provides the best natural boundary for a topographical description of the snow climate areas. There are definite differences between the forest area to the south and the tundra to the north in snow-cover characteristics, wind conditions, animal types, and vegetation. Snow climates are comprised of two separate climate types: continental subarctic and humid continental.

8.4.1.1. Continental Subarctic Regions

The continental subarctic climate is one of vast extremes. The temperature may range from -108°F to +110°F. Temperatures may also fluctuate 40 to 50 degrees within a few hours. This area includes several climate subtypes. The largest areas run from Alaska to Labrador and Scandinavia to Siberia. They are cold, snowy forest climates, moist all year, with cool, short summers. A colder climate is found in northern Siberia which has very cold winters with an average cold temperature of -36°F. Another area is found in northeastern Asia where the climate is a cold, snowy forest climate with dry winters. Winter is the dominant season of the continental subarctic climate. Because freezing temperatures occur for six to seven months, all moisture in the ground is frozen to a depth of many feet.

8.4.1.2. Humid Continental Regions

The humid continental climates are generally located between 35 and 60 degrees north latitude. For the most part, these climates are located in central and eastern parts of continents of the middle latitudes. These climates are a battle zone of polar and tropical air masses. Seasonal contrasts are strong and the weather is highly variable. In North America, this climate extends from New England westward beyond the Great Lakes region into the Great Plains and into the prairie provinces of Canada. This climate can also be found in central Asia. The summers are cooler and shorter than in any other climate in the temperate zone with the exception of the highland (Alpine) subarctic climate. The summer temperatures range from 60°F to 70°F. The winter temperatures range from -15°F to +26°F. The precipitation for the year varies from 10 to 40 inches. A higher percentage of the precipitation is snow, with less snow occurring in areas along the coasts. The weather is influenced by the polar easterly winds and the subtropical westerly winds. The effect of ocean currents on this continental climate is minimal. This climate is dominated by the high- or low-pressure cells centered in interiors of the continent.

8.4.1.3. Extreme Daylight and Darkness

Both climate regions have seasonal extremes of daylight and darkness resulting from the tilt of the Earth's axis. Snow climate nights are long, even continuous in winter; conversely, north of the Arctic Circle, the Sun is visible at midnight at least once a year. Darkness presents a number of problems to the survivor. No heat is received directly from the Sun in midwinter, thus the cold reaches extremes. Outside activities are limited to necessity, although the light from the Moon, stars, and auroras, shining on a light ground surface, is of some help. Confinement to cramped quarters adds boredom to discomfort, and depression becomes a dominant mood as time drags on. Fortunately, the period of complete darkness does not last long.

8.4.1.4. Terrain of Snow Climates

The terrain of the snow climate areas coincides with a great belt of needle-leaf forests. This region is found in the higher middle latitudes. Its pole ward side usually borders on tundra and its southern margin usually adjoins continental temperate climates. This area is like the tundra because it has poor drainage. As a result, there are an abundance of lakes and swamps. The coastlines vary from gentle plains sweeping down to the ocean to steep, rugged cliffs. Glaciers are a predominate feature of the high altitudes (6,000-feet elevation or above). These glaciers flow down to lower elevations or terminate at the ocean.

8.4.2. Ice Climates

There are three separate climates in the category of ice climates: marine subarctic climate, tundra climate, and icecap climate (Figure 8-15).

Figure 8-15 Tundra

8.4.2.1. Marine Subarctic Climate

Key characteristics of this climate are the persistence of cloudy skies and strong winds (sometimes in excess of 100 miles per hour) and a high percentage of days with precipitation. The region lies between 50 and 60 degrees north latitude and 45 to 60 degrees south latitude. The marine subarctic climate is found on the windward coasts, on islands, and over wide expanses of ocean in the Bering Sea and the North Atlantic, touching points of Greenland, Iceland, and Norway. In the Southern Hemisphere, this climate is found on small landmasses.

8.4.2.2. Tundra Climate

The tundra region lies north of 55 degrees north latitude and south of 50 degrees south latitude. The average temperature of the warmest month is below 50°F. Proximity to the ocean and persistent cloud cover keep summer air temperatures down despite abundant solar energy at this latitude near the summer solstice (Figure 8-16).

Figure 8-16 Summer and Winter Tundra

8.4.2.3. Icecap Climate

There are three vast regions of ice on the Earth. They are Greenland and Antarctic continental icecaps and the larger area of floating sea ice in the Arctic Ocean. The continental icecaps differ in various ways, both physically and climatically, from the polar sea ice and can be treated separately (Figure 8-17).

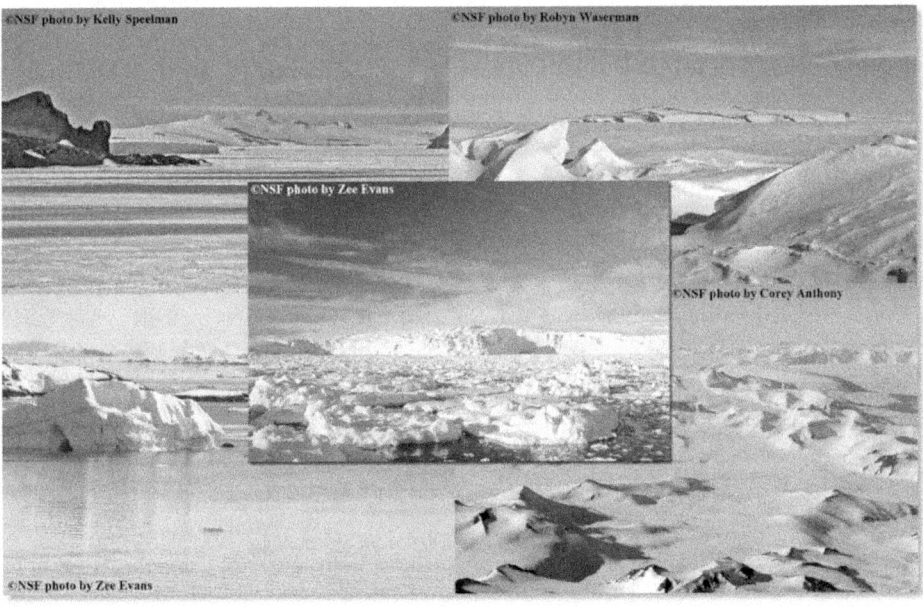

Figure 8-17 Arctic Climate

8.4.2.3.1. Greenland

The largest island in the world is Greenland. Most of the island lies north of the Arctic Circle and ice covers about 85 percent of it. The warmest region of the island is in the southwestern coast. The average summer temperature is 50°F. The coldest region is the center of the icecap. The temperature there averages -53°F in the winter.

8.4.2.3.2. The Antarctic

The Antarctic lies in a unique triangle formed by South America, Africa, and Australia. Surrounding the continent are portions of the Atlantic, Pacific, and Indian Oceans. The area is almost entirely enclosed by the Antarctic Circle. The climate is considered as one of the harshest in the world. The average temperature remains below 0°F all year. In the winter months, the mean temperature is from -40°F to -80°F. Winter temperatures inland often drop below -100°F. Great storms and blizzards (with accompanying high winds) range over the entire area due to both the continent's great elevation and by being completely surrounded by warm ocean water.

8.4.2.3.3. Sea Ice on the Arctic Ocean

Ice on the Arctic Ocean includes frozen sea water and icebergs that have broken off glaciers. This ice remains frozen near the North Pole year around. Near the coast, the sea ice melts during the summer. Currents, tides, and winds may cause it to fold and form high ridges called pressure ridges. One piece of ice may slide over another causing a formation called rafted ice. When the ice breaks into sections separated by water, these sections are called leads. Great explosions and rolling thunder are caused by the breaking and folding of the ice.

8.4.2.4. Terrain

The terrain of the true ice climates encompasses nearly every variation known. Much of the landmass is composed of tundra. In its true form, the tundra is treeless. Vast rugged mountain ranges are found in the area and rise several thousand feet above the surrounding areas. Steep terrain, snow and ice fields, glaciers, and very high wind conditions make this area a very desolate place. Continental glaciers such as the icecaps covering Greenland and the Antarctic continent are large expanses of wind-swept ice moving slowly toward the sea. Ice thickness in continental glacier areas can exceed 10,000 feet.

8.4.2.4.1. Shrub Tundra

In Russia, the area surrounding the Lena River is known as a typical shrub tundra environment. Shrubs, herbs, and mosses occur in this zone. Arctic birch predominates but other shrubs occur and several may be useful as supplementary food such as the crystal tea ledum (Labrador tea), willows, and the bog bilberry. In this same shrub zone a lower herbaceous layer occurs which is composed of black crowberry, several grasses, and the cowberry. On the ground, mosses and lichens are present in abundance. The shrubs on the open tundra reach a height of only three to four feet but in valleys and along the rivers, the same shrubs may reach the height of a person.

8.4.2.4.2. Wooded Tundra

The region immediately adjoining the treeless tundra is an extension of the coniferous areas of the south. These subarctic wooded areas include a variety of tree species of which the genus Picea (Spruce) predominates. On the Kola Peninsula of northeastern Scandinavia, these northernmost forests are birch. Siberian spruce occurs between the White Sea and the Urals. Siberian larch occurs between the Urals and the Pyasina River. Dahurian larch occurs between the Pyasina River and the upper reaches of the Anadyr River. In extreme northeastern Asia, Mongolian poplar, Korean willow, and birch are found along the rivers. The trees extending into the tundra are distinguished by their stunted growth (except in river valleys where they reach 18 to 24 feet) and sparseness. Permanent ground frost, or permafrost, penetrates most parts of the true tundra and the northern limits of the forest belt closely coincide with the southern limits of permafrost. A few different plants will cover very large areas so that extensive stands of a single variety of plant are common in the arctic tundra. All tundra plants are small in stature compared to the plants in the warmer climates of more southerly latitudes. The arctic willow and birch, for instance, spread along the ground in the tundra to form large mats. Stunted growth in all the woody plants is the rule, although there are many evergreen plants and hardy bulbous or tuberous plants. Lichens, especially reindeer moss, are widespread in the tundra. As mentioned before, the plant life of the tundra is remarkably uniform in its distribution. Some species are common to all three areas, but other species are more restricted in their distribution. The tundra also contains many species of vegetation found in the forest regions to the south.

8.4.2.4.3. Bogs

The tundra has often been classified as a continuous bog, but this is far from the truth. Many bogs do exist. There are also many hilly and even mountainous areas with considerably drier soil. The moss or sphagnum bog is less common than the sedge bog. A characteristic of more southern tundra is the development of large peat mounds 9 to 15 feet high and 15 to 75 feet in diameter. These mounds have been formed by ground upheavals caused by freezing water. Many edible plants grow on these bog mounds, such as the cloud-berry, dwarf arctic birch, bog bilberry, black crowberry, crystal tea ledum, sheathed cotton sedge, cowberry, and others.

8.4.3. Vegetation

The vegetation is similar to that found in more temperate zones; however, the cold temperatures have caused variations in the physical appearance of the plants. Dark evergreen forests thrive south of the tree line. They consist mainly of cedar, spruce, fir, and pine, mingled with birch. These subarctic forests are called taigas. A transitional zone lies between the taiga and the tundra. In this zone, the trees are sparse and seldom grow over 40 feet tall. Dwarf willow, birch, and alder mix with evergreens, and reindeer moss sometimes forms a thick carpet.

8.4.4. Animal Life

Depending on the time of year and the place, chances for obtaining animal food vary considerably. Shorelines are normally scraped clean of all animals and plants by winter ice. Inland animals are migratory.

8.4.4.1. Caribou and Reindeer

Caribou and reindeer migrate throughout northern Canada and Alaska. In northern Siberia, they migrate inland to almost 50 degrees north latitude. Some are found in west Greenland. All move close to the sea or into the high mountains in summer. In winter, they feed on the tundra. Musk oxen may be found in northern Greenland and on the islands of the Canadian archipelago. Sheep descend to lower elevations and to valley-feeding grounds in the winter. Wolves usually run in pairs or groups. Foxes are solitary and are seen most frequently when mice and lemming are abundant. Bears are dangerous, especially when wounded, startled, or with their young. They generally shun areas of human habitation.

8.4.4.2. Other Animals

Tundra animals include snowshoe and arctic hare, lemming, mice, and ground squirrels.

8.4.4.3. Poor in Species but Rich in Numbers

Compared to other parts of the world, animal life is poor in species but rich in numbers. Large animals such as caribou, reindeer, and musk oxen migrate through the tundra areas. Carnivores-wolves, foxes, lynx, wolverines, and bears-range through the landmass area and polar bears, seals, walruses, and foxes are found far out on the sea ice. Small animals are the most abundant animal life found and include hares, lemmings, marmots, mink, fishers, and porcupines.

8.4.4.4. Bird Life

Bird life is very limited during the winter months, mainly owls and ptarmigan, but during the summer months millions of migratory waterfowl nest in the arctic tundra. Species include ducks, geese, cranes, loons, and swans, nesting in and around the swamps, bogs, and lakes of the tundra. The coastal areas are home for many species of sea birds during the summer months. The coastal waters and ice flows are rich in a variety of marine life such as seals, walruses, whales, crustaceans, and fishes.

8.4.5. Insect life

Although often overlooked due to the temperature range of the Snow and Ice climates, insects can range from a nuisance to debilitating in this environment. Depending on the time of year mosquitoes and other flying insects can be as thick as to obscure vision. IP need to dress for the environment and be prepared to protect themselves.

8.4.5.1. Freshwater Fish

The freshwater rivers, lakes, and streams may be teeming with many varieties of fish-salmon, trout, and grayling dependent on the time of year. Due to the amount of surface water in the tundra area, there are a large variety of insects. Some 40 to 60 species of mosquitoes, flies, and gnats inhabit the area.

8.4.5.2. Lack of Animal Population in the Antarctic

In the Antarctic, animals are virtually nonexistent. Only the lowest forms of animal life can live mainly on mosses and lichens. Marine animals, particularly whales, seals, and penguins, are found along the coastal regions. Sea birds are abundant in the summer and nest on the coastal regions and the islands. There are a few species of wingless insects, lice, ticks, mites, etc., which live off the bird population.

8.4.6. Human Population

Density is lower due to the climate but larger groups may be found around natural resources, water sources, coastal areas, and lines of communication. IP may also see dwellings found along animal migratory routes. These dwellings may be deserted for months on end and could provide shelter.

Chapter 9
OPEN SEAS

9. Open Seas

A system of climate classification was used to describe the environmental characteristics of the landmasses in previous chapters. However, this system is not used to categorize the largest area of the world-the oceans. They are simply divided by their names and locations. All limits of oceans, seas, etc., are arbitrary, as there is only one global sea. The terms "sea" and "ocean" are often used interchangeably in reference to saltwater. However, from a geographic point of view, a sea is a body of water that is substantially smaller than an ocean or is part of an ocean.

9.1. Seas are a Large Percentage of the Earth's Surface

The seas cover 70.8 percent of the Earth's surface. The waters are not evenly distributed, covering 61 percent of the surface in the Northern Hemisphere and 81 percent in the Southern Hemisphere. Traditionally, the seas are divided into four oceans: Atlantic, Pacific, Indian, and Arctic. These oceans, with their fringing gulfs and smaller seas, make up the world's seas. The sea floor features do, to some extent, influence the surface properties of the seas; that are, currents, waves, and tides.

9.2. Seas are Geographically Aligned along Distinct Boundaries

Within each of these four major oceans, numerous subdivisions known as seas may be geographically aligned along indistinct boundaries (island chains; geography of ocean floor). Examples include:

9.2.1. The Coral Sea

The Coral Sea is an arm of the South Pacific Ocean lying east of Queensland, Australia, and west of New Hebrides and New Caledonia. It extends from the Solomon Islands on the north to the Chesterfield Islands on the south.

9.2.2. The Bearing Sea

The Bering Sea is located between Alaska and Eastern Siberia, with its southern boundary formed by the arc of the Alaskan Peninsula and the Aleutian Islands. The Bering Strait connects it with the Arctic Ocean to the North.

9.2.3. Partially Enclosed Bodies of Water

Many water bodies are partially enclosed by land and are known as gulfs. An example would be the Gulf of Mexico.

9.3. Ocean Climatic Conditions

The ocean has a complex circulation system made up of a variety of currents and countercurrents. These currents move at a rate from barely measureable to about 5.75 miles per hour. They may be relatively cold or warm currents that influence the climate and environment that exists on land and over the ocean. There is a constant movement of water from areas of high density, salinity, concentration, and pressure to areas of low density, salinity, concentration, and pressure in an attempt to establish equilibrium. These factors influence the movement of ocean currents. However, the primary influence on ocean currents is the wind. They also may be diverted by the

Coriolis force and Continental Deflection (Figure 9-1). To fully understand the general climatic conditions and seasonal variations that exist over the global sea, each major ocean must be examined separately, with the exception of the Atlantic and Pacific whose similar latitudinal references result in like characteristics (exceptions will be noted). The two physical phenomena which have the greatest impact upon climate are currents and systems of high and low air pressures.

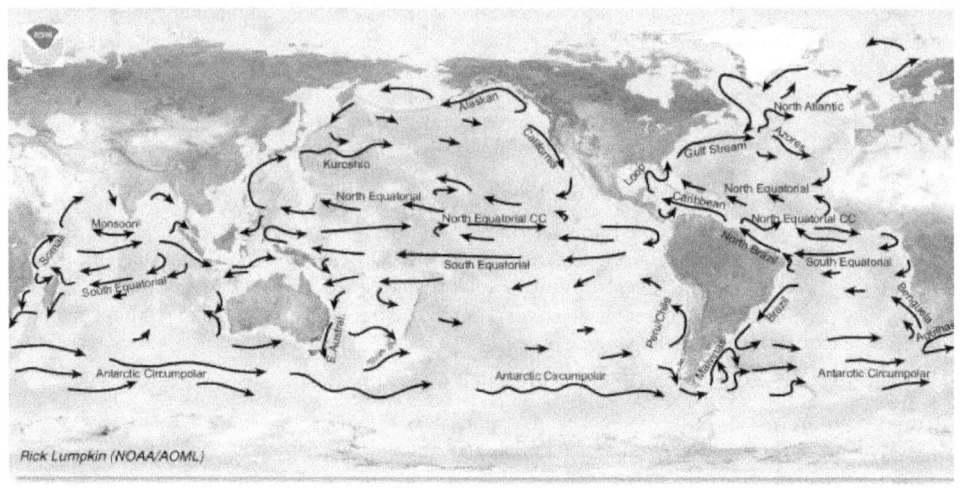

Figure 9-1 Ocean Currents

9.3.1. Currents

Currents with their basic characteristics of being either warm or cold and their inevitable convergence influence the environment of the open seas. Semi- and quasi-permanent centers of high and low atmospheric pressure have equally significant influence on typical weather sequences (for example, temperature, wind, precipitation, and storms).

9.3.2. Low Pressure Creates Storms Seen From Afar

In the Atlantic, Pacific, and Indian Oceans between five degrees north latitude and five degrees south latitude, an equatorial trough of low pressure forms a belt where no prevailing surface winds exist and is known as the doldrums. Instead, the lack of extreme pressure gradients results in shifting winds and calms which exist occur as much as one-third of the time. Intense solar heating results in violent thunderstorms associated with strong squall winds. The convergence of these equatorial winds and trade winds from the inter-tropical front can be seen at a great distance because of towering cumulus clouds rising to 30,000 feet.

9.3.3. Heavy Showers are Common near Inter-Tropical Fronts

In the vicinity of the inter-tropical front, heavy convective showers are quite common. Across the Atlantic and Eastern Pacific, the front is usually north of the Equator. Over the western Pacific, west of 180-degrees longitude, the doldrums belt oscillates considerably. Areas north of the Equator receive their heavy rainfall from June to September. Areas south of the Equator receive their heaviest precipitation between December and March. The meteorological sequence described above may be interrupted by periods of extreme weather centered on low pressure.

9.3.4. The Influence of Ocean Currents on Climate

All ocean currents have a profound influence on climate since the properties of the surface largely determine the properties of the various air masses. The following currents are examples of the currents and their influences.

9.3.4.1. The Humbolt

The cold water of the Peru or Humbolt currents has a tremendous affect on the climate of Peru and Chile. The cold air that lies over the current is warmed as it reaches land, increasing its capacity to hold moisture. The warm air does not give up the moisture until it passes over the high Andes Mountains. This accounts for the dry climate of the coast of Chili and Peru and a more temperate climate toward the Equator than is usually found in the lower latitudes.

9.3.4.2. The Labrador Current

Where the Labrador Current contacts the warm Gulf Stream, fog prevails and steep temperature gradients are present. The northeast coast of North America has much colder climates than the west coast of Europe at the same latitude.

9.3.4.3. Warm Currents in the Caribbean Sea

The warm currents account for the continually warm and pleasant weather in the Caribbean Sea and the Gulf of Mexico.

9.3.4.4. Mild Climates in Northern Europe

The winds blowing off the warm water of the Norwegian and east and west Greenland currents account for the unusually mild climates in northern Europe. At the same latitude elsewhere, the temperatures are usually much colder.

9.4. Procuring Drinking Water on the Open Seas

Seawater should never be ingested in its natural state. It will cause an individual to become violently ill in a very short period of time. When water is limited and cannot be replaced by chemical or mechanical means, it must be used efficiently. As in the desert, conserving sweat, not water, is the rule. IP should keep in the shade as much as possible and dampen clothing with seawater to keep cool. They should not over exert but relax and sleep as much as possible.

9.4.1. Collecting Rainwater for Later Use

If it rains, collect rainwater in available containers and store it for later use. Storage containers could be cans, plastic bags, or the bladder of a life preserver. Sealable containers are preferred to prevent water loss. Every item should be rinsed prior to use in order to reduce factory dyes, chemicals and residual saltwater. Fresh water is ideal over saltwater for rinsing equipment, the IP may have to weight using fresh water for rinsing to avoid "salt contamination" any stored fresh water verses not having fresh water quantities to allow for anything other than drinking. Drinking as much rainwater as possible while it is raining is advisable. If the freshwater should become contaminated with seawater discard it (Figure 9-2). At night and on foggy days, IP should try to collect dew for drinking water by using a sponge, chamois, handkerchief, etc.

Figure 9-2 Collecting Water from Spray Shield

9.4.2. Drink Only Conventional Water

Only water in its conventional sense should be consumed. The so-called "water substitutes" do little for the IP, and may do much more harm than not consuming any water at all. There is no substitute for water. Fish juices and other animal fluids are of little value in preventing dehydration. Fish juices contain protein which requires large amounts of water to be digested and the waste products must be excreted in the urine which increases water loss. IP should never drink urine - urine is body waste material and only serves to concentrate waste materials in the body and require more water to eliminate the additional waste. Blood has also often been falsely identified by IP as a potential water substitute. Blood is in fact a food source and requires additional water to digest it.

9.5. Shelter for Open Seas

Personal protection from the elements is just as important on the seas as it is anywhere else. Some rafts come equipped with insulated floors, spray shields, and canopies to protect IP from heat, cold, and water. If rafts are not equipped or the equipment has been lost, IP should try to improvise these items using parachute material, clothing, or other equipment.

9.6. Life Forms

Life forms in the seas range from one-celled animals (protozoan) to complex aquatic mammals. The fish and aquatic mammals rule the sea and are of the most concern to anyone in an isolating event on the open seas. The majority of fish and mammals can be used as food sources, but some must be considered as a hazard to life. Specific species that can be used for food will be covered in the sustenance chapter while hazardous sea life will be discussed in greater detail in survival medicine.

9.7. Traveling on Open Seas

Although accounts of sea survival incidents are often gloomy, successful survival is possible. The raft will be at the mercy of the currents and winds but IP have the ability to harness these. In the paragraphs discussing open sea travel, the environmental factors of oceans of the world will be considered as they relate to travel. The techniques for individual water rescue and swimming are important for recovering injured IP and equipment. Additionally, the problem of submersion must be considered as well as how the anti-exposure suit, life preservers, and life rafts can extend an IP's life expectancy. This is why an IP must be familiar with the individual rafts and raft procedures. The ultimate goal of an IP on the open seas is to be rescued; however, a second goal, if rescue is not made, would be to make it to land.

9.7.1. Currents

Sea currents flow in a clockwise direction in the Northern Hemisphere and counterclockwise in the Southern Hemisphere. This is caused by three factors: the Sun's heat, the winds, and the Earth's rotation (Coriolis Effect). Most sea currents travel at speeds of less than five miles per hour. Using currents as a mode of travel can be done by putting out the sea anchor and letting the current pull the raft along. IP should use caution when traveling through areas where warm and cold currents meet. It can be a storm forming area with dense fog and high winds and waves.

9.7.2. Winds

Winds also aid raft travel. In tropical areas, the winds are easterly blowing (trade winds). In higher latitudes, they blow from the west (westerlies). To use the winds as a mode of travel, the sea anchor should be pulled into the raft and, if available, a sail should be improvised from the canopy or spray shield. IP should be aware that pulling in the raft's sea anchor may cause it to flip during high winds; this is much more likely with life rafts without ballast buckets.

9.7.3. Waves

Waves can be both an asset and a hazard to raft travel. Waves are normally formed by the wind. The severity of the wind determines the size of the waves. On open seas, waves range from a few inches to over 100 feet in height. Under normal conditions, waves alone will move a life raft only a few inches at a time; therefore, using waves as a mode of propulsion is not practical. Waves are a great help in finding land or shallow areas in the sea. Ocean waves always break when they enter shallow water or when they encounter an obstruction. The force of energy of the wave depends on how abruptly the water depth decreases and the size of the waves. Breaking waves can be used as an aid to make a landfall. Storms at sea are probably the greatest hazard to IP in rafts. Aside from the waves created by a storm, the wind and rain can make life in a raft very difficult. The waves and wind can capsize a raft, or throw a person out of the raft, and then will constantly fill the raft with water; leaving some water in the raft can add to the stability during the storm. Seasickness can result from gentle to severe wave action. Additionally, rescue efforts may be severely hampered by large waves.

9.7.4. Tides

Tides are another form of wave but they are very predictable. They occur twice daily and usually cause no problems for anyone in a raft. Tides may range from one to 40 feet in height depending on the area of the world. They should be considered when planning a landing. When the tide is going in, it will help propel the raft to shore. The action of the water going away from shore when the tide is going out makes landing difficult.

9.7.4.1. Saltwater Crocodile

IP should be aware of the saltwater crocodile. It is found throughout the Southeast Asian shoreline. It is a well known man-eater, and is almost always found in salt or brackish water. It is more commonly found near river mouths and along the coasts. However, it has been known to swim as much as 40 miles out into the sea. Females with nests are likely to be vicious and aggressive. They will grow to a length of 30 feet but most specimens are less than 15 feet in length. IP should watch for this reptile while landing their raft or fishing.

9.7.4.2. Reef Fishes

IP will normally encounter reef fishes during the landing process by stepping on them. They may also be caught while fishing. Clothing and footgear must be worn at all times whether landing the raft or fishing.

9.7.4.3. Coral

Coral is normally found in warm waters, along the shores of islands and mainland. There are many different types of coral. They should be avoided since all can destroy a raft or severely injure an IP. It is best to stay in the raft when coral is encountered. If the IP must wade to shore, footgear and pants should be worn for protection. Moving slowly and watching every step may prevent serious injuries. Coral does not exist where freshwater enters the sea.

9.7.4.4. Visibility by Passing Ships

IP need to be aware that because they see a ship, that it does not mean the ship sees them. Ships can be a welcome sight to the IP, but they can also be a hazard. Since the raft is a small object in a very large sea, the IP must constantly be aware that at night or during inclement weather, the raft will be difficult to see and could be struck by a large ship. It is important to always have signaling devices ready. A lone survivor may attach reflective devices to the top of a raft to maintain signaling efforts during periods of rest. (Note: This is not ideal in non-permissive environments.)

- On the night of February 4, 1982, Steven Callahan was cruising alone in the mid-Atlantic aboard his 21-foot-4-inch sloop. In the middle of the night, his boat was struck by something (he believed was whale) and sunk. That began Callahan's 1,800-mile, two-and-a-half-month oceanic ordeal. Seven times in the course of the 76-day drift, Callahan spotted ships, twice within a mile. He fired his flare gun but there was no one on deck to see it. He wasn't bitter. "That's the way it is," he said. "My anger and frustration couldn't bring them topside to see me." Three times he turned on his emergency position-indicating radio beacon (EPIRB), hoping its signal would reach ships or aircraft, to no avail. Finally, on the night of April 20, he saw the glow of lights on the horizon before him. It was an island. By dawn he could see it, approximately 10 miles ahead, rising black-green in the rolling sea, a thin line of surf creaming at the foot of dark cliffs. Then a boat came into view—a blue-and-yellow, wood-hulled fishing boat with high flaring bows and a crew of three. The fishermen had been attracted by the swirl of birds over Callahan's raft, a sure sign of fish. As they pulled alongside, astonished to see the raft and its bearded, half-naked, emaciated cargo, one of them cried out, "Hey, mon, whatcha doin'?" Callahan was rescued in sight of shore.

9.7.5. Early Considerations for Ditching

With a ditching situation, the aircraft may not immediately sink. This may allow time to gather equipment prior to egress from the sinking aircraft. Historically, pre-planning and training have made the critical difference in IP taking equipment with them during any emergency. Should the aircraft sink prior to egress, IP should be aware that portable emergency oxygen or Helicopter Emergency Egress Device (HEEDS) may be available on the aircraft to assist in underwater egress.

9.7.5.1. Preparation for Ditching

The plan of action for possible aircraft emergencies to include ditching should be reviewed prior to the mission. During preparation for ditching stow lose equipment on aircraft and on body, don any protective clothes under the LPU, if not already on, turn off instrumentation (if applicable), adjust/position seat facing forward or aft (aft being better), have emergency gear close at hand or attached to body (i.e. Emergency Breathing Device (EBD), HEEDS bottle, medical kit, flashlight), if time, review plan of action after impact, and use the buddy system.

9.7.5.2. Bracing for Impact

When bracing for impact, position body as instructed for specific location and seat. Personnel should generally have their lap belt snug and low, grip or locate a point of reference for an exit, and remain seated until the aircraft stabilizes in the water. When aircraft ditch, they tend to impact more than once, and the second impact will normally be more severe than the first. Low wing airframes will have the majority of bottom damage occurring in the mid-section of the aircraft. Double decked aircraft will usually have flooding in the lower compartment. On aircraft with long after bodies (KC-135, EC-135, EC-18), the fuselage may break a few feet forward of the tail. The tail section may break off completely. Low-wing aircraft due to floatation support from the wings tend to ditch better then high wing (C-17 and C-130) aircraft. Low wing aircraft usually will have severe damage just aft or at the nose/cockpit area of the aircraft. High wing aircraft will normally sink rapidly after impact until the wings settle in the water. As a result most of the fuselage is submerged. Externally mounted equipment exerts variations on the aircrafts ditching characteristics.

9.7.5.3. Egress the Aircraft

Using the reference point, the IP should follow their pre-mission brief. Accomplishing assigned tasks and or immediately moves towards their ditching exit. Studies show that in controlled ditching only a small percentage of individuals are seriously injured from the impact, while most fatalities result from failure to board the life rafts. It is always safer for the IP to go from the aircraft directly to the life raft, avoiding any contact with the water if possible. Activate LPU only after exiting aircraft to avoid being pinned or trapped inside the aircraft by the activated floatation device.

9.7.5.4. Launching the Life rafts

In high-wing aircraft life rafts may have only one accessory kit and a lanyard attachment w/free float pull ring for life raft separation from aircraft. In low-wing aircraft, over-wing launching of life rafts should be launched over the leading edge of the wing to avoid jagged edges of debris and flaps. When aircraft lands into the wind, it may be more advantageous to launch life rafts to the side or rear of wing. Due to the crest and trough breaking against the wing, the life raft may get pinned under or dragged over the wing when launching from the leading edge. When the wings are wet they are extremely slick, immediately upon exiting the aircraft activate the LPUs. If possible use safety line, egress ropes/straps, or lanyards to maintain placement while on the wing.

9.7.5.5. Manual Activation of Liferafts

When activating a life raft from a hatch or over wing, have a person at each side of the life raft, one holding the activation lanyard with someone holding onto the accessory kit, when possible.

9.7.5.6. Activating Liferafts with Burning Fuel

IP should stay upwind and clear of debris and fuel covered water. IP may have to activate Life Rafts/LPUs a distance from burning fuel or aircraft wreckage. This can be done by swimming underwater or splashing burning fuel away on the way to the surface, before activation of LPUs and/or life rafts.

9.7.5.7. The Chain-Up

IP will find themselves floating in the water as they wait to board the life raft. To avoid the potential of an IP floating away they should chain-up to maintain safety of the group. The Chain-up is initiated by the IP communicating (shouting) the instruction to "chain up". IP "chain-up" by wrapping their legs around the torso of the next individual and paddling towards the life raft. IP's visibility may be reduced during the "chain-up" process by splashing water. Certain LPUs (such as the Adult/Child) have lights that will activate with water exposure and aid in identification/location during nighttime and inclement weather. Prior to boarding any type of raft, the IP needs to perform the **SWIM**:

- **S** – Separate or soften the LPUs and push them behind their body. IP soften their LPU by letting air out of the oral inflation tube.
- **W** – Wet the boarding ramp/area to include the upper cells.
- **I** – Inspect yourselves for any sharp objects that could cause damage to the raft (i.e., parachute releases, pens, pencils, etc.)
- **M** – Mount the raft.

9.7.5.8. The Five A's

A search for IP is usually started at the last known position or ditching site. Missing personnel may be unconscious and floating low in the water. Rescue procedures are illustrated in Figure 9-3. IP Immediate Actions within the Raft – Once inside of the raft, the IP needs to take care of some immediate action needs. The order of the FIVE A's is situational dependent based on the immediate needs of the IP:

- **A** – Anchor – In all rafts, IP must deploy the sea anchor, crest to trough (if your raft is at the bottom of the wave, you would place your anchor at the top and vice versa) to help reduce drift and increase stability.
- **A** – Assist yourself, as well as others into the raft.
- **A** – Air – Ensure the life raft is fully inflated at all times and make necessary adjustments based on the type of life raft (i.e., 20-man raft equalizer tube). It is difficult to enter a life raft that not fully inflated. IP may need to inflate the life raft before entering.
- **A** – Accessory Kit – All rafts, look to see if the accessory kit is floating or part of the bottom of the life raft (i.e., 46-man and 25-man), if not, it could be under the raft (i.e., 20-man, 7-man).
- **A** – Analyze – As with all isolating events, you must analyze your situation.

9.7.5.8.1. Best Technique for Water Rescue

The best technique for rescuing IP from the water is to throw them a line with a life preserver attached (Figure 9-3). The second best technique is to send a swimmer (rescuer) wearing LPUs from the raft with a line. This will help to conserve energy while recovering the IP. The least acceptable technique is to send an attached swimmer without floatable devices to retrieve an IP. In all cases, the rescuer must wear a life preserver. The strength of a person in a state of panic in the water should not be underestimated. A careful approach can prevent injury to the rescuer. Rescuers must always remain attached to the life raft; they can accomplish this by securing the heaving line or a safety line to themselves. Rafts will travel faster than an IP floating in the water, so once separated, the rescuer may never get back to the life raft.

Figure 9-3 Rescue from Water

9.7.5.8.2. Approaching the IP from Behind

When the rescuer is approaching an IP in trouble from behind, there is little danger of being kicked, scratched, or grabbed. The rescuer should swim to a point directly behind the IP and grasp the back strap of the life preserver. A sidestroke may then be used to drag the IP to the raft.

9.7.5.8.3. Salvaging Debris

All debris from the aircraft should be inspected and salvaged (rations, canteens, thermos and other containers, parachutes, seat cushions, extra clothing, maps, etc). Secure equipment to the raft to prevent loss. Special precaution should be taken with battery operated and easily corroded material to keep them dry so they will function when needed.

9.7.5.8.4. Checking Rafts for Leaks and Inflation

Rafts should be checked for inflation. Leaks and points of possible chafing should be repaired as required. Holes may be identified by water bubbling and/or the sound of hissing. All water should be removed from inside the raft. Care should be taken to avoid snagging the raft with shoes or sharp objects. Placing the sea anchor out will slow the rate of drift. If there is more than one raft, they should be connected with at least one 25 foot line. Donning the anti-exposure suit prior to the aircraft ditching is essential in cold climates. Erecting a wind break, spray shields, and canopies will protect IP from the elements. IP should huddle together and exercise regularly to maintain body heat.

9.7.5.8.5. Monitoring the Physical Condition of the IP

Monitoring the physical condition of IP and administering first aid to survivors is essential. If available, sea sickness pills will help prevent vomiting and resulting dehydration.

9.7.5.8.6. Signaling Equipment

IP should have signaling equipment available. Immediate use should be weighed against inoperability due to wetness. An example would be water getting into the day or night end of a prepared MK-13 flare.

9.7.5.8.7. Keeping Items Dry

IP need to ensure that items affected by salt water are kept dry, i.e., compasses, watches, lighters, etc.

9.7.5.8.8. Securing Raft Repair Plugs

The raft repair plugs should be secured but made easily accessible.

9.7.5.8.9. The Special Aspect of the Open Sea Environment

IP need to consider the special aspect of the open sea environment i.e., reflected sunlight, salt water chafing/sores, immersion, etc (see SERE Medicine chapter 5).

9.7.5.8.10. Senior Ranking IPs Responsibility at Sea

The senior ranking individual should calmly analyze the situation and plan a course of action to include duty assignments (watch duty, procuring and rationing food and water, etc.). All IP, except those who are badly injured or completely exhausted, are expected to perform watch duty, which should not exceed two hours. The IP on watch should look for signs of land, passing vessels or aircraft, wreckage, seaweed, schools of fish, birds, and signs of chafing/ leaking of the raft.

9.7.5.8.11. Conserving Resources

Food and water can be conserved by saving energy. IP should remain calm.

9.7.5.8.12. Maintaining a Positive Mental Attitude

Maintaining a sense of humor and a positive mental attitude will help keep morale high.

9.7.5.8.13. Rescue at Sea is a Cooperative Effort
IP should remember that rescue at sea is a cooperative effort. Search aircraft contacts are limited by the visibility of IP based on the availability of visual or electronic signaling devices. Visual and electronic communications can be increased by using all available signaling devices (signal mirrors, radios, signal panels, dye marker, and other available devices) when an aircraft is in the area.

9.7.5.8.14. Maintaining Records in a Permissive Environment

In a permissive environment, a log should be maintained with a record of last known position, time of ditching, names and physical condition of IP, ration (food and water) schedule, winds, weather, direction of swells, times of sunrise and sunset, and other navigation data.

9.8. Physical Considerations

The greatest problem an IP is faced with when submerged in cold water is death due to hypothermia. When an IP is immersed in cold water, hypothermia occurs rapidly due to the decreased insulating quality of wet clothing and the fact that water displaces the layer of still air which normally surrounds the body. Water causes a rate of heat exchange approximately 25 times greater than air at the same temperature.

9.8.1. Anti-Exposure Garments

It is critical that IP dons an anti-exposure garment prior to ditching. The best protection for an IP against the effects of cold water is to get into the life raft, stay dry, and insulate the body from the cold surface of the bottom of the life raft. Wearing the anti-exposure suit will extend an IP's life expectancy considerably even if it is donned inside the life raft. It is important to keep the head and neck out of the water and well insulated from the cold water effects. Wearing life preservers increases the predicted survival time just as the body position in the water increases the probability of survival.

9.8.1.1. HELP Body Position

Remaining still and assuming the fetal position or the "heat escape lessening posture" (HELP) (Figure 9-4) will increase an IP's survival time. About 50 percent of the heat is lost from the head. It is important to keep the head out of the water. Other areas of high heat loss are the neck, the sides, and the groin. Some side inflation LPUs can be used to bring the knees/legs higher and holding them against the body by reconnecting the LPU under the knees. This allows the IP to take a "Cannon Ball" position. IP should place their arms/hands underneath their thighs/knees in the cannon ball position to lessen fatigue.

Figure 9-4 HELP Position

9.8.1.2. Huddling

If there are several IP in the water, huddling close, side to side in a circle, will preserve body heat (Figure 9-5).

Figure 9-5 Huddling for Temperature Conservation

9.9. Life Preserver Use

An IP who knows how to relax in the water is in little danger of drowning, especially in saltwater where the body is of lower density than the water. Trapped air in clothing will help buoy the IP in the water. If IP are in the water for long periods, the IP will have to rest from treading water. The IP may do this by floating on the back. If this is not possible, the following technique should be used: Rest erect in the water and inhale; put the head face-down in the water and stroke with the arms; rest in this facedown position until there is a need to breathe again; raise the head and exhale; support the body by kicking arms and legs and inhaling; then repeat the cycle. During high sea states, an IP may ingest considerable amounts of seawater from "freestyle" swimming. The action of side breathing can cause illness and or constant spitting that leads to dehydration. The back stroke is much easier and efficient for the average swimmer.

9.9.1. Using Clothing as a Flotation Device

Clothes can be used as a flotation device. Get the head above water. Float on the stomach with arms outstretched, remembering to turn the head to the side to breathe regularly. Maintaining any improvised flotation device's buoyancy is a continual process of refilling the air and securing the bottom seal.

- Use the blouse/shirt as a temporary flotation device. Button closed; tie closed the collar/neck to create a seal. Open bottom of the shirt and vigorously slap the opening down into the water, trapping air. Once air is trapped, tie off the bottom of the shirt. The air will settle into the shoulders of the shirt, adding buoyancy.

- Use the pants/flight suit as a temporary flotation device. Remove the boots and pants/flight suit. Button the pants closed or zip flight suit to the waist and tie each pants leg closed. Care should be taken not to lose boots or other vital equipment which the IP may need later on. Orient the waist of the pants/flight suit fly towards the IP, swing the pants over the IP's head and down into the water in a quick slapping motion to trap air into the legs. Kick hard during this process to avoid toppling over. When the pants legs are full of air, hold the waist closed below the water level. Place an inflated leg under each arm with the waistband in front of the

IP. The pants should now resemble an under-arm life preserver. Keep the waistband submerged and lean into the crotch of the pants/flight suit. Splash the pants legs with water frequently to avoid drying. If the pants dry, air will be able to escape.

9.9.2. Swimming with a Life Preserver

The bulkiness of clothing, equipment, and (or) any personal injuries will necessitate the immediate need for flotation. Normally, a life preserver will be available for donning before entering the water.

9.9.2.1. Inflating the Life Preserver

Proper inflation of the life preserver must be done after clearing the aircraft but, preferably, before entering the water. Upon entering the water, the two cells of the life preserver should be fastened together. Limited swimming may be done with the life preserver inflated by cupping the hands and taking strong strokes deep into the water. The life preserver may be slightly deflated to permit better arm movement.

9.9.2.2. Swimming Strokes

If a group must swim, they should try to have the strongest swimmer in the lead with injured persons intermingled within the group. It is best to swim in a single file. Swimming is easier with LPUs when a modified swimming stroke is incorporated.

- To travel long distances with the least expenditure of energy, the backstroke will serve the IP well. To perform this stroke, the IP should lay on their back, kick their feet, and reach above the shoulders with cupped hands then bring the hands to the sides as if doing jumping jacks. The main drawback is the difficulty seeing where the IP is going.
- The sidestroke is best for pulling along an injured IP or equipment. To perform this stroke, the IP should lay on their side, grasp the object they are carrying with their top hand, kick their feet, and reach out with the opposite hand. Cup the hand and make a downward motion pulling the hand down to their side.
- When the IP needs to approach an object (i.e., raft, rescue devices, etc.), the breaststroke is the best method to get there. To perform this stroke with side inflation LPUs, ensure the LPUs are reconnected in front of the IP, and then push them under the body so they are at about stomach level. Roll forward on to the belly, reach the arms out in front with hands cupped, kick the feet, and bring the hands back to the IP's sides. For horse-collar style LPUs the process is the same minus the reconnect.

9.10. Raft Procedures

There are three needs which can be satisfied by most rafts; they are (1) personal protection, (2) mode of travel, and (3) evasion/camouflage.

9.10.1. One-Man Raft

The one-man raft has a main cell inflation chamber. If the CO_2 bottle malfunctions and does not inflate the raft or if the raft develops a leak, it can be inflated orally. The spray shield acts as a shelter from the cold, wind, and water. In some cases, this shield serves as insulation. The insulated bottom plays a significant role in the IP's protection from hypothermia by limiting the conduction of the cold through the bottom of the raft (Figure 9-6).

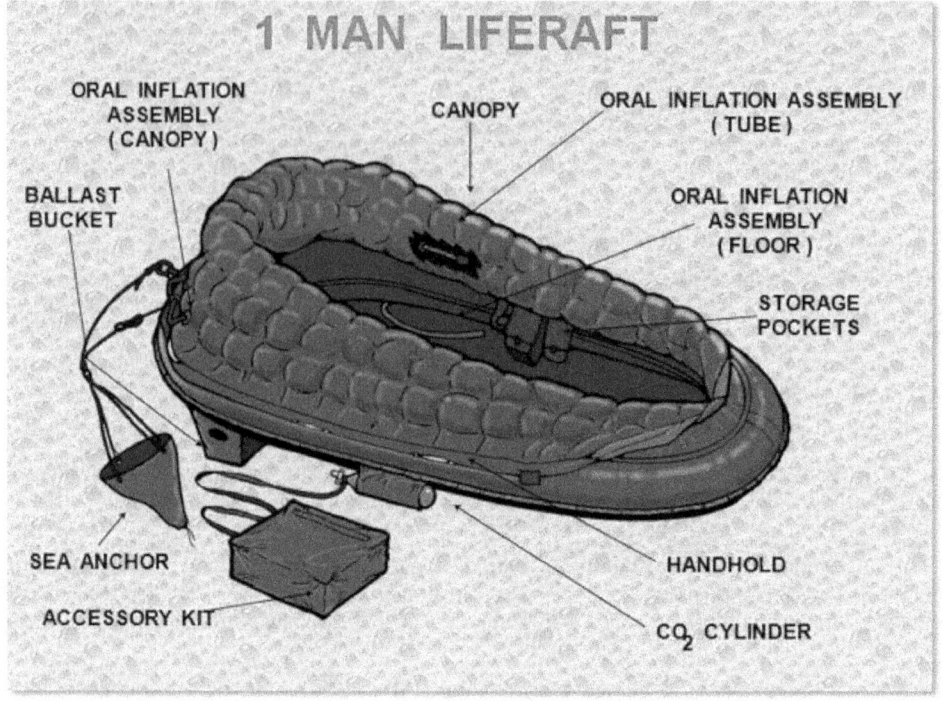

Figure 9-6 One-Man Raft with Spray Shield

9.10.1.1. Taking Advantage of the Wind

Travel is more effectively made by inflating or deflating the raft to take advantage of the wind or current. The spray shield can be used as a sail while the ballast buckets serve to increase raft drag in the water. The last device which may be used to control the speed and direction of the raft is the sea anchor. The primary purpose of the sea anchor is to stabilize the raft.

9.10.1.2. Using the Lanyard

The one-man raft is connected to the aircrew member by a lanyard parachuting to the water. IP should not swim to the raft, but pull it to their position via the lanyard. The parachute J-1 releases should be closed and the life preserver separated before boarding the raft (Figure 9-7). The raft may hit the water upside down, but may be righted by approaching the bottle side and flipping it over. The spray shield must be in the raft to expose the boarding handles.

Figure 9-7 Boarding One-Man Raft

9.10.1.3. Boarding the Raft with an Arm Injury

If the IP has an arm injury, boarding is best done by turning the back to the small end of the raft, pushing the raft under the buttocks and lying back. Another method of boarding is to push down on the small end until one knee is inside and lie forward.

9.10.1.4. Boarding the Raft in Rough Seas

In rough seas, it may be easier for the IP to grasp the small end of the raft and, in a prone position, kick and pull into the raft. Once in the raft, lying face down, the sea anchor should be deployed and adjusted. To sit upright in the raft, one side of the seat kit might have to be disconnected and the IP should roll to that side. The spray shield is then adjusted. There are two variations of the one-man raft, with the improved model incorporating an inflatable spray shield and floor for additional insulation. The spray shield is designed to help keep the IP dry and warm in cold oceans, and protect them from the Sun in the hot climates. The IP's seat kit may puncture the raft if attached to both sides of the harness. One side of the seat kit should be disconnected and the IP should then roll towards the disconnected side, so the seat kit ends up in the IP's lap as they enter the raft.

9.10.1.5. Using the Sea Anchor to Adjust the Rate of Travel

The sea anchor can be adjusted to either act as a drag by slowing down the rate of travel with the current or as a means of traveling with the current. When deployed the sea anchor (Figure 9-8) will act as a drag and the IP will stay in the general area. When the sea anchor is tied off or closed (Figure 9-8) it will form a pocket for the current to strike and propel the raft in the direction of the current. Additionally, the sea anchor should be adjusted so that when the raft is on the crest of a wave, the sea anchor is in the trough of the wave (Figure 9-9).

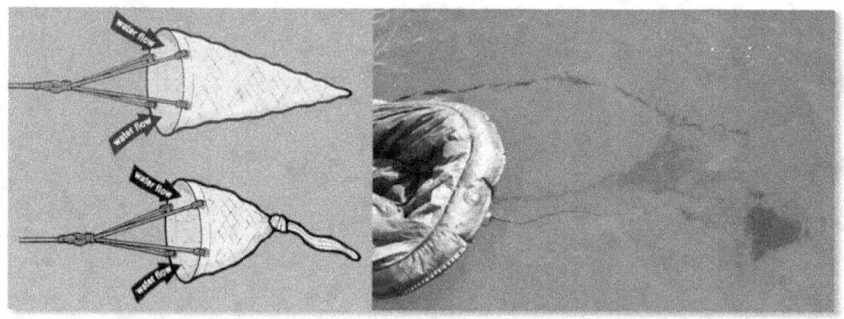

Figure 9-8 Sea Anchor

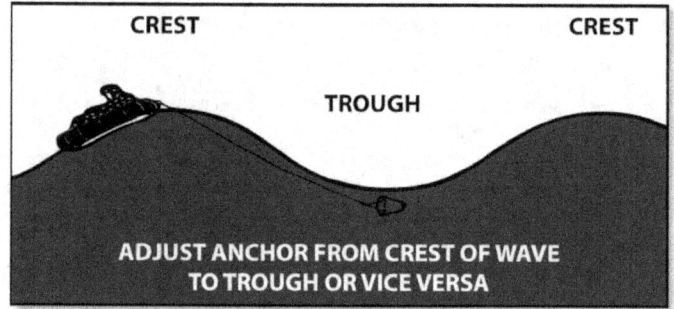

Figure 9-9 Deployment of the Sea Anchor

9.10.2. Seven-Man Raft

The seven-man raft is found in the survival drop kit (MA-1 Kit), used by USAF rescue forces. This type of raft (Figure 9-10) may inflate upside down and may, therefore, require the IP to right the raft before boarding. The IP should ensure that nothing is attached which may puncture or tear the life raft when boarding (i.e., J-1 releases closed before boarding). IP should always work from the bottle side to prevent injury if the raft turns over. Facing into the wind provides additional assistance in righting the raft. The handles on the inside bottom of the raft are used for boarding (Figure 9-10). The handles on the underside are used to right the raft (Figure 9-11).

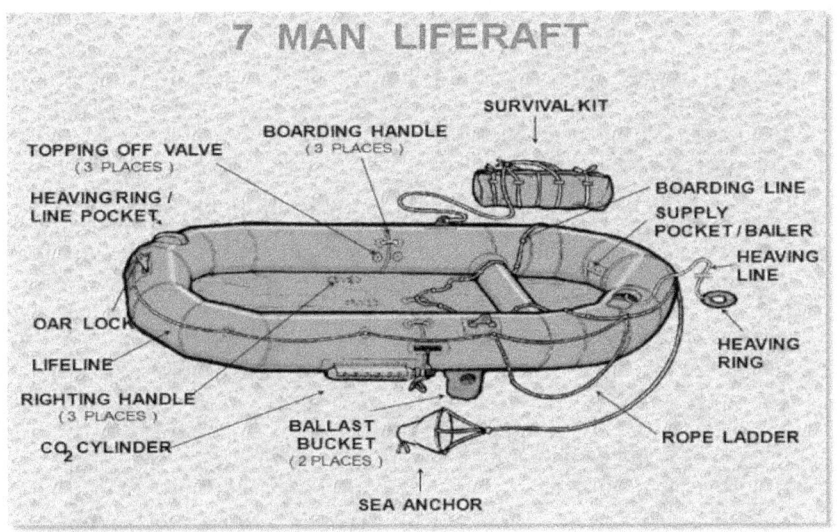

Figure 9-10 Seven-Man Raft

Figure 9-11 Method of Righting Raft

9.10.2.1. The Boarding Ladder

The boarding ladder is used to board if someone assists in holding down the opposite side. If assistance is unavailable, the IP should work from the bottle side with the wind at the back to help hold down the raft. The IP should separate the life preserver, grasp an oarlock and boarding handle, kick the legs to get the body prone on the water, and then kick and pull into the raft. If IP are weak or injured, the raft may be partially deflated to make boarding easier (Figure 9-12).

Figure 9-12 Method of Boarding Seven-Man Raft

9.10.2.2. Manual Inflation of the Raft

Manual inflation can be done by using the pump to keep buoyancy chambers and cross-seat firm, but the raft should not be over inflated. The buoyancy chambers and cross-seat should be rounded but not drum tight. Hot air expands, so on hot days, some air may be released while air may be added on cold days (Figure 9-13).

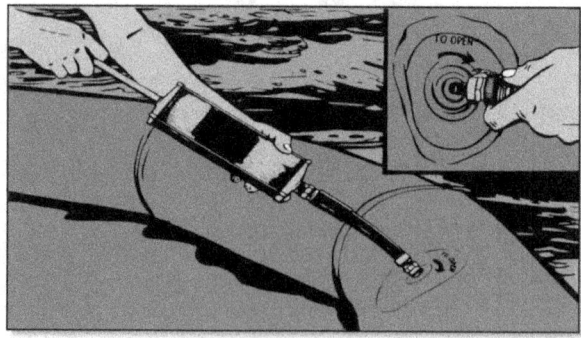

Figure 9-13 Inflating the Raft

9.10.3. Sailing a Raft into the Wind

Rafts are not equipped with keels, so they cannot sail into the wind. However, anyone can sail a raft downwind, and multi-place (except 20/25-man) rafts can be successfully sailed 10 degrees off from the direction of the wind. An attempt to sail the raft should not be made unless land is near. If the decision to sail is made and the wind is blowing toward a desired destination, IP should fully inflate the raft, sit high, rig a sail, and use an oar as a rudder. Taking in the sea anchor will make the liferaft more responsive to the wind instead of the current. Bringing in the sea anchor during strong winds may cause the liferaft to flip over.

9.10.4. Liferaft: 7-Man
A square sail should be erected in the bow using oars with their extensions as the mast and crossbar (Figure 9-14). A waterproof tarpaulin or available material may be used for the sail. If the raft does not have regular mast socket and step, the mast may be erected by tying it securely to the front cross-seat using braces. The bottom of the mast must be padded to prevent it from chafing or punching a hole through the floor whether or not a socket is provided. The heel of a shoe, with the toe wedged under the seat, makes a good improvised mast step. The corners of the lower edge of the sail should not be secured to the raft, instead the lines attached to the lower corners should be held with the hands so that a gust of wind will not rip the sail, break the mast, or capsize the raft. Every precaution must be taken to prevent the raft from turning over. In rough weather, the sea anchor is kept out away from the bow. The passengers should sit low in the raft with their weight distributed to hold the upwind side down. They should also avoid sitting on the sides of the raft or standing up to prevent falling out. Sudden movements (without warning the other passengers) should be avoided. When the sea anchor is not in use, it should be tied to the raft and stowed in such a manner that it will hold immediately if the raft capsizes.

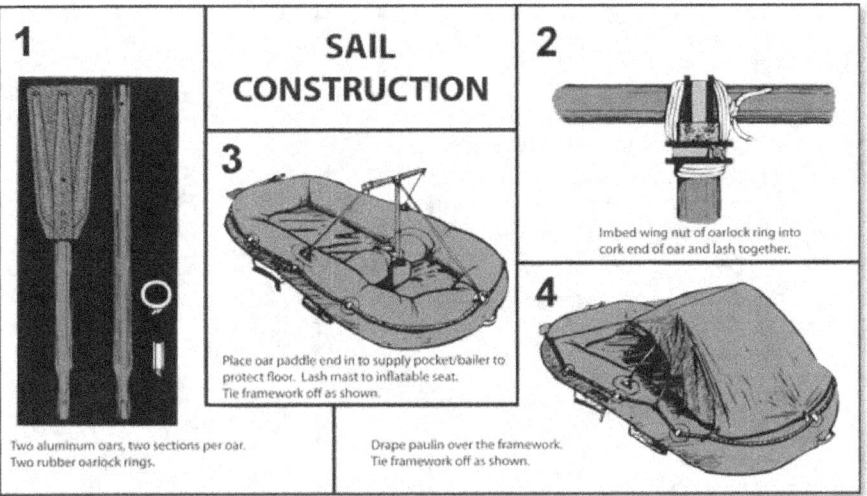

Figure 9-14 Sail Construction

9.10.5. Liferafts: 20, 25, and 46-Man No matter how the raft lands in the water, it is ready for boarding. The accessory kit is attached by a lanyard and is retrieved by hand. The center chamber can be inflated manually with the hand pump. The 20/25/46 man raft (Figure 9-15 and Figure 9-16) should be boarded from the aircraft if possible. If the raft is not boarded from the aircraft then follow the steps below.

- Approach lower boarding ramp, steps, or stirrup.
- Separate the life preserver. Perform SWIM procedures.
- Grasp the boarding handles and kick the legs to get the body into a prone position on the water's surface; then kick and pull until inside the raft (Figure 9-17).
- If for any reason the raft is not completely inflated, boarding will be made easier by approaching the intersection of the raft and ramp, grasping the upper boarding handle, swinging one leg up onto the cells of the liferaft as in mounting a horse. This leg will be partially inside the life raft, and then using the leg in the water and your hands holding the boarding handles kick and pull until your body is over the partially inflated life raft cells and inside.
- Should the raft be equipped with an equalizer tube, it should be clamped upon entering the raft to prevent deflating the entire raft, in case of puncture.
- The 20/25/46-man raft can be inflated by using the pump to keep the chambers and center ring firm. They should be well rounded but not drum tight.

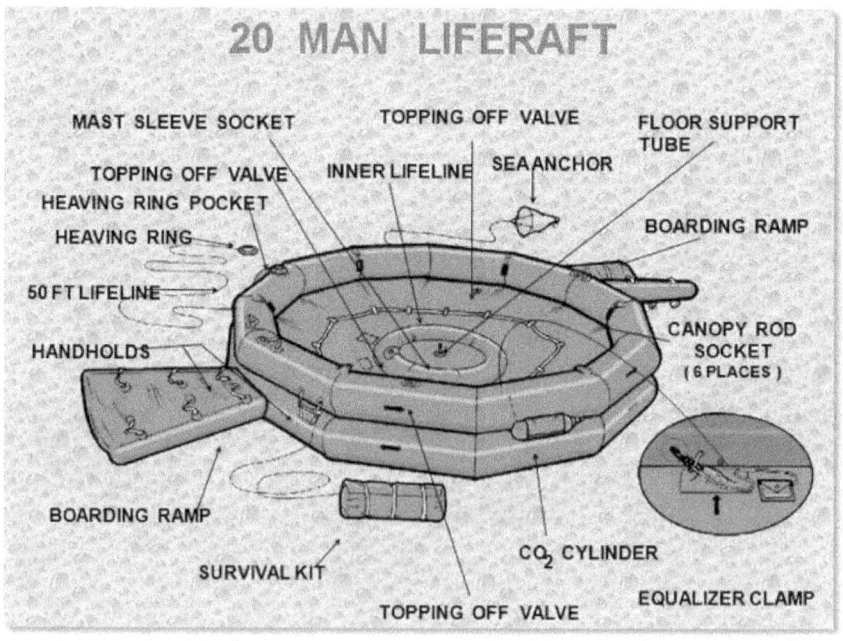

Figure 9-15 20-Man Raft

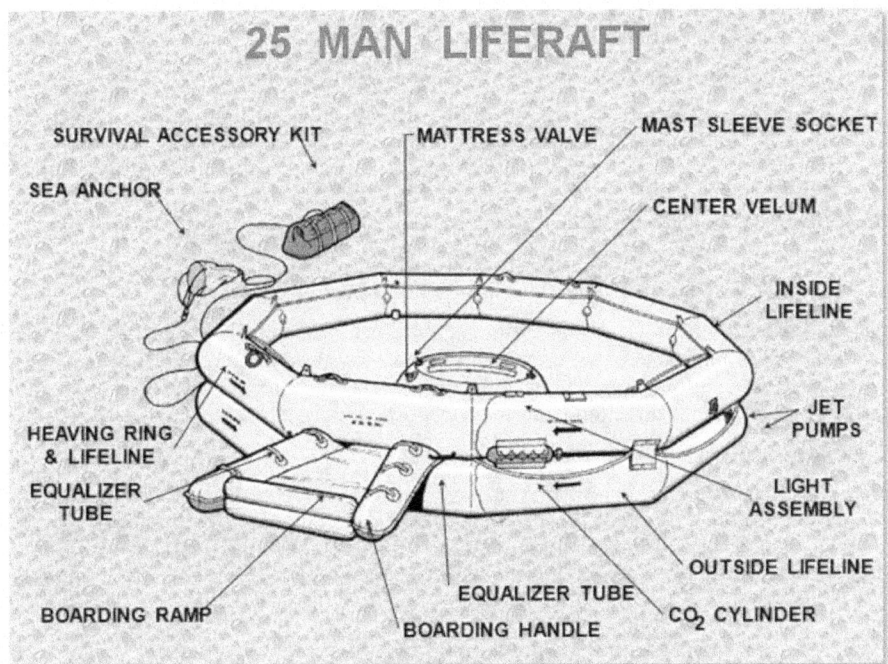

Figure 9-16 25-Man Raft

Figure 9-17 Boarding 20-Man Raft

9.11. Making Landfall

The lookout should watch carefully for signs of land. Some indications of land include:

9.11.1. Clouds can Help Determine Land Location

A fixed cumulus cloud in a clear sky, or in a sky where all other clouds are moving, often hovers over or slightly downwind from an island.

9.11.2. Appearance of Sky can Help Determine Land Location

In the tropics, a greenish tint in the sky is often caused by the reflection of sunlight from the shallow lagoons or shelves of coral reefs.

9.11.3. Appearance of Ice Fields can Help Determine Land Location

In the arctic, ice fields or snow-covered land are often indicated by light-colored reflections on clouds, quite different from the darkish gray reflection caused by open water.

9.11.4. Color of Water can Help Determine Land Location

Deep water is dark green or dark blue. Lighter color indicates shallow water, which may mean land is near.

9.11.5. Fog, Mist, or Rain can Help Determine Land Location

In fog, mist, rain, or at night, when drifting past a nearby shore, land may be detected by characteristic odors and sounds. The musty odor of mangrove swamps and mudflats and the smell of burning wood carry a long way. The roar of surf is heard long before the surf is seen. Continued cries of sea birds from one direction indicate their roosting place on nearby land.

9.11.6. Birds can Help Determine Land Location

Birds are usually more abundant near land than over the open sea. The direction from which flocks fly at dawn and to which they fly at dusk may indicate the direction of land. During the day, birds are searching for food and the direction of flight has no significance unless there is a storm approaching. While birds may be an indication of land, sighting them does not always mean land is near.

9.11.7. Wave Patterns can Help Determine Land Location

Land may be detected by the pattern of the waves, which are refracted as they approach land. Figure 9-18 shows the form the waves assume. Land should be located by observing this pattern and turning parallel to the slightly turbulent area marked "X" on the illustration and following its direction.

9.11.8. Securing Equipment before Landfall is Essential

Care should be taken in all circumstances to lanyard off and secure equipment prior to making landfall.

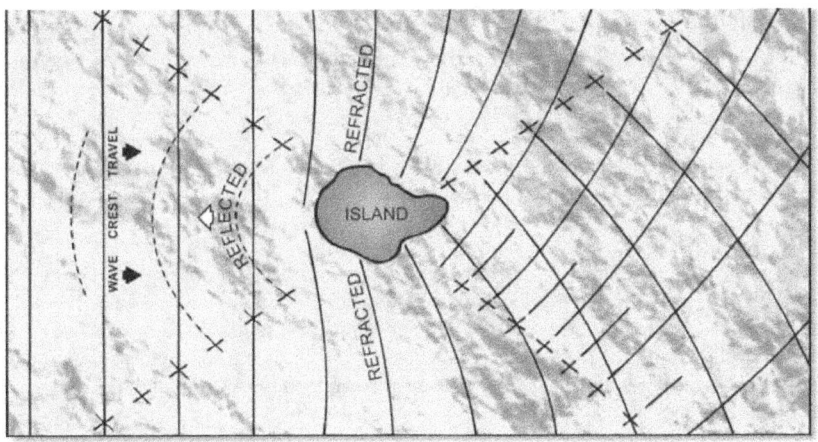

Figure 9-18 Diagram of Wave Patterns about an Island

9.12. Methods of Getting Ashore

No matter how the IP plans on getting ashore, they should observe and utilize the wave patterns to their advantage. Waves occur in patterns, and the IP should observe for the largest wave. The IP should immediately make their effort to get to shore following the largest wave to avoid getting pummeled into the sea shore.

9.12.1. Swimming Ashore

Distances on the water are very deceptive. In most instances, staying with the raft is the best course of action. If the decision is made to swim, a life preserver or other flotation aid should be used. Shoes and at least one thickness of clothing should be worn. The side or breast stroke will help conserve strength.

9.12.1.1. Using Small Waves to Assist the IP

If surf is moderate the IP can ride in on the back of a small wave by swimming forward with it and making a shallow dive to end the ride just before the wave breaks. The swimmer should stay in the trough between waves in high surf, facing the seaward wave and submerging when the wave approaches. After the wave passes, the swimmer should work shoreward in the next trough.

9.12.1.2. What to do if Caught in an Undertow

If the swimmer is caught in the undertow of a large wave, push off the bottom and swim to the surface and proceed shoreward. A place where the waves rush up onto the rocks should be selected if it is necessary to land on rocky shores and avoid places where the waves explode with a high white spray. After selecting the landing point, the swimmer should advance behind a large wave into the breakers. The swimmer should face shoreward and take a sitting position with the feet in front, two or three feet lower than the head, so the knees are bent and the feet will absorb shocks when landing or striking submerged boulders or reefs. If the shore is not reached the first time, the IP should swim with hands and arms only. As the next wave approaches, the sitting position with the feet forward should be repeated until a landing is made.

9.12.1.3. Grasping

Water is quieter in the lee of a heavy growth of seaweed. This growth can be very helpful. The swimmer should crawl over the top by grasping the vegetation with overhand movements.

9.12.1.4. Crossing a Rocky Reef

A rocky reef should be crossed in the same way as landing on a rocky shore. The feet should be close together with knees slightly bent in a relaxed sitting posture to cushion blows against coral.

9.12.2. Rafting Ashore

In most cases, any life raft can be used to make a shore landing with no danger. Going ashore in strong surf is dangerous. Survey the shore for a safe location to make landfall. The IP should look for gaps in the surf line and head for them while avoiding coral reefs and rocky cliffs. These reefs don't occur near the mouths of freshwater streams. Stay out of rip currents and strong tidal currents which may carry the raft out to sea. If caught in a rip current paddle perpendicular from the current until free from its forces.

9.12.2.1. Preparing for Landfall

- Take down the mast.
- Don clothing and shoes to avoid injuries.
- Fully inflate, adjust, and securely fasten life preserver.
- Stow equipment.
- Use paddles to maintain control.
- Ensure the sea anchor is deployed to help prevent the sea from throwing the stern of the raft around and capsizing it. The sea anchor should not be deployed when traveling through coral.

9.12.2.2. Handling the Raft in a Medium Surf with No Wind

In a medium surf with no wind, IP should keep the raft from passing over a wave so rapidly that it drops suddenly after topping the crest. If the raft turns over in the surf, every effort should be made to grab and hold onto the raft.

9.12.2.3. Landing the Raft on a Beach

The IP should ride the crest of a large wave as the raft nears the beach, staying inside until it has grounded. If there is a choice, a night landing should not be attempted. If signs of people are noted, it might be advantageous to wait for assistance.

9.12.2.4. Landing a Raft on Sea-Ice

Sea-ice landings should be made on large stable floes only. Icebergs, small floes, and disintegrating floes could cause serious problems. The edge of the ice can cut, and the raft deflated. Use paddles and hands to keep the raft away from the sharp edges of the iceberg. The raft should be stored a considerable distance from the ice edge. It should be fully inflated and ready for use in case the floe breaks up.

Chapter 10

LOCAL PEOPLE

10. Local People

One evader concluded with the following advice: "My advice is, 'When in Rome, do as the Romans do!' Show interest in their country, and they will go overboard to help you!" One of the most frequently given bits of advice is to accept, respect, and adapt to the ways of the people among whom IP find themselves. Expect challenges as you put this advice into practice (Figure 10-1 Local People).

Figure 10-1 Local People

10.1. Contact with People

The IP must give serious consideration to people. Are they farmers, fishermen, friendly people, or enemies? To the IP, "cross-cultural contact" can vary quite radically in scope. It could mean interpersonal relationships with people of an extremely different culture, or contacts with people who are culturally similar by our standards. A culture is identified by standards of behavior that are considered proper and acceptable for the members and may or may not conform to our idea of propriety. Regardless of whom these people are, the IP can expect they will have different laws, social and economic values, and political and religious beliefs.

10.1.1. Respecting Unknown (Local) People

People will be friendly, unfriendly, or unknown to the IP. If the people are known to be friendly, the IP must make every attempt to keep them that way by being courteous and respecting the religion, politics, social customs, habits, and all other aspects of their culture. If the people are known to be enemies or are unknowns, the IP should make every effort to avoid any contact and leave no sign of presence. Therefore, a basic knowledge of the daily habits of the local people can be extremely important in this attempt. An exception might be, if after careful and covert observation, it is determined an unknown people are friendly, contact might be made if assistance is absolutely necessary.

10.1.2. Thoughtful Contact with Local People

Generally, there is little to fear and everything to gain from thoughtful contact with the local peoples of friendly or neutral countries. Familiarity with local customs, displaying common decency, and most importantly, showing respect for their customs should help an IP avoid trouble and possibly gain needed assistance. Prior to making contact the IP should take into account cultural norms in selecting the person to contact. To make contact, an IP should wait until only one person is near and, if possible, let that person make the initial approach. Most people will be willing to help an IP who appears to be in need; however, political attitudes and training or propaganda efforts can change the attitudes of otherwise friendly people. Conversely, in nominally unfriendly countries, many people, particularly in remote areas, may feel abused or ignored by their politicians, and may be friendlier toward outsiders.

10.1.3. Displaying Patience when Contacting Local People

Friendly, courteous, and patient behavior is important for successful contact with local peoples. Displaying fear, displaying weapons, and making sudden or threatening movements can cause a local person to fear an IP which can prompt a hostile response. When attempting contact, smile frequently. Many local peoples may be shy and seem unapproachable or they may ignore the IP. Approach them slowly and don't rush matters.

10.2. IP Behavior

IP may use Tobacco, money, blood chit, and personal items to discreetly trade for necessities in an isolated situation. IP should be cautious and not overpay for needed items.

10.2.1. Non-Verbal Communication of Local People

Non-verbal language or acting out needs or questions can be very effective. Many people are accustomed to it and communicate using nonverbal sign language. Potential IP should learn a few words and phrases of the local language in and around their area of operations. Attempting to speak someone's language is an excellent way to show respect for their culture. Since English is widely used, some of the local people may understand a few words of English.

10.2.2. Learning the Rules of Local People

Certain areas may be taboo, and they range from religious or sacred places to diseased or danger areas. In some areas, certain animals must not be killed. An IP must learn what the rules are and follow them. The IP must be observant and learn as much as possible. This will not only help in strengthening relations, but new knowledge and skills may be very important later. IP should seek advice on local hazards and attempt to remain clear of hostile people. Keep in mind though, that frequently, people, as in our culture, insist others are hostile because they also do not understand different cultures and distant peoples. The people that generally can be trusted, in their opinion, are their immediate neighbors-much the same as in our own neighborhood. Local people may suffer from diseases which are contagious. The IP should build a separate dwelling, if possible, and avoid physical contact without seeming to do so. Personal preparation of food and drink is desirable if it can be done without giving offense. Frequently, the use of "personal or religious custom" as an explanation for isolationist behavior will be accepted by the local people.

10.2.3. Exercising Caution with Local People

The IP must be very cautious when touching people. Many people consider "touching" taboo and such actions may be dangerous. Sexual contact should be avoided.

10.2.4. Hospitality

Hospitality among some people is such a strong cultural trait they may seriously reduce their own supplies to make certain a stranger or visitor is fed. What is offered should be accepted and shared equally with all present. The IP should eat in the same way the local people eat and, most importantly, attempt to eat what is offered to avoid offense. If any promises are made, they must be kept. Personal property and local customs and manners, even if they seem odd, must be respected. Some kind of payment (money, trade items, blood chit, etc) for food, supplies, etc., may be offered depending on culture.

10.2.5. Respecting Privacy

Privacy must be respected and an IP should not enter a house or dwelling unless invited.

10.3. Political Allegiance

In today's world of fast-paced international politics and "shuttle diplomacy," political attitudes and commitments within nations are subject to rapid change. The population of many countries, especially politically hostile countries, must not be considered friendly just because they do not demonstrate open hostility. Pre-mission briefings should cover interaction with local population and those in that country who are in positions of authority such as the police, government, and military.

10.3.1. Norms of Interaction with Local People While initial contact is best done on a one-to-one basis, IP may encounter locals in larger groups in built up areas that represent multiple cultures. Aspects of culture include the norms, values, religion, language, race, ethnicity, and heritage associated with a specific group. Determining the expected interaction between the human societies, with consideration for the built-up areas, the natural environments, and support networks (transportation, food resources, religions, and other countless features of the landscape), may help the IP anticipate the types of interactions they might encounter and prepare accordingly.

10.4. Population in Built-Up Areas

A built-up area is a concentration of structures, facilities, and people that form the economic and cultural focus for the surrounding area. There is no "standard" urban terrain. Built-up areas are classified into four categories:

- Villages (population up to 3,000).
- Strip areas (urban areas built along roads connecting towns or cities).
- Towns or small cities (population up to 100,000 and not part of a major urban complex).
- Large cities with associated urban sprawl (population up to the millions, covering hundreds of square kilometers).

10.4.1. Determination of Friendly and Un-Friendly Local People

While the physical characteristics of the built-up areas are of great importance, the key variable is the population. The issue is simply whether the citizenry is friendly, unfriendly, or unknown. Too often, the evaluation of the flesh-and-blood terrain, of the human high-ground, ends there. Yet few populations are ever exclusively hostile, or truly indifferent, or unreservedly welcoming.

10.4.2. Human Terrain

Analyzing the "cultural architecture" of the built-up area begins with the recognition that there are three broad types of "human terrain." For IP purposes, the human terrain of built-up location can be classified as hierarchical, multicultural, or tribal.

- Hierarchical human terrain - chains-of-command operate within broadly accepted rule of law.
- Multicultural human terrain - contending systems of custom and belief, often aggravated by ethnic divisions, struggle for dominance.
- Tribal human terrain - Based upon differences in blood, but not in race or, necessarily, in religion, ethnic conflicts can be intractable and merciless.

10.5. Summary

These classifications provide a general understanding of the human architecture and the operational environment that an IP may be thrust into. It can provide assistance in developing a general course of action when establishing contact, define potential problems when dealing with local populations, identify possible scenarios with law-enforcement, paramilitary, and military agencies, and help in the problem solving and decision-making during an isolating event.

Chapter 11

PROPER BODY TEMPERATURE

11. Proper Body Temperature

In an isolating event, the two key requirements for personal protection are maintenance of proper body temperature and prevention of injury. The means for providing personal protection are many and varied, and include clothing, shelter, equipment, and fire (which are all covered in detail in separate chapters). These individual items are not necessary for survival in every situation; however, all four are essential in some environments. Conditions which affect the body temperature, the physical principles of heat transfer, and the methods of coping with these conditions are addressed below.

11.1. Optimum Core Temperature

The body functions at its best when core temperatures range from 96°F to 102°F. Preventing too much heat loss or gain should be a major concern for IP. Factors causing changes in body core temperature (excluding illness) are the climatic conditions of temperature, wind, and moisture.

11.1.1. Exposure to Extreme Temperatures

As a general rule, exposure to extreme temperatures can result in substantial decreases in physical efficiency. In the worst case, incapacitation and death can result.

11.1.2. Wind Increase the Chill Effect

Wind increases the chill effect (Figure 11-1), causes dissipation of heat, and accelerates loss of body moisture.

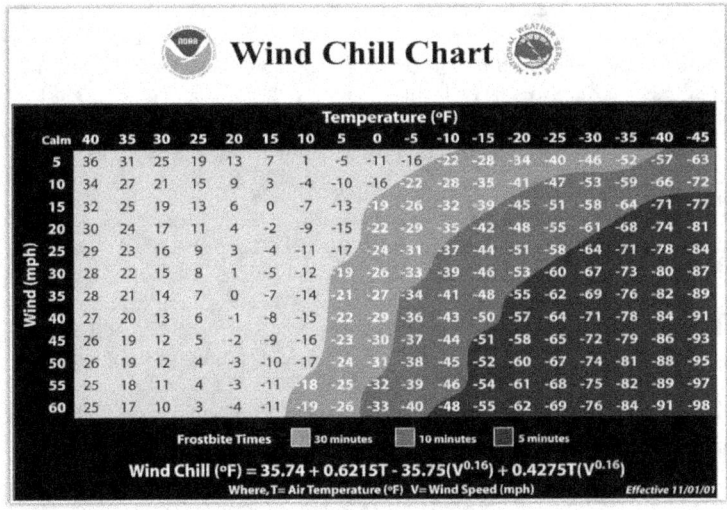

Figure 11-1 Wind Chill Chart

11.2. Water as a an Effective Way to Transfer Body Heat

Water provides an extremely effective way to transfer heat to and from the body. When a person is hot, the whole body may be immersed in a stream or other body of water to be cooled. On the other hand, in cold temperatures, a hot bath can be used to warm the body. When water is around the body, it tends to bring the body to the temperature of the water. An example is when a hand is burned and then placed in cold water to dissipate the heat. One way to lower body temperature is by applying water to clothing and exposing the clothed body to the wind. This action causes heat to leave the body 25 times faster than when wearing dry clothing. This rapid heat transfer is the reason IP must always guard against getting wet in cold and windy environments. Consider how long a person could survive after their body is totally submerged in water at a temperature of 50°F (Figure 11-2).

Water Temperature Degrees C	Degrees F	Loss of Dexterity with no protective clothing	Exhaustion or Unconsciousness	Expected Time of Survival
0.3	32.5	Under 2 min.	Under 15 min.	Under 15 to 45 min.
0.3 to 4.5	32.5 - 40	Under 3 min.	15 to 30 min.	30 to 90 min.
4.5 to 10	40 - 50	Under 5 min.	30 to 60 min.	1 to 3 hrs.
10 to 15.5	50 - 60	10 to 15 min.	1 to 2 hrs.	1 to 6 hrs.
15.5 to 21	60 - 70	30 to 40 min.	2 to 7 hrs.	2 to 40 hrs.
21 to 26.5	70 - 80	1 to 2 hrs.	2 to 12 hrs.	3 hrs. to indefinite
Over 26.5	Over 80	2 to 12 hrs.	Indefinite	Indefinite

Figure 11-2 Life Expectancy Following Cold-Water Immersion (without protective clothing)

11.3. Factors Which Affect Survival Time

Many factors, such as body size, body fat, protective clothing, water temperature, currents, and winds, affect survival time when an individual is immersed in water. Individuals who know they are going to be operating near, on, over, or even in the water should dress to protect themselves from the dangers of cold water and hypothermia. Recent studies by the US Navy provide some insight into predicted immersed survival times that can be used as a guide. Understanding that increasing body mass, body fat, or thermal clothing will, in most cases, increase predicted survival times while decreasing body mass, body fat, or thermal clothing will subsequently decrease predicted survival times.

- Consider a 170 pound male aircrew member with 19.2 percent body fat, dressed for the environment. His first layer is standard underwear, boxers or briefs, and a T-shirt. Over this first layer he is dressed in a commercial thermal insulating garment (TIG), long sleeve top and full length bottom, made of polyester batting sandwiched between layers of woven polyester microfiber fabric. This is a new version of what many people will remember as "Chinese underwear." Next, the aircrew member has on standard wool socks and an aircrew constant-wear anti exposure suit similar to the US Air Force CWU-74/P or the US Navy CWU-62B/P. Finally, the aircrew member has on a standard aircrew flight helmet and jump into the water.

11.3.1. Predicted Survival Time

Based on the US Navy study, this aircrew member would have a predicted immersed survival time of 175 minutes or 2.9 hours in 35 degree Fahrenheit water. The same individual dressed in standard flight clothing of underwear, flight suit, socks, boots and flight helmet would have a predicted immersed survival time of well under 30 minutes. As you can see, dressing appropriately could easily be the difference between survival and fatality (Figure 11-3).

Weight/Body Fat	Water Temperature F	Predicted immersed survival time wearing standard issue USAF constant-wear anti-exposure suit, TIG, standard wool socks, flight helmet and boots	Predicted immersed survival time wearing standard issue USAF constant-wear anti-exposure suit, USAF issue full-length constant wear anti-exposure suit liner, aramid long underwear top and bottom, standard wool socks, flight helmet and boots	Predicted immersed survival time wearing standard issue USAF flight suit, Xcel Icon Shorty Wetsuit, standard wool socks, flight helmet and boots
175 lbs, 19.2%	35	175 minutes	175 minutes	30 minutes
	40	220 minutes	220 minutes	40 minutes
	45	295 minutes	280 minutes	60 minutes
	50	360 minutes (indefinite; test ended)	360 minutes (indefinite; test ended)	80 minutes
	55	360 minutes (indefinite; test ended)	360 minutes (indefinite; test ended)	100 minutes
	60	360 minutes (indefinite; test ended)	360 minutes (indefinite; test ended)	160 minutes
	68	360 minutes (indefinite; test ended)	360 minutes (indefinite; test ended)	340 minutes

Figure 11-3 Life Expectancy Following Cold Water Immersion

11.4. Factors that can Transfer Body Heat

Body heat can be transferred by radiation, conduction, convection, evaporation, and respiration. Understanding how heat is transferred and the methods by which that transfer can be controlled can help IP keep the body's core temperature in the 96°F to 102°F range (Figure 11-4 Heat Transfer).

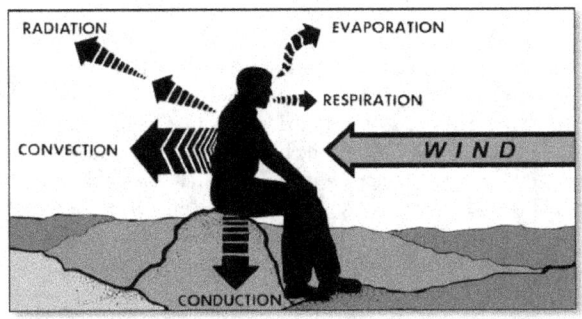

Figure 11-4 Heat Transfer

11.4.1. Radiation as the Primary Cause of Heat Loss

Radiation is the primary cause of heat loss. It involves the transfer of heat waves from the body to the environment and/or from the environment back to the body. For example, at a temperature of 50°F, 50 percent of the body's total heat loss can occur through an exposed head and neck. As the temperature drops, the situation intensifies. At 5°F, the loss can be 75 percent under the same circumstances. Not only is heat lost from the head, but also from the other extremities of the body. The hands and feet also radiate significant amounts of heat. To minimize the amount of heat lost, make sure areas of exposed skin are covered, including the head, face, neck, hands, and feet.

11.4.2. Conduction

Conduction is the movement of heat from one molecule to another molecule by direct contact within a solid object. Extreme examples of how heat is lost and gained quickly are deep frostbite and third-degree burns, both of which can happen by touching the same piece of metal at opposite extremes of cold and heat. Heat is also lost from the body in this manner by touching objects in the cold with bare hands, by sitting on a cold log, or by kneeling on snow to build a shelter. These are things which IP should avoid since they can lead to over chilling the body.

11.4.3. Handling of Liquid Fuel at Low Temperatures is Dangerous

Handling of liquid fuel at low temperatures is especially dangerous. Unlike water which freezes at 32°F, fuel exposed to outside temperatures will reach the same temperature as the air. The temperature of the fuel may be 10°F to 30°F below zero or colder. Spilling the fluid on exposed skin will cause instant frostbite, not only from the conduction of heat by the cold fluid, but by the further cooling effects of rapid evaporation of the liquid as it hits the skin.

11.4.4. Convection

Heat movement by means of air or wind to or from an object or body is known as convection. The human body is always warming a thin layer of air next to the skin by radiation and conduction. The temperature of this layer of air is nearly equal to that of the skin. The body stays warm when this layer of warm air remains close to the body. However, when this warm layer of air is removed by convection, the body cools down. A major function of clothing is to keep the warm layer of air close to the body; by removing or disturbing this warm air layer, wind can reduce body temperature. Therefore, wind can provide beneficial cooling in dry, hot conditions, or be a hazard in cold, wet conditions.

11.4.5. Evaporation

Evaporation is a process by which liquid changes into vapor, and during this process, heat within the liquid escapes to the environment. An example of this process is how a desert water bag works on the front of a jeep while driving in the hot desert. The wind created by the jeep helps to accelerate evaporation and causes the water in the bag to be cooled. The body also uses this method to regulate core temperature when it perspires and air circulates around the body. The evaporation method works any time the body perspires regardless of the climate. For this reason, it is essential that people wear fabrics that breathe in cold climates. If water vapor cannot evaporate through the clothing, it will condense, freeze, and reduce the insulation value of the clothing and cause the body temperature to go down.

11.4.6. Respiration

The respiration of air in the lungs is also a way of transferring heat. It works on the combined processes of convection, evaporation, and radiation. When breathing, the air inhaled is rarely the same temperature as the lungs. Consequently, heat is either inhaled or expelled with each breath. A person's breath can be seen in the cold as heat is lost to the outside.

Chapter 12

CLOTHING

12. Clothing

Clothing is an important asset to IP and is the most immediate form of protection. Clothing is important in staying alive, especially if food, water, shelter, and fire are limited or unobtainable. This is especially true in the first stages of an emergency situation because IP must work to satisfy other needs. If IP are not properly clothed, they may not survive long enough to build a shelter or fire, to find food, or to be rescued.

12.1. Protection

People have worn clothing for protection since they first put on animal skins, feathers, and other coverings. In most parts of the world, people need clothing for protection from harsh climates. Clothing also provides protection from physical injuries caused by vegetation, terrain features, environmental extremes, and animal life which may cause bites, stings, and cuts.

12.2. Clothing Materials

Clothing is made from a variety of natural and synthetic materials such as nylon, wool, and cotton. The type of material used has a significant effect on protection. When two or three layers of material are worn, a layer of air is trapped between each layer of material creating another layer of dead-air or insulation. The ability of these different fibers to hold dead-air is responsible for differing insulation values. Potential IP must be aware of both the environmental conditions and the effectiveness of these different materials in order to select the best type of clothing for a particular geographic region.

12.2.1. Natural Materials

Natural materials include fur, leather, and cloth made from plant and animal fibers such as wool and cotton.

- Fur and leather are made into some of the warmest and most durable clothing. Fur is used mainly for coats and coat linings. Animal skin has to be treated to make it soft and flexible and to prevent it from rotting.

- Although wool is somewhat absorbent, it retains most of its insulating qualities when wet. It contains natural lanolin oils which helps garments shed water increasing its thermal qualities.

- Cotton is a common plant fiber widely used to manufacture clothing. It absorbs moisture quickly and, with heat radiated from the body, will allow the moisture to pass away from the body. It does not offer much insulation when wet. It can be used as an inner layer against the skin with insulation (for example, wool, Dacron pile, synthetic batting) sandwiched between. The cotton protects the insulation and, therefore, provides warmth.

12.2.2. Synthetic Materials

Many synthetic materials are stronger, more shrink-resistant, and less expensive than natural materials. Synthetic fibers are manufactured from chemicals derived from water, coal, and petroleum. The fibers are of different lengths, diameters, and strengths, and sometimes have hollow cores. These fibers, woven into materials such as nylon, Dacron, polypropylene and polyester, make very strong long-lasting clothing, tarps, tents, etc. Some fibers are spun into a batting type material with air space between the fibers, providing excellent insulation when used inside clothing.

12.2.2.1. Nylon

Nylon covered with rubber is durable and waterproof but is also heavy. There are other coverings on nylon which are waterproof but somewhat lighter and less durable. However, most coated nylon has one drawback - it will not allow for the evaporation of perspiration. Therefore, individuals may have to change the design of the garment to permit adequate ventilation (for example, wearing the garment partially unzipped).

12.2.2.2. Synthetic Fibers

Synthetic fibers are generally lighter in weight than most natural materials and have much of the same insulating qualities. They work well when partially wet and dry out easily. However, they generally do not compress as well as down insulation.

12.2.2.3. Protecting Synthetic Fibers

Many synthetic fibers should be protected from heat or open flames as they can melt or catch fire.

12.2.3. Types of Insulation

There are also a variety of natural and synthetic insulating materials.

12.2.3.1. Natural Insulation

12.2.3.1.1. Down Insulation

Down is the soft plumage found between the skin and the contour feathers of birds. Ducks and geese are good sources for down. If used as insulation in clothing, remember that down will absorb moisture (either precipitation or perspiration) quite readily. Because of the light weight and compressibility of down, it has wide application in cold-weather clothing and equipment. It is one of the warmest natural materials available when kept clean and dry. It provides excellent protection in cold environments; however, if the down gets wet it tends to get lumpy and loses its insulating value.

12.2.3.1.2. Cattail Plants

Cattail plants have a worldwide distribution, with the exception of the extreme north and south latitudes. The cattail is a marshland plant found along lakes, ponds, and the backwaters of rivers. The seed head on the tops of the stalks forms dead-air spaces and makes a good down-like insulation when broken away from the stem, fluffed up, and placed between two pieces of material.

12.2.3.1.3. Leaves from Deciduous Trees

Leaves from deciduous trees (those that lose their leaves each autumn) also make good insulation. To create dead-air space, leaves should be placed between two layers of material.

12.2.3.1.4. Grass and Mosses

Grasses, mosses, and other natural resources can also be used as insulation when placed between two pieces of material.

12.2.3.2. Synthetic Insulation

12.2.3.2.1. Polyesters and Acrylics

Synthetic filaments such as polyesters and acrylics absorb very little water and dry quickly. Spun synthetic filament is lighter than an equal thickness of wool. It is also an excellent replacement for down in clothing since, unlike down, it does not collapse when wet.

12.2.3.2.2. Nylon Parachute Material

The nylon material in a parachute insulates well if used in the layer system because of the dead-air space. Use caution when using the parachute in cold climates as nylon may become "cold soaked;" that is, the nylon will take on the temperature of the surrounding air. People have been known to receive contact frostbite when placing cold nylon against bare skin.

12.3. Insulation Measurement

Consideration should be given to how well the material's fibers insulate from the heat or cold. This can be described in terms of its "Clo value", which is a relative measurement of the ability of insulation to provide warmth.

12.3.1. The Clo Value

The Clo value is a numerical representation of a clothing ensemble's thermal resistance. One "Clo" is defined as the amount of clothing required by a resting (sedentary) person to be comfortable at ambient conditions where temperature is 70°F, relative humidity is less than 50 percent, and wind velocity is just over half a mile per hour. However, the Clo value alone is not sufficient to determine the amount of clothing required. Variables such as metabolic rate, wind conditions, and the physical makeup of the individual must be considered.

12.3.2. When Less Insulation is Needed

The body's rate of burning or metabolizing food and producing heat varies among individuals. Therefore, some IP may need more insulation than others even though food intake is equal, and consequently the required Clo value must be increased. Physical activity causes an increase in the metabolic rate and the rate of blood circulation through the body. When a person is physically active, less clothing or insulation is needed than when standing still or sitting.

12.3.3. When Shelter is Mandatory for Sustaining Life

When the combination of temperature and wind drops the chill factor to minus 100°F or lower, the prescribed Clo for protecting the body may be inapplicable (over a long period of time) without relief from the wind. For example, when the temperature is minus 60°F, the wind is blowing 60 to 70 miles per hour, and the resultant chill factor exceeds minus 150°F, clothing alone is inadequate to sustain life. Shelter is essential.

12.3.4. The Affect of Physical Build on Temperature Endurance

The physical build of a person also affects the amount of heat and cold that can be endured. For example, a very thin person will not be able to endure as low a temperature as one who has a layer of fat below the skin. Conversely, heavy people will not be able to endure extreme heat as effectively as thinner people.

12.3.5. The Air Force Clothing Inventory

In the Air Force clothing inventory, there are many items which fulfill the need for insulating the body. They are made of the different fibers previously mentioned, and when worn in layers, provide varying degrees of insulation. The following average zone temperature chart is a guide in determining the best combination of clothing to wear.

TEMPERATURE RANGE	CLO REQUIRED
86 to 68°F	1 - Lightweight
68 to 50°F	2 - Intermediate Weight
50 to 32°F	3 - Intermediate Weight
32 to 14°F	3.5 - Heavyweight
14 to -4°F	4.0 - Heavyweight
-4 to -40°F	4.0 - Heavyweight

Figure 12-1 CLO Chart

12.3.6. Clo Value Per Fabric Layer

The amount of Clo value per layer of fabric is determined by the loft (distance between the inner and outer surfaces) and the amount of dead air held within the fabric. Some examples of the Clo values and some items of clothing include:

LAYERS	CLO Value
1 - Fire Retardant underwear (1 layer)	0.6 CLO
2 – Fire Retardant underwear (2 layers)	1.5 CLO
3 – Quilted liners	1.9 CLO
4 – Nomex Flight suit	0.6 CLO
6 – Nomex jacket	1.9 CLO

Figure 12-2 CLO Value Per Layer

12.3.6.1. Dressing Layers for Added Protection

This total amount of insulation should keep the average person warm at a low temperature. Comparing items one and two above shows that doubling the layer of underwear more than doubles the Clo value. This is true for all layers of any clothing system. Therefore, one gains added protection by using several very thin layers of insulation rather than two thick layers. The air held between these thin layers increases the insulation value. The use of many thin layers also provides (through removal of desired number of layers) the ability to closely regulate the amount of heat retained inside the clothing. The ability to regulate body temperature helps to alleviate the problem of overheating and sweating, and preserves the effectiveness of the insulation.

12.3.7. Using Layers for IPs Sleeping System

The principle of using many thin layers of clothing can also be applied to the "sleeping system" (sleeping bag, liner, and bed). This system uses many layers of synthetic material, one inside the other, to form the amount of dead air needed to keep warm. To improve this system, an IP should wear clean and dry clothing in layers (the layer system) in cold climates. While discussing the layer system, it is important to define the "COLDER" principle (Figure 12-3). This acronym is used to aid in remembering how to use and take care of clothing.

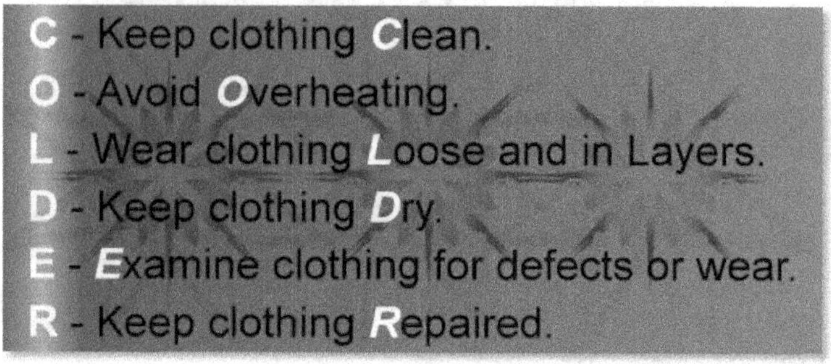

> **C** - Keep clothing **C**lean.
> **O** - Avoid **O**verheating.
> **L** - Wear clothing **L**oose and in **L**ayers.
> **D** - Keep clothing **D**ry.
> **E** - **E**xamine clothing for defects or wear.
> **R** - Keep clothing **R**epaired.

Figure 12-3 Colder Principle

- Clean. Dirt and other materials inside fabrics will cause the insulation to be ineffective, wear away and cut the fibers which make up the fabric, and cause holes. Washing clothing in the field may be impractical; therefore, IP should concentrate on using proper techniques to prevent soiling clothing. This should include avoiding, if possible, kneeling in dirt or wiping dirty hands or equipment on clothing.

- Overheating. Clothing best serves the purpose of preserving body heat when worn in layers as follows: absorbent material next to the body, insulating layers, and outer garments to protect against wind and rain. Because of the rapid change in temperature, wind, and physical exertion, garments should allow donning and removal quickly and easily. Ventilation is essential when working because enclosing the body in an airtight layer system results in perspiration which wets clothing, thus reducing its insulating qualities.

- Loose. Garments should be loose fitting to avoid reducing blood circulation and restricting body movement. Additionally, the garment should extend beyond the waist, wrists, ankles, and neck to reduce body heat loss.

- Dry. A small amount of moisture in the insulation fibers will cause heat losses up to 25 times faster than dry clothing. Internally produced moisture is as damaging as is externally dampened clothing. The outer layer should protect the inner layers from moisture as well as from abrasion of fibers. For example, wool rubbing on logs or rocks, etc. The outer shell keeps dirt and other contaminants out of the clothing. Clothing can be dried in many ways. Fires are often used however; take care to avoid burning the items. The "bare hand" test is very effective. Place one hand near the fire in the approximate place the wet items will be and count to three slowly. If this can be done without feeling excessive heat, it should be safe to dry items there. Never leave any item unattended while it is drying. Leather boots, gloves, and mitten shells require extreme care to prevent shrinkage, stiffening, and cracking. The best way to dry boots is upright beside the fire (not upside down on sticks because the moisture does not escape the boot), or simply walk them dry in the milder climates. The Sun and wind can be used to dry clothing with little supervision other than checking occasionally on the incoming weather and making sure the article is secure. Freeze-drying is used in subzero temperatures with great success. Let water freeze on or inside the item and then shake, bend, or beat it to cause the ice particles to fall free from the material. Tightly woven materials work better with this method than do open fibers.

- Examine. All clothing items should be inspected regularly for signs of damage or soil.
- Repair. When damage is detected, immediately repair it.

12.3.8. Body Parts that Need Greater Protection

The neck, head, hands, armpits, groin, and feet lose more heat than other parts of the body and require greater protection. Work with infrared film shows tremendous heat loss in those areas when not properly clothed. IP in a cold environment are in a real emergency situation if they do not have proper clothing. Figure 12-4 shows some examples of how military clothing works to hold body heat.

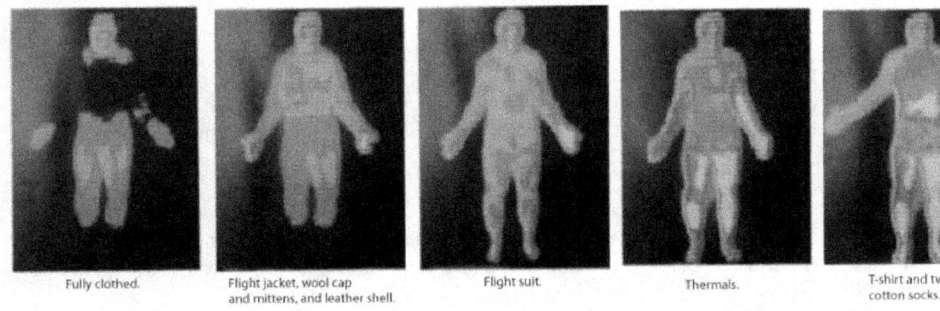

NOTE: Dark blue indicates no heat loss; the lighter the color, the greater the heat loss.

Figure 12-4 Thermogram of Body Heat Loss

12.3.8.1. Thermogram of Heat Loss Explained

Models wearing samples of aircrew attire appear as spectral figures in a thermogram; an image revealing differences in infrared heat radiated from their clothing and exposed skin. White is warmest; red, yellow, green, blue, and magenta form declining temperatures scale spanning about 15 degrees; while black represents all lower temperatures. Almost the entire scale is seen on the model in boxer shorts. Warm, white spots appear on the underarm and neck. Only the shorts block radiation from the groin. Temperatures cool along the arm to dark blue fingertip far from the heat-producing torso. The addition of the next layer of clothing (Fire retardant long underwear) prevents heat loss except where it is tight against the body. As more layers are added, it is easy to see the areas of greatest concern are the head, hands, and feet. These areas are difficult for crewmembers to properly insulate while flying an aircraft. Mittens are ineffective due to the degraded manual dexterity. Likewise, it is difficult to feel the rudder pedal action while wearing bulky warm boots. These problems require inclusion of warm hats, mittens, and footgear (mukluk type) possibly located in survival kits during cold weather operation. Research has shown when a Clo value of 10 is used to insulate the head, hands, and feet and the rest of the body is only protected by one Clo, the average individual can be exposed to low temperatures (-10°F) comfortably for a reasonable period of time (30 to 40 minutes). When the amount of Clo value placed on the individual is reversed, the amount of time an IP can spend in cold weather is greatly reduced due to the heat loss from their extremities. This same principle works in reverse in hot parts of the world if one submerges the head, hands, or feet in cold water, it lets the most vascular parts of the body lose heat quickly.

12.4. Clothing Wear in Snow and Ice Areas

IP should use the following guidelines for clothing in snow and ice areas.

12.4.1. Avoid Restricting Circulation

Clothing should not be worn so tight that it restricts the flow of blood which distributes the body heat and helps prevent frostbite. When wearing more than one pair of socks or gloves, ensure that each succeeding pair is large enough to fit comfortably over the other. Do not wear three or four pairs of socks in a shoe fitted for only one or two pairs. Release any restriction caused by twisted clothing or a tight parachute harness.

12.4.2. Prevent Perspiration by Opening Clothing

When exerting the body, prevent perspiration by opening clothing at the neck and wrists and loosening it at the waist. If the body is still warm, comfort can be obtained by taking off outer layers of clothing, one layer at a time. When work stops, the individual should put the clothing back on to prevent chilling.

12.4.3. Keep Clothing as Dry as Possible

Snow must be brushed from clothing before entering a shelter or going near a fire. Beat or shake the frost out of garments before warming them, and dry them on a rack near a fire. Socks should be dried thoroughly.

12.4.4. Keep Moving

If IP fall into water, they should keep moving and should not remove footwear until they are in a shelter or beside a fire.

12.4.5. Arrange Spare Clothing Loosely

At night, IP should arrange dry spare clothing loosely around and under the shoulders and hip to help keep the body warm. Wet clothes should never be worn into the sleeping bag. The moisture destroys the insulation value of the bag.

12.4.6. Use Wool as a Protective Inner Layer

All clothing made of wool offers good protection when used as an inner layer. When wool is used next to the face and neck, IP should be cautioned that moisture from the breath will condense on the surface and cause the insulating value to decrease. The use of a wool scarf wrapped around the mouth and nose is an excellent way to prevent cold injury, but it needs to be de-iced on a regular basis to prevent freezing flesh adjacent to it. An extra shell is generally worn over the warming layers to protect them and to act as a windbreak.

12.4.7. Keep Head and Ears Covered

At 50°F, as much as 50 percent of total body heat can be lost from an unprotected head.

12.4.8. Using the Hood to Funnel Radiant Heat

The hood is designed to funnel the radiant heat rising from the rest of the body and to recycle it to keep the neck, head, and face warm (Figure 12-5). The individual's ability to tolerate cold should dictate the size of the front opening of the hood. The "tunnel" of a parka hood is usually lined with fur of some kind to act as a protecting device for the face. This same fur also helps to protect the hood from the moisture expelled during breathing. The closed tunnel holds heat close to the face longer; the open one allows the heat to escape more freely. As the frost settles on the hair of the fur, it should be shaken from time to time to keep it free of ice buildup.

Figure 12-5 Proper Wear of Parka

12.4.9. Other Headgear

Other headgear includes the watch cap, pile cap, and hood. These items are most effective when used with a covering for the face in extreme cold. The pile cap is extremely warm where it is insulated, but it offers little protection for the face and back of the neck.

12.4.10. Eye Protection

To help prevent sun or snow blindness, an IP should wear sun glasses, snow goggles, or improvise a shield with a small horizontal slit opening (Figure 12-6).

Figure 12-6 Improvised Goggles

12.4.11. Wool Gloves

One or two pairs of wool gloves and (or) mittens should be worn inside a waterproof shell (Figure 12-7). If IP have to expose their hands, they should warm them inside their clothing.

Figure 12-7 Layer System for Hands

12.4.12. Adding Insulation to Foot Wear

If boots are big enough, add insulation around the feet by using dry grass, moss, or other material. Footgear can be improvised by wrapping parachute cloth or other fabric lined with dry grass or moss for insulation.

12.4.13. Felt Booties and Mukluks

Felt booties and mukluks with the proper socks and insoles are best for dry, cold weather. Rubber-bottomed boot shoepacs with leather tops are best for wet weather. Mukluks should not be worn in wet weather. The vapor-barrier rubber boots can be worn under both conditions and are best at extremely low temperatures. The air release valve should be closed at ground level. These valves are designed to release pressure when airborne. Air should not be blown into the valves as the moisture could decrease insulation.

12.4.14. Parachute Material in Extreme Cold

In strong wind or extreme cold, as a last resort, an IP should wrap up in parachute material, if available, and get into some type of shelter or behind a windbreak. Extreme care should be taken with hard materials, such as synthetics, as they may become cold soaked and require more time to warm.

12.4.15. Sleeping Systems

Sleeping systems (sleeping bag, liner, and bed) are the transition "clothing" used between normal daytime activities and sleep (Figure 12-8).

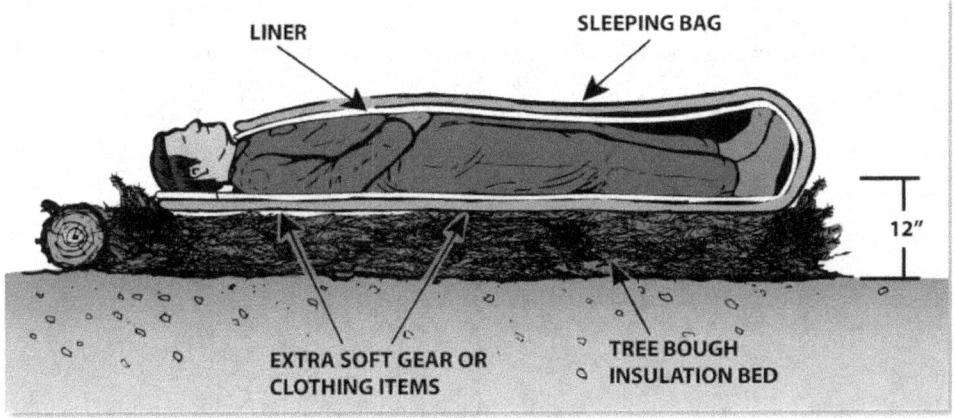

Figure 12-8 Sleeping System

12.4.15.1. Extra Insulation for Sleeping Bags

The insulating material in the sleeping bag may be synthetic or it may be down and feathers. Note that, while the covering is nylon, feathers and down lining require extra protection from moisture. Sleeping bags are compressed when packed and must be fluffed before use to restore insulation value. Clean and dry socks, mittens, and other clothing can be used to provide additional insulation.

12.5. Clothing in the Summer Arctic

In the summer arctic, there are clouds of mosquitoes and black flies so thick a person can scarcely see through them. IP can protect themselves by wearing proper clothing to ensure no bare skin is exposed. A good head net and gloves should be worn. Smoky clothing may also help to keep insects away. Treat all clothing with insect repellent at night.

12.5.1. Head Nets

Head nets must stand out from the face so they do not touch the skin (Figure 12-9). Issued head nets are either black or green. If one needs to be improvised they can be sewn to the brim of the hat or can be attached with an elastic band that fits around the crown. Black is the best color, as it can be seen through more easily than green or white. A heavy tape encasing a drawstring should be attached to the bottom of the head net for tying snugly at the collar. Hoops of wire fastened on the inside will make the net stand out from the face and at the same time allow it to be packed flat. The larger they are, the better the ventilation, but very large nets will not be as effective in wooded country where they may become snagged on brush.

12.5.1.1. Protection when a Head Net is Not Available

If the head net is lost or none is available, some protection can be obtained by wearing sunglasses with improvised screened sides, plugging ears lightly with cotton, and tying a handkerchief around the neck.

12.5.2. Gloves to Protect Against Mosquitoes

Gloves are a necessity where mosquitoes and flies are found in abundance, as in swampy areas. Gloves with a six-inch gauntlet closing the gap at the wrist and ending with an elastic band halfway to the elbow are best. Cotton/Nomex work gloves are better than no protection at all, but mosquitoes will bite through them. Treating the gloves with insect repellent will help.

Figure 12-9 Insect Protection

12.5.3. Clothing Protection against Mosquitoes

Mosquitoes do not often bite through two layers of cloth. Therefore, a lightweight undershirt and long underwear will help. To protect ankles, blouse the bottoms of trousers around boots, or wear some type of leggings or gaiters.

12.6. Clothing at Sea

In cold oceans, IP must try to stay dry and keep warm. IP should put on any available extra clothing. If anti-exposure suits are not provided, drape extra clothing around shoulders and over heads. Clothes should be loose and comfortable.

12.6.1. Decreasing the Cooling Effects of the Wind

If wet, IP should use a wind screen to decrease the cooling effects of the wind. They should also remove, wring out, and replace outer garments or change into dry clothing if available. Hats, socks, and gloves should also be dried. An IP, who is dry, should share extra clothes with those who are wet. Wet personnel should be given the most sheltered positions in the raft, and warm their hands and feet against those who are dry.

12.6.2. Keeping the Floor of the Raft Dry If Possible

For insulation, covering the floor with any available material will help. IP should huddle together on the floor of the raft and spread extra tarpaulin, sail, or parachute material over the group. If in a multi-place raft, canopy sides can be lowered.

12.6.3. Exercising to Restore Circulation

IP should exercise fingers, toes, shoulders, and buttock muscles. Performing mild exercises will help restore circulation keep the body warm, stave off muscle spasms, and possibly prevent medical problems. IP should warm hands under armpits and periodically raise feet slightly and hold them up for a minute or two. They should also move face muscles frequently to prevent frostbite.

12.6.4. Generating Heat through Shivering

Shivering is the body's way of quickly generating heat and is considered normal. However, persistent shivering may lead to uncontrollable muscle spasms. This can be avoided by exercising muscles.

12.6.5. Providing Water to those Suffering from Exposure

If water is available, additional rations should be given to those suffering from exposure to cold. IP should eat small amounts frequently rather than one large meal.

12.7. Anti-exposure Garments

The anti-exposure assemblies, both quick donning and constant wear, are designed for personnel operating in climatic cold climates where unprotected or prolonged exposure to the conditions of cold air and/or cold water as a result of ditching or abandoning an aircraft or vehicle would be dangerous or could prove fatal. The suit provides protection from the wind and insulation against the chill of the ocean. Exposure time varies depending on the particular anti-exposure assembly worn, the cold sensitiveness of the person, and survival procedures used.

12.7.1. Quick-Donning Anti-Exposure Flying Coverall

Some anti-exposure coveralls are designed for quick donning (approximately one minute) in emergencies. After ditching the aircraft or vehicle, the coverall protects the wearer from exposure while swimming in cold water, and from exposure to cold air, wind, spray, and rain.

12.7.2. The Coverall

The coverall is a one-size garment made from waterproof nylon cloth. It has two expandable-type patch pockets, an adjustable waist belt, and attached boots with adjustable ankle straps. One pair of insulated, adjustable wrist strap mittens, each with a strap attached to a pocket, is provided. A hood, also attached with a strap, is in the left pocket. The gloves and hood are normally worn after boarding the life raft. A carrying case with instructions and a snap fastener closure is furnished for storage.

- To use the coverall, personnel should wear it over regular clothing. It is large enough to wear over the usual gear. Individuals with large boots, typically size 12 or over, should remove boots prior to donning the anti-exposure suit.

- IP should be extremely careful when donning the coverall to prevent damage by snagging, tearing, or puncturing it on projecting objects. After donning the coverall, the waist band and boot ankle straps should be adjusted to take up fullness. If possible, IP should stoop while pulling the neck seal to expel air trapped in the suit (Figure 12-10). When jumping into the water, they should leap feet first with hands and arms close to sides or brought together above the head.

12.7.3. The Constant Wear Exposure Suit

Note there is a constant wear exposure suit designed to be worn continuously during overwater flights where the water temperature is 60 degrees or below. The MAJCOM may waiver it to 51 degrees.

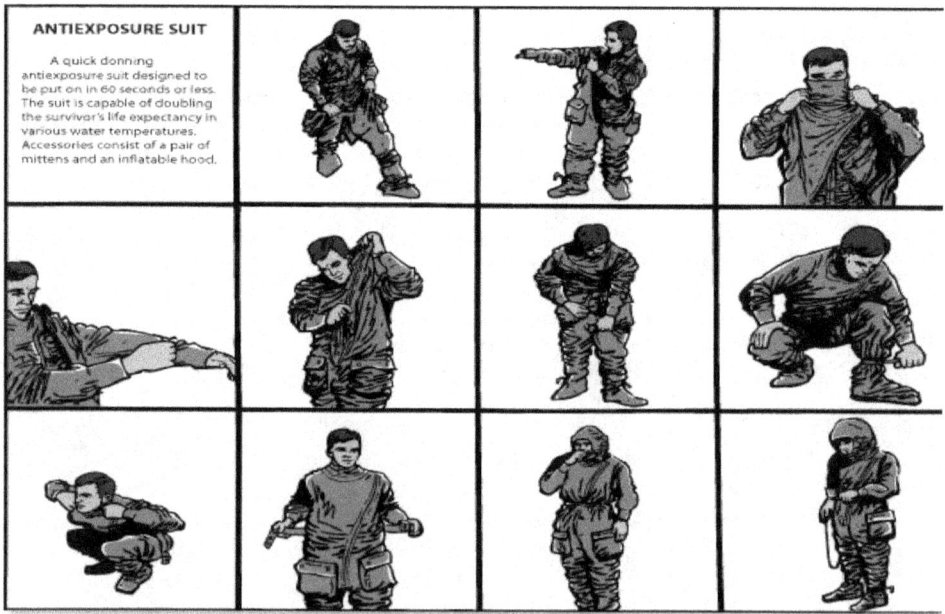

Figure 12-10 Donning Anti-Exposure Suit

12.8. Warm Oceans

Protection against the Sun and securing drinking water are the most important items in warm ocean climates. An IP should keep the body covered as much as possible to avoid sunburn. For shade, the canopy provided with life rafts may be used, or a sunshade can be improvised out of any materials available. If the heat becomes too intense, IP may dampen clothing with sea water to promote evaporation and cooling. The use of sunburn preventive cream or a lip balm is advisable. Exposure to the Sun increases thirst, wastes precious water, reduces the body's water content, and causes serious burns and so the body must be kept covered completely. IP should roll down their sleeves, pull up their socks, close their collars, wear a hat or improvised headgear, use a piece of cloth as a shield for the back of the neck, and wear sunglasses or improvise eye covers.

12.9. Tropical Climates

In tropical areas, the body should be kept covered for prevention of insect bites, scratches, and sunburn.

12.9.1. Protection when Moving through Vegetation

When moving through vegetation, IP should roll down their sleeves, wear gloves, and blouse the legs of their pants or tie them over their boot tops. Improvised gaiters can be made to protect legs from ticks and leeches.

12.9.2. Wear Loose Clothing as Protection from the Sun

Loosely worn clothing will keep an IP cooler, especially when subjected to the direct rays of the Sun.

12.9.3. Headgear to Protect Against Insects

IP should wear a head net or tie material around the head for protection against insects. The most active time for insects is at dawn and dusk. An insect repellent should be used at these times.

12.9.4. Headgear to Protect Against Sunburn

In open country or in high grass, IP should wear a neck cloth or improvised head covering for protection from sunburn and/or dust. They should also move carefully through tall grass, as some sharp-edged grasses can cut clothing to shreds. IP should dry clothing before nightfall. If an extra change of clothing is available, effort should be made to keep it clean and dry.

12.10. Dry Climates

In dry climates, people wear clothing made of lightweight materials, such as cotton or linen, which have an open weave. These materials absorb perspiration and allow air to circulate around the body.

12.10.1. Protective Clothing in Dry Climates

In the dry climates of the world, clothing will be needed for protection against sunburn, heat, sand, and insects. In some deserts there can be a temperature difference of over 60 degrees between day and night, so IP should not discard any clothing. They should keep their head and bodies covered and blouse the legs of pants over the tops of footwear during the day. IP should not roll up sleeves, but keep them rolled down and loose at the cuff to stay cool.

12.10.2. Emulating Local People's Clothing

IP should emulate what the people who live in the hot dry areas of the world often wear. Wearing heavy white or light colored flowing robes, leaving only the face and the eyes exposed to the sun, will protect almost every inch of the body. This type of dress will help keep the IP cooler and conserve perspiration because it produces an area of higher humidity between the body and the clothing. White clothing also reflects the sunlight.

12.10.3. Neckpiece for Sun Protection

IP should wear a cloth neckpiece to cover the back of the neck and protect it from the sun. A t-shirt makes an excellent neck drape, with the extra material used as padding under the cap. If hats are not available, IP can make headpieces like those worn by desert dwellers, as shown in Figure 12-11. During dust storms, they should wear a covering for the mouth and nose; parachute cloth or other fabric will work.

Figure 12-11 Protective Desert Clothing

12.11. Care of the Feet

Foot care and footgear are important in isolating situations in all climates because walking is the only means of mobility. Therefore, care of footgear is essential both before and during an isolating event. Recommendations for care include:

1. Ensure footgear is properly broken-in before deploying.
2. Treat footgear to ensure water-repellency (follow manufacturer's recommendations).
3. Keep leather boots as dry as possible.

12.11.1. Air Force Mukluks

Mukluks have been around for thousands of years and have proven their worth in extremely cold weather. Air Force mukluks are made of cotton duck with rubber-cleated soles and heels (Figure 12-12). They have slide fasteners from instep to collar, laces at instep and collar, and are 18 inches high. They are used by flying and ground personnel operating under dry, cold conditions in temperatures below +15°F.

12.11.1.1. Mukluk Liners

IP should change mukluk liners daily when possible.

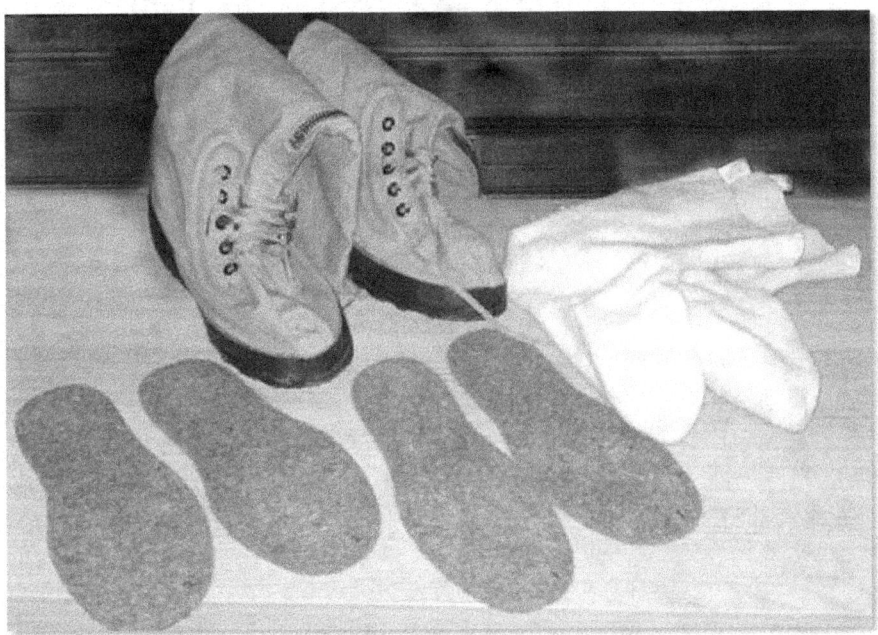

Figure 12-12 Issued Mukluks

12.11.2. Using Grass to Construct Inner Soles

Grass can be used to construct inner soles. Grass is a good insulator and will collect moisture from the feet. To prepare grass for use as inner soles, grasp a sheaf of tall grass, about one-half inch in diameter, with both hands. Rotate the hands in opposite directions. The grass will break up or fluff into a soft mass. Form this fluff into oblong shapes and spread it evenly throughout the shoes. The inner soles should be about an inch thick. Remove these inner soles at night and make new ones the following day.

12.11.3. Improvising Foot Gear

Improvising foot gear may be essential to care for feet. A combination of two or more types of improvised footwear may be more desirable and more efficient than any single type.

12.11.4. Gaiters

Gaiters can be made from parachute material or other cloth, webbing, or canvas. Gaiters help keep sand and snow out of shoes and protect the legs from bites and scratches (Figure 12-13).

Figure 12-13 Gaiters

12.11.5. The Hudson Bay Duffel Bag

A "Hudson Bay Duffel" is a triangular piece of material used as a foot covering. To improvise this foot covering, do the following:

1. Cut two to four layers of cloth into a 30-inch square.
2. Fold this square to form a triangle.
3. Place the foot on this triangle with the toes pointing at one corner.
4. Fold the front cover up over the toes.
5. Fold the side corners, one at a time, over the instep (Figure 12-14). This completes the foot wrap.

Figure 12-14 Hudson Bay Duffel

12.11.6. Making Improvised Socks

Parachute material or other cloth material can be used to improvise "Russian Socks." The material should be cut into strip approximately two feet long and four inches wide, and then wrapped bandage fashion around the feet and ankles. Socks made in this fashion will provide comfort and protection for the feet.

12.11.7. Double Socks

Double Socks are made by inserting cushion padding, feathers, dry grass, or fur between the layers of socks, and the wrapping parachute or other fabric around the feet, and tying above the ankles (Figure 12-15).

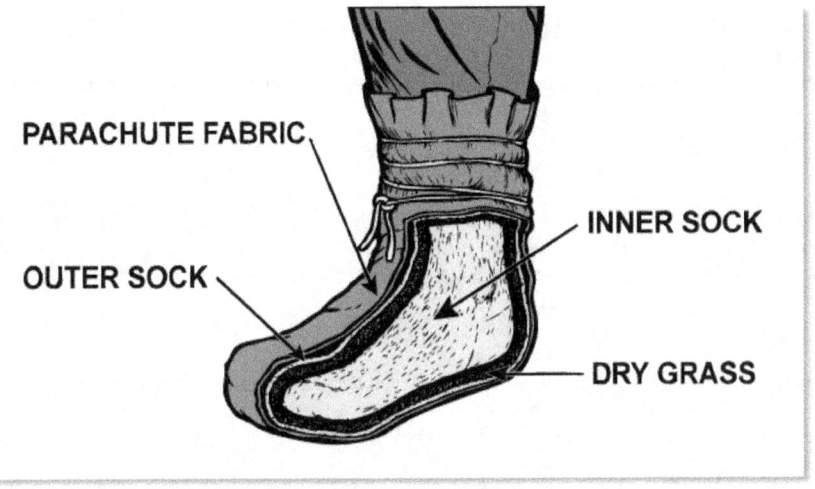

Figure 12-15 Double Socks

Chapter 13

SHELTER

13. Shelter

Shelter is anything that provides protection from environmental hazards. The environment influences shelter site selection and factors which IP must consider before constructing an adequate shelter. The techniques and procedures for constructing shelters for various types of protection are also presented.

13.1. Shelter Considerations

The location and type of shelter built by IP vary with each situation. There are many things to consider when picking a site, such as the time and energy required to establish an adequate shelter, weather conditions, life forms (human, plant, and animal), terrain, and time of day. Every effort should be made to use as little energy as possible and yet attain maximum protection from the environment.

13.1.1. Time

Late afternoon is not the best time to look for a site which will meet that day's shelter requirements. IP should not wait until the last minute to build shelter, as they may be forced to use poor materials in unfavorable conditions. They must constantly be thinking of ways to satisfy their needs for protection from environmental hazards.

13.1.2. Weather

Weather conditions are a key consideration when selecting a shelter site. Failure to consider the weather could have disastrous results. Weather factors such as temperature, wind, and precipitation can influence the IP's choice of shelter type and site selection.

13.1.2.1. Temperature

Situating a shelter site in low areas such as a valley in cold regions can expose the IP to low night temperatures and wind-chill factors. Colder temperatures are found along valley floors and are sometimes referred to as "cold air sumps." It may be advantageous to situate shelter sites to take advantage of the Sun. The shelter site could be situated in an open area during the colder months for added warmth, or in a shaded area for protection from the Sun during periods of hotter weather. In some areas, a compromise may have to be made. For example, in many deserts the daytime temperatures can be very high while low temperatures at night can turn water to ice. Protection from both heat and cold are needed in these areas. Shelter type and location should be chosen to provide protection from the existing temperature conditions.

13.1.2.2. Wind

Wind can be either an advantage or a disadvantage depending upon the temperature of the area and the velocity of the wind. During the summer or on warm days, IP can take advantage of the cool breezes and protection the wind provides from insects by locating their camps on knolls or spits of land. Conversely, wind can become an annoyance or even a hazard as blowing sand, dust, or snow can cause skin and eye irritation as well as damage to clothing and equipment. On cold days or during winter months, IP should seek shelter sites which are protected from the effects of wind-chill and drifting snow.

13.1.2.3. Precipitation

The many forms of precipitation (rain, sleet, hail, or snow) can also present problems for IP. Shelter sites should be out of major drainages and other low areas to avoid potential flash floods or mud slides resulting from heavy rains. Snow can also be a great danger if shelters are placed in potential avalanche areas.

13.1.3. Life Forms

All life forms (human animal, and plant,) must be considered when selecting the shelter site and the type of shelter that will be used. Human life forms may mean the enemy or other groups from whom IP wish to remain undetected. For a shelter to be adequate, various factors must be considered, especially if an extended isolating situation is expected.

13.1.3.1. Personal Discomfort from Insects

Insect life can cause personal discomfort, disease, and injury. By locating shelters on knolls, ridges, or other areas that have a breeze or steady wind, IP can reduce the number of flying insects in their area. Staying away from standing water sources will help to avoid mosquitoes, bees, wasps, and hornets. Ants can be a major problem, and some species will vigorously defend their territories with painful stings, bites, or pungent odors.

13.1.3.2. Issues with Animals near a Shelter Site

Animals can also be a problem, especially if the IP's shelter site is situated near their trails or waterholes.

13.1.3.3. Hazardous Plants

Some plants may be hazardous to the IP. Thorn bushes, trees, and vines can cause injuries. Dead trees that are standing, and trees with dead branches, should be avoided. Wind may cause them to fall, causing injuries or death. Poisonous plants, such as poison oak or poison ivy, must also be avoided when locating a shelter.

13.1.4. Terrain

Avalanche, rock, or mud-slide areas should be avoided. These areas can be recognized by either a clear path or a path of secondary vegetation, such as one to 15-foot tall vegetation or other new growth which extends from the top to the bottom of a hill or mountain. Dry streambeds should also be avoided due to the danger of flash flooding. Rock overhangs must be checked for safety before being used as a shelter.

13.2. Principles of Shelter Locations and Types

The basic principles of shelter requirements and planning are:

- Be near areas to meet the IPs' needs such as water, food, fuel, and a signal or recovery site.
- Be a safe area, providing natural protection from environmental hazards.
- Be near sufficient materials to construct the shelter.

- If possible, IP should try to find a shelter which needs little work to be adequate. In some cases, the shelter may already be present. IP limit themselves if they assume shelters must be a fabricated framework having predetermined dimensions and covered with material. IP must consider the amount of energy required to build the shelter.

- If nature has provided a natural shelter nearby, which will satisfy the IPs' needs, use what is already there, a complete construction of a shelter is not necessary. This saves time and energy. This does not rule out shelters with a fabricated framework and parachute or other manufactured material covering; it simply enlarges the scope of what can be used as a survival shelter. For example, rock overhangs, caves, large crevices, fallen logs, root buttresses, or snow banks can all be modified to provide adequate shelter. Modifications may include adding snow blocks to finish off an existing tree well shelter, increasing the insulation of the shelter by using vegetation or parachute material, or building a fire reflector in front of a rock overhang or cave. See Figure 13-1 for examples of naturally occurring shelters.

- Be large enough and level enough for the IP to sit up, with adequate room to lie down and to store all personal equipment. An adequate shelter provides physical and mental well-being for rest which is vital if IP are to make sound decisions. The need for rest becomes more critical as time passes and rescue or return is delayed.

Figure 13-1 Natural Shelters

13.2.1. Determining Purpose of the Shelter

Before actually constructing a shelter site, IP must determine the specific purpose of the shelter. The type of shelter to be fabricated is influenced by: climate, weather, insects, availability of nearby materials (manufactured or natural), length of expected stay, enemy presence, number and physical conditions of IP.

13.2.2. Shelter Specifications

The size limitations of a shelter are important only if there is either a lack of material on hand or if it is cold. Otherwise, the shelter should be large enough to be comfortable yet not so large as to cause an excessive amount of work. Any shelter, naturally occurring or otherwise, in which a fire is to be built, must have a ventilation system which will provide fresh air and allow smoke and carbon monoxide to escape. Even if a fire does not produce visible smoke (such as heat tabs), the shelter must still be vented. See figure 13-24 for placement of ventilation holes in a snow cave. If a fire is to be placed outside the shelter, the opening of the shelter should be placed 90 degrees to the prevailing wind. This will reduce the chances of sparks and smoke being blown into the shelter if the wind should reverse direction in the morning and evening, which frequently occurs in mountainous areas. The best fire to shelter distance is approximately three feet. One place where it would not be wise to build a fire is near an aircraft or vehicle wreckage, especially if it is being used as a shelter. The possibility of igniting spilled lubricants or fuels is great. IP may decide instead to use materials from the aircraft or vehicle to add to a shelter located a safe distance from the crash or abandon site.

13.2.2.1. Remembering Pitch and Tightness to Protect Against Rain and Snow

If IP are going to use a parachute or other porous materials, they should remember that "pitch and tightness" apply to shelters designed to shed rain and snow. Porous materials will not shed moisture unless they are stretched tightly at an angle of sufficient pitch which will encourage run-off instead of penetration. An angle of 40 to 60 degrees is recommended for the pitch of the shelter. The material stretched over the framework should be wrinkle-free and tight. IP should not touch the material when water is running over it as this will break the surface tension at that point and allow water to drip into the shelter.

13.2.2.2. Collecting the Necessary Materials

Select, collect, and prepare all materials needed before the actual construction; this includes framework, covering, bedding, or insulation, and implements used to secure the shelter ("deadmen", lines, stakes, etc.).

1. For shelters with a wooden framework, the poles or wood selected should have all rough edges and stubs removed. This will reduce the chances of the fabric being ripped, as well as eliminate the chances of injury to IP.
2. When a tree is selected as natural shelter, leave some or all of the branches in place on the outer side of the tree, as they will make a good support structure for the rest of the shelter parts.
3. Many materials can be used as framework coverings. Some can provide both framework and covering, such as bark peeled off dead trees; boughs cut off trees; bamboo, palm fronds, grasses, and other vegetation cut or woven into desired patterns.
4. If parachute material is to be used, it must be modified slightly. IP should remove all of the lines from the parachute and then cut it to size. This will eliminate bunching and wrinkling and reduce leakage.

13.2.2.3. Site Preparation

Site preparation should include brushing away rocks and twigs from the sleeping area and cutting back overhanging vegetation.

13.2.2.4. Beginning with the Framework

To actually construct the shelter, begin with the framework. The framework is very important. It must be strong enough to support the weight of the covering and precipitation buildup of snow. It must also be sturdy enough to resist strong wind gusts.

1. For natural shelters, branches may be securely placed against trees or other natural objects. The support poles or branches can then be placed and/or attached depending on their function.
2. The pitch of the shelter is determined by the framework. A 60-degree pitch is optimum for shedding precipitation and providing shelter room.
3. After the basic framework has been completed, IP can apply and secure the framework covering. The care and techniques used to apply the covering will determine the effectiveness of the shelter in shedding precipitation.
4. If using parachute material, stretch the center seam tight; then work from the back of the shelter to the front, alternating sides and securing the material to stakes or framework by using buttons and lines. Stretch the material tight by pulling the material 90 degrees to the wrinkles. If material is not stretched tight, any moisture will pool in the wrinkles and leak into the shelter.
5. If natural materials are to be used for the covering, the shingle method should be used. Start at the bottom and work towards the top of the shelter, overlapping the bottom of each piece over the top of the preceding piece. This will allow water to drain off. The material should be placed on the shelter in sufficient quantity so that it cannot be seen through from the inside.

13.3. Types of Shelters

13.3.1. Immediate Action Shelters

An immediate action shelter is one which can be erected quickly with minimal effort; for example, from raft, aircraft parts, parachutes, paulin, or plastic material. Natural formations can also shield IP from the elements immediately, to include overhanging ledges, fallen logs, caves, and tree wells (Figure 13-2). Regardless of type, the shelter must provide whatever protection is needed and, with a little ingenuity, it should be possible for IP to protect themselves and do so quickly. In some instances, immediate action shelters may have to serve as permanent shelters for IP. Situate the shelter in areas which afford maximum protection from precipitation and wind and use the basic shelter principles noted above.

13.3.1.1. Existing Shelters

If shelter is needed, IP should remember to use an existing shelter if at all possible. Improvise on natural shelters and construct new shelters only if necessary.

13.3.1.2. Using a Multi-Person Raft for Shelter

Multi-person life rafts may be the only immediate or long-term shelter available. In this situation, multi-person life rafts must be deployed in the quickest manner possible to ensure maximum advantages are attained. Anchor the raft for retention during high winds and use additional boughs, grasses, etc., for ground insulation.

Figure 13-2 Immediate Action Shelters

13.3.2. Sleeping Bag

Immediate action should be to use the whole parachute, cloth material, padding, or other manmade items, until conditions allow for improvising. A sleeping bag can be improvised by using four gores of parachute material or an equivalent amount of other materials (Figure 13-3). The material should be folded in half lengthwise and sewn at the foot. To measure the length, the IP should allow an extra six to 10 inches in addition to the IPs' height. The two raw edges can then be sewn together. The two sections of the bag can be filled with cattail down, goat's beard lichen, dry grass, insulation from aircraft walls, etc. The stuffed sleeping bag should then be quilted to keep the insulation from shifting. The bag can be folded in half lengthwise and the foot and open edges sewn. The length and width can be adjusted for the IP.

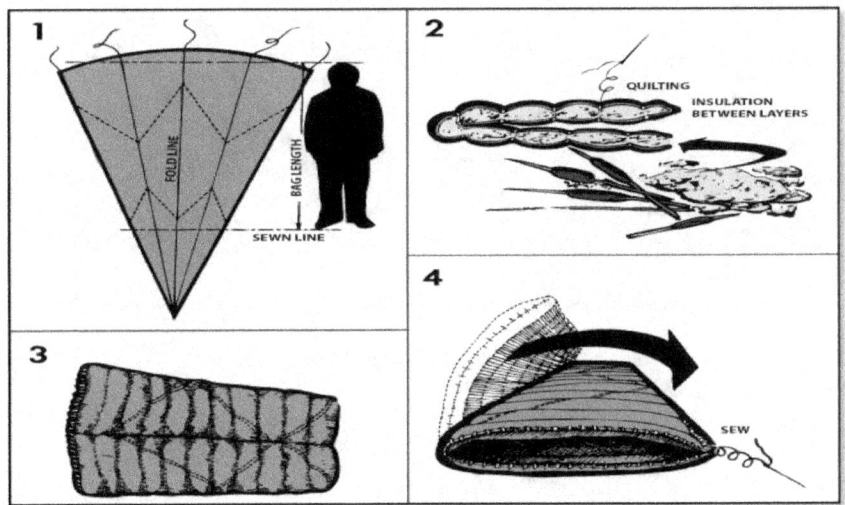

Figure 13-3 Improvised Sleeping Bag

13.3.3. Improvised Shelters

Shelters of this type should be easy to construct and/or dismantle in a short period of time. IP should consider this type of shelter only when they are not immediately concerned with getting out of the elements. Shelters of this type include A-frames, simple shade shelters, tepees, sod shelters, and snow shelters; including tree-pit shelters.

13.3.3.1. A-Frame

One way to build an A-frame shelter is to use a parachute or other material for the covering. There are as many variations of this shelter as there are builders. The procedures here will, if followed carefully, result in the completion of a safe shelter that will meet IP's needs. For an example of this and other A-frame shelters, see Figure 13-4.

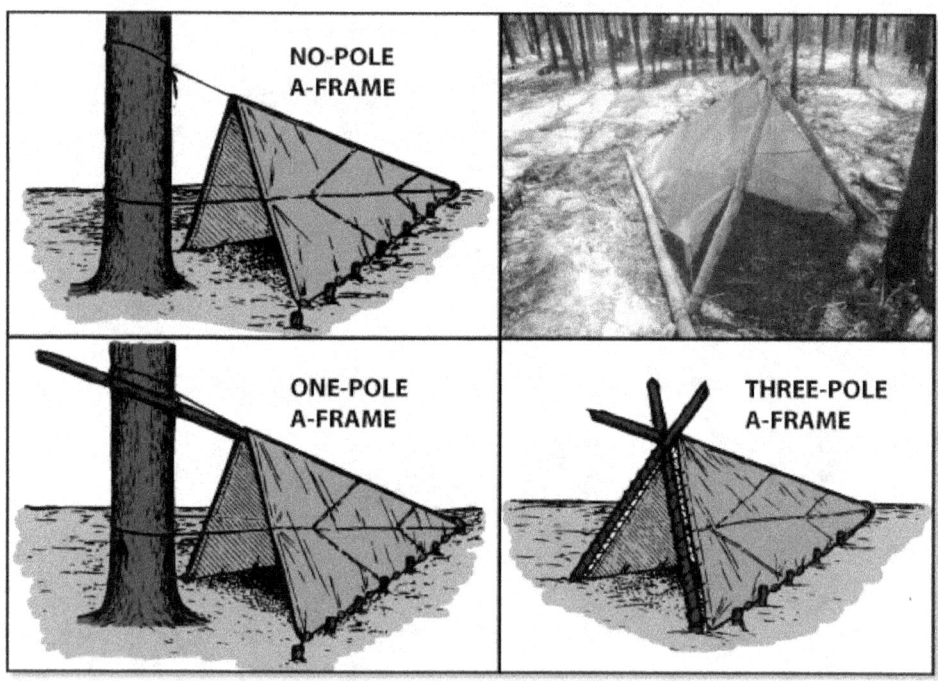

Figure 13-4 A-Frame Shelters

13.3.3.1.1. Materials Needed

- One 12 to 18 foot long sturdy ridge pole with all projections cleaned off.

- Two bipod poles, approximately seven feet long.

- Parachute material, normally five or six gores, or other natural or manmade material

- Suspension lines or cord.
- Buttons or small objects placed behind gathers of material to provide a secure way of affixing suspension line to the material.
- Approximately 14 stakes, approximately 10 inches long.

13.3.3.1.2. Assembling the Framework

1. Lash (See chapter 15 - Equipment) the two bipod poles together at eye-level height.
2. Place the ridge pole, with the large end on the ground, into the bipod formed by the poles and secure with a square lash.
3. The bipod structure should be 90 degrees to the ridge pole and the bipod poles should be spread out to an approximate equilateral triangle of a 60-degree pitch. A piece of line can be used to measure this.

13.3.3.1.3. Application of Fabric

1. Tie off approximately two feet of the apex in a knot and tuck this under the butt end of the ridge pole. Use half hitches and clove hitches to secure the material to the base of the pole.
2. Place the center radial seam of the parachute piece or the center of the fabric on the ridge pole. After pulling the material taut, use half hitches and clove hitches to secure the fabric to the front of the ridge pole.
3. Scribe or draw a line on the ground from the butt of the ridge pole to each of bipod poles. Stake the fabric down, starting at the rear of the shelter and alternately staking from side to side to the shelter front. Use a sufficient number of stakes to ensure the parachute material is wrinkle-free.
4. Stakes should be slanted or inclined away from the direction of pull. When tying off with a clove hitch, the line should pass in front of the stake first and then pass under itself to allow the button and line to be pulled 90 degrees to the wrinkle.

13.3.3.2. Lean-ToFor an example of lean-to shelters, see Figure 13-5.

13.3.3.2.1. Materials Needed

- A sturdy, smooth ridge pole long enough to span the distance between two sturdy trees.

- Support poles, 10 feet long.
- Stakes, lines, and buttons.
- Covering material.

13.3.3.2.2. Assembling the Framework

1. Lash the ridge pole (between two suitable trees) on the shelter side, about chest or shoulder high.
2. Lay the roof support poles against ridge pole from the ridge pole side of shelter so the roof support poles and the ground are at approximately a 60-degree angle. The roof support poles should push the ridge pole against the trees. Lash the roof support poles to the ridge pole.

13.3.3.2.3. Application of Fabric

1. Place the middle seam of the fabric on the middle support pole with lower lateral band along the ridge pole.
2. Tie-off the middle and both sides of the lower lateral band approximately 8 to 10 inches from the ridge pole.
3. Stake the middle of the rear of the shelter first, and then alternate from side to side.
4. The stakes that go up the sides to the front should point to the front of the shelter.

Figure 13-5 Lean-To Shelters

13.3.3.3. Raised Platform Shelter

This shelter type has many variations (Figure 13-6).

13.3.3.3.1. Example

One example is to use four trees or vertical poles in a rectangular pattern a little taller and a little wider than the IP, keeping in mind the IP will also need protection for equipment.

1. Select four trees or vertical poles in a rectangular pattern a little taller and a little wider than the IP, keeping in mind the IP will also need protection for equipment.
2. Square-lash two long, sturdy poles between the trees or vertical poles, one on each side of the intended shelter.
3. Secure cross pieces across the two horizontal poles at six- to 12-inch intervals. This forms the platform on which a natural mattress may be constructed.
4. Cover with an insect net of parachute material or other fabric.
5. Construct a roof over the structure using A-frame building techniques.
6. The roof should be waterproofed with thatching laid bottom to top in a thick shingle fashion.

13.3.3.3.2. Using Trees to Help Build Shelters

Shelters can also be built using three trees in a triangular pattern. At the foot of the shelter, two poles are joined to one tree.

Figure 13-6 Raised Platform Shelter

13.3.3.4. Paraplatform

A variation of the platform-type shelter is the paraplatform. A quick and comfortable bed is made by simply wrapping material around the two "frame" poles. Another method is to roll poles in the material in the same manner as for an improvised stretcher (Figure 13-7).

Figure 13-7 Raised Paraplatform Shelter

13.3.3.5. Hammocks

Various parahammocks and hammocks made of other fabrics can also be made. However, they are more involved than a simple framework wrapped with material and not quite as comfortable (Figure 13-8).

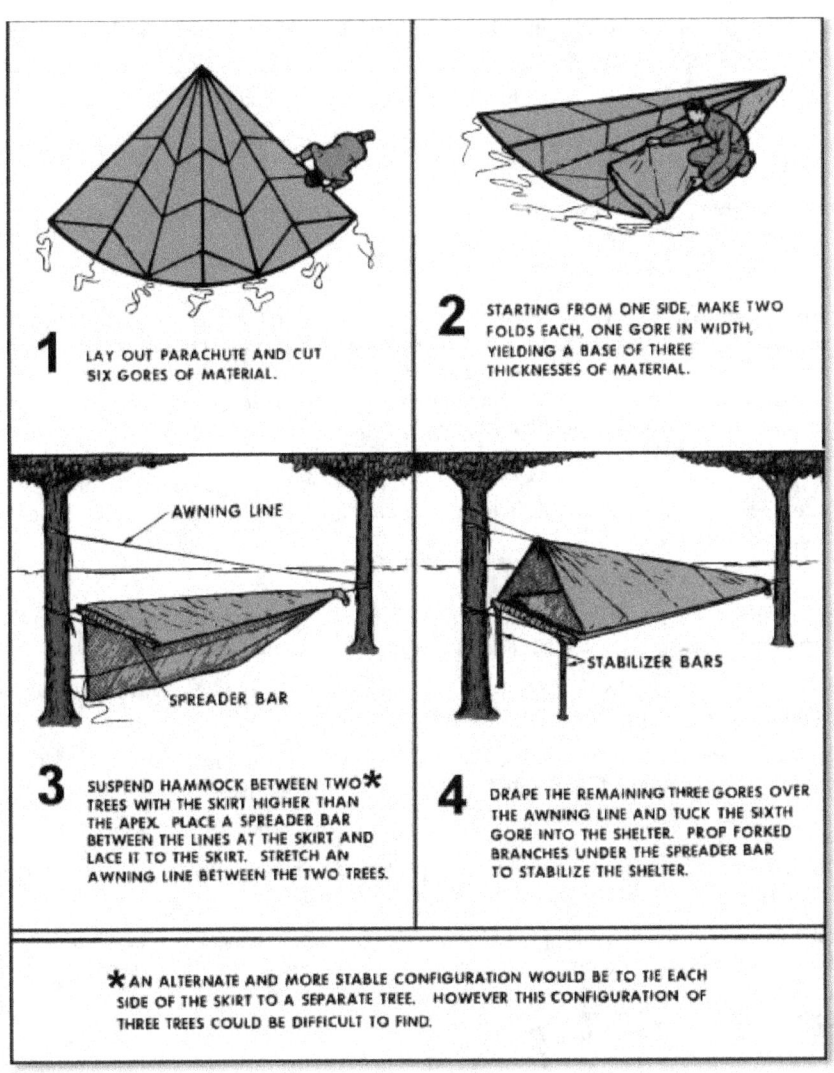

Figure 13-8 Parahammock

13.3.3.6. Debris Nest as a Sleeping System

The debris nest is a small lifesaving sleeping system, designed to protect individual(s) from the elements (Figure 13-9). It uses the old principle "do as the animals do". Although primarily designed as a sleeping system, the debris nest can be used an emergency shelter. With only a water-proof tarp the IP can survive in either a permissive or non-permissive isolating event. The key principles of a good debris nest are:

- Cold protection – nest is built small so it can be warmed by body heat.
- Precipitation protection – nest has a waterproof covering or is constructed in a natural protected location.
- Wind protection – nest is designed to allow no air from outside to move into or through except for the head area.

13.3.3.6.1. Debris Nest

The debris nest consists of dry insulation in a protected covering that an individual can crawl into. It is constructed in such a way as to be warmed primarily by the occupant's body heat. (Although heated stones or other heated heavy objects can be added for additional warmth.) There are many ways an IP can construct a debris nest depending on the environmental conditions and the materials available. If conditions are cold and the IP is in a permissive isolating event a fire should be built before beginning the shelter construction. If the situation allows choose locations to build a debris nest that provide protection from precipitation and air movement. Examples are dumpsters (use care in extreme cold), the crawl spaces in buildings, crevice of rocks, an eroded embankment, large fallen tree, or in a tree well.

13.3.3.6.2. Dry Materials as Insulation

Possible dry materials that may be used for insulation include grasses, ferns, evergreen boughs, mosses and lichens, clothing and other cloth type materials. Some examples include:

- If an IP is in a coniferous forest, the soft bow and dry pine needles can be piled up or in a deciduous forest, the dry leaves can be used.
- In an urban environment, IP may be able find a large heavy duty plastic bag, cardboard box, or container that could be filled with dry shredded papers, cloth, or available dry vegetation.
- In a cold desert, a nest may be constructed using dry grasses and available shrubs.
- Possible nest cover materials should be non-porous to provide protection from the cold (body heat can be captured in the nest), wind, and precipitation.
- Examples are tarp, EVCs (multiple EVCs can be duct taped together), large peeled bark from dead trees, thatched palm fronds, fallen logs, dumpsters (use care in extreme cold), wax coated or plastic covered card board, and the crawl spaces in buildings.

 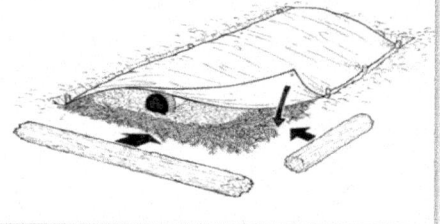

Figure 13-9 Debris Nest

13.3.3.7. Tepee, 0 to 9-Pole

The tepee is an excellent shelter for protection from wind, rain, cold, and insects. Cooking, eating, sleeping, resting, signaling, and washing can all be done without going outdoors. The tepee, whether 9-pole, 1-pole, or no-pole improvised shelter provides adequate ventilation to build an inside fire. With a small fire inside, the shelter also serves as a signal at night. The potential drawbacks of the tepee that must be considered by the IP are the large amount of time and materials required in its construction. Refer to Figure 13-10.

13.3.3.7.1. Materials Needed

- Suspension line.
- Parachute material, normally 14 gores are suitable.
    - Spread out the 14-gore section of parachute and cut off all lines at the lower lateral band, leaving about 18 inches of line attached. All other suspension lines should be stripped from the parachute.
    - Sew two smoke flaps, made from two large panels of parachute material, at the apex of the 14-gore section on the outside seams. Attach suspension line with a bowline in the end to each smoke flap. The ends of the smoke flap poles will be inserted in these.
- Stakes.
- Although any number of poles may be used, 11 poles, smoothed off, each about 20 feet long, will normally provide adequate support.

13.3.3.7.2. Assembling the Framework

(Assume 11 poles are used. Adjust instructions if different numbers are used.)

1. Lay three poles on the ground with the butts even. Stretch the canopy along the poles. The lower lateral band should be four to six inches from the bottoms of the poles before the stretching takes place. Mark one of the poles at the apex point.
2. Lash the three poles together, five to 10 inches above the marked area. A shear lash is effective for this purpose. These poles will form the tripod.
3. Scribe a circle approximately 12 feet in diameter in the shelter area and set the tripod so the butts of the poles are evenly spaced on the circle. Five of the remaining eight poles should be placed so the butts are evenly spaced around the 12-foot circle and the tops are laid in the apex of the tripod to form the smallest apex possible.

13.3.3.7.3. Application of Fabric

1. Stretch the parachute material or other fabric along the tie pole. Using the suspension line attached to the middle radial seam, tie the lower lateral band to the tie pole six inches from the butt end. Stretch the material along the middle radial seam and tie it to the tie pole using the suspension line at the apex. Lay the tie pole onto the shelter frame with the butt along the 12-foot circle and the top in the apex formed by the other poles. The tie pole should be placed directly opposite the proposed door.
2. Move the canopy material (both sides of it) from the tie pole around the framework and tie the lower lateral band together and stake it at the door. The front can now be sewn or

pegged closed, leaving three to four feet for a door. A sewing "ladder" can be made by lashing steps up the front of the tepee.

3. Enter the shelter and move the butts of the poles outward to form a more perfect circle and until the fabric is relatively tight and smooth.

4. Tighten the fabric and remove remaining wrinkles. Start staking directly opposite the door and alternate from side to side, pulling the material down and to the front of the shelter. Use clove hitches or similar knots to secure the material to the stakes.

5. Insert the final two poles into the loops on the smoke flaps. The tepee is now finished.

6. One improvement which could be made to the tepee is the installation of a liner. This will allow a draft for a fire without making the occupants cold, since there may be a slight gap between the lower lateral band and the ground. A liner can be affixed to the inside of the tepee by taking the remaining 14-gore piece of material and firmly staking the lower lateral band directly to the ground all the way around, leaving room for the door. The area where the liner and door meet may be sewn up. The rest of the material is brought up the inside walls and affixed to the poles with buttons.

Figure 13-10 9-Pole Tepee

13.3.3.8. Tepee, 1–Pole For an example of this type of tepee, refer to Figure 13-11.

13.3.3.8.1. Materials Needed

- Use a 14-gore section of canopy; strip the shroud lines leaving 16- to 18-inch lengths at the lower lateral band, or another type of material.
- Stakes.

- Inner core and needle.

13.3.3.8.2. Construction

1. Select a shelter site and scribe a circle about 14 feet in diameter on the ground.
2. The parachute material or other fabric is staked to the ground using the lines attached at the lower lateral band. After deciding where the shelter door will be located, stake the first line (from the lower band) down securely. Proceed around the scribed line and stake down all the lines from the lateral band, making sure the material is stretched taut before the line is staked down.
3. Once all the lines are staked down, loosely attach the center pole, and, through trial and error, determine the point at which the parachute material will be pulled tight once the center pole is placed upright - securely attach the material at this point.
4. Using a suspension line (or inner core), sew the end gores together leaving 3 or 4 feet for a door.

Figure 13-11 1-Pole Tepee

13.3.3.9. Tepee, No-Pole

For this shelter, the 14 gores of material are prepared the same way. A line is attached to the apex and thrown over a tree limb, etc., and tied off. The lower lateral band is then staked down starting opposite the door around a 12- to 14-foot circle. See Figure 13-12.

Figure 13-12 No-Pole Tepee

13.3.3.10. Sod Shelter

A framework covered with sod provides a shelter which is warm in cold weather and easily made waterproof and insect-proof in the summer. The framework for a sod shelter must be strong, and it can be made of driftwood, poles, willow, etc. Sod, with a heavy growth of grass or weeds, should be used since the roots tend to hold the soil together. Cutting about two inches of soil along with the grass is sufficient. The size of the blocks is determined by the strength of the individual. A sod shelter is strong and fireproof.

13.4. Shelter for Tropical Areas

In tropical areas, IP should establish a shelter site on a knoll or high spot in an open area set well back from any swamps or marshy areas. The ground in these areas is drier, and there may be a breeze which will result in fewer insects.

13.4.1. Cover to Protect from Smoke

A thick bamboo clump or matted canopy of vines for cover reflects the smoke from the shelter site and discourages insects. This cover will also keep the extremely heavy early morning dew off the bedding.

13.4.2. Fires to Protect from Insects

Insects are discouraged by smudge fires. In addition, a hammock made from parachute or other material will keep the IP off the ground and help ward off ants, spiders, leeches, scorpions, and other pests.

13.4.3. Shelter from Dampness in the Jungle

In the wet jungle, shelter from dampness is needed. If the IP stays with the aircraft or vehicle, it could be used for shelter. Covering openings with netting, parachute, or other cloth will provide protection from mosquitoes.

13.4.4. Rain Shelter

A good rain shelter can be made by constructing an A-frame type shelter and shingling it with a thickness of palm or other broad leaf plants, pieces of bark, and mats of grass (Figure 13-13).

Figure 13-13 Banana Leaf A-Frame

13.5. Shelters for Hot and Dry Climates

The extremes of heat and cold must be considered in hot areas, as most can become very cold during the night. The major problem for IP will be escaping the heat and sun rays.

13.5.1. Natural Shelter

Natural shelters in these areas are often limited to the shade of cliffs and the lee sides of hills, dunes, or rock formations. In some desert mountains, it is possible to find good rock shelters or cave-like protection under tumbled blocks of rocks which have fallen from cliffs. Use care to ensure that these blocks are in areas void of future rock falling activity and free from animal hazards.

13.5.2. Using Vegetation as Shade

Vegetation, if any exists, is usually stunted and armed with thorns. It may be possible to stay in the shade by moving around the vegetation as the Sun moves. The hottest part of the day may offer few shadows because the Sun is directly overhead. Parachute material or other fabric draped over bushes or rocks will provide some shade.

13.5.3. Materials Used in Constructing Desert Shelters

Materials which can be used in the construction of desert shelters include:

- Sand, though difficult to work with when loose, may be made into pillars by using sandbags made from parachute or any available cloth.
- Rock.
- Vegetation such as sage brush, creosote bushes, juniper trees, and desert gourd vines are valuable building materials.
- Parachute canopy and suspension lines.

13.5.4. Avoid Dense Materials when Making Shelters

The shelter should be made of dense material or have numerous layers to reduce or stop dangerous ultraviolet rays. The colors of the parachute materials used make a difference as to how much protection is provided from ultraviolet radiation. As a general rule, the order of preference should be to use as many layers as practical in the order of orange, green, tan, and white.

13.5.5. Proximity of Material from IP

The material should be kept approximately 12 to 18 inches above the individual. This allows the air to cool the underside of the material.

13.5.6. Aircraft or Vehicle Parts

Aircraft or vehicle parts and life rafts can also be used for shade shelters. IP may use sections of the wing, tail, or fuselage to provide shade. However, the interior of the aircraft or vehicle will quickly become superheated and should be avoided as a shelter. An inflatable raft can be tilted against a raft paddle or natural object such as a bush or rock to provide relief from the Sun (Figure 13-14).

Figure 13-14 Improvised Natural Shade Shelters

13.5.7. Multilayered Desert Shelter

The roof of a desert shelter should be multilayered so the resulting airspace reduces the inside temperature of the shelter. The layers should be separated 12 to 18 inches apart (Figure 13-15).

Figure 13-15 Shade Shelter

13.5.8. Placement of the Shelter Floor to Increase Cooling Effect

The floor of the shelter should be placed about 18 inches above or below the desert surface to increase the cooling effect.

13.5.9. Movable Sides

The sides of shelters should be movable in order to protect IP during cold and/or windy periods and to allow for ventilation during hot periods.

13.5.10. Using Light Colored Material in Warmer Areas

In warmer deserts, white parachute or other light-colored material should be used as an outer layer to reflect heat. Orange or sage green material should be used as an inner layer for protection from ultraviolet rays.

13.5.11. Using Multiple Layers of Material in Cooler Areas

In cooler areas, multiple layers of material should be used with sage green or orange material as the outer layer to absorb heat.

13.5.12. Location of Shelter in Hot Deserts

In a hot desert, shelters should be built away from large rocks which store heat during the day. IP may need to move to the rocky areas during the evening to take advantage of the warmth heated rocks radiate.

13.5.13. Location of Shelters near Dunes

Build shelters on the windward sides of dunes for cooling breezes.

13.5.14. Time of Day for Building a Shelter

Build shelters during early morning, late evening, or at night. Keep in mind that being in a desert area during daylight hours brings immediate concern for protection from the Sun and loss of water. In this case, any available material, i.e., parachute canopy material, can be draped over a liferaft, vegetation, or a natural terrain feature for quick shelter.

13.6. Shelters for Snow and Ice Areas

The differences in arctic and arctic-like environments create the need for different shelters. There are two types of environments which may require special shelter characteristics or building principles before IP will have adequate shelter:

1. Barren lands, which include some seacoasts, icecaps, sea ice areas, and areas above the tree line (Figure 13-16 to Figure 13-18).
2. Tree-line areas (Figure 13-19 to Figure 13-23).

13.6.1. Barren Lands

Barren lands offer a limited variety of materials for shelter construction. These are snow, small shrubs, and grasses. Ridges formed by drifting or wind-packed snow may be used for wind protection (build on the lee side). In some areas, such as sea ice, windy conditions usually exist and cause the ice to shift forming pressure ridges. These areas of unstable ice and snow should be avoided at all times. Shelters which are suitable for barren-type areas include:

13.6.1.1. Molded Snow Dome or Quinzee

1. Determine the size of the shelter based off the number of persons in your party. If alone, use the considerations discussed earlier for the appropriate size.
2. Dig out the chosen area, leaving 12 inches of snow on the ground. This will provide insulation.
3. Fill in the hole with the snow that was just dug up and mound the snow to chest height.
4. Allow the snow to set up for a minimum of ninety minutes. Longer times may be required based off the ambient air temperature.
5. Hollow out the interior of the shelter with a snow shovel, seat pan, or improvised digging tool, until the walls are 10 to 12 inches thick. As a general rule snow looks bluish when the 10-12 inches of thickness is achieved.
6. If possible create a low area at the entrance to act as a cold well or cold sump. A cold well or cold sump is the lowest point in the shelter, allowing cold air to drop below the sleeping area of the IP.
7. Ensure there is proper ventilation by creating a vent hole in the roof of the shelter. Use a ski pole, knife, or stick to puncture a hole through the snow. The hole should be two inches in diameter.
8. To finish out the shelter, create a door plug by mounding snow in a piece of material large enough to seal the opening, tie off the material so snow doesn't fall out, and pull into the entrance.

Figure 13-16 Molded Dome

13.6.1.2. Snow Cave

Refer to Figure 13-17.

Figure 13-17 Snow Cave

13.6.1.3. Fighter Trench

Refer to Figure 13-18.

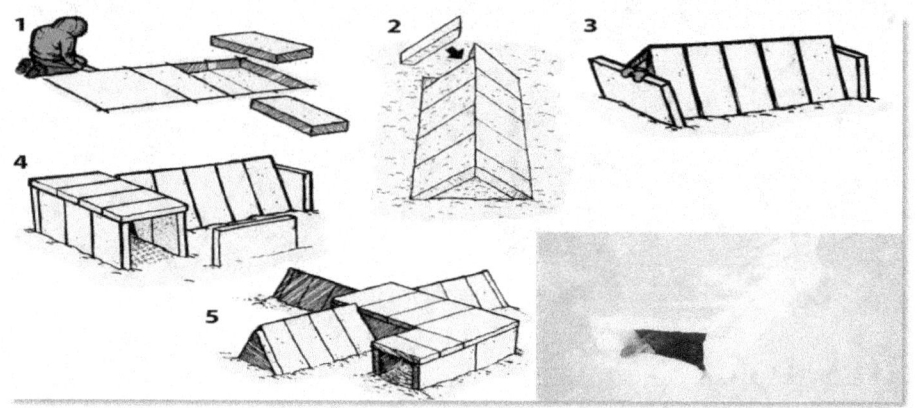

Figure 13-18 Fighter Trench

13.7. Tree-Line Areas

13.7.1. Sufficient Natural Shelter in Tree Covered Areas

In tree-covered areas, sufficient natural shelter building materials are normally available. Caution is required. Shelters built near rivers and streams may get caught should the river banks overflow. Shelters which are suitable for tree line-type areas include:

13.7.1.1. Thermal A-Frame

Refer to Figure 13-19.

Figure 13-19 Thermal A-Frame

13.7.1.2. Double Lean-To

Refer to Figure 13-20.

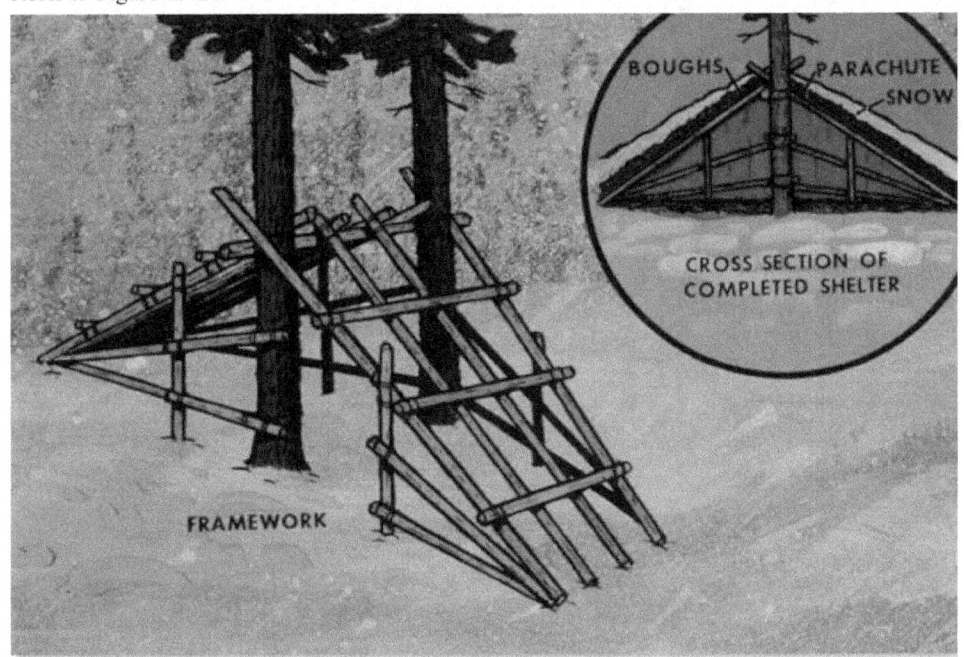

Figure 13-20 Double Lean-To

13.7.1.3. Fan

Refer to Figure 13-21.

Figure 13-21 Fan Shelter

13.7.1.4. Willow Frame

Refer to Figure 13-22.

Figure 13-22 Willow Frame Shelter

13.7.1.5. Tree Well Refer to Figure 13-23.

Figure 13-23 Tree Well Shelter

13.7.2. Thermal Principles and Insulation No matter what type of shelter is used, the use of thermal principles and insulation in arctic shelters is required. Heat radiates from bare ground and from ice masses over water. This means that shelter areas on land should be dug down to bare earth if possible. A minimum of six to 12 inches of insulation above the IP is needed to retain heat. All openings except ventilation holes should be sealed to avoid heat loss. Leaving vent holes open is especially important if heat producing devices are used. Candles, Sterno, or small oil lamps produce carbon monoxide. In addition to the ventilation hole through the roof, another may be required at the door to ensure adequate circulation of the air. As a general rule, if IP cannot see their breath, the snow shelter is too warm and should be cooled down to preclude melting and dripping.

13.7.3. Snow Caves as Shelter

Regardless of how cold it may get outside, the temperature inside a small well-constructed snow cave will probably not be lower than -10 F. Body heat alone can raise the temperature of a snow cave 45 degrees above the outside air. A burning candle will raise the temperature 4 degrees. Burning Sterno (small size, 2-1/2 oz) will raise the cave temperature about 28 degrees. However, since they cannot be heated many degrees above freezing, snow shelters provide a rather rugged life. Once the inside of the shelter "glazes" over with ice, this layer of ice should be removed by chipping it off or a new shelter built since ice reduces the insulating quality of a shelter. Maintain the old shelter until the new one is constructed. It will provide protection from the wind.

13.7.4. Conditions when Aircraft Should Not be Used as Shelter

An aircraft or vehicle should not be used as a shelter when temperatures are below freezing except in high wind conditions. Even then a thermal shelter should be constructed as soon as the conditions improve. The aircraft or vehicle will not provide adequate insulation, and the floor will usually become icy and hazardous.

13.8. General Construction Techniques

13.8.1. Construction of Thermal Shelters

Thermal shelters use a layering system consisting of the frame, parachute or other material, boughs or shrubs, and snow. The framework must be sturdy enough to support the cover and insulation. A door block should be used to minimize heat loss. Insulation should be added on sleeping areas.

13.8.2. Constructing Barren Land-Type Shelters

If a barren land-type shelter is being built with snow as the only material, a long knife or digging tool is a necessity. It normally takes two to three hours of hard work to dig a snow cave.

13.8.3. Dressing Appropriately While Constructing Shelters

IP should dress lightly while digging and working. They can easily become overheated and dampen their clothing with perspiration, which will rapidly turn to ice.

13.8.4. Position of Opening

If possible, construct all shelter types to have their openings 90 degrees to the prevailing wind. The entrance to the shelter should also be screened with snowblocks stacked in an L-shape.

13.8.5. Cutting Snow Blocks

Snow on the sea ice, suitable for cutting into blocks, will usually be found in the lee of pressure ridges or ice hummocks. The packed snow is often so shallow that the snowblocks have to be cut out horizontally.

13.8.6. Using a Digging Tool in the Event of Snowfall

No matter which shelter is used, IP should take a digging tool into the shelter at nighttime to cope with the great amount of snow which may block the door during the night.

13.9. Arctic or Cold Weather Shelter Living Considerations

13.9.1. Limit Shelter Entrances

An IP should limit the number of shelter entrances to help conserve heat. Fuel is generally scarce in the arctic. To conserve fuel, it is important to keep the shelter entrance sealed as much as possible (Figure 13-24). When it is necessary to go outside the shelter, activities such as gathering fuel, snow or ice for melting, etc., should be done. To expedite matters, a trash receptacle may be kept inside the door, and equipment may be stored in the entry way. Necessities which cannot be stored inside may be kept just outside the door.

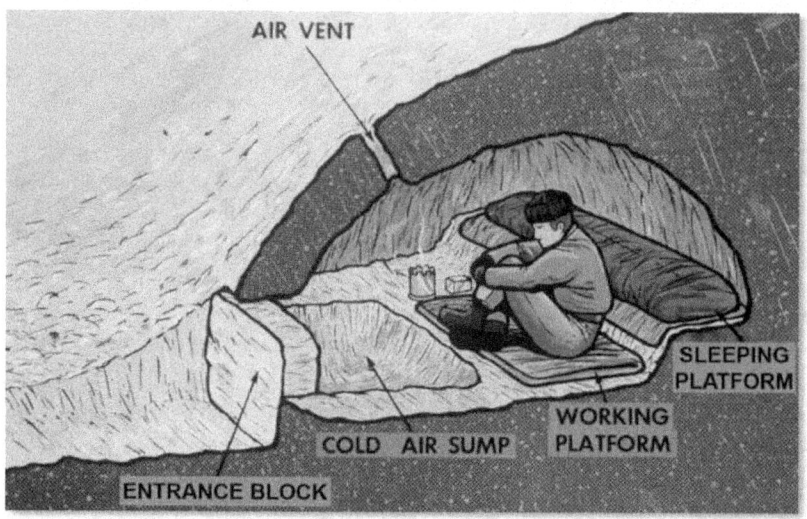

Figure 13-24 Snow Cave Shelter Living

13.9.2. Personal Hygiene in Snow Shelter Living

A standard practice in snow shelter living is for people to relieve themselves indoors when possible. This practice conserves body heat. If the snowdrift is large enough to dig connecting snow caves, one may be used as a toilet room. If not, aluminum cans may be used for urinals, and snow-blocks for solid waste (fecal) matter.

13.10. Summer Considerations for Arctic and Arctic-Like Areas

IP need shelter against rain and insects. Modify shelters discussed to provide additional protection as needed.

13.11. Maintenance and Improvements

Once a shelter is constructed, it must be maintained. Additional modifications may make the shelter more effective and comfortable. With parachute or fabric material shelters, Indian lacing the front of the shelter to the bipod will tighten the shelter. A door may help block the wind and keep insects out. Other modifications may include a fire reflector, porch or work area, or another whole addition such as an opposing lean-to.

13.11.1. Indian Lacing

Indian lacing is the sewing or lacing of the lower lateral band of a parachute with inner core or line which is secured to the bipod poles. Small holes, six to eight inches apart, should be cut in either the lower lateral band material or at least 1" from the lower edge of other material. Parachute suspension line or other cord can then be girth hitched to the first hole, run around the frame and laced through the next hole. Continue until the line is laced around frame and through each hole. Starting at the first hole, pull tension around the frame and draw slack through next hole. Continue process until line is tight and slack/wrinkles are removed. Tie off end of line. A rain fly can also be constructed to reduce the amount of precipitation that penetrates shelter material. Two layers of material, four to six inches apart, will create a more effective water repellent covering. Even during hard rain, the outer layer only lets a mist penetrate if it is pulled tight. The inner layer will then channel off any moisture which may penetrate. This layering of material also creates a dead-air space that covers the shelter. This is especially beneficial in cold areas when the shelter is enclosed.

13.11.2. Insulation Beds

Insulation beds can be constructed using natural or manmade materials. As a general rule beds should be six to 12 inches thick if being constructed of natural materials. When using manmade thermal or non-conductive materials it may not be necessary to use as much.

13.11.2.1. Ground Insulation

In addition to the sleeping bag, some form of ground insulation is advisable. An insulation mat will help insulate the survivor from ground moisture and the cold. Any nonpoisonous plants such as ferns and grasses will suffice. Leaves from a deciduous tree make a comfortable bed. If available, extra clothing, seat cushions, aircraft insulation, rafts, and parachute material may be used. In a coniferous forest, boughs from the trees would do well if the bed is constructed properly.

13.11.2.2. Making a Bed

The survivor should start at the foot of the proposed bed and stick the cut ends in the ground at about a 45-degree angle and very close together. The completed bed should be slightly wider and longer than the body. If the ground is frozen, a layer of dead branches can be used on the ground with the green boughs placed in the dead branches, similar to sticking them in the ground.

13.11.2.3. Bough Bed

A bough bed should be a minimum of 12 inches thick before use. This will allow sufficient insulation between the survivor and the ground once the bed is compressed. The bough bed should be fluffed up and boughs added daily to maintain its comfort and insulation capabilities.

13.11.2.4. Spruce Boughs

Spruce boughs have many sharp needles and can cause some discomfort. Also the needles on various types of pines are generally located on the ends of the boughs, and it would take an abundance of pine boughs to provide comfort and insulation. Fir boughs on the other hand, have an abundance of needles all along the boughs and the needles are rounded. These boughs are excellent for beds, providing comfort and insulation (Figure 13-25).

Figure 13-25 Boughs

13.11.3. Outer Clothing as Mattress Material

Outer clothing makes good mattress material. A parka makes a good foot bag. The shirt and inner trousers may be rolled up for a pillow. Socks and insoles can be separated and aired in the shelter. Drying may be completed in the sleeping bag by stowing around the hip. This drying method should only be used as a last resort.

13.11.4. Maintaining the Bag for Maximum Warmth

Keeping the sleeping bag clean, dry, and fluffed will give maximum warmth. To dry the bag, it should be turned inside out, frost beaten out, and warmed before the fire-taking care that it doesn't burn.

13.11.5. Keeping Moisture from Sleeping Bag

To keep moisture (from breath) from wetting the sleeping bag, a moisture cloth should be improvised from a piece of clothing, a towel, or parachute fabric. It can then be lightly wrapped around the head in such a way that the breath is trapped inside the cloth. A piece of fabric dries easier than a sleeping bag.

13.11.6. Exercising in the Shelter to Keep Warm

If cold is experienced during the night, IP should exercise by fluttering their feet up and down or by beating the inside of the bag with their hands. Food or hot liquids can be helpful.

Chapter 14

FIRECRAFT

14. Firecraft

Fire is used for warmth, light, drying clothes, signaling, making tools, cooking, and water purification. When using fire for warmth, the body uses fewer calories for heat and consequently requires less food. Having a fire to sit by can also be used as a morale booster. Also, smoke from a fire can be used to discourage insects.

14.1. Considerations

Avoid building a very large fire. Small fires require less fuel, are easier to control, and their heat can be concentrated. Never leave a fire unattended unless it is banked or contained. Banking a fire is done by scraping cold ashes and dry earth onto the fire, leaving enough air coming through the dirt at the top to keep the fuel smoldering. This will keep the fire safe and allow it to be rekindled from the saved coals.

14.2. Elements of Fire

The three essential elements for successful fire building are fuel, heat, and oxygen. These combined elements are referred to as the "fire triangle." By limiting fuel, only a small fire is produced. If the fire is not fed properly, there can be too much or too little fire. Fresh cut, or green fuel is difficult to ignite, and the fire must be burning well before it is used for fuel. Heat and oxygen must be accessible to ignite any fuel.

14.2.1. Preparing for Firecraft

The IP must take time and prepare well. Preparing all of the stages of fuel and all of the parts of the fire starting apparatus is the key. To be successful at firecraft, preparation, practice, and patience are required by the fire builder.

14.2.2. Three Categories of Fuels Used for Firecraft

The fuels used in building a fire normally fall into three categories (Figure 14-1) relating to their size and flash point: tinder, kindling, and fuel.

Figure 14-1 Stages of a Fire

14.2.2.1. Tinder

Tinder is any type of small material having a low flash point. It is easily ignited with a minimum of heat, such as from a spark. Tinder must be arranged to allow air (oxygen) between the hair-like, bone-dry fibers. The preparation of tinder for fire is one of the most important parts of firecraft. Dry tinder is so critical that a potential IP should keep it available at all times. It may be necessary to have two or three stages of tinder to get the flame to a useful size. Examples of tinder include:

- The shredded bark from some trees and bushes, such as Cedar, birch bark, or palm fiber.
- Crushed fibers from dead plants.
- Fine, dry wood shavings and straw/grasses.
- Resinous sawdust.
- Very fine pitch wood shavings (resinous wood (Fat Lighter) from pine or sappy conifers).
- Bird or rodent nest linings.
- Seed down (milkweed, cattail, thistle).
- Charred cloth.
- Cotton balls or lint.
- Steel wool.

- Dry powdered sap from the pine tree family (also known as pitch).
- Paper.
- Foam rubber.

14.2.2.2. Kindling

Kindling is the next larger stage of fuel material. Kindling should also have a high combustible point. It is added to, or arranged over, the tinder in such a way that it ignites when the flame from the tinder reaches it. Kindling is used to bring the burning temperature up to the point where larger and less combustible fuel material can be used. Examples of kindling include:

- Dead dry small twigs or plant fibers.
- Dead dry thinly shaved pieces of wood, bamboo, or cane (always split bamboo as sections can explode).
- Coniferous seed cones and needles.
- "Squaw wood" from the underside of coniferous trees; dead, small branches next to the ground sheltered by the upper live part of the tree.
- Pieces of wood removed from the insides of larger pieces.
- Some plastics such as the spoon from an in-flight ration.
- Wood which has been soaked or doused with flammable materials; that is, wax, insect repellent, petroleum fuels, and oil.
- Strips of petrolatum gauze from a first aid kit.
- Dry split wood burns readily because it is drier inside. Also the angular portions of the wood burn easier than the bark-covered round pieces because it exposes more surfaces to the flame. The splitting of all fuels will cause them to burn more readily.

14.2.2.3. Fuel

Fuel, unlike tinder and kindling, does not have to be kept completely dry as long as the fire is well established to dry the wet fuel prior to burning. It is recommended that all fine materials be protected from moisture to prevent excessive smoke production. (Highly flammable liquids should not be poured on an existing fire. Even a smoldering fire can cause the liquids to explode and cause serious burns). The type of fuel used will determine the amount of heat and light the fire will produce. Dry split hardwood trees (oak, hickory, ash) are less likely to produce excessive smoke and will usually provide more heat than soft woods. They may also be more difficult to break into usable sizes. Pine and other conifers are fast-burning and produce smoke unless a large flame is maintained. Rotten wood is of little value since it smolders and smokes. The weather plays an important role when selecting fuel. Standing or leaning wood is usually dry inside even if it is raining. In tropical areas, avoid selecting wood from trees that grow in swampy areas or those covered with mosses. Tropical soft woods are not usually a good fuel source. Trial and error is sometimes the best method to determine which fuel is best. After identifying the burning properties of available fuel, a selection can be made of the type needed. Recommended fuel sources include:

- Dry standing dead wood and dry dead branches (those that snap when broken). Dead wood is easy to split and break. It can be pounded on a rock or wedged between other objects and bent until it breaks.
- The insides of fallen trees and large branches may be dry even if the outside is wet. The heart wood is usually the last to rot.
- As a last resort, green wood which can be made to burn is found almost anywhere, if finely split and mixed evenly with dry dead wood.
- In treeless areas, other natural fuels can be found. Dry grasses can be twisted into bunches. Dead cactus and other plants are available in deserts. Dry peat moss can be found along the surface of undercut stream banks. Dried animal dung, animal fats, and sometimes even coal can be found on the surface. Oil impregnated sand can also be used when available.

14.3. Fire Site Preparation

The location of a fire should be carefully selected. Locate and prepare the fire carefully.

14.3.1. Building a Platform for a Fire Site

After a site is located, twigs, moss, grass, or duff should be cleaned away and scraped to bare soil if possible. If the fire must be built on snow, ice, or wet ground, IP should build a platform of green logs or rocks. Beware of wet or porous rocks, they may explode when heated.

14.3.2. Rocks for the Purpose of Holding the Platform

There is no need to dig a hole or make a circle of rocks in preparation for fire building. Rocks may be placed in a circle and filled with dirt, sand, or gravel to raise the fire above the moisture from wet ground. The purpose of these rocks is to hold the platform only.

14.3.3. Getting the Most of Warmth from a Fire

To get the most warmth from the fire, it should be built against a rock or log reflector (Figure 14-2). This will direct the heat into the shelter. Cooking fires can be walled-in by logs or stones. This will provide a platform for cooking utensils and serve as a windbreak to help keep the heat confined.

Figure 14-2 Fire Reflector

14.3.4. Fire Fuel Organization

After preparing the fire, all materials should be placed together and arranged by size (tinder, kindling, and fuel). As a rule of thumb, IP should have three times the amount of tinder and kindling than is necessary for one fire. It is to their advantage to have too much than not enough. Having plenty of material on hand will prevent the possibility of the fire going out while additional material is gathered.

14.4. Firecraft Tips

14.4.1. Matches

Conserve matches by only using them on properly prepared fires. They should never be used to light cigarettes or for starting unnecessary fires.

14.4.2. Preserving Dry Tinder

Keep some dry tinder in a waterproof container. Tinder should be exposed to the Sun on dry days. Adding a little powdered charcoal will improve its ability to burn. Cotton cloth is good tinder, especially if scorched or charred. It works well with a magnifying glass or flint and steel.

14.4.3. Difficulty of Building Fires in the Arctic Environment

Fire making can be a difficult job in an arctic environment. The main problem is the availability of fire making materials. Making a fire starts well before the match is lit. The fire must be protected from the wind. In wooded areas, standing timber and brush usually make a good windbreak but in open areas, some type of windbreak may have to be constructed. A row of snow blocks, the shelter of a ridge, or a pile of brush will work as a windbreak. The wind break must be high enough to shield the fire from the wind, and it can also act as a heat reflector if it is of solid material.

14.4.4. Platform for Fire

A platform will be required to prevent the fire from melting down through the deep snow and extinguishing it. A platform is also needed if the ground is moist or swampy. The platform can be made of green logs, metal, or any material that will not burn through very readily. Care must be taken when selecting an area for fire building. If the area has a large accumulation of humus material and (or) peat, a platform is needed to avoid igniting the material as it will tend to smolder long after the flames of the fire are extinguished. A smoldering peat fire is almost impossible to put out and may burn for years.

14.4.5. Ignition Source

The ignition source used to ignite the fire must be quick and easily operated with hand protection such as mittens. Any number of devices will work well-matches, candle, lighter, fire starter, metal matches, etc.

14.5. Fire Making with Matches (or Lighter)

14.5.1. Arranging Kindling

IP should arrange a small amount of kindling in a low pyramid, close enough together so flames can jump from one piece to another. A small opening should be left for lighting and air circulation.

14.5.2. Using a Shave Stick

Matches can be conserved by using a "shave or feather stick," or by using a loosely tied fagot or bundle of thin, dry twigs. The match must be shielded from wind while igniting the shave stick. The stick can then be applied to the lower windward side of the kindling (Figure 14-3).

Figure 14-3 Feather Stick

14.5.3. Small Pieces of Wood to Assist with Kindling

Small pieces of wood or other fuel can be laid gently on the kindling before lighting or can be added as the kindling begins to burn. The IP can then place smaller pieces first, adding larger pieces of fuel as the fire begins to burn. To avoid smothering the fire by crushing the kindling with heavy wood, IP should use a soda can diameter brace to lean the additional fuel against. This will ensure adequate ventilation of the fire.

14.5.4. Fire Bundles

IP only have a limited number of matches or other instant fire-starting devices. In a long-term situation, they should use these devices sparingly or carry fire with them when possible. Many primitive cultures carry fire (fire bundles) by using dry punk or fibrous barks (cedar) encased in a bark. Others use torches. Natural fire bundles also work well for holding the fire (Figure 16-3).

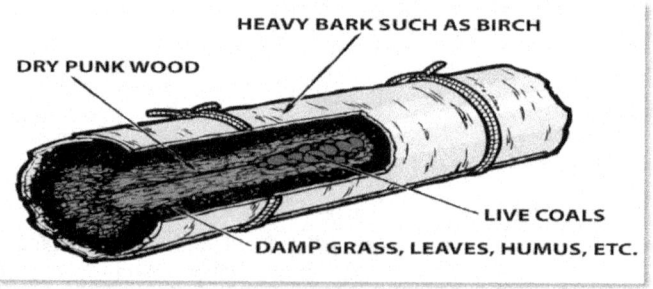

Figure 14-4 Fire Bundles

14.5.5. Keeping the Fire Bundle Alive

The amount of oxygen must be just enough to keep the coals inside the dry punk burning slowly. This requires constant vigilance to control the rate of the burning process. The natural fire bundle is constructed in a cross section as shown in Figure 14-4.

14.6. Heat Sources

A supply of matches, lighters, and other such devices will only last a limited time. Once the supply is depleted, they cannot be used again. If possible, before the need arises, IP should become skilled at starting fires with more primitive means, such as friction, heat, or a sparking device. It is essential that they continually practice these procedures. The need to start a fire may arise at the most inopportune times. One of the greatest aids an IP can have for rapid fire starting is the "tinder box" previously mentioned. Using friction, heat, and sparks are very reliable methods for those who use them on a regular basis. Therefore, an IP must practice these methods. IP must be aware of the problems associated with the use of primitive heat sources. If the humidity is high in the immediate area, a fire may be difficult to ignite even if all other conditions are favorable. For primitive methods to be successful, the materials must be bone dry. Preparation, practice, and patience in the use of primitive fire-building techniques cannot be over emphasized. A key point in all primitive methods is to ensure that the tinder is not disturbed.

14.6.1. Flint and Steel

Flint and steel is one way to produce fire without matches. To use this method, IP must hold a piece of flint in one hand above the tinder. Grasp the steel in the other hand and strike the flint with the edge of the steel in a downward glancing blow (Figure 14-5).

Figure 14-5 Fire Starting with Flint and Steel

14.6.1.1. Using Iron Pyrite and Quartz if Flint is Unavailable

True flint is not necessary to produce sparks. Iron pyrite and quartz will also give off sparks even if they are struck against each other. Check the area and select the best spark-producing stone as a backup for the available matches. The sparks must fall on the tinder and then be blown or fanned to produce a coal and subsequent flame.

14.6.1.2. Scratching Flint with a Knife-Blade to Create a Spark

Flint, such as the so-called metal match, consists of the same type material used for flints in commercial cigarette lighters. Some contain magnesium which can be scraped into tinder and into which the spark is struck. The residue from the "match" burns hot and fast and will compensate for some moisture in tinder. If issued survival kits do not contain this item and the IP choose to make one rather than buy it, lighter flints can be glued into a groove in a small piece of wood or plastic. The IP can then practice striking a spark by scratching the flint with a knife blade. A 90-degree angle between the blade and flint works best. The device must be held close enough for the sparks to hit the tinder, but enough distance must be allowed to avoid accidentally extinguishing the fire. Cotton balls dipped in petroleum jelly make excellent tinder with flint and steel. When the tinder ignites, additional tinder, kindling, and fuel can be added. (Note: Care should be taken to protect synthetic flint from water, and especially salt water.)

14.6.2. Batteries

Another method of producing fire is to use the battery of the aircraft, vehicle, storage batteries (AA, C, D, CR123, etc). Using two insulated wires, connect one end of a wire to the positive post of the battery and the end of the other wire to the negative post. Touch the two remaining ends to the ends of a piece of non-insulated wire. This will cause a short in the electrical circuit and the non-insulated wire will begin to glow and get hot. Material coming into contact with this hot wire will ignite. IP should use caution when attempting to start a fire with a battery. They should ensure that sparks or flames are not produced near the battery because explosive hydrogen gas may be produced and could result in serious injury (Figure 14-6).

Figure 14-6 Fire Starting With Batteries

14.6.2.1. Fine Grade Steel Wool

If fine grade steel wool is available, a fire may be started by stretching it between the positive and negative posts until the wire itself makes a red coal.

14.6.3. Magnifying Glass

If IP have sunlight and a burning glass, a fire can be started with very little physical effort (Figure 14-7). Concentrate the rays of the sun on tinder by using the lens of a lensatic compass, a camera lens, or the lens of a flashlight which magnifies; even a convex piece of bottle glass may work. Hold the lens so that the brightest and smallest spot of concentrated light falls on the tinder. Once a wisp of smoke is produced, the tinder should be fanned or blown upon until the smoking coal becomes a flame. Powdered charcoal in the tinder will decrease the ignition time. Add kindling carefully as in any other type of fire. Practice will reduce the time it takes to light the tinder.

Figure 14-7 Fire Starting With Burning Glass

14.6.4. Flashlight Reflector

A flashlight reflector can also be used to start a fire (Figure 14-8). Place the tinder in the center of the reflector where the bulb is usually located. Push it up from the back of the hole until the hottest light is concentrated on the end and smoke results. If a cigarette is available, use it as tinder for this method.

Figure 14-8 Fire Starting With Flashlight Reflector

14.6.5. Bamboo Fire Saw

The bamboo fire saw is constructed from a section of dry bamboo with both end joints cut off. The section of bamboo, about 12 inches in length, is split in half lengthwise. The inner wall of one of the halves (called the "running board") is scraped or shaved thin. This is done in the middle of the running board. A notch to serve as a guide is cut in the outer sheath opposite the scraped area of the inner wall. This notch runs across the running board at a 90-degree angle (Figure 14-9).

14.6.5.1. Bamboo Joints

The other half of the bamboo joint is further split in half lengthwise, and one of the resultant quarters is used as a "baseboard." One edge of the baseboard is shaved down to make a tapered cutting edge. The baseboard is then firmly secured with the cutting edge up. This may be done by staking it to the ground in any manner which does not allow it to move.

14.6.5.2. Tinder

Tinder is made by scraping the outer sheath of the remaining quarter piece of the bamboo section. The scrapings (approximately a large handful) are then rubbed between the palms of the hands until all of the wood fibers are broken down and dust-like material no longer falls from the tinder. The ball of scrapings is then fluffed to allow maximum circulation of oxygen through the mass.

14.6.5.3. Finely Shredded and Fluffed Tinder

The finely shredded and fluffed tinder is placed in the running board directly over the shaved area, opposite the outside notch. Thin strips of bamboo should be placed lengthwise in the running board to hold the tinder in place. These strips are held stationary by the hands when grasping the ends of the running board.

Figure 14-9 Bamboo Fire Saw

14.6.5.4. Bamboo Pick

A long, very thin sliver of bamboo (called the "pick") should be prepared for future use. One end of the running board is grasped in each hand, making sure the thin strips of bamboo are held securely in place. The running board is placed over the baseboard at a right angle, so that the cutting edge of the baseboard fits into the notch in the outer sheath of the running board. The running board is then slid back and forth as rapidly as possible over the cutting edge of the baseboard, with sufficient downward pressure to ensure enough friction to produce heat.

14.6.5.5. Using the Pick to Push the Embers Toward the Timber

As "billows" of smoke rise from the tinder, the running board is picked up. The pick is used to push the glowing embers from the bottom of the running board into the mass of tinder. While the embers are being pushed into the tinder, they are gently blown upon until the tinder bursts into flame.

14.6.5.6. Slowly Add Kindling

When the tinder bursts into flame, slowly add kindling in small pieces to avoid smothering the fire. Fuel is gradually added to produce the desired size fire. If the tinder is removed from the running board as soon as it flames, the running board can be reused by cutting a notch in the outer sheath next to the original notch and directly under the scraped area of the inner wall.

14.6.5.7. Bow and Drill

This is a friction method which has been used successfully for thousands of years. A spindle of yucca, elm, basswood, or any other straight grain wood (not softwood) should be made. The IP should make sure that the wood is not too hard or it will create a glazed surface when friction is applied. The spindle should be 12 to 18 inches long and three-fourths inch in diameter. The sides should be octagonal, rather than round, to help create friction when spinning. Round one end and work the other end into a blunt point. The round end goes to the top upon which the socket is placed. The socket is made from a piece of hardwood large enough to hold comfortably in the palm of the hand with the curved part up and the flat side down to hold the top of the spindle. Carve or drill a hole in this side and make it smooth so it will not cause undue friction and heat production. Grease or soap can be placed in this hole to prevent friction (Figure 14-10).

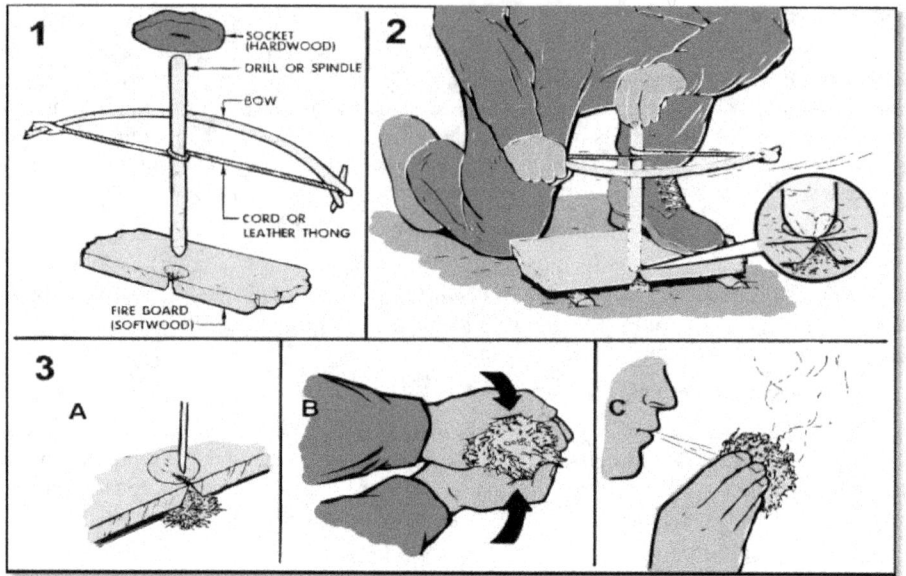

Figure 14-10 Bow and Drill

14.6.5.8. Constructing a Bow

The bow is made from a stiff branch about three feet long and about one inch in diameter. This piece should have sufficient flexibility to bend. It is similar to a bow used to shoot arrows. Tie a piece of suspension line or leather thong to both ends so that it has the same tension as that of a bow. There should be enough tension for the spindle to twist comfortably.

14.6.5.9. Fireboard

The fireboard is made of the softwood and is about 12 inches long, three-fourths inch thick, and three to six inches wide. A small hollow should be carved in the fireboard. A V-shaped cut can then be made in from the edge of the board. This V-shape should extend into the center of the hollow where the spindle will make the hollow deeper. The object of this "V" cut is to create an angle which cuts off the edge of the spindle as it gets hot and turns to charcoal dust. This is the critical part of the fireboard and must be held steady during the operation of spinning the spindle.

14.6.5.10. How to use a Fireboard

While kneeling on one knee, the other foot can be placed on the fireboard and the tinder placed under the fireboard just beneath the V-cut. Care should be taken to avoid crushing the tinder under the fireboard. Space can be obtained by using a small, three-fourths inch diameter stick to hold up the fireboard. This will allow air into the tinder where the hot powder (spindle charcoal dust) is collected.

14.6.5.11. The Bow String

The bow string should be twisted once around the spindle. The spindle can then be placed upright into the spindle hollow (socket). The IP may press the socket down on the spindle and fireboard. The entire apparatus must be held steady with the hand on the socket braced against the leg or knee. The spindle should begin spinning with long even slow strokes of the bow until heavy smoke is produced. The spinning should become faster until the smoke is very thick. At this point, hot powder that can be blown into a glowing ember has been successfully produced. The bow and spindle can then be removed from the fireboard and the tinder can be placed next to the glowing ember making sure not to extinguish it. The tinder must then be rolled gently around the burning ember, and blow into the embers, starting the tinder to burn. This part of the fire is most critical and should be done with care and planning.

14.6.5.12. Placing the Burning Tinder

The burning tinder is then placed into the waiting fire "lay" containing more tinder and small kindling. At no time in this process should the IP break concentration or change sequence. The successful use of these primitive methods of fire starting will require a great deal of patience. Success demands dedication and practice.

14.6.6. The Fire Thong

The fire thong, another friction method, is used in only those tropical regions where rattan is found. The system is simple and consists of a twisted rattan thong or other strong plant fiber, four to six feet long, less than one inch in diameter, and a four-foot length of dry wood which is softer than rattan (deciduous wood) (Figure 14-11). Rub with a steady but increasing rhythm.

14.6.6.1. Using Wire as a Fire Thong

A variation of this would be to use a wire as your thong. As the wire is pulled back and forth across the length of wood it will get hot with friction. Once hot, the wire can be applied to accelerants such as liquid fuel or gun powder from an issued bullet. To work best, mix accelerant with another, more traditional tinder.

Figure 14-11 Fire Thong

14.7. Other Methods of Fire Starting

14.7.1. Flares

The night end of the day-night flare can be used as a fire starter. To work effectively the flare should be used as it is starting to burn out rather than at the start of the ignition. Using the flare at the start of the ignition cycle will quickly expend available fuel due to the extreme heat and exhaust of the flare. Additionally, IP must weigh the importance of a fire against the loss of a night flare.

14.7.2. Fire Starters in Emergency Kits

Some emergency kits contain small fire starters, cans of special fuels, windproof matches, and other aids. IP should save the fire starters for use in extreme cold and damp (moist) weather conditions.

14.7.3. Plastic Items

Plastic items may be the type that burn readily. As an example, plastic spoon handles can be pushed deep enough into the ground to support the spoon in an upright position. Light the tip of the spoon. It will burn for about 10 minutes (long enough to dry out and ignite small tinder and kindling). If a candle is available, it should be ignited to start a fire and thus prevent using more than one match. As soon as the fire is burning, the candle can be extinguished and saved for future use. If a candle is packed in a personal survival kit (PSK) the wick should be prepped prior to packing.

14.7.4. Adding Fuel to Tinder

Tinder can be made more combustible by adding a few drops of flammable fuel/material. An example of this would be mixing the powder from an ammunition cartridge with the tinder. After preparing tinder in this manner, it should be stored in a waterproof container for future use. Care must be used in handling this mixture because the flash at ignition could burn the skin and clothing.

14.7.5. Improvising a Stove to Burn Refined Petroleum

IP can improvise a stove from a container and burn either refined petroleum products or rendered animal fats for fuel. Animal fat can be rendered through heating to produce flammable oil. Refined petroleum or rendered animal fat oil can be collected in a container and an improvised wick placed in it. When this is lit it will provide both heat and light for the IP.

14.8. Burning Aircraft/Vehicle Fuel

On barren lands in the arctic, fuel may be the only materials IP have available for fire.

14.8.1. Improvising a Stove to Burn Lubricating Oil

A stove can be improvised to burn fuel, lubricating oil, or a mixture of both. The IP should place one or two inches of sand or fine gravel in the bottom of a can or other container and add fuel. Care should be used when lighting the fuel because it may explode. Slots should be cut into the top of the can to let flame and smoke out, and holes punched just above the level of the sand to provide a draft. A mixture of fuel and oil will make the fire burn longer. If there is not a can available, a hole can be dug and filled with sand. Fuel is then poured on the sand and ignited. The IP should not allow the fuel to collect in puddles.

14.8.2. Using Wick

Lubricating oil can be burned as fuel by using a wick arrangement. The wick can be made of string, rope, rag, sphagnum moss, or even a cigarette and should be placed on the edge of a receptacle filled with oil. Rags, paper, wood, or other fuel can be soaked in oil and thrown on the fire.

14.8.3. Making a Stove out of a Waxed Carton

A stove can be made of any empty waxed carton by cutting off one end and punching a hole in each side near the unopened end. IP can stand the carton on the closed end and loosely place the fuel inside the carton. The stove can then be lit using fuel material left hanging over the end. The stove will burn from the top down.

14.8.4. Burning Seal Blubber

Seal blubber makes a satisfactory fire without a container if gasoline or heat tablets are available to provide an initial hot flame. The heat source should be ignited on the raw side of the blubber while the fur side is on the ice. A square foot of blubber burns for several hours. Once the blubber catches fire, the heat tablets can be recovered. Eskimos light a small piece of blubber and use it to kindle increasingly larger pieces. The smoke from a blubber fire is dirty, black, and heavy and will penetrate clothing and blacken the skin, but the flame is very bright and can be seen for several miles.

14.9. Fire Lays

Most fires are built to meet specific needs or uses, either heat, light, or preparing food and water. The following configurations are the most commonly used for fires and serve one or more needs (Figure 14-12).

Figure 14-12 Fire Lays

14.9.1. Tepee

The Tepee Fire can be used as a light source and has a concentrated heat point directly above the apex of the tepee which is ideal for boiling water. To build a Tepee Fire, place a large handful of tinder on the ground in the middle of the fire site, push a stick into the ground slanting over the tinder, and then lean a circle of kindling sticks against the slanting stick (like a tepee) with and opening toward the windward side for draft. To light the fire, crouch in front of the fire laying with your back to the wind and use an ignition source to light the tinder. Feed the fire from the downwind side, first with thin pieces of kindling, then gradually with thicker pieces of fuel creating a teepee shape of fuel above the burning kindling. Continue feeding until the fire has reached the desired size. The Tepee Fire has one big drawback, it tends to fall over easily; however, it serves as an excellent starter fire.

14.9.2. Log Cabin

As the name implies, this lay looks similar to a log cabin. Log Cabin Fires give off a great amount of light and heat primarily because of the amount of oxygen which enters the fire. The Log Cabin Fire creates a quick and large bed of coals and can be used for cooking or as the basis for a signal fire.

14.9.3. Long Fire

The Long Fire begins as a trench, the length of which is laid to take advantage of existing wind. The Long Fire can also be built above ground by using two parallel green logs to hold the coals together. These logs should be at least 6 inches in diameter and situated so the cooking utensils will rest upon the logs. Two one-inch thick sticks can be placed under both logs, one at each end of the Long Fire. This is done to allow the coals to receive more air.

14.9.4. Pyramid Fire

The Pyramid Fire looks similar to a Log Cabin Fire except there are layers of fuel in place of a hollow framework. The advantage of a Pyramid Fire is that it burns for a long period of time resulting in a large bed of coals. This fire could possibly be used as an overnight fire when placed in front of a shelter opening.

14.9.5. Star Fire

The Star Fire is used when conservation of fuel is necessary or a small fire is desired. It burns at the center of the "wheel" and must be constantly tended. Hardwood fuels work best with this type of fire.

Chapter 15
EQUIPMENT

15. Equipment

IP in a survival situation have needs which must be met including food, water, clothing, shelters, etc. The survival kit contains equipment which can be used to satisfy these needs. Quite often, however, this equipment may not be available due to damage or loss. This chapter will address the care and use of issued equipment and improvising the needed equipment when not available. The uses of some issued items are covered in appropriate places throughout this handbook. The care and use of equipment (not covered elsewhere) will be addressed here.

15.1. Issued Equipment

15.1.1. Electronic Equipment

Electronic signaling devices are by far the IP's most important signaling devices. Therefore, it is important for IP to properly care for them and any spare batteries to ensure their continued effectiveness. In cold temperatures, electronic devices must be kept warm to prevent the batteries from becoming cold soaked, decreasing their lives.

15.1.1.1. Speaking into a Microphone in a Cold Environment

In a cold environment, if IP speak directly into a microphone, the moisture from their breath may condense and freeze on the microphone, creating communication problems.

15.1.1.2. Exercising Caution when Using Radios in a Cold Environment

Caution must be used when using radios in a cold environment. If the radio is placed against the side of the face to communicate, frostbite could result.

15.1.1.3. Keeping Electronics Dry in a Wet Environment

In a wet environment, IP should make every effort to keep their electronic signaling devices dry. In an open-sea environment, the only recourse may be to shake the water out of the microphone before transmitting.

15.1.1.4. Care of Batteries in an Extreme Heat Environment

In an extreme heat environment, batteries have a tendency to overheat and IP need to take steps to keep the radio cool. Keep electronic devices out of direct sun and next to the body while not in use.

15.1.2. Firearms

A firearm is a precision tool. It will continue functioning only as long as it is cared for. Saltwater, perspiration, dew, and humidity can all corrode or rust a firearm until it is inoperable. If immersed in saltwater, the IP should wash the parts in freshwater and then dry and oil them if possible. Also, any firearms should be stored outside of shelters to prevent condensation from building on the weapon which could lead to malfunctions.

- Many petroleum-based lubricants used in cold environments will stiffen or freeze causing the firearm to become inoperative. It would be better to thoroughly clean the firearm and remove all lubricant. Metal becomes brittle from cold and internal firing mechanisms may be more prone to breakage.

- A firearm was not intended for use as a club, hammer, or pry bar. To use it for any purpose other than for which it was designed, would only result in damage to the firearm.

15.1.3. Cutting Tools

One of the most valuable items in any survival situation is a knife, since it has numerous uses. A file or sharpening stone should be packed in the personal survival kit. The file is normally used for axes, and the stone is normally used for knives.

15.1.3.1. Control of a Cutting Tool

Control of a cutting tool is easier to maintain if it is sharp, and the possibility of accidental injury is reduced.

15.1.3.2. Sharpening Knives with a Stone Removes the Steel

A knife should be sharpened only with a stone as repeated use of a file rapidly removes steel from the blade. In some cases, it may be necessary to use a file to remove plating from the blade before using the stone.

15.1.3.3. Sharpening a Knife Using a Stone (Draw)

One of two methods should be used to sharpen a knife. One method is to push the blade down the stone in a slicing motion. Then turn the blade over and draw the blade toward the body (Figure 15-1).

Figure 15-1 Knife Sharpening (Draw)

15.1.3.4. Sharpening a Knife Using a Stone (Circular)

The other method is to use a circular motion the entire length of the blade; turn the blade over and repeat the process. What is done to one side of the cutting edge should also be done to the other to maintain an even cutting edge (Figure 15-2).

Figure 15-2 Knife Sharpening (Circular)

15.1.3.5. Natural Whetstone

If a commercial whetstone is not available, a natural whetstone can be used. Water should be applied to these stones. The water will help to float away the metal removed by sharpening and make cleaning of the stone easier. Any sandstone will sharpen tools, but gray, clay-like sandstone gives better results while quartzite should be avoided. IP can recognize quartzite instantly by scratching the knife blade with it-the quartz crystals will bite into steel. If no sandstone is available, granite or crystalline rock can be used. If granite is used, two pieces of the stone should be rubbed together to smooth the surface before use.

15.1.3.6. A Sharp Axe Saves Time and Energy

As with a knife, a sharp axe will be safer and save time and energy.

15.1.3.7. Using a File to Sharpen and Axe or Hatchet

A file should be used on an axe or hatchet. The IP should file toward the cutting edge, but care must be taken to avoid slipping and cutting yourself. The file should be worked from one end of the cutting edge to the other. The opposite side should be worked to the same degree. This will ensure that the cutting edge is even. After using a file, the stone may be used to hone the axe blade (Figure 15-3).

Figure 15-3 Sharpening Axe

15.1.3.8. How to Cut a Tree with an Axe

When using an axe, don't try to cut through a tree with one blow. Rhythm and aim are more important than force. Too much power behind a swing interferes with aim. When the axe is swung properly, its weight provides all the power needed.

15.1.3.9. Care in Preserving the Axe Handle

Carving a new axe handle and mounting the axe head takes a great deal of time and effort. For this reason, IP should avoid actions which would require the handle to be changed. Using aim and paying attention to where the axe falls will prevent misses which could result in a cracked or broken handle. IP should not use an axe as a pry bar and should avoid leaving the axe out in cold weather where the handle may become brittle.

15.1.3.10. Removing a Broken Axe Handle

A broken handle is difficult to remove from the head of the axe. Usually the most convenient way is to burn it out (Figure 15-4). For a single-bit axe, bury the bit in the ground up to the handle, and build a fire over it. For a double-bit, an IP should dig a small trench, lay the middle of the axe head over it, cover both "bits" with earth, and build the fire. The covering of earth keeps the flame from the cutting edge of the axe and saves its temper. A little water added to the earth will further ensure this protection.

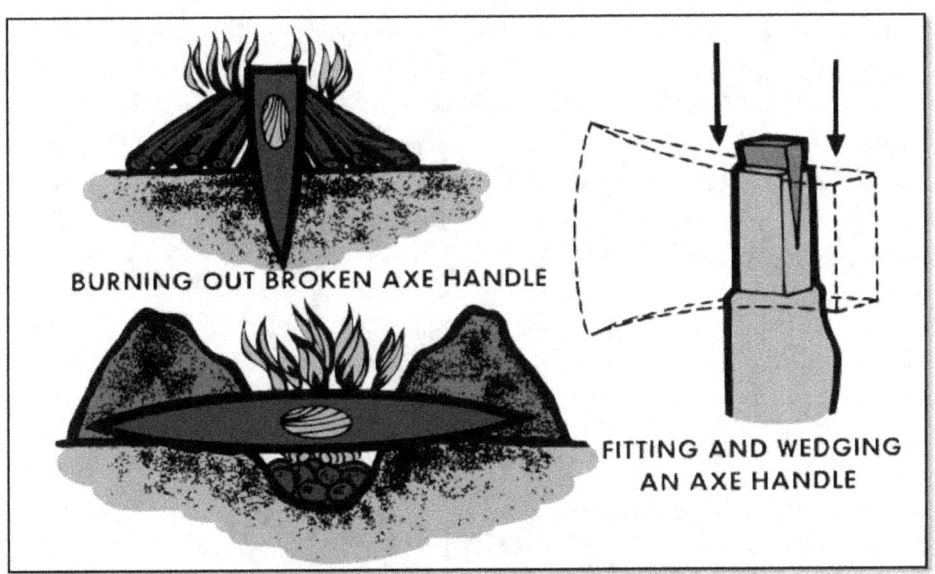

Figure 15-4 Removing Broken Axe Handle

15.1.3.11. Improvising a New Handle

When improvising a new handle, an IP can save time and trouble by making a straight handle instead of a curved one like the original. IP should use a young, straight piece of hardwood without knots. The wood should be whittled roughly into shape and finished by shaving. A slot should be cut into the axe-head end of the handle. After it is fitted, a thin, dry wooden wedge can then be pounded into the slot. IP should use the axe for a while, pound the wedge in again, and then trim it off flush with the axe. The handle must be smoothed to remove splinters. The new handle can be seasoned to prevent shrinkage by "scorching" it in the fire.

15.1.4. Knife Use

When whittling, IP must hold the knife firmly and cut away from the body (Figure 15-5). Wood should be cut with the grain. Branches should be trimmed as shown in Figure 15-6.

Figure 15-5 Whittling

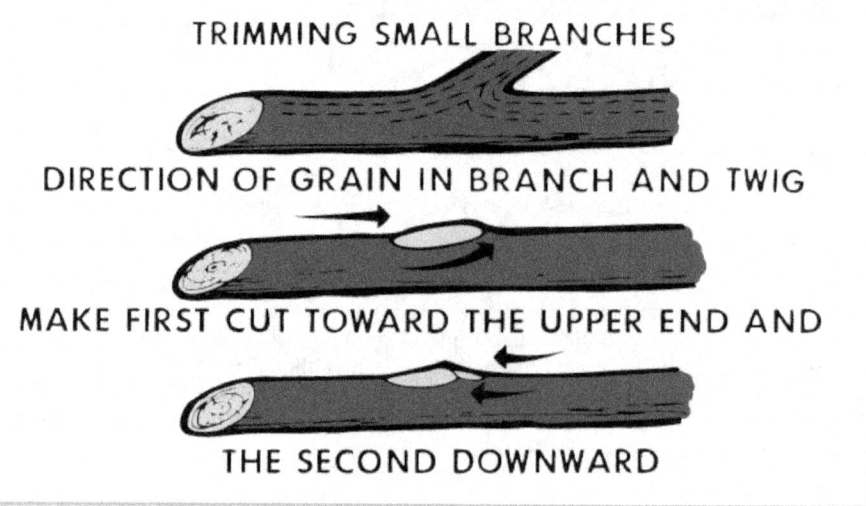

Figure 15-6 Trimming Branches

15.1.4.1. Using a Knife to Cut Wood

To cut completely through a piece of wood, a series of V-cuts should be made all the way around as in Figure 15-7. Once the piece of wood has been severed, the pointed end can then be trimmed.

Figure 15-7 Cutting Through a Piece of Wood

15.1.4.2. Using the Thumb to Steady the Hand

The thumb can be used to help steady the hand. Be sure and keep the thumb clear of the blade. To maintain good control of the knife, the right hand is steadied with the right thumb while the left thumb pushes the blade forward (Figure 15-8). This method is very good for trimming.

Figure 15-8 Fine Trimming

15.1.5. Felling Trees

To fell a tree, the IP must first determine the direction in which the tree is to fall. It is best to fell the tree in the direction in which it is leaning. The lean of the tree can be found by using the axe as a plumb line. The IP should then clear the area around the tree from underbrush and overhanging branches to prevent injury (Figure 15-9).

Figure 15-9 Clearing Brush from Cutting Area

15.1.5.1. Making Two Cuts

The IP should make two cuts. The first cut should be on the leaning side of the tree and close to the ground and the second cut on the opposite side and a slightly higher than the first cut (Figure 15-10).

Figure 15-10 Felling Cuts

15.1.5.2. IP Should Be Careful of Falling Trees

Falling trees often kick back and can cause serious injury, so IP must ensure they have a clear escape route. When limbing a tree, start at the base of the tree and cut toward the top. This procedure will allow for easier limb removal and results in a smoother cut. For safety, the IP should stand on one side of the trunk with the limb on the other.

15.1.5.3. Preventing Damage and Physical Injury

To prevent damage to the axe head and possible physical injury, any splitting of wood should be done on a log as in Figure 15-11. The log can also be used for cutting sticks and poles (Figure 15-12).

Figure 15-11 Splitting Wood

Figure 15-12 Cutting Poles

15.1.6. Improvised Equipment

If issued equipment is inoperative, insufficient, or nonexistent, IP will have to rely upon their ingenuity to manufacture the needed equipment. IP must determine whether the need for the item outweighs the work involved to manufacture it. They will also have to evaluate their capabilities. If they have injuries, will the injuries prevent them from manufacturing the item(s)?

15.1.6.1. Five Basic Rules of Improvising

IP should follow five basic rules of improvising to meet their needs using a logical problem solving technique. The five rules are:

1. Determine the need. Needs are determined by the IP's situation and prioritization (see chapter 1, 2, 3).
2. Inventory materials, both manmade and natural. Consider using abundant natural materials and avoid wasting limited materials.
3. Consider alternatives (brainstorm and list possible solutions).
4. Select the best alternative based on time, energy, and materials.
5. Plan and construct making the best use of materials and insuring it is safe and durable.

15.1.6.2. The IP Should Have a Plan

Undue haste may not only waste materials, but also waste the IP's time and energy. Before manufacturing equipment, they should have a plan in mind.

15.1.6.3. Ways to Meet IP Equipment Needs

The IP's equipment needs may be met in two different ways. They may alter an existing piece of equipment to serve more than one function, or they may construct a new piece of equipment from available materials. Since the items IP can improvise are limited only by their ingenuity, all improvised items cannot be covered in this document.

15.1.7. Manufacturing Equipment

The methods of manufacturing the equipment referred to in this handbook are only ideas and do not have to be strictly adhered to.

15.2. Parachute

Aircrew members may have a parachute. This device can be used to improvise a variety of needed equipment items. Refer to Figure 15-13.

15.2.1. The Pilot Chute

The pilot chute deploys first and pulls the rest of the parachute out.

15.2.2. Parachute Canopy

The parachute canopy consists of the apex (top) and the skirt or lower lateral band. A typical C-9 parachute canopy is divided by radial seams into 28 sections called gores. Each gore measures about 3 feet at the skirt and tapers to the apex. Each gore is further subdivided into four sections called panels. The canopy is normally divided into four colors. These colored areas are intended to aid the IP in shelter construction, signaling, and camouflage.

15.2.3. Suspension Lines

Fourteen suspension lines connect the canopy material to the harness assembly. Each piece of suspension line is 72 feet long from riser to riser and 22 feet long from riser to skirt and 14 feet from skirt to apex. The tensile strength of each piece of suspension line is 550 pounds. Each piece of suspension line contains seven to nine pieces of inner core with a tensile strength of 35 pounds. The harness assembly contains risers and webbing, buckles, snaps, "D" rings, and other hardware which can be used in improvisation.

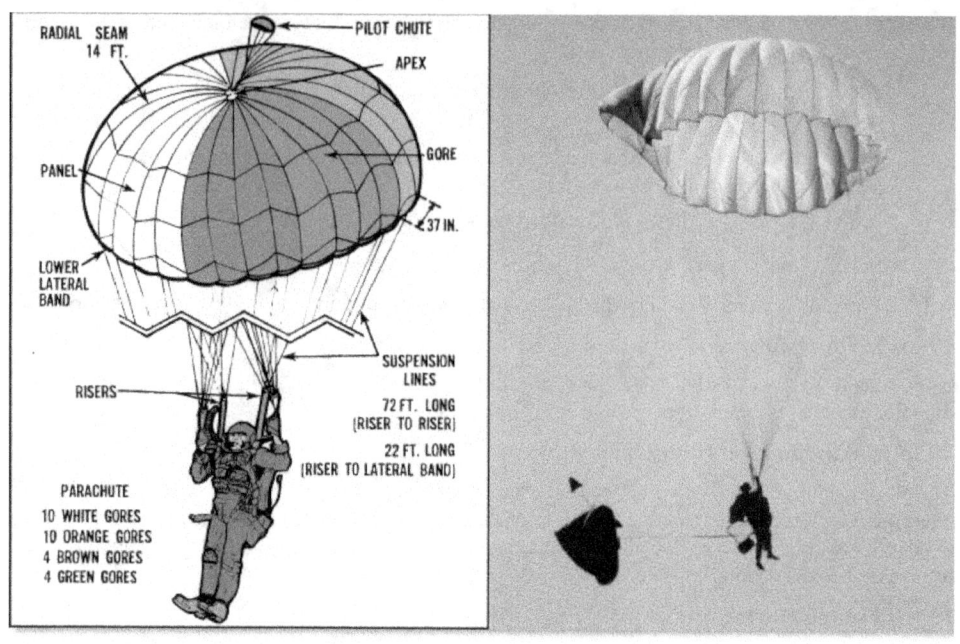

Figure 15-13 Parachute Diagram

15.2.4. The Parachute as a Resource

The whole parachute assembly should be considered as a resource. Every piece of material and hardware can be used.

15.2.4.1. Obtaining the Suspension Lines

To obtain the suspension lines, an IP should cut them at the risers or, if time and conditions permit, consider disassembling the connector links. Cut the suspension lines about two feet from the skirt of the canopy. When cutting suspension lines or dismantling the canopy/pack assembly; it will be necessary to maintain a sharp knife for safety and ease of cutting.

15.2.4.2. Obtaining all Available Suspension Lines

IP should obtain all available suspension line due to its many uses. Even the line within the radial seams of the canopy should be stripped for possible use. The suspension line should be cut above the radial seam stitches next to the skirt end of the canopy (two places). The cut should not go all of the way through the radial seam (Figure 15-14). At the apex of the canopy, and just below the radial seam stitching, a horizontal cut can be made and the suspension line extracted. The line can then be cut.

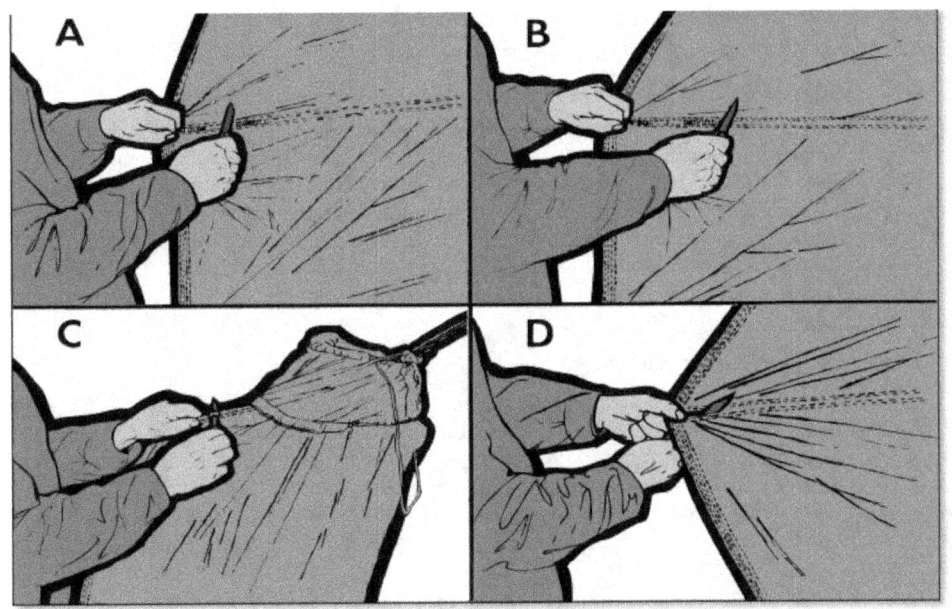

Figure 15-14 Cutting the Parachute

15.2.4.3. Maximum Canopy Usage

For maximum use of the canopy, IP must plan its disassembly. The quantity requirements for shelter, signaling, etc., should be thought out and planned for. Once these needs have been determined, the canopy may be cut up. The radial seam must be stretched tightly for ease of cutting. The radial seam can then be cut by holding the knife at an angle and following the center of the seam. With proper tension and the gentle pushing (or pulling) of a sharp blade there will be a controlled splitting of the canopy at the seam (Figure 15-14). It helps to secure the apex either to another individual or to an immobile object such as a tree.

15.2.4.4. Having the Necessary Material to Improvise

One requirement in improvising is having material available. When stripping the harness assembly, the seams of the webbing should be split so the maximum usable webbing is obtained. The harness material and webbing should not be randomly cut as it will waste much needed materials. Parachute fabric, harness, suspension lines, etc., can be used for clothing. Needles are helpful for making any type of emergency clothing. Wise IP should always have extra sewing needles hidden somewhere on themselves. If no sewing needles are available, a good needle or sewing awl can be made from a sliver of bone. Thread is usually available in the form of inner core. It will benefit the IP to collect small objects which may "come in handy." Wire, nails, buttons, a piece of canvas, or animal skin should not be discarded. Any such object may be worth its weight in gold when placed in a hip pocket or a sewing kit. Any kind of animal skin can be used for making clothing such as gloves or mittens or making a ground cover to keep the sleeping bag dry and clean. Small skins can be used for mending and for boot insoles. Mending and cleaning clothes when possible will pay dividends in health, comfort, and safety.

15.2.4.5. Sewing Improvised Equipment

The improvised equipment IP may need to make will probably involve sewing. IP knowledge of hand stitching may be critical in improvising or repairing articles of clothing and making multi-use equipment. Knowing and understanding the proper stitch to use is important to the item's function and longevity. Depending on the stitch, its use, and the type of thread some techniques require double thread sewing on an improvised button or securing an item to a piece of material), but most are worked with a single thread. A well-made hand stitch will be secured at beginning/ending with no loose threads and be smooth and not create puckers on the outer or inner sides of the fabric.

15.2.4.5.1. Places to Obtain Thread

Thread can be obtained from untwining line, inner core of a parachute line, pulling a line apart from ripped material, tree bark, or even repurposed from fishing line, dental floss, metal wire, etc. Sinew can also be used especially with natural materials such as rawhide and brain-tanned buckskin. Stronger than any thread, sinew is used in both fine and heavy applications; from sewing, to making bows and rope. Genuine sinew used as thread can be prepared wet or dry. This same technique for separating a strand a strand can be applied to a dry piece of sinew, taking only the thread that you will use at the moment. However soaking in water until pliable and preparing threads ahead of time may be an easier method of improvising.

1. Soak the sinew piece in room temperature water until soft and pliable.
2. Grasp the new sinew by the sides and begin working it with your hands in a twisting motion back and forth until the fibers begin to separate. This will take some work as the fibers are held together by natural glues.
3. Peel strands one at a time from the piece in the thickness you desire.
4. While still wet and pliable, hold the end of a strand in the fingertips of one hand while rolling the strand down the top of your thigh to twist it. Put a knot in one end and form the other end into a pointed tip. Twisting and knotting the strand while still wet allows you to work the sinew easily.
5. These dried threads are now ready for sewing. Techniques include pre-punching the material with an awl then pushing the sinew's pointed tip through, or threading the strand through a conventional/improvised needle. If pre-punching the material with an awl then pushing the thread through the material will not work, the IP will have to use some type of needle.

15.2.4.5.2. Making Needles of Bone or Wood

The first needles were made of bone or wood; needles for hand sewing have a hole, called the eye, at the non-pointed end to carry thread or cord through the fabric after the pointed end pierces it. IP can improvise a needle or an awl-type tool from scrap metal, bone, and/or wood. Yucca leaves can be pulled sharply up and then peeled down leaving the plant fibers attached to the tip of the leaf creating a "threaded" needle which can be used. When putting needles in personal survival kits it is always recommended to pack them already threaded.

15.2.4.5.3. Sewing with Fur

Sewing with fur, whether real or faux, is a bit different from sewing with fabric. Cut the fabric in a single layer; cut only through the backing. Smooth the fur or pile away from the stitching line toward the fabric right side. A whipstitch is recommended.

15.2.4.5.4. Knotting the Thread to Fabric

To sew anything by hand, begin by knotting the thread to your fabric: make a slip knot, run the thread through a small section of the fabric, slip the needle through the slip knot hole, and then pull everything tight. The thread should now be fastened. Another way to knot your fabric is to slip the needle into the fabric to create a loop, and then slip the needle through the loop. Pull the loop tight, and repeat to make a second knot.

15.2.4.5.5. Backstitch

Backstitch is the strongest hand stitch for two pieces of standard cloth material. Work backstitch from right to left (Figure 15-15).

1. Lay the two pieces of fabric you wish to join with either the right or the wrong sides facing. Right sides facing will create an invisible stitch; wrong sides facing will create a visible stitch.
2. Thread a sewing needle and attach the thread to the right-hand edge of the first piece of fabric using whatever method you prefer.
3. Begin with a couple of stitches worked on the spot, and then take a stitch and a space.
4. Take the needle back over the space and bring it out the same distance in front of the thread.
5. Continue to the end of the seam. Fasten off with a couple of stitches on the spot.

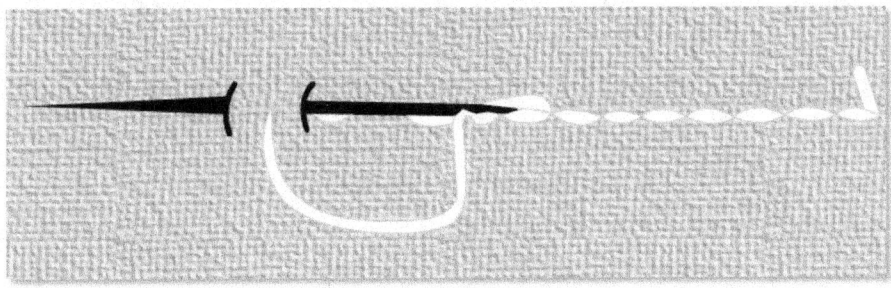

Figure 15-15 Backstitch

15.2.4.5.6. Whipstitch

The whipstitch is commonly used to attach two pieces of material or finish hem something (Figure 15-16). It works very well on thicker or stiffer material. The whipstitch is an overhand technique in which the needle enters the fabric from below and exits from above for every stitch worked.

1. Lay the two pieces of fabric you wish to join with either the right or the wrong sides facing. Right sides facing will create an invisible stitch; wrong sides facing will create a visible stitch.
2. Thread a sewing needle and attach the thread to the right-hand edge of the first piece of

fabric using whatever method you prefer.
3. Insert the needle through the second piece of fabric and pull the thread tight.
4. Bring the needle back to the first piece of fabric by laying the thread over the seam.
5. Insert the needle through both pieces of fabric, beside and approximately 1/16 of an inch over from where the needles was inserted the first time. The second stitch will enter the first piece of the fabric beside the first stitch, with the thread wrapping over the top of the seam with each stitch.
6. Repeat Steps four and five until you come to the end of the row, "whipping" the thread over the top of the seam with every stitch.
7. Tie off your thread at the end of the row by inserting your needle back under the previous whipstitch and pulling the thread almost all the way through. When you have only a small loop of thread left, insert the needle through the center of the loop and pull it tight to create a knot. Cut off any remaining thread just past the knot.

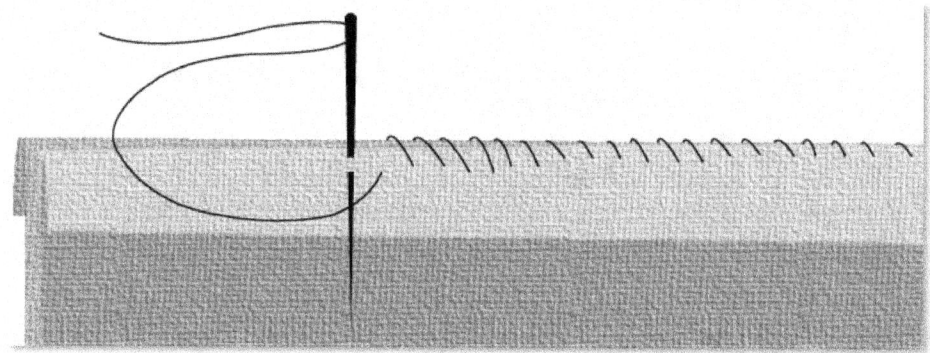

Figure 15-16 Whipstitch

15.2.4.5.7. Difficult to Sew Material

The material to be sewn may be quite thick and hard to sew. If this is the case, a palm type thimble can be improvised to keep from stabbing fingers and hands (Figure 15-17). A piece of webbing, leather, or other heavy material, with a hole for the thumb, can be used. A flat rock, metal, or piece of wood can be used as the thimble and this is held in place by a doughnut-shaped piece of material sewn onto the palm piece. To use, the end of the needle with the eye is placed on the thimble and the thimble is then used to push the needle through the material to be sewn.

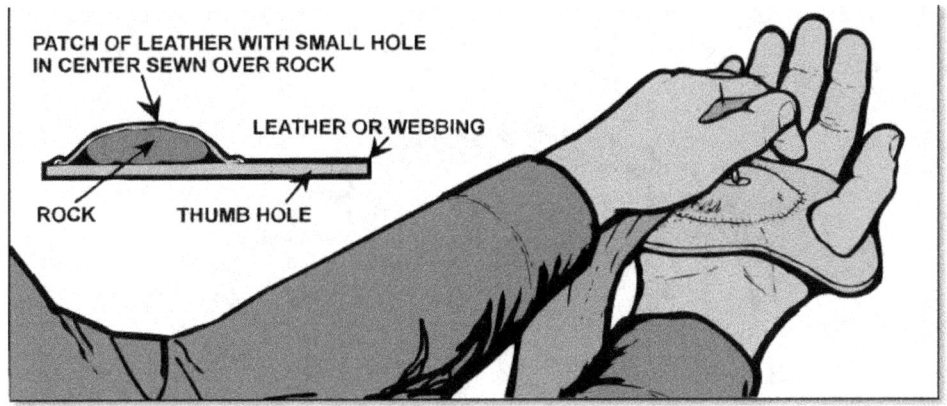

Figure 15-17 Palm Thimble

15.3. Other Improvised Equipment

15.3.1. Improvised Trail-Type Snowshoes

The snowshoe frame can be made from a sapling one inch in diameter and five feet long. The sapling should be bent and spread to 12 inches at the widest point. The IP can then include the webbing of suspension lines (Figure 15-18). The foot harness, for attaching the snowshoe to the boot, is also fashioned from suspension line or other cord.

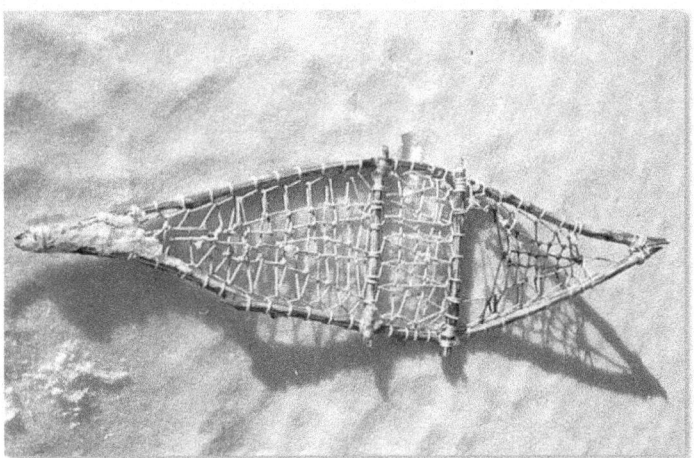

Figure 15-18 Improvised Trail Snowshoes

15.3.2. Improvised Bear Paw-Type Snowshoes

A sapling can be held over a heat source and bent to the shape shown in Figure 15-19. Wire from the aircraft or vehicle, parachute suspension line or other cord can be used for lashing and for making webbing. Snowshoes can also be quickly improvised by cutting a few pine boughs and lashing them together at the cut ends. The lashed boughs positioned with the cut ends forward can then be tied to the feet (Figure 15-20).

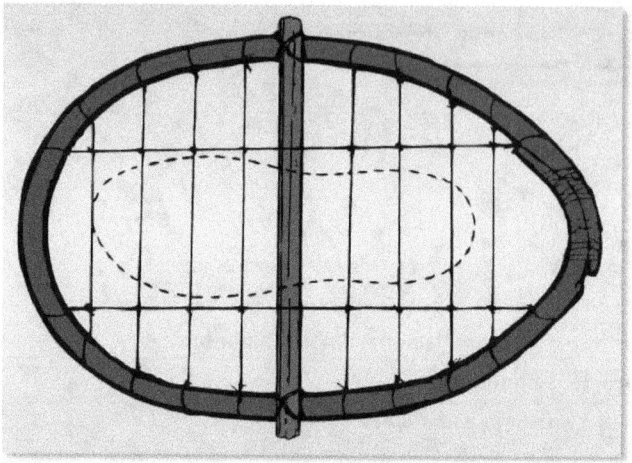

Figure 15-19 Improvised Bear Paws

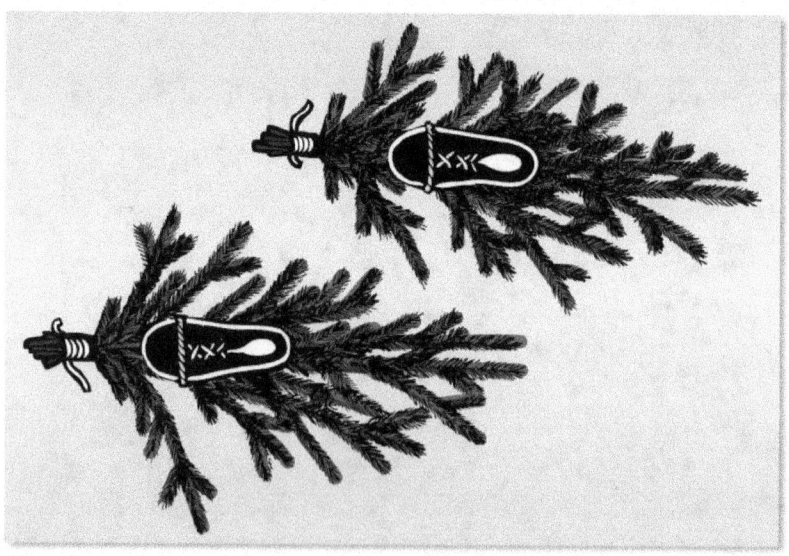

Figure 15-20 Bough Snowshoes

15.3.2.1. Guarding Against Frostbite while Snowshoeing

IP should guard against frostbite and blistering while snowshoeing. Due to the design of the harness, the circulation of the toes is usually restricted, and the hazard of frostbite is greater. IP should check the feet carefully by stopping often, taking off the harness, and massaging the feet when they seem to be getting cold.

15.3.2.2. Avoiding Blisters Between Toes

Blistering between the toes or on the ball of the foot is sometimes unavoidable if a good deal of snowshoeing is done. To make blisters less likely, the IP should keep socks and insoles dry and change them regularly.

15.3.3. Rawhide

Rawhide is a very useful material which can be made from any animal hide. Processing rawhide is time consuming but the material obtained is strong and very durable. It can be used for making sheaths for cutting tools, lashing materials, ropes, etc.

- The first step in making rawhide is to remove all of the fat and muscle tissue from the hide. The large pieces can be cut off and the remainder scraped off with a dull knife or similar instrument.
- The next step is to remove the hair. This can be done by applying a thick layer of wood ashes to the hair side. Ashes from a hardwood fire work best. Thoroughly sprinkle water all over the ashes. This causes lye to leach out of the ash. The lye will remove the hair. The hide should be rolled with the hair side in and stored in a cool place for several days. When the hair begins to shed (check by pulling on the hair), the hide should be unrolled and placed over a log. Remove the hair by scraping it off with a dull knife. Once the hair is removed, the hide should be thoroughly washed, stretched inside a frame, and allowed to dry slowly in the shade. When dry, rawhide is extremely hard. It can be softened by soaking in water.

15.3.4. Wire Saws

Wire or pieces of metal can be used to replace broken issued saws. With minor modifications, the IP can construct a usable saw (Figure 15-21). A bucksaw arrangement will help to prevent the blade from flexing. If a more durable saw is required and time permits, a bucksaw may be improvised. Blade tension can be maintained by use of a tightening device known as a "windlass".

Figure 15-21 Buck Saw

15.3.5. Cooking Utensils

Tins can serve as adequate cooking utensils. If the end has been left intact, use a green stick long enough to prevent burning the hand while cooking. If the side has been left intact, a forked stick may be used to add support to the container (Figure 15-22).

Figure 15-22 Cooking Utensils

15.4. Ropes and Knots

15.4.1. Basic Knowledge of Tying a Knot

A basic knowledge of correct rope and knot procedures will aid the IP in doing many necessary actions. Such actions as improvising equipment, building shelters, assembling packs, and providing safety devices entail the use of proven techniques. Tying a knot incorrectly could result in ineffective improvised equipment, injury, or death.

15.4.2. Rope Terminology For a depiction of the following rope terminology, refer to Figure 15-23.

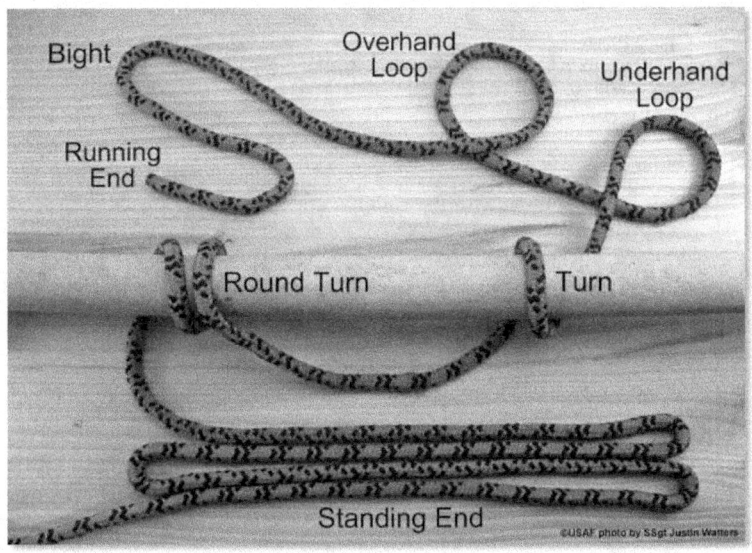

Figure 15-23 Elements of Ropes and Knots

15.4.2.1. Bend

A bend (called a knot in this handbook) is used to fasten two ropes together or to fasten a rope to a ring or loop.

15.4.2.2. Bight

A bight is a bend or U-shaped curve in a rope.

15.4.2.3. Hitch

A hitch is used to tie a rope around a timber, pipe, or post so that it will hold temporarily but can be readily untied.

15.4.2.4. Knot

A knot is an interlacement of the parts of bodies, as cordage, forming a lump or knot or any tie or fastening formed with a cord, rope, or line, including bends, hitches, and splices. It is often used as a stopper to prevent a rope from passing through an opening.

15.4.2.5. Line

A line (sometimes called a rope) is a single thread, string, or cord.

15.4.2.6. Loop

A loop is a fold or doubling of the rope through which another rope can be passed. A temporary loop is made by a knot or a hitch. A permanent loop is made by a splice or some other permanent means.

15.4.2.7. Overhand Turn or Loop

An overhand loop is made when the running end passes over the standing part.

15.4.2.8. Rope

A rope (often called a-line) is made of strands of fiber twisted or braided together.

15.4.2.9. Round Turn

A round turn is the same as a turn, with the running end leaving the circle in the same general direction as the standing part.

15.4.2.10. Running End

The running end is the free or working end of a rope.

15.4.2.11. Standing End

The standing end is the balance of the rope, excluding the running end.

15.4.2.12. Turn

A turn describes the placing of a rope around a specific object such as a post, rail, or ring with the running end continuing in the opposite direction from the standing end.

15.4.2.13. Underhand Turn or Loop

An underhand turn or loop is made when the running end passes under the standing part.

15.4.2.14. Whipping the Ends of a Rope

The raw, cut end of a rope has a tendency to untwist and should always be knotted or fastened in some manner. Whipping is one method of fastening the end of the rope. This method is particularly satisfactory because it does not increase the size of the rope. The whipped end of a rope will still thread through blocks or other openings. Before cutting a rope, place two whippings on the rope one or two inches apart and make the cut between the whippings. This will prevent the cut ends from untwisting immediately after they are cut. A rope is whipped by wrapping the end tightly with a small cord. Make a bight near one end of the cord and lay both ends of the small cord along one side of the rope. The bight should project beyond the end of the rope about one-half inch. The running end of the cord should be wrapped tightly around the rope and cord starting at the end of the whipping which will be farthest from the end of the rope. The wrap should be in the same direction as the twist of the rope strands. Continue wrapping the cord around the rope, keeping it tight, to within about one-half inch of the end. At this point, slip the running end through the bight of the cord. The standing part of the cord can then be pulled until the bight of the cord is pulled under the whipping and cord is tightened. The ends of cord should be cut at the edge of the whipping, leaving the rope end whipped (Figure 15-24).

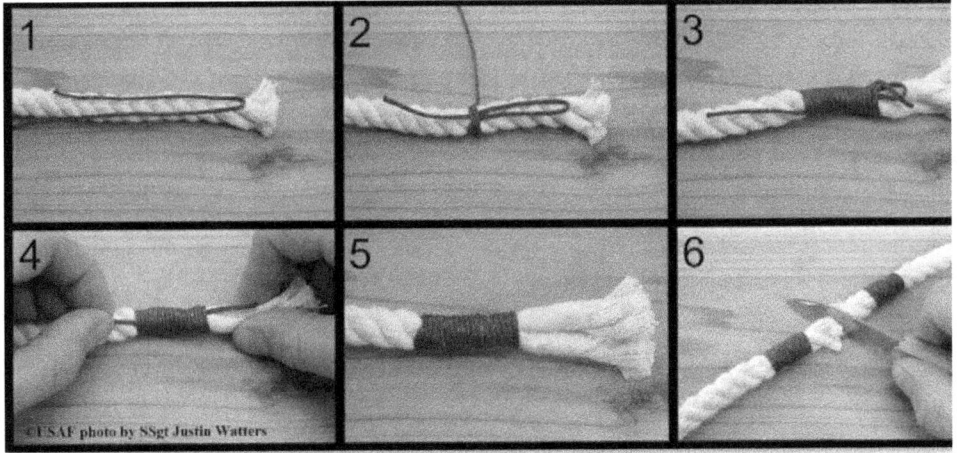

Figure 15-24 Whipping the End of a Rope

15.4.3. Knots at End of the Rope

15.4.3.1. Overhand Knot

The overhand knot (Figure 15-25) is the most commonly used and the simplest of all knots. An overhand knot may be used to prevent the end of a rope from untwisting, to form a knot at the end of a rope, or as a part of another knot. To tie an overhand knot, make a loop near the end of the rope and pass the running end through the loop, pulling it tight.

Figure 15-25 Overhand Knot

15.4.3.2. Figure-Eight Knot

The figure-eight knot (Figure 15-26) is used to form a larger knot than would be formed by an overhand knot at the end of a rope. A figure-eight knot is used in the end of a rope to prevent the ends from slipping through a fastening or loop in another rope. To make the figure-eight knot, make a loop in the standing part, pass the running end around the standing part back over one side of the loop, and down through the loop. The running end can then be pulled tight.

Figure 15-26 Figure-Eight Knot

15.4.4. Knots for Joining Two Ropes

15.4.4.1. Square Knot

The square knot (Figure 15-27) is used for tying two ropes of equal diameter together to prevent slippage. To tie the square knot, lay the running end of each rope together but pointing in opposite directions. The running end of one rope can be passed under the standing part of the other rope. Bring the two running ends up away from the point where they cross and crossed again. Once each running end is parallel to its own standing part the two ends can be pulled tight. If each running end does not come parallel to the standing part of its own rope, the knot is called a "granny knot" (Figure 15-28). Because it will slip under strain, the granny knot should not be used. A square knot can also be tied by making a bight in the end of one rope and feeding the running end of the other rope through and around this bight. The running end of the second rope is routed from the standing side of the bight. If the procedure is reversed, the resulting knot will have a running end parallel to each standing part but the two running ends will not be opposite each other. This knot is called a "thief" knot (Figure 15-28). It will slip under strain and is difficult to untie. A true square knot will draw tighter under strain. A square knot can be untied easily by grasping the bends of the two bights and pulling the knot apart.

Figure 15-27 Square Knot

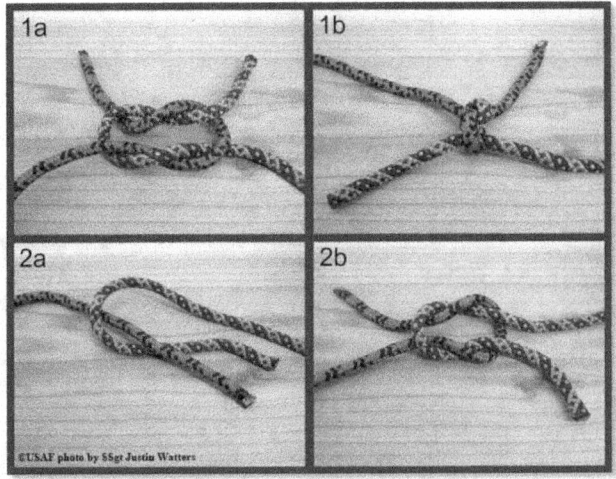

Figure 15-28 Granny and Thief Knots

15.4.4.2. Single Sheet Bend

The use of a single sheet bend (Figure 15-29) sometimes called a weaver's knot, is limited to tying together two dry ropes of unequal size. It is also useful for joining different materials like cloth or plastic sheeting to make cordage which could be useful to an IP in an urban evasion, escape or isolating event. To tie the single sheet bend, the running end of the smaller rope should pass through a bight in the larger rope. The running end should continue around both parts of the larger rope and back under the smaller rope. The running end can then be pulled tight. This knot will draw tight under light loads but may loosen or slip when the tension is released.

Figure 15-29 Single Sheet Bend

15.4.4.3. Double Sheet Bend

The double sheet bend (Figure 15-30) works better than the single sheet bend for joining ropes of equal or unequal diameter, joining wet ropes, or for tying a rope to an eye. It will not slip or draw tight under heavy loads. To tie a double sheet bend, a single sheet bend is tied first. However, the running end is not pulled tight. One extra turn is taken around both sides of the bight in the larger rope with the running end for the smaller rope. Then tighten the knot.

Figure 15-30 Double Sheet Bend

15.4.5. Knots for Making Loops

15.4.5.1. Bowline

The bowline (Figure 15-31) is a useful knot for forming a loop in the end of a rope. It is also easy to untie. To tie the bowline, the running end of the rope passes through the object to be affixed to the bowline and forms a loop in the standing part of the rope. The running end is then passed through the loop from underneath and around the standing part of the rope, and back through the loop from the top. The running end passes down through the loop parallel to the rope coming up through the loop. The knot is then pulled tight.

Figure 15-31 Bowline

15.4.5.2. Double Bowline

The double bowline (Figure 15-32) with a slip knot is a rigging used by tree surgeons who work alone in trees for extended periods. It can be made and operated by one person and is comfortable as a sling or boatswain's chair (Figure 15-33). A small board with notches as a seat adds to the personal comfort of the user. To tie a double bowline, the running end of a line should be bent back about 10 feet along the standing part. The bight is formed as the new running end and a bowline tied as described and illustrated in Figure 15-31. The new running end or loop is used to support the back and the remaining two loops and support the legs.

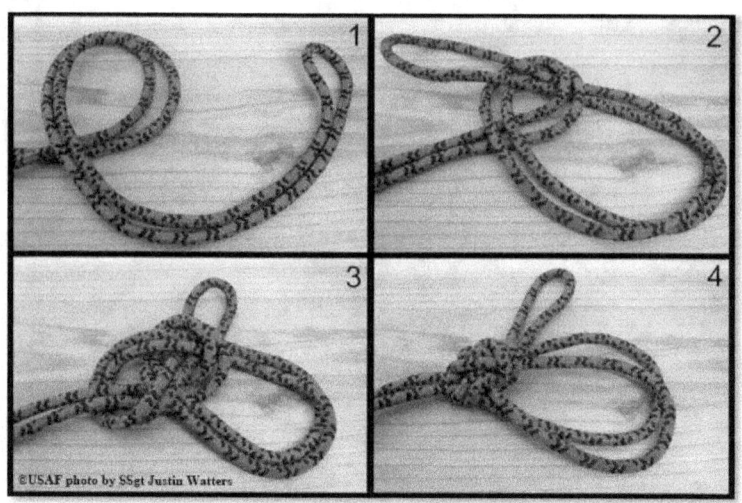

Figure 15-32 Double Bowline

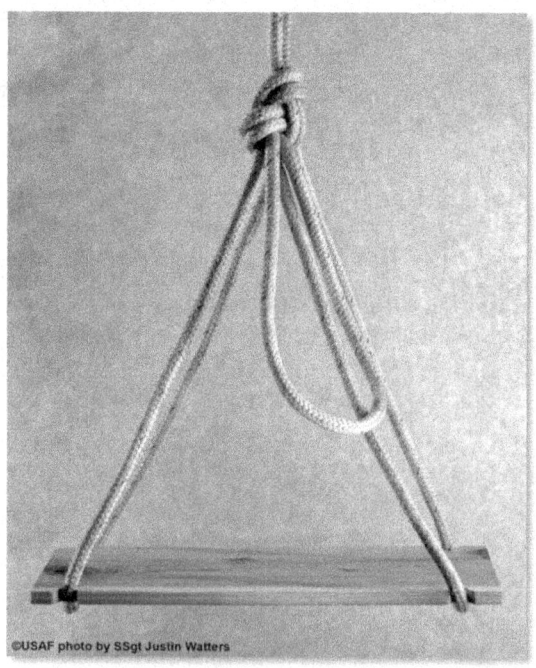

Figure 15-33 Boatswain's Chair

15.4.5.3. Butterfly Knot

The butterfly (Figure 15-34) is used to form a non-slipping loop in a rope. To make the butterfly, form a bight in the running end of the rope. Hold this bight in the left hand and form a second bight in the standing part of the rope. The right hand is used to pass bight over bight. Holding all loops in place with the left hand, the right hand is inserted through bight behind the upper part of bight. The bottom of the first loop is grasped and pulled up through the entire knot pulling it tight.

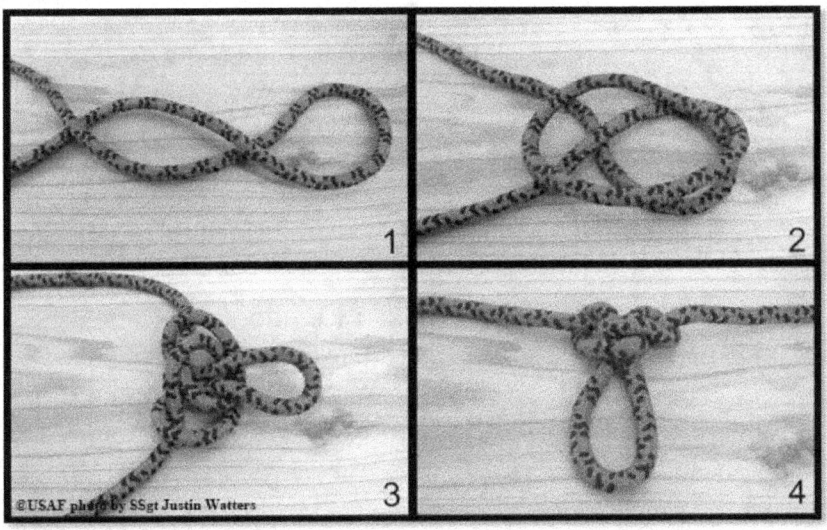

Figure 15-34 Butterfly

15.4.6. Hitches

15.4.6.1. Half Hitch

The half hitch (Figure 15-35) is used to tie a rope to a timber or to another larger rope. It is not a very secure knot or hitch and is used for temporarily securing the free end of a rope. To tie a half hitch, the rope is passed around the timber, bringing the running end around the standing part, and back under itself.

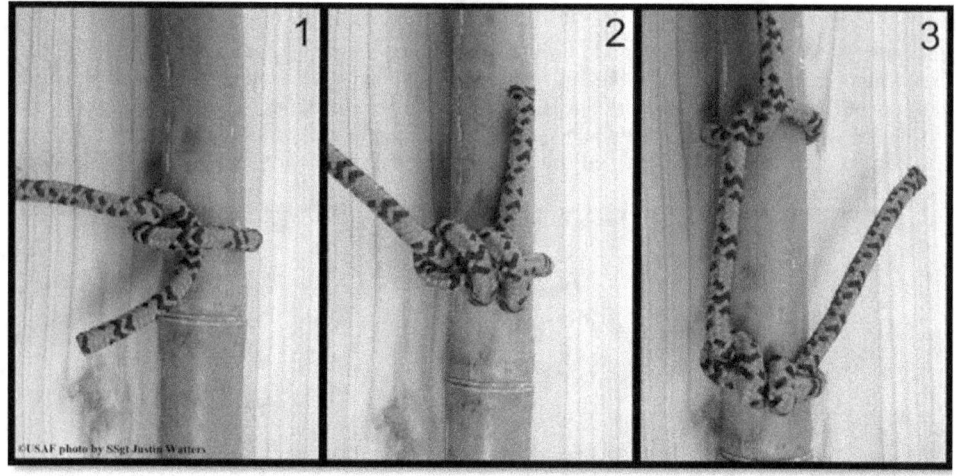

Figure 15-35 Half Hitch, Timber Hitch

15.4.6.2. Timber Hitch

The timber hitch (Figure 15-35) is used for moving heavy timbers or poles. To make the timber hitch, a half hitch is made and similarly the running end is turned about itself at least another time. These turns must be taken around the running end itself or the knot will not tighten against the pull.

15.4.6.3. Timber Hitch and Half Hitch

To get a tighter hold on heavy poles for lifting or dragging a timber hitch and half hitch are combined (Figure 15-35). The running end is passed around the timber and back under the standing part to form a half hitch. Further along the timber, a timber hitch is tied with the running end. The strain will come on the half hitch and the timber hitch will prevent the half hitch from slipping.

15.4.6.4. Clove Hitch

A clove hitch (15-36) is used to fasten a rope to a timber, pipe, or post. It can be tied at any point in a rope. To tie a clove hitch in the center of the rope, two turns are made in the rope close together. They are twisted so that the two loops lay back-to-back. These two loops are slipped over the timber or pipe to form the knot. To tie the clove hitch at the end of a rope, the rope is passed around the timber in two turns so that the first turn crosses the standing part and the running end comes up under itself on the second turn.

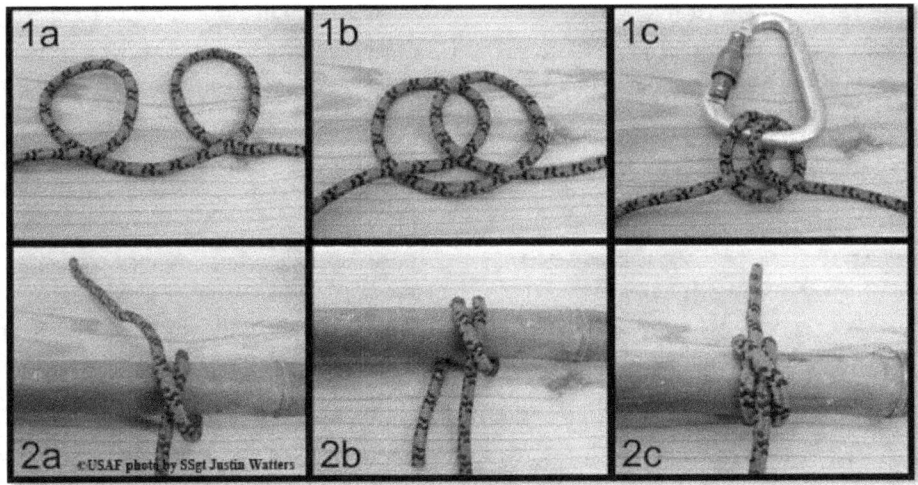

15-36 Clove Hitch

15.4.6.5. Two Half Hitches

A quick method for tying a rope to a timber or pole is the use of two half hitches. The running end of the rope is passed around the pole or timber, and a turn is taken around the standing part and under the running end. Doing this creates one half hitch. The running end is passed around the standing part of the rope and back under itself again.

15.4.6.6. Round Turn and Two Half Hitches

Another hitch used for fastening a rope to a pole, timber, or spar is the round turn and two half hitches(Figure 15-37). The running end of the rope is passed around the pole or spar in two complete turns, and the running end is brought around the standing part and back under itself to make a half hitch. A second half hitch is made. For greater security, the running end of the rope should be secured to the standing part.

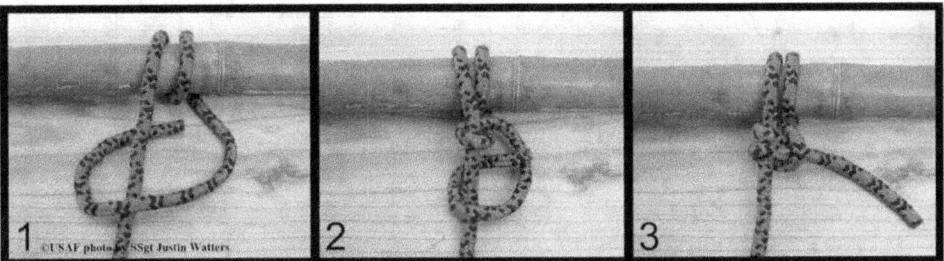

Figure 15-37 Round Turn and Two Half Hitches

15.4.6.7. Fisherman's Bend

The fisherman's bend (Figure 15-38) is used to fasten a cable or rope to an anchor, or for use where there will be a slackening and tightening motion in the rope. To make this bend, the running end of the rope is passed in two complete turns through the ring or object to which it is to be secured. The running end is passed around the standing part of the rope and through the loop which has just been formed around the ring. The running end is then passed around the standing part in a half hitch. The running end should be secured to the standing part.

Figure 15-38 Fisherman's Bend

15.4.6.8. Sheepshank

A sheepshank (Figure 15-39) is a method of shortening a rope, but it may also be used to take the load off a weak spot in the rope. To make the sheepshank (which is never made at the end of a rope), two bights are made in the rope so that three parts of the rope are parallel. A half hitch is made in the standing part over the end of the bight at each end.

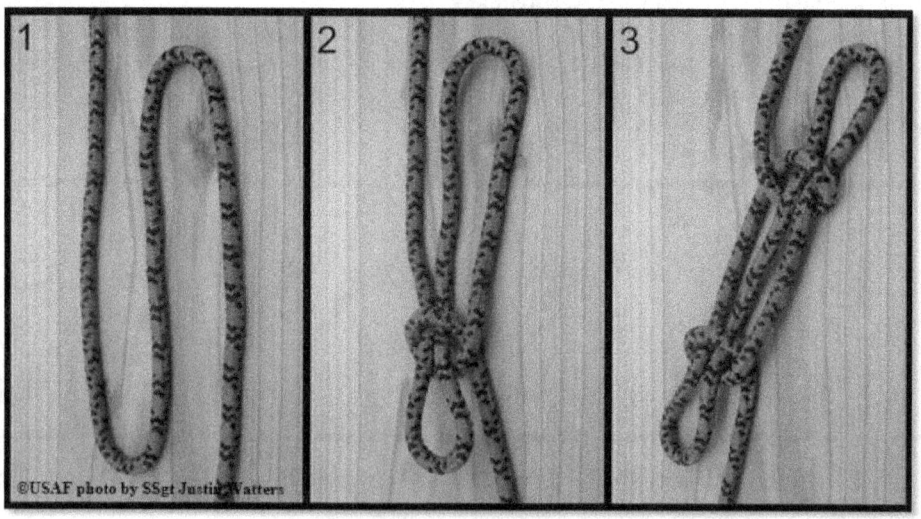

Figure 15-39 Sheep Shank

15.4.7. Lashing

There are numerous items which require lashings for construction; for example, shelters, equipment racks, and smoke generators. Three types of lashings will be discussed here; the square lash, the diagonal lash, and the shear lash.

15.4.7.1. Square Lash

The square lash (Figure 15-40) is started with a clove hitch around the log, immediately under the place where the crosspiece is to be located. In laying the turns, the rope goes on the outside of the previous turn around the crosspiece, and on the inside of the previous turn around the log. The rope should be kept tight. Three or four turns are necessary. Two or three "frapping" turns are made between the crosspieces. The rope is pulled tight; this will bind the crosspiece tightly together. It is finished with a clove hitch around the same piece that the lashing was started. The square lash is used to secure one pole at right angles to another pole. Another lash that can be used for the same purpose is the diagonal lash.

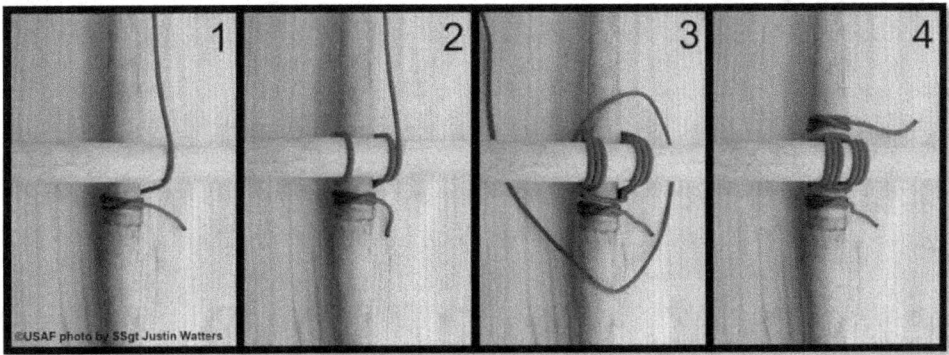

Figure 15-40 Square Lash

15.4.7.2. Diagonal Lash

The diagonal lash (Figure 15-41) is started with a clove hitch around the two poles at the point of crossing. Three turns are taken around the two poles. The turns lie beside each other, not on top of each other. Three more turns are made around the two poles, this time crosswise over the previous turns. The turns are pulled tight. A couple of frapping turns are made between the two poles, around the lashing turns, making sure they are tight. The lashing is finished with a clove hitch around the same pole the lash was started on.

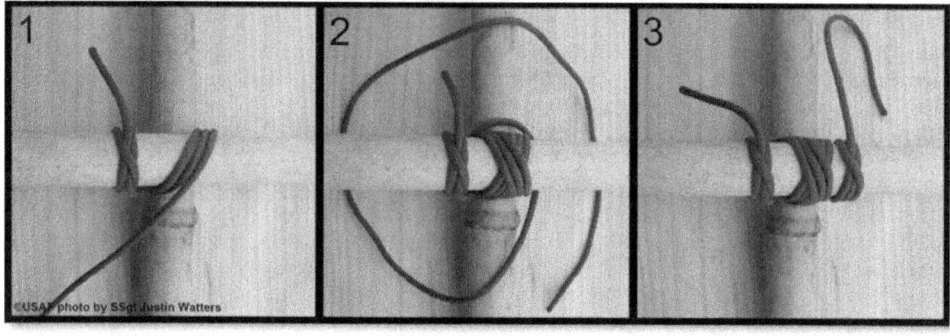

Figure 15-41 Diagonal Lash

15.4.7.3. Shear Lash

The shear lash (Figure 15-42) is used for lashing two or more poles in a series. The desired numbers of poles are placed parallel to each other and the lash is started with a clove hitch on an outer pole. The poles are then lashed together, using seven or eight turns of the rope laid loosely beside each other. Make frapping turns between each pole. The lashing is finished with a clove hitch on the pole opposite that on which the lash was started.

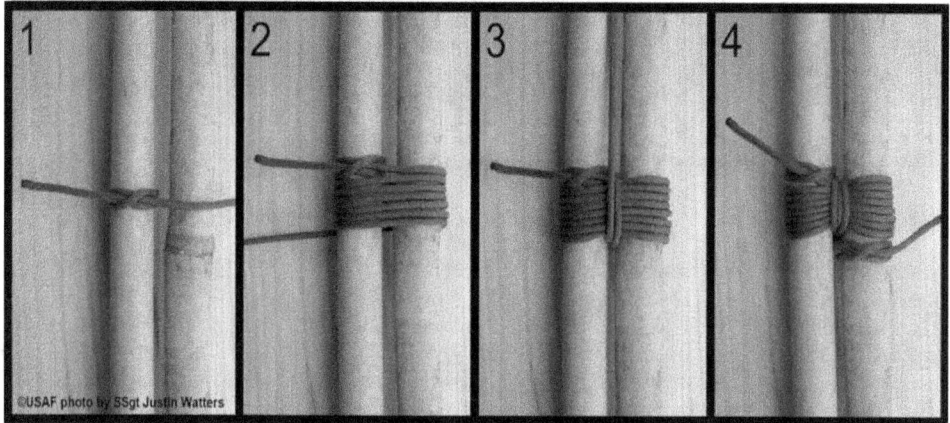

Figure 15-42 Shear Lash

15.4.8. Making Ropes and Cords

Almost any natural fibrous material can be spun into good serviceable rope or cord, and many materials which have a length of 12 to 24 inches or more can be braided. Ropes up to 3 and 4 inches in diameter can be "laid" by four people, and tensile strength for bush-made rope of 1-inch diameter range from 100 pounds to as high as 3,000 pounds.

15.4.8.1. Tensile Strength

Using a three-lay rope of one-inch diameter as standard, the following table of tensile strengths may serve to illustrate general strengths of various materials (Figure 15-43). For safety's sake, the lowest figure should always be regarded as the tensile strength.

MATERIAL	TENSILE STRENGTH
Green Grass	–100 lbs to 250 lbs
Bark Fiber	200 lbs to 1,500 lbs
Palm Fiber	650 lbs to 2,000 lbs
Sedges	2,000 lbs to 2,500 lbs
Monkey Rope (Lianas)	–56 lbs to 700 lbs
Lawyer Vine (Calamus) ⅜-inch diameter	–1,200 lbs

NOTE: Doubling the diameter quadruples the tensile strength; half the diameter reduces the tensile strength to one-fourth.

Figure 15-43 Tensile Strength

15.4.8.2. Principles of Rope-making Materials

To discover whether a material is suitable for rope making, it must have four qualities:

1. It must be reasonably long in the fiber.
2. It must have "strength."
3. It must be pliable.
4. It must have "grip" so the fibers will "bite" onto one another.

15.4.8.3. Determining Suitability of Material

There are simple tests to determine if a material is suitable. First, pull on a length of the material to test for strength. Second, twist it between the fingers and "roll" the fibers together; if it will withstand this and not "snap" apart, an overhand knot is tied and gently tightened. If the material does not cut upon itself, but allows the knot to be pulled taut and will "bite" together and is not smooth and slippery then it is suitable for rope making

15.4.8.4. Where to Find Suitable Material

These qualities can be found in various types of plants, in ground vines, in most of the longer grasses, in some of the water reeds and rushes, in the inner barks of many trees and shrubs, and in the long hair or wool of many animals.

15.4.8.5. Obtaining Fibers for Making Ropes

Some green freshly gathered materials may be stiff or unyielding. When this is the case, it should be passed through hot flames for a few moments. The heat treatment should cause the sap to burst through some of the cell structure, and the material thus becomes pliable. Fibers for rope making may be obtained from many sources such as:

- Surface roots of many shrubs and trees with strong fibrous bark.
- Dead inner bark of fallen branches of some species of trees and in the new growth of many trees such as willows.
- The fibrous material of many water and swamp growing plants and rushes.
- Many species of grass and weeds.
- Some seaweed.
- Fibrous material from leaves, stalks, and trunks of many palms.
- Many fibrous-leaved plants such as the aloes.

15.4.8.6. Gathering and Preparing Materials

There may be a high content of vegetable gum in some plants. This can often be removed by soaking the plants in water, by boiling, or by drying the material and "teasing" it into thin strips.

15.4.8.6.1. Using Green Materials

Some of the materials have to be used green if any strength is required. The materials that should be green include the sedges, water rushes, grasses, and lianas.

15.4.8.6.2. Palm Fiber

Palm fiber is harvested in tropical or subtropical regions. It is found at the junction of the leaf and the palm trunk, or it will be found lying on the ground beneath many palms. Palm fiber is a "natural" for making ropes and cords.

15.4.8.6.3. Fibrous Matter

Fibrous matter from the inner bark of trees and shrubs is generally more easily used if the plant is dead or half dead. Much of the natural gum will have dried out and when the material is being teased, prior to spinning, the gum or resin will fall out in fine powder.

15.4.8.7. Making a Cord by Spinning with the Fingers

15.4.8.7.1. Use Fibers that have been Tested for Strength

Use any material with long strong threads or fibers which have been previously tested for strength and pliability. The fibers are gathered into loosely held strands of even thickness. Each of these strands is twisted clockwise. The twist will hold the fibers together. The strands should be formed one-eighth inch diameter. As a general rule, there should be about 15 to 20 fibers to a strand. Two, three, or four of these strands are later twisted together, and this twisting together or "laying" is done with a counterclockwise twist, while at the same time, the separate strands which have not yet been laid up are twisted clockwise. Each strand must be of equal twist and thickness.

15.4.8.7.2. Bonding the Fibers into Strands

Figure 15-44 shows the general direction of twist and the method whereby the fibers are bonded into strands. In a similar manner, the twisted strands are put together into lays, and the lays into ropes.

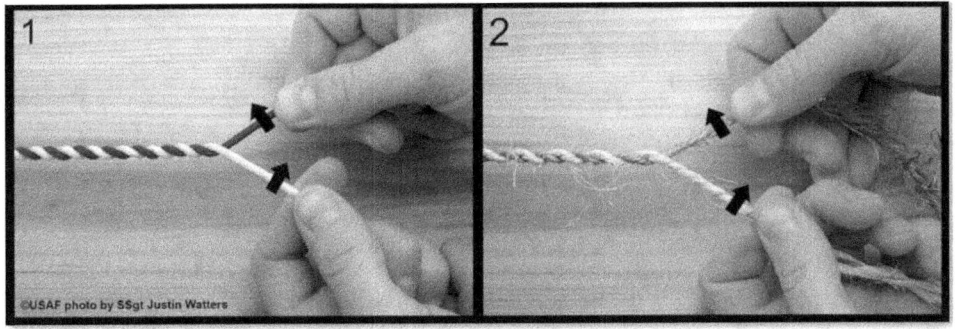

Figure 15-44 Twisting Fibers

15.4.8.7.3. The "Layer"

The person who twists the strands together is called the "layer" and must see that the twisting is even, the strands are uniform, and the tension on each strand is equal. In "laying," care must be taken to ensure each of the strands is evenly "laid up"; that is, one strand does not twist around the other one.

15.4.8.7.4. Spinning Fine Cords

When spinning fine cords for fishing lines, snares, etc., considerable care must be taken to keep the strands uniform and the lay even. Fine thin cords of no more than 1/32-inch thickness can be spun with the fingers and are capable of taking a breaking strain of 20 to 30 pounds or more.

15.4.8.7.5. Using Two People to Spin

Normally two or more people are required to spin and lay up the strands for cord. However, many native people spin cord unaided. They twist the material by running the flat of the hand along the thigh, with the fibrous material between hand and thigh; and with the free hand, they feed in fiber for the next "spin." Using this technique, one person can make long lengths of single strands. This method of making cord or rope with the fingers is slow if any considerable length of cord is required.

15.4.8.8. Thickness of Strands

Equal thickness and twist for each of the strands throughout their length are important. The thickness should not be greater than is necessary with the material being used. For a grass rope, the strand should not be more than one-fourth inch diameter; for coarse bark or palm, not more than one-eighth or three-sixteenth inch; and for fine bark, hair, or sisal fiber, not more than one-eighth inch.

15.4.8.9. Common Errors in Rope Making

- There is a tendency with beginners to feed unevenly. Thin wispy sections of strand are followed by thick portions. Such feeding degrades the quality of rope. Rope made from such strands will break with less than one-fourth of the tensile strain on the material.

- Beginners are wise to twist and feed slowly. Speed, with uniformity of twist and thickness, comes with practice.

- Thick strands do not help. It is useless to try and spin a rope from strands an inch or more in thickness. Such a rope will break with less than half the tensile strain on the material. Spinning "thick" strands does not save time in rope making.

15.4.8.10. Lianas, Vines, and Canes

Lianas and ground vines are natural ropes, and grow in subtropical and tropical scrub and jungle. Many are of great strength and useful for braiding, tree climbing, and other purposes. The smaller ground vines, when "braided", give great strength and flexibility. Canes and stalks of palms provide excellent material if used properly. Only the outer skin is strong, and will split off easily if the main stalk is bent away from the skin. This principle also applies to the splitting of lawyer cane (calamus), palm leaf stalks, and all green material. If the split starts to run off, bend the material away from the thin side, and it will gradually gain in size and come back to an even thickness with the other split side.

15.4.8.11. Bark Fibers

The fibers in many barks which are suitable for "rope making" are located near the innermost layers. This is the bark next to the sap wood. When seeking suitable barks of green timber, cut a small section about three inches long and one inch wide. Cut this portion from the wood to the outer skin of the bark.

- The specimen should be peeled and the different layers tested. Green bark fibers are generally difficult to spin because of "gum" and it is better to search around for windfall dead branches and try the inner bark of these. The gum has likely leached out, and the fibers should separate easily.

- Many shrubs have excellent bark fiber. It is advisable to cut the end of a branch and peel off a strip of bark for testing. Thin bark from green shrubs is sometimes difficult to spin into fine cord and is easier to use as braid for small cords.

- Where it is necessary to use green bark fiber for rope spinning the gum will generally wash out when the bark is teased and soaked in water for a day or so. After removing from the water, the bark strips should be allowed to dry before shredding and teasing into fiber.

15.4.9. Braiding

One person can braid and make suitable rope. The three-strand braid makes a flat rope, and while quite good, it does not have finish or shape, nor is it as "tight" as the four-strand braid. On other occasions, it may be necessary to braid broad bands for belts or for shoulder straps. There are many fancy braids which can be developed from these, but these three are basic, and essential for practical woodcraft work. A general rule for all braids is to work from the outside into the center.

15.4.9.1. Three Plait

1. The right-hand strand is passed over the strand to the left.

2. The left-hand strand is passed over the strand to the right.

3. This is repeated alternately from left to right (Figure 15-45).

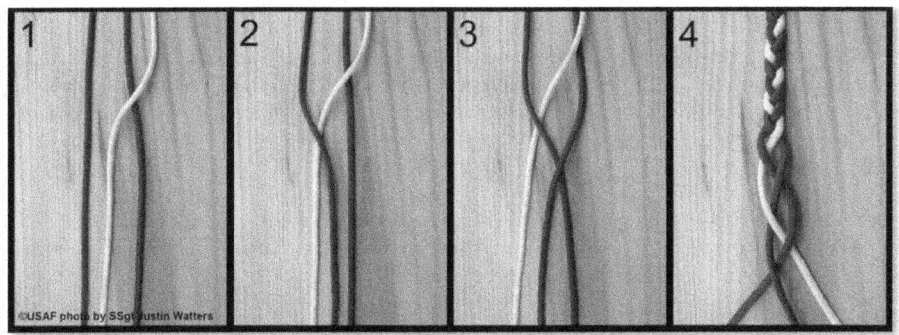

Figure 15-45 Three-Stranded Braid

15.4.9.2. Flat Four-Strand Braid

1. The four strands are placed side by side. The right-hand strand is taken and placed over the strand to the left.
2. The outside left-hand strand is laid under the next strand to itself and over what was the first strand.
3. The outside right-hand strand is laid over the first strand to its left.
4. The outside left strand is placed under and over the next two strands, respectively, moving toward the right.
5. Thereafter, the right-hand strand goes over one strand to the left, and the left-hand strand under and over to the right (Figure 15-46).

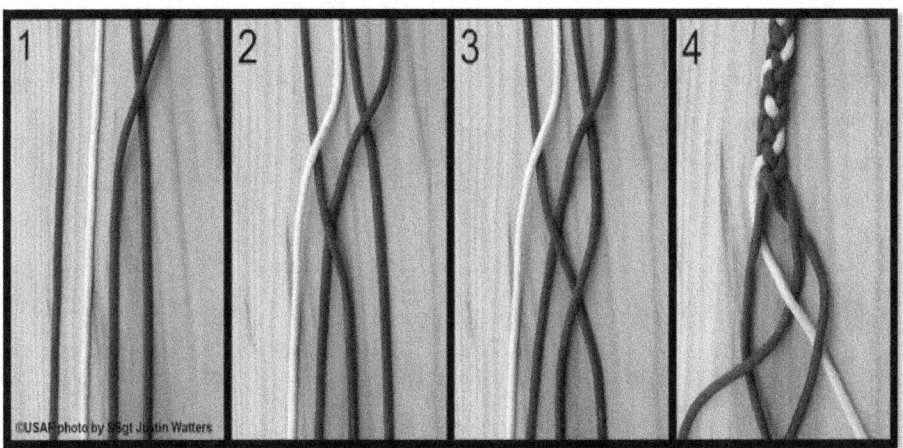

Figure 15-46 Four-Strand Braid

15.5. Personal Survival Kit

Potential IP should consider assembling and carrying personal survival kits. The shock and fear associated with some isolating events have caused IP to hit the ground running, leaving their survival kits behind. If IP have a personal survival kit in their pocket, it may improve their survival chances considerably.

15.5.1. Preparing Personal Survival Kits

A great deal of thought should go into preparing personal survival kits. The potential needs of the IP must be a consideration, such as the impact of the environmental elements, type of mission, availability of recovery, and distance to friendly forces.

15.5.2. Ways to Carry

There are two basic ways to carry a personal survival kit. One way is to pack all items into one or two waterproof containers. The other way is to scatter the items throughout personal clothing. Any type of small container can be used to encase the contents of the personal survival kit. Plastic cigarette cases, soap dishes, and even metal Band-Aid boxes are excellent containers.

15.5.2.1. Items that can be Packed in a Small Container

Examples of items which can be packed into a small container include:

- Matches.
- Safety pins (varied sizes).
- Fishhooks and line.
- Knife or Multi-tool (small, multi-bladed).
- Button compass.
- Prophylactic (for water container).
- Snare wire.
- Water purification tablets.
- Signal mirror.
- Needles with thread.
- Band-Aids.
- Aluminum foil.
- Insect repellent stick.
- Lip Balm.
- Soap (Antiseptic).
- Super Glue
- Duct Tape

Chapter 16
WATER

16. Water

Nearly every isolating event details the need IP have for water. Many ingenious methods of locating, procuring, purifying, and storing water are included in the recorded experiences of IP. Whether isolated in temperate, tropic, or dry climates, water may be one of their first and most important needs. The priority of finding water over that of obtaining food must be emphasized to potential IP. Individuals may be able to live for weeks without food, depending on the temperature and amount of energy being exerted, but a person who has no water can be expected to die within days. Even in cold climate areas or places where water is abundant, IP should attempt to keep their body fluids at a level that will maintain them in the best possible state of health. Even in relatively cold climates, the body needs two quarts of water per day to remain efficient.

16.1. Water Sources

IP should be aware of both the water sources available to them and the resources at their disposal for producing water.

16.1.1. Obtaining Water

IP may obtain water from carried storage, reverse osmosis pumps or flex packs found in various survival kits. Flex packs provide approximately four ounces of water per package. Knowledge of issued water procurement systems and the amount of water they may provide is extremely important.

16.1.2. Carrying Extra Water

Potential IP should always carry extra water during their missions. The initial shock of the isolating event can produce feelings of confusion and thirst and having additional water containers available will benefit the IP. The issued items containing water should be stored and protected by the IP for times when natural sources of freshwater are not available.

16.1.3. Naturally Occurring Water Sources

Naturally occurring water sources include:

- Surface water, including streams, lakes, springs, ice, and snow.
- Precipitation, such as rain, snow, dew, and sleet.
- Subsurface water, such as underground springs and streams.

16.1.4. Indicators of Possible Water

Several indicators of possible water include:

- Presence of abundant vegetation of a different variety such as deciduous growth in a coniferous area.
- Drainages and low-lying areas.
- Large clumps of plush grass and/or vibrantly colored vegetation.

- Wells or cisterns, for indication of underground water sources.

16.1.5. Location of Water in the Libyan and Sahara Desert

In the Libyan Sahara Desert, donut-shaped mounds of camel dung often surround wells or other water sources. Bird flights can indicate direction to or from water. Pigeons and doves fly from water in the morning and to it in the evening. Large flocks of birds may also congregate around or at areas of water. Bird droppings around cracks or crevices in rocks may indicate water sources. Bees require a nearby water source to support their hives; water can usually be found within ¼ mile.

16.1.6. Swarming Insects as an Indication of Water

The presence of swarming insects indicates water is near. In some places, IP should look for signs of animal presence. For example, in damp places, animals may have scratched depressions into the ground to obtain water. Animal trails may lead to water; the "V" formed by intersecting trails often point toward water sources. There are typically more intersections closer to the water source.

16.1.7. Presence of People as an Indication of Water

The presence of people will indicate water. The location of this water can take many forms - stored water in containers that are carried with people who are traveling, wells, irrigation systems, pools, etc. IP who are evaders should be extremely cautious when approaching any water source, especially if they are in dry areas; these places may be guarded or inhabited.

16.1.8. Tapping the Earth's Supply of Water

When surface water is not available, IP may have to tap the Earth's supply of ground water. Access to this depends upon the type of ground-rock or loose material, clay, gravel, or sand.

16.1.8.1. Potential Water in Rocky Ground

In rocky ground, IP should look for springs and seepages. Limestone and lava rocks will have more and larger springs than any other rocks. Most lava rocks contain millions of bubble holes; ground water may seep through them. IP can also look for springs along the walls of valleys that cross a lava flow. Some flows will have no bubbles but do have "organ pipe" joints - vertical cracks that part the rocks into columns a foot or more thick and 20 feet or more high. At the foot of these joints, IP may find water creeping out as seepage, or pouring out in springs.

16.1.8.2. Cracks in Rocks as an Indication of Water

Most common rocks, like granite, contain water only in irregular cracks. A crack in a rock with bird dung around the outside may indicate a water source that can be reached by a piece of surgical hose used as a straw or siphon.

16.1.8.3. Water in Loose Sediments

Water is more abundant and easier to find in loose sediments than in rocks. Springs are sometimes found along valley floors or down along their sloping sides. The flat benches or terraces of land above river valleys usually yield springs or seepages along their bases, even when the stream is dry. IP should not waste time digging for water unless there are signs that water is available. Digging in the floor of a valley under a steep slope, especially if the bluff is cut in a terrace, can produce a water source. A lush green spot where a spring has been during the wet season is a good place to dig for water. Water moves slowly through clay, but many types of clay contain strips of sand which may yield springs. IP should look for a wet place on the surface of clay bluffs and try digging it out.

16.1.8.4. Drinking Water at the Coast

Along coasts, water may be found by digging beach wells (Figure 16-1). Locate the wells behind the first or second pressure ridge. Wells can be dug three to five feet deep and should be lined with driftwood to prevent sand from refilling the hole. Rocks should be used to line the bottom of the well to prevent stirring up sand when procuring the water. The average well may take as long as two hours to produce four to five gallons of water. Do not be discouraged if the first try is unsuccessful - dig another.

Figure 16-1 Beach Well

16.2. Locating and Procuring Water

Precipitation may be procured by laying a piece of nonporous material such as a poncho, piece of canvas, plastic, or metal material on the ground. If rain or snow is being collected, it may be more efficient to create a bag or funnel shape with the material so the water can be easily gathered. Dew can be collected by wiping it up with a sponge or cloth first, and then wringing it into a container (Figure 16-2). If the Sun is shining, snow or ice may be placed on a dark surface to melt (dark surfaces absorb heat, whereas light surfaces reflect heat). Ice can be found in the form of icicles on plants and trees, sheet ice on rivers, ponds, and lakes, or sea ice. If snow must be used, IP should use snow closest to the ground. This snow is packed and will provide more water for the amount of snow than will the upper layers.

- Consideration should be given to the possibility of contaminating the water with dyes, preservatives, or oils on the surfaces of the objects used to collect the precipitation. Ice will yield more water per given volume than snow and requires less heat to do so.

- When snow is to be melted for water, place a small amount of snow in the bottom of the container being used and place it over or near a fire. Snow should be added a little at a time. Snow absorbs water, and if packed, forms an insulating airspace at the bottom of the container. When this happens, the bottom may burn out. IP should allow water in the container bottom to become warm so that when more snow is added, the mixture remains slushy. This will prevent burning the bottom out of the container.

Figure 16-2 Methods of Procuring Water

16.3. Water in Snow and Ice Areas

Due to the extreme cold of arctic areas, water requirements are greatly increased. Increased body metabolism, respiration of cold air, and extremely low humidity all act to reduce the body's water content. The processes of heat production and digestion in the body also increase the need for water in colder climatic zones. Constructing shelters, signals, and gathering firewood are extremely demanding tasks for IP. Physical exertion and heat production in extreme cold place the water requirements of an IP close to five or six quarts per day to maintain proper hydration levels. The diet of IP will often be dehydrated rations and high protein food sources. For the body to digest and use these food sources effectively, increased water intake is essential.

16.3.1. Obtaining Water in the Arctic

Obtaining water need not be a serious problem in the arctic because an abundant supply of water is available from streams, lakes, ponds, snow, and ice. All surface water should be purified by some means, if possible. In the summer, surface water may be discolored but is drinkable when purified. Water obtained from glacier-fed rivers and streams may contain high concentrations of dirt or silt. By letting the water stand for a period of time, most silt will settle to the bottom; the remaining water can be strained through porous material for further filtration.

16.3.2. Constructing a Water Machine

A "water machine" can be constructed which will produce water while IP are doing other tasks. It can be made by placing snow on any porous material (such as parachute or cotton), gathering up the edges, and suspending the "bag" of snow from any support near the fire. Radiant heat will melt the snow and the water will drip from the lowest point on the bag. A container should be placed below this point to catch the water (Figure 16-3).

Figure 16-3 Water Machine

16.3.3. Using Body Heat to Melt Snow for Water

In some arctic areas, there may be little or no fuel supply with which to melt ice and snow for water. In this case, body heat can be used to do the job. The ice or snow can be placed in a waterproof container like a water bag and placed between clothing layers next to the body. This cold substance should not be placed directly next to the skin as it will cause chilling and lowering of the body temperature.

16.3.4. Icebergs as a Source for Drinking Water

Since icebergs are composed of freshwater, they can be a readily available source of drinking water. IP should use extreme caution when trying to obtain water from this source. Even large icebergs can suddenly roll over and dump IP into the frigid sea water. If sea ice is the primary source of water, IP should recall that like seawater itself, saltwater ice should never be ingested. To obtain water in Polar Regions or sea ice areas, select old sea ice - a bluish or blackish ice which shatters easily and generally has rounded corners. This ice will be almost salt-free. New sea ice is milky or gray colored with sharp edges and angles. This type of ice will not shatter or break easily. Snow and ice may be saturated with salt from blowing spray; if it tastes salty, IP should select different snow or ice sources.

16.3.5. Ingesting Un-Melted Snow

The ingesting of un-melted snow or ice is not recommended. Eating snow or ice lowers the body's temperature, induces dehydration, and causes minor cold injury to lips and mouth membranes. Water consumed in cold areas should be in the form of warm or hot fluids. The ingestion of cold fluids or foods increases the body's need for water and requires more body heat to warm the substance.

16.4. Water in Tropical Areas

Surface water is normally available in the form of streams, ponds, rivers, and swamps. In the savannas during the dry season, it may be necessary for the IP to resort to digging for water in the places previously mentioned. Water obtained from these sources may need filtration and should be purified. Jungle plants can also provide IP with water.

16.4.1. Plants as Rainfall Collectors

Many plants have hollow portions which can collect rainfall, dew, etc. (Figure 16-4). Since there is no absolute way to tell whether this water is pure, it should be purified. The stems or the leaves of some plants have a hollow section where the stem meets the trunk. Look for water collected here. This includes any Y-shaped plants (palms or air plants). The branches of large trees often support air plants (relatives of the pineapple) whose overlapping, thickly growing leaves may hold a considerable amount of rainwater. Trees may also catch and store rainwater in natural receptacles such as cracks or hollows.

Figure 16-4 Water Collectors

16.4.2. Freshwater from Plant Sources

Pure freshwater needing no purification can be obtained from numerous plant sources. There are many varieties of vines which are potential water sources. The vines are from 50 feet to several hundred feet in length and one to six inches in diameter. They also grow like a hose along the ground and up into the trees. The leaf structure of the vine is generally high in the trees. Water vines are usually soft and easily cut. The smaller species may be twisted or bent easily and are usually heavy because of the water content. The water from these vines should be tested for potability.

- The first step in testing the water from vines is to nick the vine and watch for sap running from the cut. If milky sap is seen, the vine should be discarded; if no milky sap is observed, the vine may be a safe water vine.

- Cut out a section of the vine, hold that piece vertically, and observe the liquid as it flows out. If it is clear and colorless, it may be a drinkable source. If it is cloudy or milky-colored, discard the vine.

- Let some of the liquid flow into the palm of the hand and observe it. If the liquid does not change color, IP can now taste it. If it tastes like water or has a woody or sweet taste, it should be safe for drinking. Liquid with a sour or bitter taste should be avoided.

16.4.2.1. Water Trapped within a Vine

Water trapped within a vine is easily obtained by cutting out a section of the vine. The vine should first be cut high above the ground and then near the ground. This will provide a long length of vine and, in addition, will tend to hide evidence of the cuts which is especially important if the IP are in an evasion situation. When drinking from the vine, it should not touch the mouth as the bark may contain irritants which could affect the lips and mouth (Figure 16-5). The pores in the upper end of the section of vine may reclose, stopping the flow of water. If this occurs, cut off the end of the vine opposite the drinking end. This will reopen the pores allowing the water to flow.

Figure 16-5 Water Vines and Bamboo

16.4.3. Water from the Rattan Palm

Water from the rattan palm and spiny bamboo may be obtained in the same manner as from vines. It is not necessary to test the water if positive identification of the plant can be made. The slender stem (runner) of the rattan palm is an excellent water source. The joints are overlapping in appearance, as if one section is fitted inside the next.

16.4.4. Water Trapped within Sections of Green Bamboo

Water may be trapped within sections of green bamboo. To determine if water is trapped within a section of bamboo, it should be shaken. If it contains water, a sloshing sound can be heard. An opening may be made in the section by making two 45-degree angle cuts, both on the same side of the section, and prying loose a piece of the section wall. The end of the section may be cut off and the water drunk or poured from the open end. The inside of the bamboo should be examined before consuming the water. If the inside walls are clean and white, the water will be safe to drink. If there are brown or black spots, fungus growth, or any discoloration, the water should be purified before consumption. Sometimes water can also be obtained by cutting the top off certain types of green bamboo, bending it over, and staking it to the ground (Figure 16-5). A water container should be placed under it to catch the dripping water. This method has also proven effective on some vines and the rattan palm.

16.4.5. Water from Banana Plants

Water can also be obtained from banana plants. Cut a banana plant down into a long section which can be easily handled. The section is taken apart by slitting from one end to the other and pulling off the layers one at a time. A strip three inches wide, the length of the section, and just deep enough to expose the cells should be removed from the convex side. This section is folded toward the convex side to force the water from the cells of the plant. The layer must be squeezed gently to avoid forcing out any tannin (an astringent which has the same effect as alum) into the water. Another technique for obtaining water from the banana plant is by making a "banana-well." This is done by making a bowl out of the plant stump, fairly close to the ground, by cutting out and removing the inner section of the stump (Figure 16-6). Water which first enters the bowl may contain a concentration of tannin. A leaf from the banana or other plant should be placed over the bowl while it is filling to prevent contamination by insects, etc.

Figure 16-6 Water from Banana Plant

16.4.6. Water Trees

Water trees can also be a valuable source of water in some jungles. It is important to properly identify the tree as the sap of some trees can be very dangerous. Water trees can be identified by their blotched bark which is fairly thin and smooth. The leaves are large, leathery, fuzzy, and evergreen, and may grow as large as eight or nine inches. The trunks may have short outgrowths with fig-like fruit on them or long tendrils with round fruit comprised of corn kernel-shaped nuggets.

- In a non-tactical situation, the tree can be tapped in the same manner as a rubber tree, with either a diagonal slash or a "V." When the bark is cut into, it will exude a white sap which if ingested causes temporary irritation of the urinary tract. This sap dries up quite rapidly and can easily be removed. The cut should be continued into the tree with a spigot (bamboo, knife, etc.) at the bottom of the tap to direct the water into a container. The water flows from the leaves back into the roots after sundown, so water can be procured from this source only after sundown or on overcast (cloudy) days.

- If IP are in a tactical situation, they can obtain water from the tree and still conceal the procurement location. If the long tendrils are growing thickly, they can be separated and a hole bored into the tree. The white sap should be scraped off and a spigot placed below the tap with a water container to catch the water. Moving the tendrils back into place will conceal the container. Instead of boring into the tree, a couple of tendrils can be cut off or snapped off if no knife is available. The white sap should be allowed to dry and then be removed. The ends of the tendrils should be placed in a water container and the container concealed (Figure 16-7).

Figure 16-7 Water Tree

16.4.7. Coconut Water

Coconuts contain a refreshing fluid. Where coconuts are available, they may be used as a water source. The fluid from a mature coconut contains oil, which when consumed in excess can cause diarrhea. There is little problem if used in moderation or with a meal and not on an empty stomach. Green unripe coconuts about the size of a grapefruit are the best for use because the fluid can be taken in large quantities without harmful effects. There are more fluids and less oil so there is less possibility of diarrhea.

16.4.8. Water from Mud

Water can also be obtained from liquid mud. Mud can be filtered through a piece of cloth. Water taken by this method must be purified.

16.4.9. Collecting Water from a Tree

Rainwater can be collected from a tree by wrapping a cloth around a slanted tree and arranging the bottom end of the cloth to drip into a container (Figure 16-8).

Figure 16-8 Collecting Water from Slanted Tree

16.5. Water in Dry Areas

Some of the ways to find water in this environment have been explored, such as locating a concave bend in a dry riverbed and digging for water (Figure 16-9). If there is any water within a few feet of the surface, the sand will become slightly damp. Dig until water is obtained.

Figure 16-9 Dry Stream Bed

16.5.1. Dew in Deserts

Some deserts become humid at night. The humidity may collect in the form of dew. This dew can then be collected by digging a shallow basin in the ground about three feet in diameter and lining it with a piece of canvas, plastic, or other suitable material. A pyramid of stones taken from a minimum of one foot below the surface should then be built in this basin. Dew will collect on and between the stones and trickle down onto the lining material where it can be collected and placed in a container.

16.5.2. Roots Near the Surface

Plants and trees having roots near the surface may be a source of water in dry areas. Water trees of dry Australia are a source of water, as their roots run out 40 to 80 feet at a depth of two to nine inches under the surface. IP may obtain water from these roots by locating a root four to five feet from the trunk and cutting the root into two- or three-foot lengths. The bark can then be peeled off and the liquid from each section of root drained into a container. The liquid can also be sucked out. The trees growing in hollows or depressions will have the most water in their roots. Roots that are one to two inches thick are an ideal size. Water can be carried in these roots by plugging one end with clay.

16.5.3. Cactus as a Water Source

Cactus-like or succulent plants may be sources of water, but IP should recall that no plants should be used for water procurement which has a milky sap. The barrel cactus of the United States provides a water source. To obtain it, cut off the top of the plant. The pulpy inside portions of the plant should then be mashed to form a watery pulp. Water may ooze out and collect in the bowl; if not, the pulp may be squeezed through a cloth directly into the mouth.

16.5.4. Vegetation Bag

A vegetation bag can be constructed by cutting foliage from trees or herbaceous plants, sealing it in a large clear plastic bag, and allowing the heat of the Sun to extract the fluids contained within. A large, heavy-duty clear plastic bag should be used. The bag should be filled with about one cubic yard of foliage, sealed, and exposed to the Sun. The average yield for one bag tested was 320 ml/bag per five-hour day. This method is simple to set up. The vegetation bag method of water procurement does have one primary drawback. The water produced is normally bitter to taste, caused by biological breakdown of the leaves as they lay in the water produced and superheated in the moist "hothouse" environment. This method can be readily used in a survival situation, but before the water produced by certain vegetation is consumed, it should undergo the taste test. This is to guard against ingestion of cyanide-producing substances and other harmful toxins, such as plant alkaloids (see Figure 16-10).

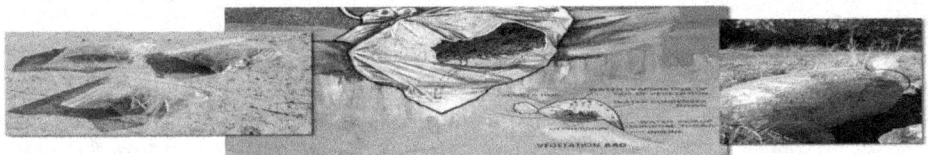

Figure 16-10 Vegetation Bag

16.5.5. Transpiration Bag

The transpiration bag method is simple to use and therefore enhances your chance of survival. This method is the vegetation bag process taken one step further. A large plastic bag is placed over a living limb of a medium-size tree or large shrub. The bag opening is sealed at the branch, and the limb is then tied down to allow collected water to flow to the corner of the bag. For a diagram of the water transpiration method, see Figure 16-11.

Figure 16-11 Transpiration Bag

16.5.5.1. Amount of Transpired Water

The amount of water yielded by this method will depend on the species of trees and shrubs available. Transpired water has a variety of tastes depending on whether or not the vegetation species is allowed to contact the water.

16.5.5.2. Minimal Effort Expended

The effort expended in setting up water transpiration collectors is minimal. It takes about five minutes' work and requires no special skills once the method has been described or demonstrated. Collecting the water necessitate that the IP dismantling the plastic bag at the end of the day, draining the contents and setting it up again the following day. The same branch may be reused (in some cases with almost similar yields); however, as a general rule, when vegetation abounds, a new branch should be used each day.

16.5.5.3. Water Transpiration as a Superior Method

Without a doubt, the water transpiration bag method surpasses other methods (such as vegetation bag, cutting roots, barrel cactus) in yield; ease of assembly, and in most cases, taste. The benefits of having a simple plastic bag can't be over-emphasized. As a water procurer, in dry, semi-dry, or desert environments where low woodlands predominate, it can be used for transpiration; in scrubland, steppes, or treeless plains, as a vegetation bag. Up to three large, heavy-duty bags may be needed to sustain one IP in certain situations.

16.6. Preparation of Water for Consumption

The following are ways IP can possibly determine the presence of harmful agents in the water:

- Strong odors, foam, or bubbles in the water.
- Discoloration or turbid (muddy with sediment).
- Lake water found in desert areas may be salty if the lake has been without an outlet for an extended period of time. Magnesium or alkali salts may produce a laxative effect; if not too strong, it is drinkable.
- If the water gags the IP or causes gastric disturbances, drinking should be discontinued.
- The lack of healthy green plants growing around any water source.

16.6.1. Rendered Potable

Because of IP's potential aversion to water from natural sources, it should be rendered as potable as possible through filtration. Filtration only removes the solid particles from water - it does not purify it. One simple and quick way of filtering is to dig a sediment hole or seepage basin along a water source and allow the soil to filter the water. The seepage hole should be covered while not in use. Another way is to construct a filter - layers of parachute material stretched across a tripod (Figure 16-13). Charcoal is used to eliminate bad odors and foreign materials from the water. Activated charcoal (obtained from freshly burned wood is used to filter the water). If a solid container is available for making a filter, use layers of fine-to-coarse sand and gravel along with charcoal and grass.

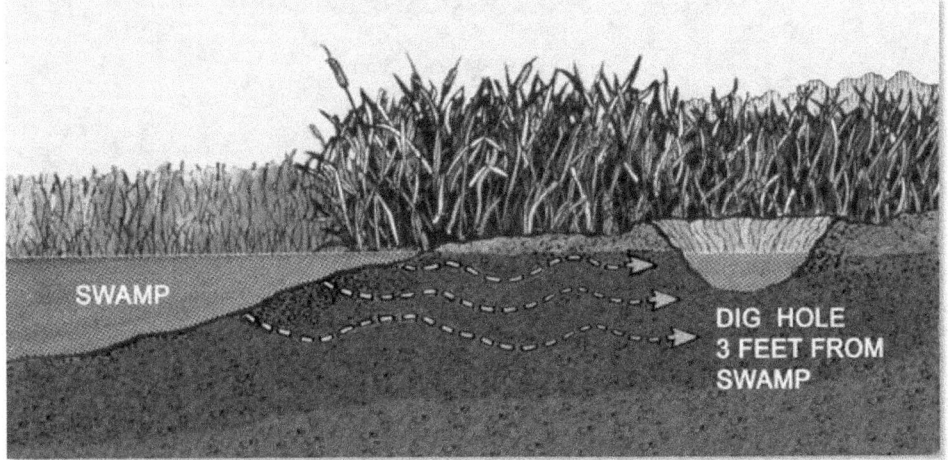

Figure 16-12 Sediment Hole

Figure 16-13 Water Filter

16.6.2. Situational Water Purification

Methods of water purification will be dictated by the situation. Some examples include:

16.6.2.1. Boiling Water

Boil for at least one minute at sea level. If above 8,000 feet boil for at least three minutes.

16.6.2.2. Purification Tablets

To use purification tablets, IP should follow instructions on the bottle. If purification tablets are not available, plant sources or non-stagnant, running water obtained from a location upstream from habitation should be consumed.

16.6.3. Solar Purification

Solar purification, also known as "safe drinking water in six hours" (SODIS), is a method of purifying water using only sunlight and plastic polyethylene terephthalate (PET) bottles. Colorless, transparent PET water or soda pop bottles (two quart or smaller size) with few surface scratches are chosen for use (Polycarbonate bottles block all UVA and UVB rays and should not be used.) Remove labels and wash bottles before the first use. Fill the bottles with water from the contaminated source. To improve oxygen saturation, bottles can be filled three-quarters, shaken for 20 seconds (with the cap on), then filled completely and recapped. Very cloudy water must be filtered prior to exposure to the sunlight. Filled bottles are then exposed to the sun.

16.6.3.1. SODIS

At a water temperature of about 90°F (33°C), it takes at least six hours for SODIS to be efficient. If the water temperatures rise above 122°F (50°C), the purification process is three times faster.

16.6.4. Heating Bottles

Bottles will heat faster and to higher temperatures if they are placed on a sloped sun-facing corrugated metal roof or black material as compared to thatched roofs. The treated water can be consumed directly from the bottle.

Suggested Treatment Schedule	
Weather Conditions at 90°F	Minimum Treatment Duration
sunny (less than 50% cloud cover)	6 hours
cloudy (50-100% cloudy, little to no rain)	2 days

16.6.5. Storing Water for Later Consumption

After water is found and purified, IP may wish to store it for later consumption. The following make good containers, but may need to be washed or rinsed out before using:

- Water bag and Canteen.
- Tarp, sheet of plastic, or an EVC.
- Plastic Bag
- Prophylactic (placed inside a sock for protection of bladder).
- Segment of bamboo.
- Birch bark and pitch canteen.
- LPU bladder.
- Hood from anti-exposure suit.

Chapter 17
FOOD

17. Food

Except for the water they drink and the oxygen they breathe, IP must meet their physical needs through the intake of food. This chapter will explore the relationship of proper nutrition to physical and mental efficiency. It is extremely important that IP maintain a proper diet at all times. A nutritionally sound body stands a much better chance of surviving. Improper diet over a long period of time may lead to a lack of stamina, slower reactions, lower resistance to illness, and reduced mental alertness, all of which can cost IP their lives in a survival situation. Knowledge of the body's nutritional requirements will help IP select foods to supplement their rations.

17.1. Nutrition

IP expend much more energy during isolating situations (especially while evading or traveling) than they would in the course of their normal everyday jobs and life. Basal metabolism is the amount of energy expended by the body when it is in a resting state. The rate of basal metabolism will vary slightly with regard to the sex, age, weight, height, and race of a person. The basic energy expended or number of calories consumed by the hour will change as a person's activity level changes. For example, a person who is simply sitting in a warm shelter, may consume anywhere from 20 to 100 calories per hour, while that same person in evasion situation maneuvering through thick undergrowth with a heavy pack, would expand a greater amount of energy. In an isolating event, proper food can make the difference between success and failure.

17.1.1. Carbohydrates, Fats, and Proteins

The three major categories of food are carbohydrates, fats, and proteins. Vitamins and minerals are also important as they keep certain essential body processes in good working order. It is also necessary for an IP to maintain proper water and salt levels in their bodies, as they aid in the body's ability to break down and use food, as well as preventing certain heat and cold disorders.

17.1.1.1. Carbohydrates

Carbohydrates are composed of very simple molecules which are easily digested. Carbohydrates lose little of their energy to the process of digestion and are therefore efficient energy suppliers. Because carbohydrates supply easily used energy, many nutritionists recommend that, if possible, individuals should try to use them for up to half of their calorie intake. Obviously, since an IP in an isolating event is likely to expend more than the individual faced with a normal day, they should increase their intact of carbohydrates to 60-70 percent of their total calories from carbohydrates. It is important that IPs, if possible, eat before, during, and after the isolating event to make sure that they will have enough energy in the muscles for the next obstacle or task they must overcome or accomplish. Examples of carbohydrates include: starches, sugars, and cellulose. These can be found in fruits, vegetables, candy, milk, cereals, legumes, and baked goods. Cellulose cannot be digested by humans, but it does provide needed roughage for the diet.

17.1.1.2. Fats

Fats are more complex than carbohydrates. The energy contained in fats is released slower than the energy in carbohydrates. As a result, it is a longer lasting form of energy. Fats supply certain fat-soluble vitamins. Sources of these fats and vitamins include butter, cheese, oils, nuts, egg yolks, margarine, and animal fats. If IP eat fats before sleeping, they will be warmer during sleep. If fats aren't included in the diet of IP, they can become run down and irritable. This can lead to both physical and psychological breakdown.

17.1.1.3. Digestive Process

The digestive process breaks protein down into various amino acids. These amino acids are formed into new body tissue protein, such as muscles. Some protein gives the body the exact amino acids required to rebuild itself. These proteins are referred to as "complete." Protein that lacks one or more of these essential amino acids is referred to as "incomplete." Incomplete protein examples include cheese, milk, cereal grains, and legumes. Incomplete protein when eaten in combination with milk and beans for example, can supply an assortment of amino acids needed by the body. Some complete protein is found in fish, meat, poultry, and blood. No matter which type of protein is consumed, it will contain the most complex molecules of any food type listed.

17.1.1.4. Protein

If possible, a recommended daily allowance of two and a half to three ounces complete protein should be consumed by each IP each day. If only incomplete protein is available, two, three, or even four types of foods may need to be eaten in combination so enough amino acids are combined to form complete protein.

17.1.1.5. Amino Acids

If amino acids are introduced into the body in a large amount and some of them are not used for the rebuilding of muscle, they are changed into fuel or stored in the body as fat. Because protein contains the more complex molecules, over fats and carbohydrates, protein supplies energy after other forms of energy have been used up. A lack of protein causes malnutrition, skin and hair disorders, as well as muscle atrophy.

17.1.2. Vitamins

Vitamins occur in small quantities in many foods, and are essential for normal growth and health. Their chief function is to regulate the body processes. Vitamins can generally be placed into two groups: fat-soluble and water-soluble. The body only stores slight amounts of the water-soluble type. During a long isolating event where a routinely balanced diet is not available, IP must overcome food aversions and eat as much of a variety of vitamin-rich foods as possible. Often one or more of the four basic food groups (meat, fish, poultry, vegetables and fruits, grain and cereal, milk and milk products) are not available in the form of familiar foods, and vitamin deficiencies such as beriberi or scurvy result. If IP can overcome aversions to local foods high in vitamins, these diseases as well as signs and symptoms such as depression and irritability can be warded off.

17.1.3. Minerals

Adequate minerals can also be provided by a balanced diet. Minerals build and (or) repair the skeletal system and regulate normal body functions. Minerals needed by the body include iodine, calcium, iron, and salt. A lack of minerals can cause problems with muscle coordination, nerves, water retention, and the ability to form or maintain healthy red blood cells.

17.1.4. Required Calories

For IP to maintain their efficiency due to the extremes of an isolating event and work involved, the following number of calories per day is recommended. These figures will vary because of individual differences in basal metabolism, weight, etc. During warm weather, IP should consume anywhere from 3,000 to 5,000 calories per day. In cold weather, the calorie intake should rise from 4,000 to 6,000 calories per day. A familiarity with the calorie and fat amounts in foods is important for IP to meet their nutritional needs. For example, it would take quite a few mussels and dandelion greens to meet those requirements. IP should attempt to be familiar enough with foods that they can select or find foods that provide a high calorie intake (Figure 17-1).

Food/Amount	Calories	Grams of Fat	Grams of Protein	Grams of Carbohydrates
Pheasant, roasted (3.5 oz)	133	3.25g	24.4g	
Venison, roasted (3.5 oz)	157	7g	21.8g	
Rabbit (3.5 oz)	144	2.3g		
Stinging Nettles; cooked, boiled, drained (1 cup)	37			6g
Quail Egg, cooked (1 egg)	14	1g	1g	
Duck Egg, cooked (1 egg)	130	9.6g	9g	1g
Breadfruit, raw (1 cup)	227	1g	2g	60g
Cattail, Narrow Leaf Shoots, raw (3.5 oz)	24.5			4g
Mixed Species Trout, cooked (1 fillet/79g)	117	5g	16g	
Dandelion greens, cooked, boiled, drained (1 cup)	35	1g	2g	7g
Prickly Pears, cooked (1 cup)	61	1g	1g	14g
Crayfish, raw or cooked (3.5 oz)	65-70	1g	14g	
Snail, raw (3.5 oz)	87.5		17.5g	4g
Fiddlehead Ferns, raw (3.5 oz)	35		3.5g	7g
Crickets, dried (100g)	562	5.5g	6.7g	
Hairless Caterpillars, dried (100g)	3701		28.2g	
Ant Eggs, raw (3.5 oz)	126	3.5g	17.5g	4g

Figure 17-1 Food and Caloric Chart

17.1.4.1. Rationing Food

IP should also be familiar with the number of calories supplied by the food in issued rations. In most situations, rations will have to be supplemented with other foods procured by IP. If possible, IP should limit their activities to save energy. Rationing food is a good idea because IP never know when their survival ordeal will end. They should eat when they can, keeping in mind that they should maintain at least a minimum calorie intake to satisfy their basic activity needs.

17.1.4.2. Caloric and Fat Values

Caloric and fat values of selected foods are shown in the chart, and unless otherwise specified, the foods listed are raw. Depending on how IP cook the food, the usable food value can be increased or decreased.

17.2. Food

IP should be able to find something to eat wherever they are. One of the best places to find food is along the seacoast, between the high and low watermark. Other likely spots are the areas between the beach and a coral reef: Marshes, mud flats, or mangrove swamps where a river flows into the ocean or into a larger river; riverbanks; inland waterholes, shores of ponds and lakes, margins of forests, natural meadows, protected mountain slopes, and abandoned cultivated fields are also good food sources.

17.2.1. Rations

Rations have been developed especially to provide some of the proper sustenance needed. When eaten as directed on the package, they will keep IP relatively nourished. If enough other food can be found, rations should be conserved for emergency use.

17.2.2. Environmental Conditions

Consideration must be given to available food and water and how long the isolating event may last. Environmental conditions must also be considered. If the IP are in a cold environment, more of the proper food will be required to provide necessary body heat. Available food must be rationed based on the estimated time which will elapse before recovery and being able to supplement issued rations with natural foods. If IP should seek help, each traveler should be given twice as much food as those staying behind. The IP resting at the encampment and those seeking help will stay in about the same physical condition for about the same length of time.

17.2.2.1. Dry, Starchy, and Highly Salted Foods

Avoid dry, starchy, and highly salted foods and meats if less than a quart of water is available per day. Keep in mind that eating increases thirst. For water conservation, the best foods to eat are those with high carbohydrate content, such as hard candy and fruit. All physical activity requires additional food and water. When physical activity is being performed, the IP must increase food and water consumption to maintain physical efficiency. If food is available, it is acceptable to eat several times throughout the day. It is preferable to have at least two meals a day, with one meal being hot. Cooking usually makes food safer, more digestible, and palatable. The time spent cooking will provide a good rest period. On the other hand, some foods are not palatable unless eaten raw.

17.2.3. Native Foods

Native foods may be more appetizing if they are eaten by themselves. In many countries, vegetables are often contaminated by feces (animal or human) which the natives use as fertilizer. Dysentery is transmitted in this way. If possible, the IP should try to select and prepare their own meals. If necessary to avoid offending the native peoples, indicate that religious beliefs or taboos require self-preparation of food.

17.2.4. Food Aversions

Learn to overcome food aversions. Foods that may not look good are often a part of the local's regular diet. Wild foods are high in mineral and vitamin content. With a few exceptions, all animals are edible when freshly killed. Avoid strange looking fish and fish with flesh that remains indented when depressed as it is possibly spoiled and should not be eaten. With proper knowledge (pre-mission study, EVCs, etc.) and the ability to overcome food aversions, IP can eat and sustain themselves in strange or hostile environment.

17.3. Animal Food

Animal food gives the most food value per pound. Anything that creeps, crawls, swims, or flies is a possible source of food. People eat grasshoppers, hairless caterpillars, wood-boring beetle larvae and pupae, ant eggs, spider bodies, and termites. Such insects are high in fat, carbohydrates, and should be cooked until dried. Everyone has probably eaten insects contained in flour, cornmeal, rice, beans, fruits, and greens in their daily foods.

17.3.1. Hunting

To become successful in hunting, IP must go through a behavioral change and reorganize personal priorities. This means the one and only goal for the present is to kill an animal to eat. To kill this animal, the IP must mentally become a predator. The IP must be willing and prepared to forgo stress in order to hunt down and kill an animal. Because of the type of weapons IP are likely to have, it will be necessary to get very close to the animal to immobilize or kill it. This is going to require large amounts of stealth and cunning. Knowledge of the animal being hunted is also very important. If in an unfamiliar area, IP may learn much about the animal life of the area by studying signs such as trails, droppings, and bedding areas.

17.3.2. Animal Characteristics

IP should understand the general characteristics of the animals in the area. The size of the tracks will give a good idea of the size of the animal. The depth of the tracks will indicate the weight of the animal. Animal droppings can tell the IP much. For example, if it is still warm or slimy, it was made very recently; if there is a large amount scattered around the area, it could well be a feeding or bedding area. The droppings may also indicate what the animal feeds on. Carnivores often have hair and bone in their droppings. Herbivores have coarse portions of the plants they have eaten. Many animals mark their territory by urinating or scraping areas on the ground or trees. These signs could indicate good trap or ambush sites. By following those signs, (tracks, droppings, etc.), IP may learn about the feeding, watering, and resting areas of the animals they are hunting. Well worn trails will often lead to the animal's watering place. Having made a careful study of all the signs of the animal, the IP is in a much better position to capture it, whether electing to stalk, trap, snare, or lie in wait to shoot it.

17.3.3. Success of the Hunt

If IP elect to hunt, there are some basic techniques which will be helpful and improve chances of success. Wild animals rely entirely upon their senses for their preservation. These senses are smell, vision, and hearing. Humans have lost the keenness of some of their senses like smelling, hearing, etc. To overcome this loss, humans have the ability to reason. As an example, some animals have a fantastic sense of smell, but this can be overcome by approaching the quarry from a downwind direction. The best times to hunt are at dawn and dusk as animals are either leaving or returning to their bedding areas. Both diurnal and nocturnal animals are active at this time. There are five basic methods of hunting:

17.3.3.1. Ambush

Ambush is the best method for inexperienced hunters as it involves less skill. The main principle of this method is to wait along a well-used game trail, until the quarry approaches within killing range. Morning and evening are usually the best times to still hunt. Care should be taken not to disturb the area; always wait downwind. Patience and self-control are necessary to remain motionless for long periods of time.

17.3.3.2. Stalking

Stalking refers to the stealthily approach toward game. This method is normally used when an animal has been sighted and the IP proceeds to close the distance using all available cover. Stalking must be done slowly so that minimum noise is made. Quick movement is easily detected by the animal. Always approach from the downwind side and move when the animal's head is down eating, drinking, or looking in another direction. The same techniques are used in blind stalking as in regular stalking, the main difference being that the IP is stalking a position where the animal is expected to be while the animal is not in sight.

17.3.3.3. Tracking

Tracking is very difficult unless conditions are ideal. This method involves reading all of the signs left behind by the animal, interpreting what the animal is doing, and how it can best be killed. The most common signs are trails, beds, urine, droppings, blood, tracks, and feeding signs.

17.3.3.4. Funneling

Some wild animals can be scared or driven in a direction where other IP or traps have been set. This method is normally used where the animals can be funneled; a valley or canyon is a good place to make a drive. More than one person is usually necessary to make a drive.

17.3.3.5. Imitating an Injured Animal

Small predators may be called in by imitating an injured animal. Ducks and geese can be attracted by imitating their feeding calls. These noises can be made by sucking on the hand, blowing on a blade of grass or paper, sucking the lip, or using specially designed devices. IP should not call animals unless they know what they are doing as strange noises may "spook" the animal.

17.3.4. Weapons

It is difficult to kill animals of any size without using some type of tool or weapon. As our technology has increased in complexity, so have our killing tools. If a firearm, spear, or club, etc, is available, a basic knowledge of shooting and hunting techniques is necessary.

17.3.4.1. Firearms

One of the limiting factors in the use of firearms is the amount of ammunition on hand. Therefore, IP cannot afford to waste ammunition on moving game or game which is beyond the effective range of the firearm being used. Wait for a pause in the animal's motions. The shot must be placed in a vital area with any firearm. Aim for the brain, spine, lungs, or heart. A hit in these areas is usually fatal.

17.3.4.2. Tracking Wounded Animals

A full-jacketed bullet often won't immediately down a larger animal hit in a vital area such as the lungs or heart. The alternative to losing the animal is tracking it to where it falls. Often it is better to wait a while before pursuing the animal. If not pursued, the animal may lie down and stiffen or perhaps bleed to death. Follow the blood trail to where the animal has gone down and kill it if it is still alive. Even though ammunition might be limited, small animals may be more productive than large ones. Although they present smaller targets and have less meat, they are less wary, more numerous, and travel less distance to escape if wounded. A large amount of edible meat on small animals can be destroyed from a bullet wound. On rodents, most of the meat is on the hindquarters and front quarters; on birds, most of the meat is in the breast and legs. IP should try to hit a vital spot that spoils the least amount of meat.

17.3.4.3. Hunting at Night

Night hunting is usually best, since most animals move at night. A flashlight or torch may be used to shine in the animal's eyes. It will be partly blinded by the light and IP can get much closer than in the daytime. If a gun is not available, the animal can be killed with either a club or a sharpened stick used as a spear.

17.3.4.4. Danger of Wounded Animals

Remember that large animals, when wounded, cornered, or with their young, can be dangerous. Be sure the animal is dead, not just wounded, unconscious, or playing "possum." Animals usually die with their eyes open and glazed-over. Poke all "dead" animals in the eye with a long sharp stick before approaching them.

17.3.5. Other Food Sources

Small freshwater turtles can often be found sunning themselves along rivers and lakeshores. If they dash into shallow water, they can still be procured with nets, clubs, etc. IP should be careful of the mouth and claws. Frogs and snakes also sun and feed along streams. Use both hands to catch a frog - one to attract it and keep it busy while grabbing it with the other. Bright cloth on a fishhook also works. All snakes are edible and can be killed with a long stick but the IP should consider the potential danger of being bitten verses needing food. Both marine and dry-land lizards are edible. A noose, small fishhook baited with a bright cloth lure, slingshot, or club can be used. A slingshot can be made with a forked stick and the elastic from the parachute pack or surgical tubing found in some survival/medical kits (Figure 17-2). With practice, the slingshot can be very effective for killing small animals.

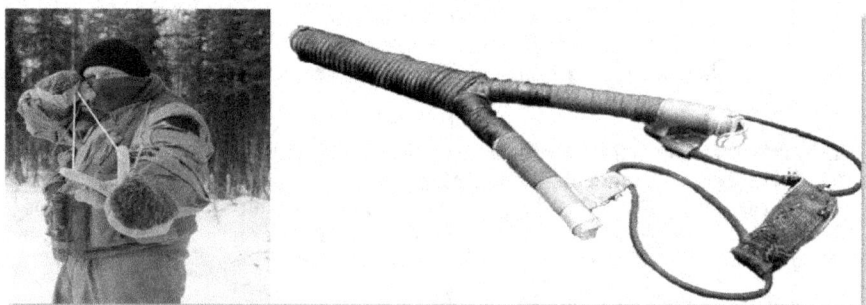

Figure 17-2 Slingshot

17.3.6. Snaring and Trapping

Snaring and trapping animals are ways IP can procure animal food to supplement issued rations. Since small animals are usually more abundant than large animals, they will probably be the IP's main source of food. Snares should be set out on a 15: 1 ratio; 15 snares should be set out for every one animal expected to be caught.

17.3.6.1. Advantages of Traps and Snares

Using traps and snares are more advantageous than going out on foot and physically hunting the animal. The most important advantage is that traps work 24 hours a day without assistance from the hunter. A large area can be effectively strapped with the possibility of catching many animals within the same period of time. IP (generally) use much less energy maintaining a trap line than is used by hunting. This means less food is required because less energy is expended.

17.3.6.2. Trap and Snare Location

The traps or snares should be set in areas where the game is known to live or travel. Look for signs such as tracks, droppings, feeding signs, or actual sightings of the animal. If snares are used, they should be set up to catch the animal around the neck. Therefore, the loop must allow the head to pass through but not the body. Loops will vary in size from one animal to another. When placing snares, try to find a narrow area of the game trail where the animal has no choice but to enter the loop. If a narrow area cannot be found, brush or other obstacles can be arranged to funnel the animal into the snare (Figure 17-3). Do not overdo the funneling; use as little as possible. Avoid disturbing the natural surroundings if possible. Do not walk on game trails, but approach 90 degrees to the trail, set the snare, and back away. Snares may also be set over holes or burrows. All snares and traps should be set during the midday because most animals are nocturnal in nature. Check snares and traps twice daily. If possible, check in the morning after the sunrises and prior to sunset. Checking the snares should be made from a distance so any animals moving at the time of checking will not be disturbed or frightened away.

Figure 17-3 Funneling

17.3.6.3. Three ways to Immobilize or Trap Animals

1. A free-sliding noose, when tightened around the neck, will restrict circulation of air and blood. The materials should be strong enough to hold the animal. For example, suspension line, string, wire, cable, or rawhide.

2. Mangle traps use a weight which is suspended over the animal's trail or over bait. When

the animal trips the trigger, the weight (log) will descend and mangle the animal (Figure 17-4).

3. Any means of holding the animal and detaining its progress would be considered a hold-type trap.

Figure 17-4 Mangle

17.3.6.3.1. The Apache Foot Snare and Box Trap

The apache foot snare is an example of a hold-type trap. It is used for large browsers and grazers like deer (Figure 17-5). It should be located along game trails where an obstruction, such as a log blocks the trail. When animals jump over this obstruction, a very shallow depression is formed where their hooves land. The apache foot snare should be placed at this depression. The box trap for birds is another example of hold-type traps (Figure 17-6).

Figure 17-5 Apache Foot Snare

Figure 17-6 Box Trap

17.3.6.3.2. The Simple Loop

The simple loop is the quickest snare to construct. All snares and traps should be simple in construction with as few moving parts as possible. This loop can be constructed from any type of bare wire, suspension line, inner core, vines, long strips of green bark, clothing strips or belt, and any other material that will not break under the strain of holding the animal. If wire is being used for snares, a figure "8" or locking loop should be used (Figure 17-7). Once tightened around the animal, the wire is locked into place by the figure "8" which prevents the loop from opening again. A simple loop snare is generally placed in the opening of a den, with the end of the snare anchored to a stake or similar object. The simple loop snare can also be used when making a squirrel pole (Figure 17-8) or with some types of trigger devices.

Figure 17-7 Locking Loop and Setting Noose

Figure 17-8 Squirrel Pole

17.3.7. Triggers

Triggers may be used with traps. The purpose of the trigger is to set the device in motion, which will eventually strangle, mangle, or hold the animal. There are many triggers. Some of the more common ones include:

17.3.7.1. Two-Pin Toggle with Counterweight

Use a two-pin toggle with a counterweight for small to medium sized animals, and lift out of the reach of predators (Figure 17-9).

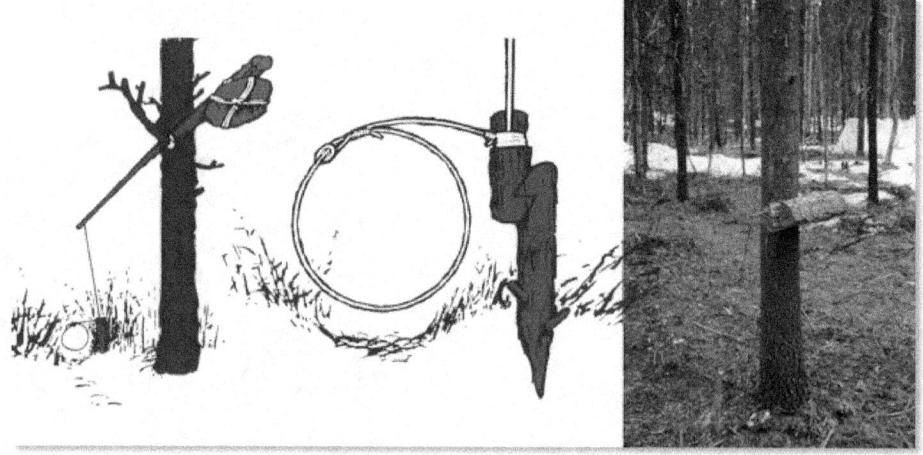

Figure 17-9 Two-Pin Toggle

17.3.7.2. Figure "H" Wire Snare

Figure "H" with wire snare for small mammals and rodents (Figure 17-10).

Figure 17-10 Figure "H"

17.3.7.3. Canadian Ace

Canadian ace for predators such as bobcat, coyote, etc. (Figure 17-11).

Figure 17-11 Canadian Ace

17.3.7.4. Three-Pin Toggle with Deadfall

The Three-pin toggle with deadfall is used for medium to large animals (Figure 17-12). Medium and large animals can be captured using deadfalls, but this type of trap is recommended only when big game exists in large quantities to justify the great expense of time and effort spent in constructing the trap.

Figure 17-12 Three-Pin Toggle

17.3.7.5. Twitch-up Snare

The twitch-up snare incorporates the simple loop for small animals (Figure 17-13). When the animal is caught, the sapling jerks it up into the air and keeps the carcass out of the reach of predators. This type of snare may not work well in cold climates, since the bent sapling will freeze in position and not spring up when released. A possible cold weather/environment modification to this trap is to use a counterweight instead of bent sapling.

Figure 17-13 Twitch Up

17.3.7.6. Long Forked Stick

A long forked stick can be used as a twist stick to procure ground squirrels, rabbits, etc. A den that has signs of activity must be located. Using the long forked stick, probe the hole with the forked end until something soft is felt. Twisting the stick will entangle the animal's hide in the stick and the animal can be extracted (Figure 17-14).

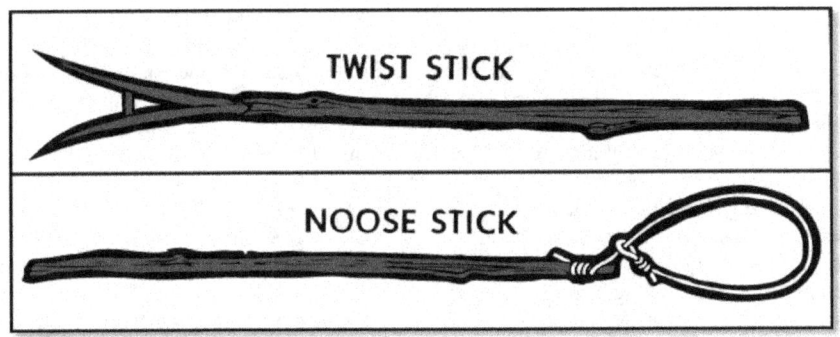

Figure 17-14 Twist Stick and Noose Stick

17.3.8. Ways to Catch Birds

There are several ways to catch birds. Birds can be caught with a gill net. The net should be set up at night vertically to the ground in some natural flyway, such as an opening in dense foliage. A small gill net on a wooden frame with a disjointed stick for a trigger can also be used. A gill net can be made by using inner core from parachute suspension line.

17.3.8.1. Baited Fishhooks

Birds can be caught on baited fishhooks (Figure 17-15) or simple slipping loop snares. A bird's nest can be a source of food. All bird eggs are edible when fresh. Large wading birds such as cranes and herons often nest in mangrove swamps or in high trees near water.

Figure 17-15 Baited Fishhook

17.3.8.2. Loss of Flight

During molting season, birds cannot fly because of the loss of their "flight" feathers therefore they can be captured by clubbing or netting.

17.3.8.3. Ojibwa Snare

Birds can be also caught in an Ojibwa snare. This snare is made by cutting a one or two inch thick sapling at a height of four and a half to five feet above the ground (Figure 17-16). A springy branch is whittled flat at the butt end and a rectangular hole is cut through the flattened end. One end of a stick, 15 inches long, is then whittled to fit slightly loose in the hole and the top corner of the whittled end is rounded off so the stick will easily drop away from the hole. The branch is then tied by its butt end to the top of the sapling. A length of inner core from suspension line is tied to the bottom end of the branch and the branch is bent into a bow with the line passing through the hole in the butt end. A knot is tied in the line and the 15-inch stick is placed in the hole to lock the line in place (just behind the knot). An eight inch loop is made at the end of the line and laid out on the 15-inch stick (spread out as well as possible). A piece of bait is placed on top of the sapling, and when a bird comes to settle on the 15-inch stick, the stick drops from the hole causing the loop to tighten around the bird's legs.

Figure 17-16 Ojibwa Bird Snare

17.3.8.4. Loop Snare

When many birds frequent a particular type of bush, some simple loop snares may be set up throughout the bush. Make the snares as large as necessary for the particular type of birds that come to perch, feed, or roost there.

- In wild, wooded areas, many larger species of birds such as spruce grouse, which has merited the name "fools hen", and ptarmigan may be approached and killed with a stick with little trouble. It often sits on the lower branches of trees and may be caught with a long stick with a loop at the end (Figure 17-14).

17.3.9. Insects

IP may turn to insects as a food source and may overcome food aversions as a result. When food is limited and insects are available, they can become a valuable food source. In some places, locusts and grasshoppers, cicadas, and crickets are eaten regularly; occasionally termites, ants, and a few species of stonefly larvae are consumed. Big beetles such as the Goliath Beetle of Africa, the Giant Water Beetles, and the big Long Horns are relished the world over. Clusters, like those of the Snipe fly Atherix (that overhang the water), and the windrows of Brine fly puparia are eaten. Aquatic water bugs of Mexico are grown especially for food. All stages of growth can be eaten, including the eggs but, the large insects must be cooked to kill internal parasites.

17.3.9.1. Termites and White Ants

Termites and white ants are also an important food source. Strangely enough, these are closely related to cockroaches. The reason they are eaten so extensively in Africa is the fact that they occur in vast numbers and are easily collected both from their nests and during flight.

17.3.9.2. Carpenter Ants

Many Native Americans made a habit of eating the large carpenter ants that are sometimes pests in many households (Figure 17-17). These ants were eaten both raw and cooked. Although the practice of eating them has not entirely disappeared, ants do not form an essential part of the diet of any of the inhabitants of this country.

Figure 17-17 Ants

17.3.9.3. Other Ants

Honey ants, leaf cutting or umbrella ants, and others are edible in all stages, both raw and cooked.

17.3.9.4. Caterpillars

It is natural that caterpillars, the larvae of moths and butterflies, form a very substantial part of the food of primitive peoples because they large size and abundant. Caterpillars with hairs should be avoided. If eaten, the hairs may become lodged in the throat causing irritation or infection. Today it is known that insects have nutritional or medicinal value. The praying mantis, for example, contains 58 percent protein, 12 percent fat, three percent ash, vitamin B complex, and vitamin A. The insect's outer skeleton is an interesting compound of sugar and amino acids.

17.3.9.5. Other Insects

Bee larvae, locusts, dragonflies, bumblebees, cockroaches and locust grasshoppers, golden June beetles, crickets, wasp larvae, and silkworm larvae are all used for food.

17.3.9.6. Preparation

Insects should have any stinging apparatus, (legs, wings, and heads) removed before they are eaten.

17.3.9.7. Food Aversions

Insects have been used as a food source for thousands of years and will undoubtedly continue to be used. If IP cannot overcome their aversion to insects as a food source, they will miss out on a valuable and plentiful supply of food.

17.3.10. Fishing

Fishing is one way to obtain food throughout the year wherever water is found (Figure 17-18). There are many ways to catch fish which include hook and line, gill nets, poisons, traps, and spearing.

17.3.10.1. Emergency Fishing Kits

If an emergency fishing kit is available, there will be a hook and line in it, however, if a kit is not available a hook and line will have to be procured or improvised. Hooks can be made from wire or carved from bone or wood. The line can be made by unraveling a parachute suspension line or by twisting threads from clothing or plant fibers. A piece of wire between the fishing line and the hook will help prevent the fish from biting through the line. Insects, smaller fish, shellfish, worms, or meat can be used as bait. Bait can be selected by observing what the fish are eating. Artificial lures can be made from pieces of brightly colored cloth, feathers, or bits of bright metal or foil tied to a hook. If the fish will not take the bait, try to snag or hook them in any part of the body as they swim by. In freshwater, the deepest water is usually the best place to fish. In shallow streams, the best places are pools below falls, at the foot of rapids, or behind rocks. The best time to fish is usually early morning or late evening. Sometimes fishing is best at night, especially in moonlight or if a light is available to attract the fish. IP should be patient and fish at different depths in all kinds of water. Fishing at different times of the day and changing bait is often rewarding.

Figure 17-18 Fishing Places

17.3.10.2. Fishing Nets

The most effective fishing method is a net because it will catch fish without having to be attended (Figure 17-19 and Figure 17-20). If a gill net is used, stones can be used for anchors and wood for floats. The net should be set at a slight angle to the current in order to clear itself of any floating refuse that comes down the stream. The net should be checked at least twice daily (Figure 17-21). A net with poles attached to each end works effectively if moved up or down a stream as rapidly as possible while moving stones and threshing the bottom or edges of the stream banks. The net should be checked every few moments so the fish cannot escape.

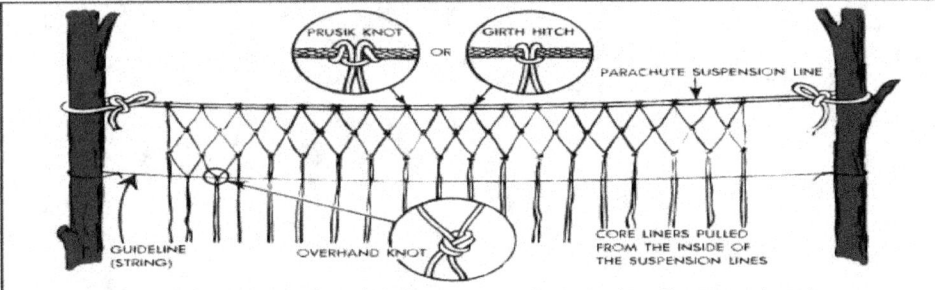

1. Suspend a suspension line casing (from which the core liners have been pulled) between two uprights, approximately at eye level.
2. Hang core liners (an even number) from the line suspended as in 1, above. These lines should be attached with a Prusik knot or girth hitch and spaced in accordance with the mesh you desire. One-inch spacing will result in a 1-inch mesh, etc. The number of lines used will be in accord with the width of the net desired. If more than one man is going to work on the net, the length of the net should be stretched between the uprights, thus providing room for more than one man to work. If only one man is to make up the net, the depth of the net should be stretched between the uprights and step 8, below, followed.
3. Start at left or right. Skip the first line and tie the second and third lines together with an overhand knot. Space according to mesh desired. Then tie fourth and fifth, sixth and seventh, etc. One line will remain at the end.
4. On the second row, tie the first and second, third and fourth, fifth and sixth, etc., to the end.
5. Third row, skip the first line and repeat step 3 above.
6. Repeat step 4, and so on.
7. You may want to use a guide line which can be moved down for each row of knots to ensure equal mesh. Guide line should run across the net on the side opposite the one you are working from so that it will be out of your way.
8. When you have stretched the depth between the uprights and get close to ground level, move the net up by rolling it on a stick and continue until the net is the desired length.
9. String suspension line casing along the sides when net is completed to strengthen it and make the net easier to set.

Figure 17-19 Making a Gill Net

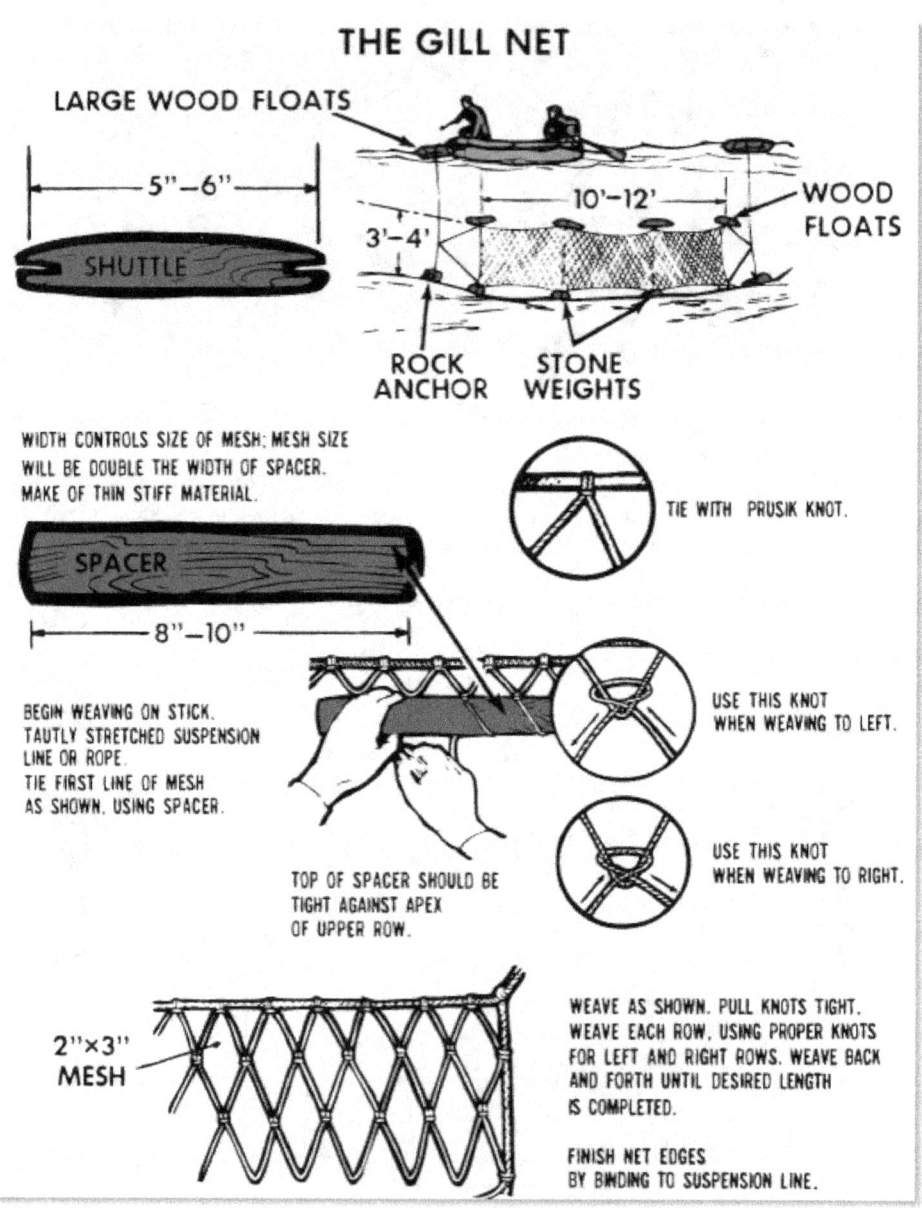

Figure 17-20 Making a Gill Net with Shuttle and Spacer

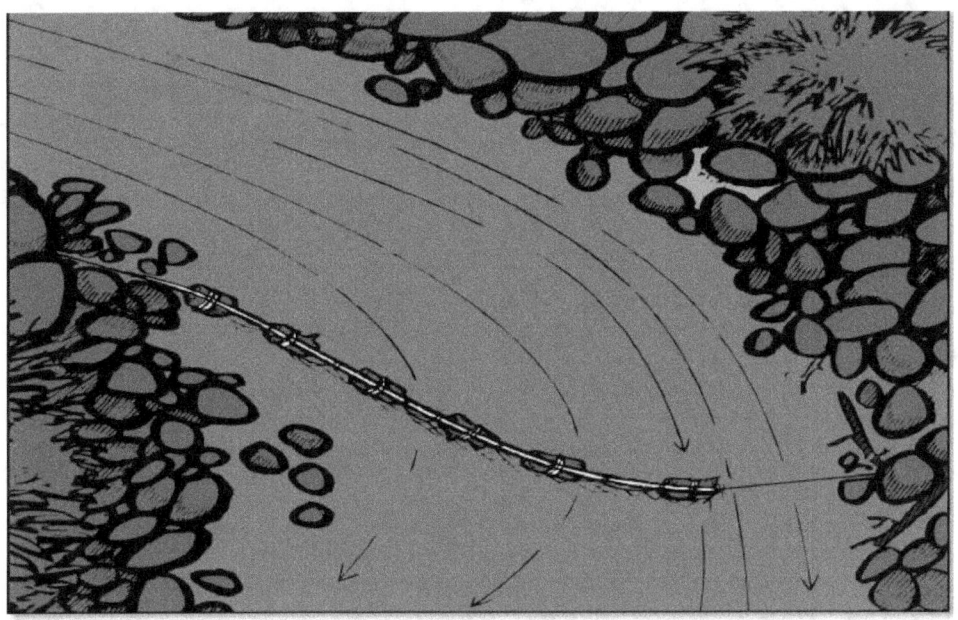

Figure 17-21 Setting the Gill Net

17.3.10.3. Dip Nets

Shrimp (prawns) live on or near the sea bottom and may be scraped up or lured to the surface with light at night. A hand net made from parachute cloth or other material is excellent for catching shrimp. Lobsters are creeping crustaceans found on or near the sea bottom. A lobster trap, jig, baited hook, or dip net can be used to catch lobster. Crabs will creep, climb, and burrow and are easily caught in shallow water with a dip net or in traps baited with fish heads or animal viscera.

17.3.10.4. Fish Traps

Fish traps (Figure 17-22) are very useful for catching both freshwater and saltwater fish, especially those that move in schools. In lakes or large streams, fish tend to approach the banks and shallows in the morning and evening. Sea fish, traveling in large schools, regularly approach the shore with the incoming tide, often moving parallel to the shore guided by obstruction in the water.

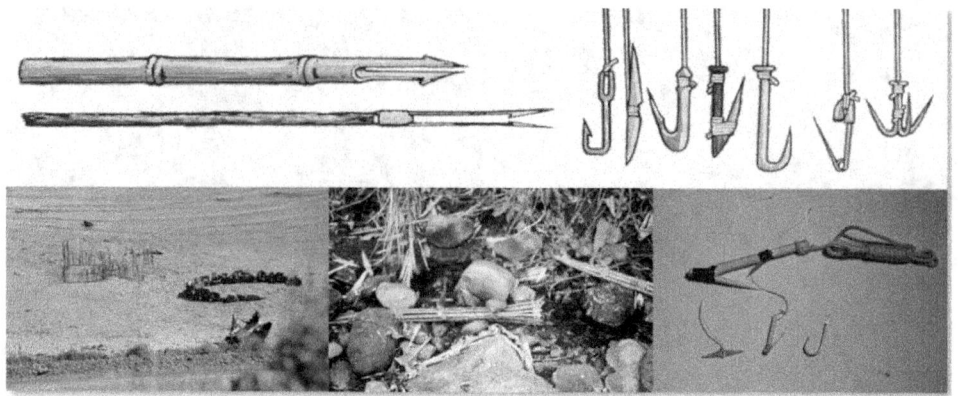

Figure 17-22 Maze-type Fish Traps

17.3.10.4.1. Fish Trap Basics

A fish trap is basically an enclosure with a blind opening where two fence-like walls extend out, like a funnel, from the entrance. The time and effort put into building a fish trap should depend on the need for food and the length of time IP plan to stay in one spot.

17.3.10.4.2. Trap Location

The trap location should be selected at high tide and the trap built at low tide. One to two hours of work should be sufficient to hold the trap. Consider the location, and try to adapt natural features to reduce the amount of effort it takes to build the trap. Natural rock pools should be used on rock shores. Natural pools on the surface of reefs should be used on coral islands by blocking the opening as the tide recedes. Sandbars, and the ditches they enclose, can be used on sandy shores. The best fishing off sandy beaches is the lee side of offshore sandbars. By watching the swimming habits of fish, a simple dam can be built which extends out into the water forming an angle with the shore. This will trap fish as they swim in their natural path. When planning a more complex brush dam, select protected bays or inlets using the narrowest area and extending one arm almost to the shore.

17.3.10.4.3. Streams

In small, shallow streams, the fish traps can be made with stakes or brush set into the stream bottom or weighted down with stones so that the stream is blocked except for a small narrow opening into a stone or brush pen or shallow water. Wade into the stream, herding the fish into the trap, and catch or club them when they get in shallow water. Mud-bottom streams can be trampled until cloudy and then netted. The fish are blinded and cannot avoid the nets. Freshwater crawfish and snails can be found under rocks, logs, overhanging bushes, or in mud bottoms.

1. Fish may be confined in properly built enclosures and kept for days. In many cases, it may be advantageous to keep them alive until needed and thus ensure there is a fresh supply without danger of spoilage. Mangrove swamps are often good fishing grounds.

2. At low tide, clusters of oysters and mussels are exposed on the mangrove "knees" or lower branches. Clams can be found in the mud at the base of trees. Crabs are very active among

branches or roots and in the mud. Fish can be caught at high tide. Snails are found on mud and clinging to roots. Shellfish which are not covered at high tide or those from a colony containing diseased members should not be eaten. Some indications of diseased shellfish are shells gaping open at low tide, foul odor, and (or) milky juice.

17.3.10.5. Poisoning Fish

Throughout the warm regions of the world, there are various plants which the natives use for poisoning fish. The active poison in these plants is harmful only to cold-blooded animals. IP can eat fish killed by this poison without ill effects.

17.3.10.5.1. Using Lime

Lime thrown in a small pond or tidal pool will kill fish in the pool. Lime can be obtained by burning coral and seashells.

17.3.10.5.2. Fish-Poison Plants

The most common method of using fish-poison plants, such as Derris root found in Southeast Asia (Figure 17-23), is to crush the plant parts (most often the roots) and mix them with water. Drop large quantities of the crushed plant into pools or the headwaters of small streams containing fish. Within a short time, the fish will rise in a helpless state to the surface. After putting in the poison, follow slowly down stream and pick up the fish as they come to the surface, sink to the bottom, or swim erratically to the bank. A stick dam or obstruction will aid in collecting fish as they float downstream. The husk of "green" black walnuts can be crushed and sprinkled into small sluggish streams and pools to act as a fish stupefying agent. In the southwest Pacific, the seeds and bark from the Barringtonia tree (Figure 17-24) are commonly used as a source of fish poison. The Barringtonia tree usually grows along the seashore.

17.3.10.5.3. Derris Plant

Derris is a climbing leguminous plant of Southeast Asia and the southwest Pacific islands, including New Guinea. Its roots contain rotenone, a strong insecticide and fish poison. Also known as Derris powder, it was formerly used as an organic insecticide used to control pests on crops such as peas. However, due to studies revealing its extreme toxicity, as well as the concentration level of rotenone to which the powder is often refined, experts in ecological and organic growing no longer consider it ecologically sound. Rotenone is still sold in the US.

Figure 17-23 Derris

Figure 17-24 Barringtonia

17.3.10.6. Tickling

Tickling can be effective in small streams with undercut banks or in shallow ponds left by receding flood waters. Place hands in the water and reach under the bank slowly, keeping the hands close to the bottom if possible. Move the fingers slightly until they make contact with a fish. Then work the hands gently along the fish's belly until reaching its gills. Grasp the fish firmly just behind the gills and scoop it onto land. In the tropics, this type of fishing can be dangerous due to hazardous marine life in the water such as piranhas, eels, and snakes.

17.3.10.7. Detecting Poisonous Fish

There is no known way to detect a poisonous fish merely by its appearance. Fish that are poisonous in one area may be safe to eat in another. In general, bottom dwellers and feeders, especially those associated with coral reefs, should be suspect. Also, unusually large predator-type fish should be eaten with caution. The internal organs and roe of all tropical marine fish should never be eaten, as those parts contain a higher concentration of poison.

17.3.10.7.1.   Marine Animal Edibility Test

Under certain conditions, where IP may be required to eat questionable fish, rules should be followed. A fish will be safer if it can be caught away from reefs or entrances to lagoons. Once the fish has been secured, the "marine animal edibility test" should be used. The fish should be cut into thin strips and boiled in successive changes of water for an hour or more. This may help since some, but not all, of the toxins are water soluble. Further, it should be noted, that normal cooking techniques and temperatures will not weaken or destroy poisons.

- If boiling is not possible, cut the meat into thin strips and soak in changes of sea water for an hour or so, squeezing the meat juices out as thoroughly as possible. IP should eat only a small portion of the flesh and wait 12 hours to see if any symptoms arise (if the fish will not spoil). Remember that the degree of poisoning is directly related to how much fish is eaten. If in doubt, do not eat it. The advice of native people on eating tropical marine fish may not be valid. In many instances they check edibility by first feeding fish portions to their dogs and cats. As soon as any symptoms arise, vomiting should be induced by administering warm saltwater or egg whites. If these procedures do not work, try sticking a finger down the person's throat. A laxative should also be given to the victim if one is available. The victim may have to be protected from injury during convulsions. If the victim starts to foam at the mouth and exhibits signs of respiratory distress, a cricothyroidotomy may have to be performed (see Chapter 5 for more information.). Morphine may help relieve pain in some cases. If the victim complains of severe itching, cool showers may give some relief. Treat any other symptoms as they arise.

17.4. Plant Food

The thought of having a diet consisting only of plant food may be distressing to IP. This is not the case if the isolating event is entered into with the confidence and intelligence based on knowledge or experience. If IP know what to look for, can identify it, and know how to prepare it properly for eating, there is no reason why they can't find sustenance.

17.4.1. Carbohydrates

Plants provide carbohydrates, which provide the body energy and calories. Carbohydrates keep weight and energy up, and include important starches and sugars.

17.4.2. Plant Availability

Another advantage of a plant diet is availability. In many instances, a situation may present itself in which procuring animal food is difficult because of injury, being unarmed, being in enemy territory, exhaustion, or being in an area which lacks wildlife. If convinced that vegetation can be depended upon for daily food needs, the next question is "where to get what and how."

17.4.2.1. Procuring Plant Foods

Experts estimate there are about 300,000 classified plants growing on the surface of the Earth, including many which thrive on mountain tops and on the floors of the oceans. There are two considerations that IP must keep in mind when procuring plant foods. The first consideration is determining if the plant is edible, and preferably, palatable. The next consideration is that plant food must be fairly abundant in the areas in which it is found. If it includes an inedible or poisonous variety in its family, the edible plant must be distinguishable to the average eye from the poisonous one. Usually a plant is selected because one special part is edible, such as the stalk, the fruit, or the nut.

17.4.2.2. Aids in Determining Plant Edibility

To aid in determining plant edibility, there are general rules which should be observed and an edibility test that should be performed. In selecting plant foods, the following should be considered. First, look for plant parts that contain high energy such as fruit, seeds, grains, nuts, roots, bulbs and tubers. Also, select plants resembling those cultivated by people. It is risky to rely upon a plant (or parts thereof) being edible for human consumption simply because animals have been seen eating it. When selecting an unknown plant as a possible food source, apply the following general rules:

17.4.2.2.1. Toxic Peptides

Mushrooms and fungi should not be selected. Fungi have toxic peptides, a protein-base poison which has no taste. There is no field test other than eating to determine whether an unknown mushroom is edible. Anyone gathering wild mushrooms for eating must be absolutely certain of the identity of every specimen picked. Some species of wild mushrooms are difficult for an expert to identify. Because of the potential for poisoning, relying on mushrooms as a viable food source is not worth the risk.

17.4.2.2.2. Plants to Avoid

Plants with umbrella-shaped flowers are to be completely avoided, although carrots, celery, dill, and parsley are members of this family. One of the most poisonous plants, poison water hemlock, is also a member of this family (Figure 17-25).

- All of the legume family should be avoided (beans and peas). They absorb minerals from the soil and cause problems. The most common mineral absorbed is selenium. Selenium is what has given locoweed its fame. (Locoweed is a vetch).
- A milky sap indicates a poisonous plant. Plants that are irritants to the skin should not be eaten, such as poison ivy.
- As a general rule, all bulbs should be avoided. Examples of poisonous bulbs are tulips and death camas.
- White and yellow berries are to be avoided as they are almost always poisonous. Approximately one-half of all red berries are poisonous. Blue or black berries are generally safe for consumption.
- Aggregated fruits and berries are always edible (for example, thimbleberry, raspberry, salmonberry, and blackberry).
- Single fruits on a stem are generally considered safe to eat.
- Plants with shiny leaves are considered to be poisonous and caution should be used.

Figure 17-25 Water Hemlock

17.4.3. Edibility Testing

A plant that grows in sufficient quantity within the local area should be selected to justify the edibility test and provide a lasting source of food if the plant proves edible.

- Plants growing in the water or moist soil are often the most palatable.
- Plants are less bitter when growing in shaded areas.

17.4.3.1. The Edibility Test

The previously mentioned information concerning plants is general. There are exceptions to every rule, but IP should only select unknown plants as a last resort. When selecting unknown plants for possible consumption, poisonous characteristics should be avoided. Apply the edibility test to only one plant at a time so if some abnormality does occur, it will be obvious which plant caused the problem. Once a plant has been selected for the edibility test, proceed as follows:

1. First, if there are any unpleasant odors such as a moldy or musty smell coming from the plant; stop testing and disregard as a possible edible plant option. Also, if the plant gives off an "almond" scent, it should be disregarded as a possible edible plant option
2. Crush or break part of the plant to determine the color of its sap. If the sap is clear, proceed to the next step.
3. Touch the plant's sap or juice to the inner forearm. If there are no ill effects, such as a rash or burning sensation to the skin, then proceed with the rest of the steps. (NOTE: Sometimes

heavy smokers are unable to taste various poisons, such as alkaloids)

4. If a reaction was not noted when touching the inner forearm, place some of the plant juice on the outer lip for eight minutes. If a reaction occurs, stop the test.
5. If still no reaction, taste a small pinch of the plant and leave in the mouth for eight minutes. If there is an unpleasant taste, such as bitterness or a numbing sensation of the tongue or lips, stop the test. If a reaction does not occur swallow the pinch of plant.
6. After swallowing, wait eight hours. If there is no reaction after eight hours, chew a handful of the plant, swallow, and wait an additional eight hours. If no reaction occurs after eight hours, consider the tested plant part edible.
7. Keep in mind that any new or strange food should be eaten with restraint until the body has become accustomed to it. The plant may be slightly toxic and harmful when large quantities are eaten.

17.4.3.2. Plant Preparation

If IP are not able to boil the plant before consumption, plant food may be prepared as follows:

17.4.3.2.1. Leaching a Plant

Leach the plant by crushing the plant material and placing it in a container. Pour large quantities of cold water over it (rinse the plant parts). Leaching removes some of the bitter elements of nontoxic plants. If leaching is not possible, IP should follow the steps they can in the edibility test.

17.4.4. Edible Plants in Temperate Environments

In the temperate environment, edible plants are widely varied with seeds, resins, cattail, chicory, fiddlehead fern, sorrel and others are available. In many places, the ground is thickly covered with mosses and there may be a few varieties of early flowering plants and many berry-bearing shrubs which invite birds and mammals into the open areas. Some of the largest herbivores (plant eaters) live in these evergreen forests - caribou, reindeer, moose, and deer. The small herbivores may include porcupines, several species of squirrels, mice, and rabbits. The carnivores (flesh eaters) which feed upon the smaller animals include black bear, gray wolf, lynx, wolverine, red fox, and weasel. Multitudes of insects provide food for the birds. (These insects may also present a menace to IP). A large variety of birds feed not only on insects but also on plants. More information is available in the sustenance chapter and through self study.

17.4.4.1.1. Winter Months

In winter, the primary edible food plants available include:

- Rootstalks
- Roots
- Seeds
- Resins (from Pines)
- Infusion (teas) from evergreen needles
- Bark (inner part)

17.4.4.2. Summer Months
During summer months, many more plants are available for food, including:
- Tree Nuts such as almonds, walnuts, pine nuts, chestnuts, hazelnuts, and acorns
- Agave (Figure 17-26)
- Sweet Acacia
- Water Plantain
- Baobab
- Wild Rhubarb
- Wild Sorrel
- Cattail
- Tree ferns, Spreading Wood Ferns, Bracken Fern, and Fiddleheads (Ferns)
- Wild Dock
- Chicory
- Juniper
- Common Jujube
- Wild Lily and Water Lily (Water Arum)

Figure 17-26 Agave

17.4.5. Edible Parts of Plants

IP will find some plants which are completely edible, but many plants which they may find will have only one or more identifiable parts having food and thirst-quenching value. The variety of plant component parts which might contain substance of food value is shown in Figure 17-27.

EDIBLE PARTS OF PLANTS

Underground Parts	Tubers
	Rots and Rootstalks
	Bulbs
Stems and Leaves (potherbs)	Shoots and Stems
	Leaves
	Pith
	Bark
Flower Parts	Flowers
	Pollen
Fruits	Fleshy Fruits (dessert and vegetable)
	Seeds and Grains
	Nuts
	Seed Pods
	Pulps
Gums and Resins	
Saps	

Figure 17-27 Edible Parts of Plants

17.4.5.1. Underground Edible Foods

17.4.5.1.1. Edible Tubers.

The potato is an example of an edible tuber. Many other kinds of plants produce tubers such as the tropical yam, the Eskimo potato, and tropical water lilies. Tubers are usually found below the ground. Tubers are rich in starch and should be cooked by roasting in an earth oven or by boiling to break down the starch for ease of digestion. The following are some plants with edible tubers:

- Arrowroot, East Indian
- Taro
- Cassava (Tapioca)

17.4.5.1.2. Edible Roots

Many plants produce roots which may be eaten. Edible roots are often several feet in length. In comparison, edible rootstalks are underground portions of the plant which have become thickened, and are relatively short and jointed. Both true roots and rootstalks are storage organs rich in stored starch. The following are some plants with edible roots or rootstalks (rhizomes):

- Baobab
- Pine, Screw
- Bean, Goa

17.4.5.1.3. Edible Bulbs

The most common edible bulb is the wild onion, which can easily be detected by its characteristic odor. Wild onions may be eaten uncooked, but other kinds of bulbs are more palatable if cooked. In Turkey and Central Asia, the bulb of the wild tulip may be eaten. All bulbs contain a high percentage of starch. (Some bulbs are poisonous, such as the death camas which has white or yellow flowers.) The following are some plants with edible bulbs:

- Wild Lily
- Wild Onion
- Tiger Lily

17.4.5.2. Shoots and Leaves

All edible shoots grow in much the same fashion as asparagus. The young shoots of ferns (fiddleheads) and especially those of bamboo and numerous kinds of palms are desirable for food. Some kinds of shoots may be eaten raw, but most are better if first boiled for 5 to 10 minutes, the water drained off, and the shoots re-boiled until they are sufficiently cooked for eating (parboiled,). The following are some plants with edible shoots:

- Palm
- Coconut
- Bamboo (Figure 17-28)
- Cattail
- Rock Tripe

Figure 17-28 Bamboo

17.4.5.2.1. Edible Leaves

The leaves of spinach-type plants (potherbs), such as wild mustard, wild lettuce, and lamb quarters, may be eaten either raw or cooked. Prolonged cooking, however, destroys most of the vitamins. Plants which produce edible leaves are perhaps the most numerous of all edible plants. The young tender leaves of nearly all nonpoisonous plants are edible. The following are only some plants with edible leaves:

- Sorrel, Wild
- Dock
- Plantain

17.4.5.2.2. Edible Pith

Some plants have edible pith in the center of the stem. The pith of some kinds of tropical plants is quite large. Pith of the sago palm is particularly valuable because of its high food value. The following are some palms with edible pith (starch):

- Coconut
- Rattan
- Sugar

17.4.5.2.3. Tree Bark

The inner bark of a tree, the layer next to the wood, may be eaten raw or cooked. It is possible in northern areas to make flour from the inner bark of such trees as the cottonwood, aspen, birch, willow, and pine. The outer bark should be avoided in all cases because this part contains large amounts of bitter tannin. Pine bark is high in vitamin C. The outer bark of pines can be cut away and the inner bark stripped from the trunk and eaten fresh, dried, cooked, or pulverized into flour. Bark is most palatable when newly formed in spring. As food, bark is most useful in the arctic regions, where plant food is often scarce.

17.4.5.3. Flower Parts

Fresh flowers may be eaten as part of a salad or to supplement a stew. The hibiscus flower is commonly eaten throughout the southwest Pacific area. In South America, the people of the Andes eat nasturtium flowers. In India, it is common to eat the flowers of many kinds of plants as part of a vegetable curry. Flowers of desert plants may also be eaten. The following are some plants with edible flowers:

- Abal
- Papaya
- Banana

17.4.5.3.1. Pollen

Pollen looks like yellow dust. All pollen is high in food value and is found in some plants, especially the cattail. Quantities of pollen may easily be collected and eaten as a kind of gruel.

17.4.5.4. Edible Fruits

Edible fruits can be divided into sweet and non-sweet (vegetable) types. Both are the seed bearing parts of the plant. Sweet fruits are often plentiful in all areas of the world where plants grow. For instance, in the far north, there are blueberries and crowberries; in the temperate zones, cherries, plums, and apples; and in the American deserts, fleshy cactus fruits. Tropical areas have more kinds of edible fruit than other areas, and a list would be endless. Sweet fruits may be cooked, or for maximum vitamin content, left uncooked. Common vegetable fruits include the tomato, cucumber, and pepper. The following are some plants with edible fruits:

- Apple, Wild
- Huckleberry
- Prickly Pear

17.4.5.4.1. Edible Vegetables

The following are some plants with edible fruits (vegetables):

- Breadfruit (Figure 17-29)
- Horseradish
- Caper, Wild

Figure 17-29 Breadfruit

17.4.5.5. Seeds

Seeds of many plants, such as buckwheat, ragweed, amaranth, and goosefoot, contain oils and are rich in protein. The grains of all cereals and many other grasses, including millet, are also extremely valuable sources of plant protein. They may either be ground between stones, mixed with water and cooked to make porridge, parched or roasted over hot stones. In this state, they are still wholesome and may be kept for long periods without further preparation (Figure 17-30). The following are some plants with edible seeds and grains:

- Rice
- Screw Pine
- Water Lily (Temperate)

Figure 17-30 Grains

17.4.5.6. Nuts

Nuts are among the most nutritious of all raw plant foods and contain an abundance of valuable protein. Plants bearing edible nuts occur in all the climatic zones of the world and in all continents except in the arctic regions (Figure 17-31). Most nuts can be eaten raw but some, such as acorns, are better when cooked. The following are some plants with edible nuts:

- Almond
- Beechnut
- Pine

Figure 17-31 Edible Nuts

17.4.5.7. Pulp

The pulp around the seeds of many fruits is the only part that can be eaten. Some fruits produce sweet pulp, others have a tasteless or even bitter pulp. Plants that produce edible pulp include the custard apple, Inga pod, breadfruit, and tamarind. The pulp of breadfruit must be cooked, whereas in other plants, the pulp may be eaten uncooked. Use the edibility rules in all cases of doubt.

17.4.5.8. Gum and Resin

Gum and resin are sap that collects and hardens on the outside surface of the plant. It is called gum if it is soft and soluble, and resin if it is hard and not soluble. Most people are familiar with the gum which exudes from cherry trees and the resin which seeps from the pine trees. These plant byproducts are edible and are a good source of nutritious food which should not be overlooked.

17.4.5.9. Vines

Vines or other plant parts may be tapped as potential sources of usable liquid. The liquid is obtained by cutting the flower stalk and letting the fluid drain into some sort of container such as a bamboo section. Palm sap with its high-sugar content is highly nutritious. The following are some plants with edible sap and drinking water:

- Sweet Acacia (water)
- Saxual (water)
- Nipa Palm (sap)

17.5. Food in Tropical Climates

There are more types of animals in the jungles of the world than in any other region. A jungle visitor who is unaware of the life style and eating habits of these animals would not observe the presence of a large number of the animals.

17.5.1. Game Trails

Game trails are the normal routes along which animals travel through a jungle. Some of the animals used as food are hedgehogs, porcupines, anteaters, mice, wild pigs, deer, wild cattle, bats, squirrels, rats, monkeys, snakes, and lizards.

17.5.1.1. Reptiles

Reptiles are located in all jungles and should not be overlooked as a food source. All snakes should be considered poisonous and extreme caution used when killing the animal for a food source. All cobras should be avoided since the spitting cobra aims for the eyes; the venom can blind if not washed out immediately. Lizards are good food, but may be difficult to capture since they can be extremely fast. A blow to the head of a reptile will usually kill it. Crocodiles and caimans are extremely dangerous on land as well as in the water.

17.5.1.2. Frogs

Frogs can be poisonous; all brilliantly colored frogs should be totally avoided. Some frogs and toads in the tropics secrete substances through the skin which has a pungent odor. These frogs are often poisonous.

17.5.1.3. Large Animals

The larger, more dangerous animals should be left alone.

17.5.2. Seafood

Seafood such as fish, crabs, lobster, crayfish, and small octopi can be poked out of holes, crevices, or rock pools (Figure 17-32). IP should be ready to spear them before they move off into deep water. If they are in deeper water, they can be teased shoreward with a baited hook, or stick.

17.5.2.1. Methods of Procuring Fish

A number of methods can be used for procuring fish. Some are listed below.

- Hook-and-line fishing on a rocky coast requires a lot of care to keep the line from becoming entangled or cut on sharp edges. Most shallow-water fish are nibblers. Unless the bait is well placed and hooked and the barb of the hook offset by bending, the bait may be lost without catching a fish. Use hermit crabs, snails, or the tough muscle of a shellfish as bait. Take the cracked shells and any other animal remains and drop them into the area to be fished. This brings the fish to the area and provides a better opportunity to catch the fish. Examine stomach contents of the first fish caught to determine what the fish are feeding on.

- Jigging is taking a baited or spooned hook dipped repeatedly beneath the surface of the water is sometimes effective. This method may be used at night.

- Spearing is difficult except when the stream is small and the fish are large and numerous during the spawning season, or when the fish congregate in pools. Make a spear by sharpening a long piece of wood, lashing two long thorns on a stick, or fashioning a bone spear point, and take a position on a rock over a fish run. Wait patiently and quietly for a fish to swim by.

Figure 17-32 Edible Invertebrates

17.5.2.2. Tropical Seafood

Examples of tropical seafood include:

- A small heap of empty oyster shells near a hole may indicate the presence of an octopus. To catch an octopus located in a hole, push a long stick into the hole where the octopus is located. When tentacle comes out of the hole, grab the tentacles and pull out the octopus and stick. Also, a baited hook placed in the hole will sometimes catch the octopus. IP should allow the octopus to surround the hook and line before lifting. Octopi are not scavengers like sharks, but they are hunters, fond of spiny lobster and other crab-like fish. At night, octopus come into shallow water and can be easily seen and speared.

- Snails and limpets cling to rocks and seaweed from the low-water mark up. Large snails called chitons adhere tightly to rocks just above the surf line.

- Mussels usually form dense colonies in rock pools, on logs, or at the bases of boulders. Mussels are poisonous in tropical zones during the summer, especially when seas are highly phosphorescent or reddish.

- Sea cucumbers and conchs (large snails) live in deep water. The sea cucumber will shoot out its stomach when excited. The stomach is not edible. The skin and the five strips of muscle can be eaten after boiling. Conches can be boiled out of their shells and have very firm flesh. Use care when picking conches up. The bottom of their "foot" has a bony covering which can severely cut IP who picks them up.

- The safest fish to eat are those from the open sea or deep water beyond the reef. Silvery fishes, river eels, butterfly fishes, and flounders from bays and rivers are good to eat.
- Land crabs are common on tropical islands and are often found in coconut groves. An open coconut can be used for bait.

17.5.2.2.1. Marine Turtles

There are over 265 species of marine turtles. Of these, only five have been reported as poisonous and dangerous to IP. Many of these species are commonly eaten, but for some unknown reason, these same turtles become extremely toxic under certain conditions. Basically, the main species to be concerned with are the green, the hawksbill, and the leatherback turtles. These turtles are found mainly in tropical and subtropical seas but can also be found in temperate waters.

- The origin of turtle poison is unknown but some investigators suggest it comes from the poisonous marine algae eaten by the turtles. It should be noted that a species of turtle may be safe to eat in one area but deadly in another. There is absolutely no way IP can distinguish between a poisonous and nonpoisonous sea turtle just by looking at it or by examining any part of it. Toxicity may occur at any time of the year; however, the most dangerous months appear to be the warmer months. The degree of freshness also has nothing to do with how poisonous the turtle is.
- The symptoms will vary with the amount of turtle ingested. Symptoms will develop within a few hours to a few days after eating the food. These symptoms include nausea, vomiting, diarrhea, pain, sweating, coldness in the extremities, vertigo, dry and burning lips and tongue, tightness of the chest, drooling, and difficulty in swallowing. Other victims reported a heavy feeling of the head, a white coating on the tongue, diminished reflexes, coma, and sleepiness. About 44 percent of the victims poisoned by marine turtles die.
- There is no known antidote for this kind of poisoning. There is no specific treatment-treat symptomatically.
- If there is the slightest suspicion about the edibility of a marine turtle, it should not be eaten, or at least the marine animal edibility test should be used. Turtle liver is especially dangerous to eat because of its high vitamin A content.

17.5.2.3. Red Tide

Red tide is a name used to describe the reddish or brownish coloration in saltwater, resulting from tiny plants and organisms called plankton, which suddenly increase tremendously in numbers. Red tides appear in waters worldwide. In the United States, they are most common off the coasts of Florida, Texas, and southern California. Although most red tides are harmless, some may kill fish and other water creatures. Still other types of red tides do not kill sea life, but cause the shellfish feeding on them to be poisonous. Some of these creatures secrete poisons which can paralyze and kill fish, or can kill fish by using nearly all of the oxygen in the water. Although the exact reason for the sudden increase of the plankton is unknown, there is evidence that shows favorable food, temperature, sunlight, water currents, and salt in the water will increase the population. It is not unusual for it to remain from a few hours to several months. IP should not eat any fish that are found dead.

17.5.3. Jungle Environments

In jungle environments, rainfall is distributed throughout the year and there is a lack of cold season. This allows plants in the humid regions to grow, produce leaves and flower year round. Some plants grow very rapidly. For example, the stem of the giant bamboo may grow more than 22 inches in a single day.

17.5.3.1. Plant Food Search

IP in search of plant food should apply some basic principles to the search. IP are lucky to find a plant that can readily be identified as edible. If a plant resembles a known plant, it is very likely to be of the same family and can be used. If a plant cannot be identified, the edibility test should be applied. IP will find many edible plants in the tropical forest, but chances of finding them in abundance are better in an area that has been cultivated in the past (secondary growth).

17.5.3.2. Some Plants IP Might Find

Citrus fruit trees may be found in uncultivated areas, but are primarily limited to areas of secondary growth. The many varieties of citrus fruit trees and shrubs have leaves two to four inches long alternately arranged. The leaves are leathery, shiny, and evergreen. The leaf stem is often winged. Small (usually green) spines are often present by the side of the bud. The flowers are small and white to purple in color. The fruit has a leathery rind with numerous glands and is round and fleshy with several cells (fruit sections or slices) and many seeds. A great number of wild and cultivated fruits (oranges, limes, lemons, etc.) native to the tropics are eaten raw or used in beverages (Figure 17-33).

Figure 17-33 Citrus Fruit Tree

17.5.3.2.1. Taro

Taro can be found in both secondary growth and in virgin areas (Figure 17-34). It is usually found in the damp, swampy areas in the wild, but certain varieties can be found in the forest. It can be identified by its large heart-shaped or arrowhead-shaped leaves growing at the top of a vertical stem. The stem and leaves are usually green and rise a foot or more from a tuber at the base of the stem. Taro leaf tips point down; poisonous elephant ear points up. All varieties of taro must be cooked to break down the irritating crystals in the plant.

Figure 17-34 Taro Plant

17.5.3.2.2. Wild Pineapple

Wild pineapple can be found in the wild, and common pineapples may be found in secondary growth areas (Figure 17-35). The wild pineapple is a coarse plant with long clustered, sword-shaped leaves with saw-toothed edges. The leaves are spirally arranged in a rosette. Flowers are violet or reddish. The wild pineapple fruit will not be as fully developed in its wild state as when cultivated. The seeds from the flower of the plant are edible as well as the fruit. The ripe fruit may be eaten raw, but the green fruit must be cooked to avoid irritation. (The leaf fibers make excellent lashing material and ropes can be manufactured from it.)

Figure 17-35 Wild Pineapple

17.5.3.2.3. Yams

Yams may be found cultivated or wild. There are many varieties of yam, but the most common has a vine with square-shaped cross section and two rows of heart-shaped leaves growing on opposite sides of the vine. The vine can be followed to the ground to locate the tuber. The tubers should be cooked to destroy the poisonous properties of the plant (Figure 17-36).

Figure 17-36 Yams

17.5.3.2.4. Ginger

Ginger grows in the tropical forest and is a good source of flavoring for food (Figure 17-37). It is found in shaded areas of the primary forest. The ginger plant grows five to six feet high. It has seasonal white snapdragon-type flowers, while some variations have red flowers. The leaves when crushed produce a very sweet odor and are used for seasoning or tea. The tea can be used to treat colds and fever.

Figure 17-37 Ginger

17.5.3.2.5. Coconut Palm

The coconut palm is found wild on the seacoast and cultivated in areas inland. The coconut palm is a tree 50 to 100 feet high, either straight or curved, marked with ring-like leaf scars. The base of the tree is swollen and surrounded by a mass of rootlets. The leaves are leathery and reach a length of 15 to 20 feet (the leaves make excellent sheathing for shelter). The fruit grows in clusters at the top of the tree. Each nut is covered with a fibered hard shell. The "heart" of the coconut palm is edible and is found at the top (the new leaves grow out of the heart). Cutting the tree down and removing the leaves allow an IP to gain access to the heart. The flower of the coconut tree is also edible and is best used as a cooked vegetable. The germinating nut is filled with a meat that can be eaten raw or cooked. There are many other varieties of palm found in the tropics which have edible hearts and fruits (Figure 17-38).

Figure 17-38 Coconut Palm

17.5.3.2.6. Papaya

The papaya is an excellent source of food and can be found in secondary growth areas. The tree grows to a height of six to 20 feet. The large, dark green, many fingered, rough-edged leaves are clustered at the top of the plant. The fruit grows on the stem clustered under the leaves. The fruit is small in the wild state, but cultivated varieties may grow to 15 pounds. The peeled fruit can be eaten raw or cooked. The peeling should never be eaten. The green fruit is usually cooked. The milky sap of the green fruit is used as a meat tenderizer; care should be taken not to get it in the eyes. Always wash the hands after handling fresh green papayas. If some of the sap does get in the eyes, they should be washed immediately (Figure 17-39).

Figure 17-39 Papaya

17.5.3.2.7. Cassava

Cassava (tapioca) can be found in secondary growth areas (Figure 17-40). It can be identified by its stalk-like leaves which are deeply divided into numerous pointed sections or fingers. The woody (red) stem of the plant is slender and at points appears to be sectioned. When found growing wild in secondary growth areas, pull the trunks to find where a root grows. When one is found, a tuber can be dug. Tubers have been found growing around a portion of the stem that was covered with vegetation. The brown tuber of the plant is white inside and must be boiled or roasted. The tuber must also be peeled before boiling (the green-stemmed species of cassava is poisonous and must be cooked in several changes of water before eating it).

Figure 17-40 Cassava

17.5.3.2.8. Ferns

Ferns can be found in the virgin tropical forest or in secondary growth areas. The new leaves (fiddle heads) at the top are the edible parts. They are covered with fuzzy hair which is easily removed by rubbing or washing. Some can be eaten raw, but as a rule, should be cooked as a vegetable (Figure 17-41).

Figure 17-41 Edible Ferns

17.5.3.2.9. Sweet Sops

Sweet sops can be found in the tropical forest (Figure 17-42). It is a small tree with simple, oblong leaves. The fruit is shaped like a blunt pine cone with thick grey-green or yellow, brittle spines. The fruit is easily split or broken when ripe, exposing numerous dark brown seeds imbedded in the cream colored, very sweet pulp.

Figure 17-42 Sweet Sops

17.5.3.2.10.   Star Apple

The star apple is common in the tropical forests (Figure 17-43). The tree grows up to a height of 60 feet and can be identified by the leaves which have shiny, silky, brown hairs on the bottom. The fruit looks like a small apple or plum with a smooth greenish or purple skin. The meat is greenish in color and milky in texture. When cut through the center, the brown, elongated seeds make a figure like a six or 10 pointed stars. The fruit is sweet and eaten only when fresh. When cut, the rind (like other parts of the tree) will emit a white sticky juice or latex (it is not poisonous, which is an exception to the milky sap rule).

Figure 17-43 Star Apple

17.5.3.3. Different Kinds of Wild Plants

Of the 300,000 different kinds of wild plants in the world, a large number of them are found in the tropics and many of them are potentially edible. Only a small number of jungle plants have been discussed. It would be of great benefit to anyone flying over or passing through a tropical environment to study the plant foods available in this type of environment.

17.6. Food in Dry Climates Although not as readily available as in the tropical climate, food is available and obtainable in dry climates.

17.6.1. Desert Plant Life

Plant life in the desert is varied due to the different geographical areas. It must be remembered, therefore, that available plants will depend on the actual desert, the time of year, and if there has been any recent rainfall. Potential IP should be familiar with plants in the area they will be operating in or flying over.

17.6.1.1. Date Palms

Date palms are located in most deserts and are cultivated by the native people around oases and irrigation ditches. They bear a nutritious, oblong, black fruit (when ripe).

17.6.1.2. Fig Trees

Fig trees are normally located in tropical and subtropical zones, however, a few species can be found in the deserts of Syria and Europe. Many kinds of figs are cultivated and the fruit can be eaten when ripe. Most figs resemble a top or a small pear somewhat squashed in shape. Ripe figs vary greatly as to palatability. Many are hard, woody, covered with irritating hairs, and worthless as survival food. The edible varieties are soft, delectable, and almost hairless. They are green, red, or black when ripe.

17.6.1.3. Millet

Millet, a grain bearing plant, is grown by natives around oases and other water sources in the Middle East deserts.

17.6.1.4. Cacti Fruits

Most cacti fruits and leaves have spines to protect them from birds and animals seeking water stored in the stems and leaves. The fruit and flower of all cacti are edible. Some fruits are red, some yellow, but all are soft when ripe. Any of the flat leaf variety, such as the prickly pear, can be boiled and eaten as greens (like spinach) if the spines are first removed. Although the cactus originates in the American deserts, the prickly pear has been introduced to the desert edges in Asia, Africa, the Near East, and Australia, where it grows profusely.

17.6.1.4.1. Prickly Pear

Prickly Pear is native and most abundant in the American deserts, but has been introduced into the Gobi, Sahara, and Australian Deserts, and other parts of the world (Figure 17-44).

Figure 17-44 Prickly Pear

17.6.1.4.2. Barrel Cactus

Barrel Cactus is found in many places, but is only native to the North American deserts. It grows up to five to six feet high.

17.6.1.4.3. Suguaro Cactus

Suguaro (Giant) Cactus is abundant in southern Arizona and in Sonora, Mexico and can grow up to 50 feet tall.

17.6.1.5. Onions

There are two types of onions in the Gobi desert. A hot, strong, scallion-type grows in the late summer. It will improve the taste of food, but should not be used as a primary food. The highland onions grow two to two and a half inches in diameter. These can be eaten like apples and the greens can also be eaten raw or cooked.

17.6.1.6. Abal

Abal grows to about four feet tall in sandy deserts. The fresh flowers can be eaten. The dry twigs can be crushed and used as a tea substitute. It is found in the Sahara and Arabian Deserts.

17.6.1.7. Acacia

Acacia is most common in the Sahara, Gobi, and Australian desert regions, and in the warmer and drier parts of America. The beans can be crushed and cooked as porridge. It is spiny with many branches and grows to 10 feet tall. Roots yield water four to five feet from the tree trunk (Figure 17-45).

Figure 17-45 Acacia

17.6.1.8. Saxaul

Saxaul can be found on the salt deserts of the Gobi Desert. The bark acts as water storage and is a good water source.

17.6.1.9. St. John's Bread

St. John's Bread can be found along the border of the Mediterranean coast of the Sahara and across the Arabian Desert. It grows 40 to 50 feet tall and seeds can be pulverized and cooked as porridge (the most nutritious plant food in the Middle East).

17.6.1.10. Wild Desert Gourds

Wild desert gourds are found in the Sahara and Arabian Deserts. They have a vine which grows from eight to 30 feet long, and produce a melon-like poisonous fruit. The seed can be eaten when roasted or boiled. The flowers are also edible (Figure 17-46).

Figure 17-46 Wild Desert Gourd

17.6.1.11. Succulent Plants

Succulent plants are filled with juices and store moisture to survive. The surface is covered with a layer of wax or a blanket of fine hairs for protection against the heat. The moisture is contained in tough cellulose that is not digestible and must be manually broken down to release the water.

- All desert flowers can be eaten except those with milky or colored sap.
- All grasses are edible. Usually the best part is the whitish tender end that shows when the grass stalk is pulled from the ground. All grass seeds are edible.

17.6.2. Animal Food Sources

Animal food sources may be used to supplement diets and provide needed protein and fats. When looking at a desert area, it is sometimes difficult to visualize an abundance of animal life existing in it. There is, however, a great quantity of animal life present. Most are edible, but some may be hazardous to IP during the procurement stage. Some of the abundant animal life includes:

17.6.2.1. Bugs

At the peak of seasonal plant growth, the desert crawls and buzzes with an enormous number and variety of beetles, ants, wasps, moths, and bugs. They appear with the first good rains and generally feed during nighttime. IP can harvest crickets, locust, grasshoppers, and caterpillars in the desert.

17.6.2.2. Freshwater Shrimp

On the playas of the Sonora and Chihuahua deserts, several species of freshwater shrimp appear every summer in warm temporary ponds. In the Mohave Desert, where summer rains are rare, they may appear only a few times in a century.

17.6.2.3. Snakes and Lizards

Snakes, lizards, etc., have adapted well to the desert environment. Care must be observed when procuring them as some are hazardous, such as the Gila monster and rattlesnake.

17.6.2.4. Desert Birds

In general, desert birds stay in areas of heavier vegetation and many need water daily. Therefore, most desert birds will be found within short flights of some type of water source. Many birds will migrate during the drought season. If an abundance of birds is seen, insects, vegetation, and a water source will normally be nearby.

17.6.2.5. Rabbits

Rabbits, prairie dogs, and rats remain in the shade or burrow into the ground protecting themselves from the direct sun and heated air as well as from the hot desert surface.

17.6.2.6. Larger Mammals

Larger mammals are also found in the desert. This group consists of but is not limited to gazelles, antelope, deer, foxes, small cats, badgers, dingoes and hyenas. Most are nocturnal and generally avoid humans. Many roam at night eating smaller game and insects and a few eat plants. Only few of these mammals can be hazardous to IP, although all of them should be approached with caution.

17.6.2.7. Available Food Sources

Only a few of the available animals and plants have been discussed. If the possibility of having to survive in a desert area exists, IP should try to become familiar with the food source available in that area.

17.7. Food in Snow and Ice Climates In the snow and ice climates, food is more difficult to find than water. Animal life is normally more abundant during the warm months, but it can still be found in the cold months. Fish are available in most waters during the warmer months but they congregate in deep waters, large rivers, and lakes during the cold months. Also, some edible plant life can be found throughout the year in most areas of the arctic.

17.7.1. Animals in the Arctic

All animals in the arctic regions are edible, but the livers of seals and polar bears must not be eaten because of the high concentration of vitamin A. Death could result from ingesting large quantities of the liver. On the open sea ice, game animals such as seal, walrus, polar bear, and fox are available. Also, many types of birds can be found during the warmer months and fish can be caught throughout the year.

17.7.1.1. Seals

Seals will probably be the main source of food when stranded on the open sea ice. They can be found in open leads, areas of thin ice, or where snow has drifted over a pressure ridge forming a cave which could have open water or very thin ice. Caution must be taken in these areas; they may also house polar bears which feed primarily on seals.

- Newborn seals have trouble staying afloat or swimming and will be found on the ice in the early summer. The seal cubs can be easily killed with a club, spear, knife, or firearm and make an excellent source of food. The meat, blubber (fat), and coagulated milk in their stomachs are edible. When killing a cub, it is best to keep a lookout for the mother as she tends to protect her offspring in any way possible.

- Seals must surface periodically to breathe. When the icepack is thin, the seals poke their noses through the ice and take a breath of air in a lead or in open water. In thick ice, the seal will chew and (or) claw a breathing hole through the ice. Normally most seals will have more than one breathing hole. In hunting seals, it is best to take a position beside a breathing hole and wait until a seal comes up to breathe, then spear or strike it on the head with a club. Seals are very sensitive to blows on or around the nose. They will often lose consciousness but not die. Also, a hook can be suspended in a breathing hole so it hangs down at least six inches below the ice. When a seal comes to breathe, it can become hooked when it tries to depart the breathing hole. Seals can be recovered by gaffing or grabbing by hand, but in some cases, the breathing hole might have to be enlarged to pull the body through. If the seal is killed in open water, a "manak" or "grapple hook" can be used to retrieve it. All seals killed in open water or those that fall into open water should be recovered immediately. During the cold months, they will float for quite awhile, but during the warm months or when a female is nursing young, they sink rapidly. This is due to the loss of body fat.

17.7.1.2. Birds

Birds are plentiful during the summer months and can be procured by spearing, clubbing, catching with a baited fishhook, or use of a weapon.

17.7.1.3. Animals in Tundra Areas

On tundra areas, there are a large variety of animals available as a food source.

17.7.1.3.1. Large Game

The large game consists of caribou, musk oxen, sheep, wolf, and bears. Even though the large game animals can be a food source, they will be difficult to procure if a firearm is not available. Therefore, they should be considered a hazard to IP without a firearm. In the spring, bears tend to congregate along rivers and streams due to the amount of food available, normally salmon. During certain seasons of the year, bears will be found feeding at berry patches in which these areas should be avoided.

17.7.1.3.2. Small Game

Small game animals of the tundra include hares, lemmings, mice, ground squirrels, marmots, and foxes. They may be trapped or killed the entire year. Most prefer some cover and can be found in shallow ravines or in groves of short willows. Both ground squirrels and marmots hibernate in the winter. In summer, ground squirrels are abundant along sandy banks of large streams. Marmots typically live among the rocks in the mountains, usually near the edge of a meadow or in deep soil. The marmot always seeks relief in the same spot not far from a hidden entrance. To find the burrow in rocky areas, look for a large patch of orange colored lichen on rocks, it grows best on animal and bird dung. When snaring, it is best to use a simple loop made of strong line or wire. The wire must be a two-strand twisted wire since metal becomes brittle in the cold and breaks very easily. Other snares and triggers will be less effective in the cold climate. A gill net can be used as a snare by spreading it across a trail so that the animal will entangle itself.

17.7.1.3.3. Arctic Birds

The arctic is the breeding ground for many birds. In summer, ducks, geese, loons, and swans build their nests near ponds on the coastal plains or bordering lakes or rivers of the low tundra. A few ducks on a small pond usually indicates that setting birds may be found and flushed from the surrounding shores. Swans and loons normally nest on small, grassy islands in the lakes. Geese crowd together near large rivers or lakes. Smaller wading birds customarily fly from pond to pond. Grouse and ptarmigan, are common in the swampy forest regions of Siberia. Sea birds may be found on cliffs or small islands off the coast. Their nesting areas can often be located by their flights to and from their feeding grounds. Jaeger gulls are common over the tundra, and frequently rest on higher hillocks. In the winter, fewer birds are available because of migratory patterns. Ravens, grouse, ptarmigan, and owls are the primary birds available. Ptarmigan are seen in pairs or flocks, feeding along grassy or willow covered slopes. The eggs and young birds are an excellent food source and can be easily procured.

17.7.1.4. Wildlife in the Arctic

As in the tundra areas, the forested areas in the arctic and arctic-like areas abound in wildlife.

- The large game species include moose, deer, caribou, and bear.
- Small game of the forests includes hares, squirrels, porcupine, muskrat, and beaver. They can be snared or trapped easily in winter or summer. Small animal trails can be found in the winter with great ease. Most animals do not like to travel in deep snow and tend to travel the same trail often. This results in a trail that resembles a small superhighway where the snow is packed down well below the normal snow level. Most trails will also be located in heavy cover and undergrowth or parallel to roads and open areas. The same trails will normally be used during the summer.

17.7.1.5. Summer Months

During the summer months, the open water provides an excellent opportunity to procure all types of seafood, both freshwater and saltwater.

17.7.1.5.1. Ocean Shores

Arctic and tom cod, sculpin, eelpout, and other fish may be caught in the ocean. The ocean shores are rich hunting grounds for edible sea life such as clams, mussels, scallops, snails, limpets, sea urchins, chitons, and sea cucumbers. The inland lakes and rivers of the surrounding coastal tundra generally have plenty of fish which are easily caught during the warmer season. In the North Pacific and in the North Atlantic extending slightly northward into the Arctic Sea, the coastal waters are rich in all seafood such as fish, crawfish, snails, clams, oysters, and the king crab. Generally clams, mussels, scallops, snails, limpets, sea urchins, chitons and sea cucumbers can be handpicked in tide pools on ocean shores. Fish can be netted, speared, clubbed, or caught with a hook and line. In the spring, king crab come close to shore and may be caught on fish lines set in the deep water or by lowering baiting lines through holes cut in the ice. After freeze up, fishing is still possible through the ice. Fish tend to congregate in the deepest water possible. A hole should be cut through the ice at the estimated deepest point. Other good locations are at outlets or where tributaries flow into lakes or ponds. The ice is normally thinner over rapid moving water and at the edges of deep streams or rivers with snowdrifts extending out from the banks. Open water is often marked by a mist or fog formed over the area by vaporizing water. All methods of procuring fish in the summer will work in the winter. Shallow lakes, rivers, or ponds can freeze completely killing off all fish life. All sea life can be eaten raw, but cooking usually makes it more palatable.

17.7.1.5.2. Shellfish

Do not eat shellfish that are not covered at high tide. Never eat any type of shellfish that is dead when found, or any that do not close tightly when touched. Poisonous fish are rarer in the arctic than in the tropics. Some fish, such as sculpins, lay poisonous eggs; but eggs of the salmon, herring, or freshwater sturgeon are safe to eat. In arctic or subarctic areas, the black mussel may be very poisonous. If mussels are the only available food, select only those in deep inlets far from the coast. Remove the black meat (liver) and eat the white meat. Arctic shark meat is also poisonous (high concentration of vitamin A).

17.7.2. Plant Life

The plant life of the arctic regions is generally small and stunted due to the effects of permafrost, low mean temperatures, and a short growing season.

17.7.2.1. Barren Tundra

On the barren tundra areas, a wide variety of small edible plants and shrubs exist. During the short summer months on the tundra, Labrador tea, fireweed, coltsfoot, dwarf arctic birch, willow, and numerous other plants and berries can be found. During the winter, roots, rootstalks, and frozen berries can be found beneath the snow. Lichens and mosses are abundant but should be selected carefully as some species are poisonous.

17.7.2.2. Bog and Swamp Areas

In bog or swamp areas, many types of water sedge, cattail, dwarf birch, and berries are available. During spring and summer, many young shoots from these plants are easily collected.

17.7.2.3. Wooded Areas

The wooded areas of the arctic contain a variety of trees (birch, spruce, poplar, aspen, and others). Berry producing plants can also be found, such as blueberries, cranberries, raspberries, cloud berries, and crow berries. Wild rose hips (Figure 17-47), Labrador tea, alder, and other shrubs are very abundant. Many wild edible plants are highly nutritious. Greens are particularly rich in carotene (vitamin A). Leafy greens, many berries, and rose hips are all rich in ascorbic acid (vitamin C). Many roots and rootstalks contain starch and can be used as a potato substitute in stews and soups.

Figure 17-47 Wild Rose Hips

17.7.2.4. Edible Mushrooms

Although there are several types of edible mushrooms, fungi, and puff-balls in the arctic, a person should avoid ingesting them because it is difficult to identify the poisonous and nonpoisonous species. During the growing season, the physical characteristics can change considerably making positive identification even more difficult.

17.7.2.5. Poisonous Plants and Berries

There are many poisonous plants and a few poisonous berries in the arctic. Very few cause death; many will cause extreme nausea, dizziness, abdominal pain, and diarrhea. Contact poisonous plants, such as poison ivy, are not found in the arctic. Some of the more common poisonous plants are shown in Figure 17-48 through Figure 17-55.

Figure 17-48 Baneberry

Figure 17-49 Buttercup

Figure 17-50 Death Camas

Figure 17-51 False Hellebore

Figure 17-52 Monkshood and Larkspur

Figure 17-53 Lupine

Figure 17-54 Vetch and Locoweed

Figure 17-55 Poison Hemlock

17.7.2.6. Edible Plants

When selecting edible plants, choose young shoots when possible, as these will be the most tender. Plants should be eaten raw to obtain the most nutritive value. Some of the more common edible plants are:

17.7.2.6.1. Dandelions

Dandelions generally grow with grasses but may be scattered over rather barren areas. Both leaves and roots are edible raw or cooked. The young leaves make good greens; the roots (when roasted) are used as a substitute for coffee.

17.7.2.6.2. Black and White Spruce Trees

Black and white spruce trees are generally the northern most evergreens. These trees have short, stiff needles that grow singularly rather than in clusters like pine needles. The cones are small and have thin scales. Although the buds, needles, and stems have a strong resinous flavor, they provide essential vitamin C by chewing them raw. In spring and early summer, the inner bark can be used for food (Figure 17-56).

Figure 17-56 White Spruce

17.7.2.6.3. Dwarf Arctic Birch

The dwarf arctic birch is a shrub with thin tooth-edged leaves and bark which peels off in sheets. The fresh green leaves and buds are rich in vitamin C. The inner bark may also be eaten (Figure 17-57).

Figure 17-57 Dwarf Birch

17.7.2.6.4. Arctic Willow

There are many different species of willow in the arctic. Young tender shoots may be eaten as greens and the bark of the roots is also edible. They have a decidedly sour taste but contain a large amount of vitamin C (Figure 17-58).

Figure 17-58 Arctic Willow

17.7.2.7. Lichens

Lichens are abundant and widespread in the far North and can be used as a source of emergency food. Many species are edible and rich in starch-like substances, including Iceland moss, peat moss and reindeer lichen. Beard lichen growing on trees has been used as food by Indians. However, some of it contains a bitter acid which causes irritation of the digestive tract. If lichens are boiled, dried, and powdered, this acid is removed and the powder can then be used as flour or made into a thick soup.

17.8. Food on the Open Seas Almost all sea life is not only edible, but is also an excellent source of nutrients essential to humans. The protein is complete because it contains all the essential amino acids, and the fats are similar to those of vegetables. Sea foods are high in minerals and vitamins. The majority of life in the sea (fish, birds, plants, and aquatic animals) is edible.

17.8.1. Seaweed

Most seaweed is edible and a good source of food, especially for vitamins and minerals. Some seaweed contains as much as 25 percent protein, while others are composed of over 50 percent carbohydrates. At least 75 different species are used for food by seacoast residents around the world. For many people, especially the Japanese, seaweeds are an essential part of the diet, and the most popular varieties have been successfully farmed for hundreds of years. The high cellulose content may require gradual adaptation because of their laxative quality if they comprise a large part of the diet. As with vegetables, some species are more flavorful than others. Generally, leafy green, brown, or red seaweeds can be washed and eaten raw or dried. The following list of edible seaweeds gives a description of the plant, tells where it may be found, and in many cases, suggests a method of preparation:

17.8.1.1. Green Seaweed

Common green seaweeds (Figure 17-59), often called sea lettuce, are in abundance on both sides of the Pacific and North Atlantic oceans. After washing it in clean water, it can be used as a garden lettuce.

Figure 17-59 Edible Green Seaweeds

17.8.1.2. Brown Seaweed

The most common edible brown seaweeds are the sugar wrack, kelp, and Irish moss (Figure 17-60).

17.8.1.2.1. Sugar Wrack

The young stalks of the sugar wrack are sweet to taste. This seaweed is found on both sides of the Atlantic and on the coasts of China and Japan.

17.8.1.2.2. Edible Kelp

Edible kelp has a short cylindrical stem and thin, wavy olive-green or brown fronds that grow one to several feet in length. It is found in the Atlantic and Pacific oceans, usually below the high-tide line on submerged ledges and rocky bottoms. Kelp should be boiled before eating. It can be mixed with vegetables or soup.

17.8.1.2.3. Irish Moss

Irish moss, a variety of brown seaweed, is quite edible, and is often sold in market places. It is found on both sides of the Atlantic Ocean and can be identified by its tough, elastic, and leathery texture, however, when dried, it becomes crisp and shrunken. It can be found at or just below the high-tide line and is sometimes found cast upon the shore. It should be boiled before eating.

Figure 17-60 Edible Brown Seaweeds

17.8.1.3. Red Seaweeds

Red seaweeds can usually be identified by their characteristic reddish tint, especially the edible varieties. The most common and edible red seaweeds include the dulse, laver, and other warm-water varieties (Figure 17-61).

17.8.1.3.1. Dulse

Dulse has a very short stem which quickly broadens into a thin, broad, fan-shaped expanse which is dark red and divided by several clefts into short, round-tipped lobes. The entire plant is from a few inches to a foot in length. It is found attached to rocks or coarser seaweeds, usually at the low-tide level, on both sides of the Atlantic Ocean and in the Mediterranean. Dulse is leathery in consistency and is sweet to the taste. If dried and rolled, it can be used as a substitute for tobacco.

17.8.1.3.2. Laver

Laver is usually red, dark purple, or purplish-brown, and has a satiny sheen or filmy luster. Common to both the Atlantic and Pacific oceans, it has been used as food for centuries. This seaweed is used as a relish, or is cleaned and then boiled gently until tender. It can also be pulverized and added to crushed grains and fried in the form of flat cakes. During World War II, laver was chewed for its thirst-quenching value by New Zealand troops. Laver is usually found on the beach at the low-tide level.

17.8.1.3.3. Warm Water Seaweed

A great variety of red, warm-water seaweed is found in the South Pacific area. This seaweed accounts for a large portion of the native diet. When found on the open sea, bits of floating seaweed may not only be edible but often contains tiny animals that can be used for food. The small fish and crabs can be dislodged by shaking the clump of seaweed over a container.

Figure 17-61 Edible Red Seaweeds

17.8.2. Plankton

Plankton includes both minute plants and animals that drift about or swim weakly in the ocean. These basic organisms in the marine food chain are generally more common near land since their occurrence depends upon the nutrients dissolved in the water. Plankton can be caught by dragging a net or sea anchor through the water. The taste of the plankton will depend upon the types of organisms predominant in the area. If the population is mostly fish larvae, the plankton will taste like fish. If the population is mostly crab or shellfish larvae, the plankton will taste like crab or shellfish. Plankton contains valuable protein, carbohydrates and fats. Because of its high chiton and cellulose content, however, plankton cannot be immediately digested in large quantities. Therefore, anyone subsisting primarily on a plankton diet must gradually increase the quantities consumed. Most of the planktonic algae are smaller than the planktonic animals and, although edible, are less palatable. Some plankton algae, for example, those that cause "red tides" and paralytic shellfish poisoning, are toxic to humans.

17.8.2.1. Plankton as a Food Source

If IP are going to use plankton as a food source, there must be a sufficient supply of fresh water for drinking. Each plankton catch should be examined to remove all stinging tentacles broken from jellyfish or Portuguese man-of-war. The primarily gelatinous species may also be selectively discarded since their tissues are predominately composed of saltwater. When the plankton is found in subtropical waters during the summer months, and the presence of poisonous dinoflagellates is suspected (due to discoloration or high luminescence of the ocean), the edibility test should be applied before eating.

17.8.2.2. Final Precaution

The final precaution before ingesting plankton is to feel or touch the plankton to check for species that are especially spiny. The catch should be sorted (visually) or dried and crushed before eating if it contains large numbers of these spiny species.

17.8.3. Fishing

If a fishing kit is available, the task of fishing will be made much easier. Small fish will usually gather under the shadow of the raft or in clumps of floating seaweed. These fish can be eaten or used as bait for larger fish. A net can also be used to procure most sea life. Also, a light source such as a flashlight or full moon can be used to attract certain types of fish. It is not advisable to secure fishing lines to the body or the raft. A large fish may pull a person out of the raft or damage the raft itself. Fish, bait, or bright objects dangling in the water can attract large dangerous fish. All large fish should be killed outside the raft by a blow to the head or by cutting off the head.

17.8.4. Sea Birds

Sea birds have proven to be a useful food source which may be more easily caught than fish. IP have reported capturing birds by using baited hooks, by grabbing, and by shooting. Freshly killed birds should be skinned, rather than plucked, to remove the oil glands. They can be eaten raw, sun dried, or cooked. The gullet contents can be a good food source. The flesh should be eaten or preserved immediately after cleaning. The viscera, along with any other unused parts, make good fish bait.

17.8.5. Marine Mammals

Marine mammals are rarely encountered by a person in the water, although they may be seen from a distance. Any large mammal is capable of inflicting injuries but they will generally avoid people. The killer whale (Orca) is rarely seen and, although large enough to feed on humans, has never been known to do so. Almost all sea mammals are a good source of food but difficult to obtain. The liver, especially of any arctic or cold-water mammal, should not be eaten because of toxic concentrations of vitamin A.

17.8.6. Sea Life

All sea life must be cleaned, cut up, and eaten as soon as possible to avoid spoilage. Any meat left over can be preserved by sun-drying or smoking. The internal parts can be used as bait. General rules should be followed to identify any features that would indicate that sea life is inedible prior to preparation.

17.9. Preparing Game Food

IP must know how to use the meat of game and fish to their advantage and how to do this with the least effort and physical exertion. Many people have died from starvation because they had failed to take full advantage of a game carcass. They abandoned the carcass on the mistaken theory that they could get more game when needed.

17.9.1. Large Animals

If the animal is large, the first impulse is usually to pack the meat to camp. In some cases, it might be easier to move the camp to the meat. A procedure often advocated for transporting the kill is to use the skin as a sled for dragging the meat. When the entire animal is dragged, this method may prove satisfactory only on frozen lakes or rivers or over very smooth snow-covered terrain. In rough or brush-covered country, however, it is generally more difficult to use this method, although it will work. Large mountain animals can sometimes be dragged down a snow-filled gully to the base of the mountain. If meat is the only consideration, and IP do not care about the condition of the skin, mountain game can sometimes be rolled for long distances. Before transporting a whole animal, it should be gutted and the incision closed. Once the bottom of the hill is reached, almost invariably the method is either to backpack the meat to camp, making several trips if no other IP are present, or to pack the camp to the animal. When the weight of the meat proves excessive and moving the camp is not practical, some of the meat could be eaten at the scene. The heart, liver, and kidneys should be eaten as soon as possible to avoid spoilage.

17.9.1.1. Skinning and Butchering

Skinning and butchering must be done carefully so that all edible meat can be saved. When the decision is made to discard the skin, a rough job can be done. However, considerations should be given to possible uses of the skin. A square of fresh skin, long enough to reach from the head to the knees, will not weigh much less when it is dried, and is an excellent ground cloth for use under a sleeping bag on frozen ground or snow. The best time to skin and butcher an animal is immediately after the kill. However, if an animal is killed late in the day, it can be gutted immediately and the other work done the next morning. An effort to keep the carcass secure from predators should be made.

17.9.1.2. Edible Fat

When preparing meat, all edible fat should be saved. This is especially important in cold climates, as the diet may consist almost entirely of lean meat. Fat must be eaten in order to provide a complete diet. Rabbit meat tends to have very little fat content. Rabbit starvation, also referred to as protein poisoning or mal de caribou, is a form of acute malnutrition caused by excessive consumption of any lean meat (e.g., rabbit and birds such as ptarmigan) coupled with a lack of other sources of nutrients in combination with other stressors, such as severe cold or dry environment. Symptoms include diarrhea, headache, fatigue, low blood pressure and heart rate, and a vague discomfort and hunger that can only be satisfied by consumption of fat or carbohydrates. Some early arctic explorers died after an extended diet consisting only of rabbit meat. The fact that a person will die after an extended diet consisting only of rabbit meat indicates the importance of fat in a primitive diet. The same is true of birds, such as the ptarmigan.

17.9.1.3. Birds

Birds should be handled in the same manner as other animals. They should be cleaned after killing and protected from flies. Birds, with the exception of sea birds, should be plucked and cooked with the skin on. Carrion-eating birds, such as vultures, must be boiled for at least 20 minutes to kill parasites before further cooking and eating. Fish-eating birds have a strong, fish-oil flavor. This may be lessened by baking them in mud or by skinning them before cooking.

17.9.2. General Ways to Skin Animals

There are two general ways to skin animals depending upon the size: the big game method, or the glove skinning method.

17.9.2.1. Big Game Skinning Method

IP should use the big game method when skinning and butchering large game.

17.9.2.1.1. First Steps in Big Game Skinning

The first step in skinning is to turn the animal on its back and with a sharp knife, cut through the skin on a straight line from the tail bone to a point under its neck as illustrated in Figure 17-62. In making this cut, pass around the anus and, with great care, press the skin open until the first two fingers can be inserted between the skin and the thin membrane enclosing the entrails. When the fingers can be forced forward, place the blade of the knife between the fingers, blade up, with knife held firmly. While forcing the fingers forward, palm upward, follow with the knife blade, cutting the skin but not cutting the membrane.

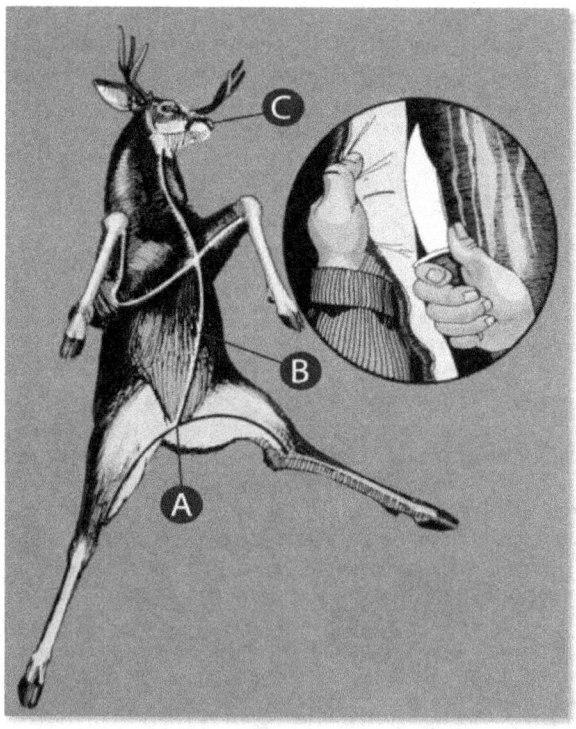

Figure 17-62 Big Game Skinning

17.9.2.1.2. Male Animals

If the animal is a male, cut the skin parallel to, but not touching, the penis. If the tube leading from the bladder is accidentally cut, a messy job and unclean meat will be the result. If the gall or urine bladders are broken, washing will help clean the meat. Otherwise, it is best not to wash the meat but to allow it to form a protective glaze.

17.9.2.1.3. Reaching the Ribs

On reaching the ribs, it is no longer possible to force the fingers forward, because the skin adheres more strongly to flesh and bone. Furthermore, care is no longer necessary. The cut to point C (figure17-62) can be quickly completed by alternately forcing the knife under the skin and lifting it. With the central cut completed, make side cuts consisting of incisions through the skin, running from the central cut (A-C figure 17-62) up the inside of each leg to the knee and hock joints. Make cuts around the front legs just above the knees and around the hind legs above the hocks. Make the final cross cut at point C, and then cut completely around the neck and in back of the ears. Once the cuts are complete, the IP can begin skinning.

17.9.2.1.4. Skinning Down the Sides

After skinning down the animal's side as far as possible, roll the carcass on its side to skin the back. Then spread out the loose skin to prevent the meat from touching the ground and turn the animal on the skinned side. Follow the same procedure on the opposite side until the skin is free.

17.9.2.1.5. The Membrane Enclosing the Entrails

In opening the membrane which encloses the entrails, follow the same procedure used in cutting the skin by using the fingers of one hand as a guard for the knife and separating the intestines from the membrane. This thin membrane along the ribs and sides can be cut away in order to see better. Be careful to avoid cutting the intestines or bladder. The large intestine passes through an aperture in the pelvis. This tube must be separated from the bone surrounding it with a knife. Tie a knot in the bladder tube to prevent the escape of urine. With these steps completed, the entrails can be easily disengaged from the back and removed from the carcass. Another method of gutting or field dressing is shown in Figure 17-63. After gutting is completed, it may be advisable to hang the animal. (Note: If it is hot, gut the animal before skinning it.)

Figure 17-63 Field Dressing

17.9.2.1.6. The Intestines

The intestines of a well-conditioned animal are covered with a lace-like layer of fat which can be lifted off and placed on nearby bushes to dry for later use. The gall bladder which is attached to the liver of some animals should be carefully removed. If it should happen to rupture, the bile will taint anything it touches. Be sure to clean the knife if necessary. The kidneys are imbedded in the back, forward of the pelvis, and are covered with fat. Running forward from the kidneys on each side of the backbone are two long strips of chop-meat or muscle called tenderloin and/or back strap. Eat this after the liver, heart, and kidneys as it is usually very tender. Edible meat can also be removed from the head, brisket, ribs, backbone, and pelvis.

17.9.2.1.7. Large Animals should be Quartered

Large animals should be quartered. To do this, cut down between the first and second rib and then sever the backbone with an axe or machete. Cut through the brisket of the front half and then chop lengthwise through the backbone to produce the front quarters. On the rear half, cut through the pelvic bone and lengthwise through the backbone. To make the load lighter and easier to transport, a knife could be used to bone the animal, thereby eliminating the weight of the bones. Butchering is the final step and is simplified for survival purposes. The main purpose is to cut the meat in manageable size portions.

17.9.2.1.8. Glove Skinning

Glove skinning is usually performed on small game (Figure 17-64).

Figure 17-64 Glove Skinning

17.9.2.1.9. Initial Cuts

The initial cuts are made down the insides of the back legs. The skin is then peeled back so that the hindquarters are bare and the tail is severed. To remove the remaining skin, pull it down over the body in much the same way a pullover sweater is removed. The head and front feet are severed to remove the skin from the body. For one-cut skinning of small game, cut across the lower back and insert two fingers under each side of the slit. By pulling quickly in opposite directions, the hide will be easily removed (Figure 17-65).

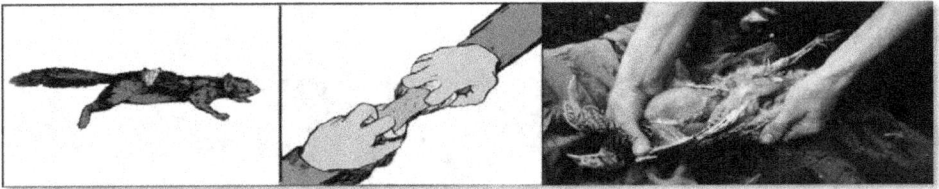

Figure 17-65 Small Animal Skinning

17.9.2.1.10. Removal of Internal Organs

To remove the internal organs, a cut should be made into the abdominal cavity without puncturing the organs. This cut must run from the anus to the neck. There are muscles which connect the internal organs to the trunk and they must be severed to allow the viscera to be removed. A rabbit may be gutted by using a knife-less method with no mess and little time lost. Squeeze the entrails toward the rear resulting in a tight bulging abdomen. Raise the rabbit over the head and sling it down hard striking the forearms against the thighs. The momentum will expel the entrails through a tear in the vent. Save the internal organs such as heart, liver, and kidneys, as they are nutritious. The liver should be checked for any white blotches and discarded if affected as these indicate tularemia (also known as rabbit fever). The disease is transmitted by rodents but also infects humans.

17.9.3. Cold-Blooded Animals

Cold-blooded animals are generally easy to clean and prepare.

17.9.3.1. Snakes and Lizards

Snakes and lizards are very similar in taste and they have similar skin. Like the mammals, the skin and viscera should be removed. The easiest way to do this is to sever the head and (or) legs. In the case of a lizard, peel back enough skin so that it may be grasped securely and simply pull it down the length of the body turning the skin inside out as it goes. If the skin does not come away easily, a cut down the length of the animal can be made. This will allow the skin to part from the body more easily. The entrails are then removed and the animal is ready to cook. As another option, after gutting the lizard, IP could also char the lizard over a fire, split it down the middle, and eat the meat without skinning first.

17.9.3.2. Amphibians

Except for the larger amphibians such as the bullfrog, the hind legs are the largest portion of the animal worth saving. Remove the hindquarters, by cutting through the backbone, leaving the abdomen and upper body. Pull the skin from the legs and they are ready to cook. With the bullfrogs and larger amphibians, the whole body can be eaten. The head, the skin, and viscera should be removed and discarded (use as bait to catch something else).

17.9.4. Fish

Most fish need little preparation before they are eaten. Scaling the fish before cooking is not necessary. A cut from the anus to the gills will expose the internal organs which should be removed. The gills should also be removed before cooking. The black line along the inside of the backbone is the kidney and should be removed by running a thumbnail from the tail to the head. There is some meat on the head and should not be discarded. See Figure 17-66 for one method of filleting a fish.

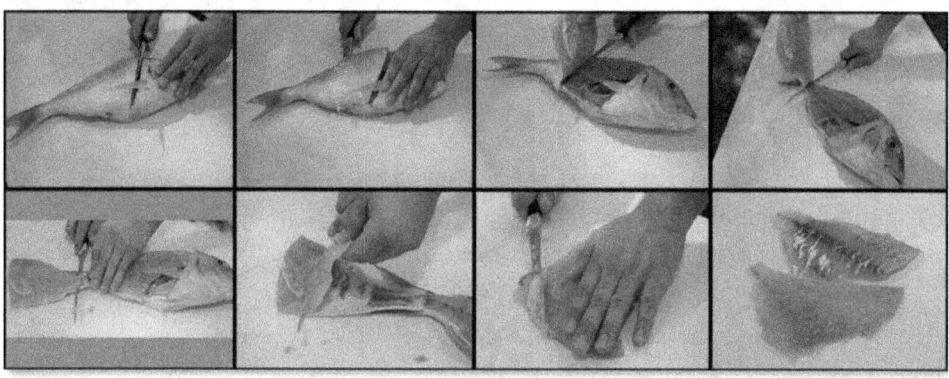

Figure 17-66 Filleting a Fish

17.9.5. Birds

All birds have feathers which can be removed in two ways: by plucking or by skinning. The gizzard, heart, and liver should be retained. The gizzard should be split open as it contains partially digested food and stones which must be discarded before being eaten.

17.9.6. Insects

Insects are an excellent food source and they require little or no preparation. The main point to remember is to remove all hard portions such as the hind legs of a grasshopper and the hard wing covers of beetles. The rest is edible.

17.10. Preparing Plant Food

Preparing plant foods can be more involved than preparing animal life.

17.10.1. Plant Foods Containing Tannin

Some plant foods, such as acorns and tree bark may be bitter because of tannin. These plants will require leaching by chopping up the plant parts, and pouring several changes of fresh water over them. This will help wash out the tannin, making the plant more palatable. Other plants such as cassava and green papaya must be cooked before eating to break down the harmful enzymes and chemical crystals within them and make them safe to eat. Plants such as skunk cabbage must undergo this cooking process several times before it is safe to eat.

17.10.2. Starchy Foods

Starchy foods should be cooked since raw starch is difficult to digest. They can be boiled, steamed, roasted, or fried and are eaten plain, or mixed with other wild foods. The maniac (cassava) must be cooked, because the bitter form (green stem) is poisonous when eaten raw. Starch is removed from sago palm, cycads, and other starch-producing trunks by splitting the trunk and pounding the soft, whitish inner parts with a pointed club. This pulp is washed with water and the white sago (pure starch) is drained into a container. It is washed a second time, and then it may be used directly as a flour. One trunk of the sago palm will supply IP starch needs for many weeks.

17.10.3. Fiddlehead Ferns

The fiddlehead ferns are the curled, young succulent fronds which have the same food value as cabbage or asparagus. Practically all types of fiddleheads are covered with hair which makes them bitter. The hair can be removed by washing the fiddleheads in water. If fiddleheads are especially bitter, they should be boiled for 10 minutes and then re-boiled in fresh water for 30 to 40 minutes. Wild bird eggs or meat may be cooked with the fiddleheads to form a stew.

17.10.4. Wild Grasses

Wild grasses have an abundance of seeds and may be eaten either boiled or roasted after separating the chaff from the seeds which can be done by rubbing. No known grass is poisonous. If the kernels are still soft and do not have large stiff barbs attached, they may be used for porridge. If brown or black rust is present, the seeds should not be eaten (Ergot Poisoning). To gather grass seeds, a cloth can be placed on the ground and the grass heads beaten with sticks.

17.10.5. Plants that Thrive in Wet Places

Plants that grow in wet places (along rivers, lakes, and ponds or directly in water) are of potential value as survival food. The succulent underground parts and stems are most frequently eaten. Various members of the calla lily family (which have arrowhead shaped leaves), often grow in very wet places in the Tropics. Jack-in-the-pulpit, calla lily, and sweet flag are members of the Arum family. To be eaten, the members of this plant family must be cooked in frequent changes of water to destroy the irritant crystals in the stems. Two kinds of marsh and water plants are the cattail and the water lily.

17.10.5.1. Cattail

The cattail is found worldwide except in tundra regions of the far north (Figure 17-67). Cattails can be found in the more moist places in desert areas of all continents as well as in the moist tropic and temperate zones of both hemispheres. The young shoots taste like asparagus. The spikes can be boiled or steamed when green and then eaten. The rootstalks, without the outer covering, are eaten boiled or raw. Cattail roots can be cut into thin strips, dried, and then ground into flour. They are 46 percent starch, 11 percent sugar, and the rest is fiber. While the plant is in flower, the yellow pollen is very abundant; this may be mixed with water and made into small cakes and steamed as a substitute for bread.

Figure 17-67 Cattails

17.10.5.2. Water Lilies

Water lilies grow on all the continents, but principally in southern Asia, Africa, North America, and South America (Figure 17-68). The two main types are: 1) Temperate water lilies which produce enormous rootstalks and yellow or white flowers which float on the water. 2) Tropical water lilies which produce large edible tubers and flowers which are elevated above the water surface.

Figure 17-68 Water Lilies

17.10.5.3. Rootstalks and Tubers

Rootstalks or tubers may be difficult to obtain because of deep water. They are starchy and high in food value. They can be eaten either raw or boiled. Stems may be cooked in a stew. Young seed pods may be sliced and eaten as a vegetable. Seeds may be bitter, but are very nourishing. They may be parched and rubbed between stones as flour. The water lily is considered an important food source by native peoples in many parts of the world.

17.10.6. Nuts

Nuts are very high in nutritional value and usually can be eaten raw. Nuts may be roasted in the fire or roasted by shaking them in a container with hot coals from the fire. They may then be ground to make flour. If IP do not wish to eat a plant or its parts raw, it can be cooked using the same methods used in cooking meat; by boiling, roasting, baking, broiling, or frying.

17.10.7. Preserving Plant Foods

If IP have been able to gather more plant foods than can be eaten, the excess can be preserved in the same manner as animal foods. Plant foods can be dried by wind, air, sun, or fire, with or without smoke. A combination of these methods can also be used. The main object is to remove the moisture. Most wild fruits can be dried. If the plant part is large, such as some tubers, it should be sliced, and then dried. Some type of protection may be necessary to prevent consumption and (or) contamination by insects. Extra fruits or berries can be carried with IP by wrapping them in leaves or moss.

17.11. Cooking

All wild game, large insects, freshwater fish, clams, mussels, snails, and crawfish must be thoroughly cooked to kill internal parasites. Mussels and large snails may have to be minced to make them tender.

17.11.1. Boiling

Boiling is the most nutritious, simplest, and safest method of cooking (Figure 17-69). Numerous containers can be used for boiling; for example, a metal container suspended above, or set beside, a heat source to boil foods. Green bamboo makes an excellent cooking container. Stone boiling is a method of boiling using super-heated rocks and a container that holds water but cannot be suspended over an open flame. Examples of these containers are survival kit containers, flying helmet, a hole in the ground lined with waterproof material, or a hollow log. The container is filled with food and water and then heated with super-hot stones until the water boils. Stones from a stream or damp area should not be used. The moisture in the stones may turn to steam and cause the stone to explode while the stones are being heated in the fire. The container should be covered and new stones added as the water stops boiling. The rocks can be removed with the aid of a wire secured to the rock before being put into the container or two sticks used in a chopstick fashion.

Figure 17-69 Boiling

17.11.2. Baking

Baking is a good method of cooking as it is slow and is usually done by putting food into a container and cooking it slowly. Baking is often used with various types of ovens. Foods may be wrapped in wet leaves (Figure 17-70) (avoid using a type of plant that will give an unpleasant flavor to what is being cooked), placed inside a metal container, or they may be packed with mud or clay and placed directly on the coals. Fish and birds packed in mud and baked must not be skinned because the scales, skin, or feathers will come off the animal when the mud or clay is removed. Clambake-style baking is done by heating a number of stones in a fire and allowing the fire to burn down to coals. A layer of wet seaweed or leaves is then placed over the hot rocks. Food such as mussels and clams in their shells are then placed on the wet seaweed and (or) leaves (Figure 17-71). More wet seaweed and (or) leaves and soil is used as a cover. When thoroughly steamed in their juices, clam, oyster, and mussel shells will open and may be eaten without further preparation.

Figure 17-70 Baking

Figure 17-71 Clam Baking

17.11.3. Improvised Ovens

Any type of food can be cooked in the ground in a rock oven. First, a hole is dug about two feet deep and two to three feet square, depending on the amount of food to be cooked. The sides and bottom are then lined with rock. Next, procure several green trees about six inches in diameter and long enough to bridge the hole. Firewood and grass or leaves for insulation should also be gathered. A fire is started in the hole. Two or three green trees are placed over the hole and several rocks are placed on the trees. The fire must be maintained until the green trees burn through. This indicates the fire has burned long enough to thoroughly heat the rocks and the oven is ready. The fallen rocks, fire, and ash are removed from the hole and a thin layer of dirt is spread over the bottom. The insulating material (grass, leaves, moss, etc.) is placed over the soil, then the food, then more insulating material on top and around the food, another thin layer of soil, and the extra hot rocks are placed on top. The hole is then filled with soil up to ground level. Small pieces of meat (steaks, chops, etc.) cook in one and a half to two hours and large pieces take five to six hours.

17.11.4. Roasting

Roasting is less desirable as it involves exposing the food to direct heat which quickly destroys the nutritional properties (Figure 17-72). Putting a piece of meat on a stick and holding it over the fire is considered roasting.

17.11.5. Broiling

Broiling is the quickest way to prepare fish. A rock broiler may be made by placing a layer of small stones on top of hot coals, and laying the fish on the top. Scaling the fish before cooking is not necessary, and small fish need not be cleaned. Cooked in this manner, fish have a moist and delicious flavor. Crabs and lobsters may also be placed on the stones and broiled.

Figure 17-72 Broiling and Roasting

17.11.6. Planking

Meat may be cooked by placing it on a flat board or stone (planking) which is propped up close to the fire (Figure 17-73). The meat will have to be turned over at least once to allow thorough cooking. The cooking time depends on how close the meat is to the fire.

Figure 17-73 Planking

17.11.7. Frying Food

Frying is by far the least favorable method of preparing food. Nearly all of the natural juices are cooked out of the meat which tends to make it tough and some of the nutritional value of the meat will destroyed. Frying can be done on any nonporous surface which can be heated. Examples are unpainted metal aircraft or vehicle parts, large seashells, flat rocks, and some survival kit parts. Turtle shells can also be considered but are often thought to be too valuable as a helmet to be used for this purpose.

17.12. Preserving Food

Finding natural foods is an uncertain aspect of survival. IP must make the best use of the available food. Food, especially meat, has a tendency to spoil within a short period of time unless it is preserved. There are many ways to preserve food including cooking, refrigeration, freezing, and dehydration.

17.12.1. Cooking Methods

Cooking will slow down the decomposition of food but will not eliminate it. This is because many bacteria are present which work to break it down. Cooking methods which are the best for immediate consumption, such as boiling, are the least effective for preserving food. Food should be re-cooked every day until all is consumed.

17.12.2. Storing Food

Cooling is an effective method of storing food for short periods of time. Heat tends to accelerate the decomposition process where cooling retards decomposition. The colder food becomes, the less the likelihood of deterioration until freezing which eliminates decomposition. Cooling devices available to IP include:

- Food items buried in snow will maintain a temperature of approximately 32°F.
- Food wrapped in waterproof material and placed in streams will remain cool in summer months. Care should be taken to ensure food is secured.
- Below the Earth's surface (particularly in shady areas or along streams), remains cooler than the surface. A hole may be dug, lined with grass, and covered to form an effective cool storage area similar to a root cellar.
- When water evaporates, it tends to cool down the surrounding area. Articles of food may be wrapped in an absorbent material such as cotton or burlap and rewetted as the water evaporates.
- Once food is frozen, it will not decompose. Food should be frozen in meal-sized portions so refreezing is avoided.
- Drying removes all moisture and preserves the food. Drying is done by sunning, smoking, or burying it in hot sand.

17.12.2.1. Sun-Drying Foods

For sun-drying, the food should be sliced very thin and placed in direct sunlight. Meat should be cut across the grain to improve tenderness and decrease drying time. If salt is available, it should be added to improve flavor and accelerate the drying process (Figure 17-74).

Figure 17-74 Sun-Drying

17.12.2.2. Smoking

Smoking is a process done through the use of non-resinous wood such as willow or aspen to produce smoke which adds flavor and dries the meat. A smoke rack is also necessary to contain the smoke. The following are the procedures for drying meat using smoke:

1. Cut meat very thin and across the grain. If the meat is warm and difficult to slice thin, cut the meat in one or two inch cubes and beat it thin with a clean wooden mallet (improvised).
2. Remove fat.
3. Hang the meat on a rack so each piece is separate.
4. Elevate meat no less than two feet above coals.
5. Coals are placed in the bottom of a smoke rack with green woodchips on top to produce smoke.

17.12.3. Preserving Fish

The method used to preserve fish through warm weather is similar to that used in preserving other meat. When there is no danger of predatory animals disturbing the fish, the fish should be placed on available fabric and allowed to cool during the night.

- Fish may be dried in the same manner described for smoking meat. To prepare fish for smoking, the heads and backbone are removed and the fish are spread flat on a grill. Thin willow branches with bark removed make skewers.
- Fish may also be dried in the Sun. They can be suspended from branches or spread on hot rocks. When the meat has dried, sea water or salt should be used on the outside, if available.

17.12.4. Protecting Food Sources

Many animals and insects will devour the IPs' food if it is not correctly stored. Protecting food from insects and birds is done by wrapping it in material, wrapping and tying brush around the bundle, and finally, wrapping it with another layer of material. This creates "dead air" space making it more difficult for insects and birds to get to the food. If the outer layer is wetted, evaporation will also cool the food to some degree. In most cases, if the food is stored several feet off the ground, it will be out of reach of most animals. This can be done by hanging the food or putting it into a "cache". If the food is dehydrated, the container must be completely waterproof to prevent re-absorption. Frozen food will remain frozen only if the outside temperature remains below freezing. Burying food is a good way to store as long as scavengers are not in the area to uncover it. Insects and small animals should also be remembered when burying the food. Food should never be stored in the shelter as this may attract wild animals and could be hazardous to IP.

Chapter 18

LAND NAVIGATION

18. Land Navigation

IP must know their location in order to intelligently decide if they should wait for rescue or if they should determine a destination and (or) route to travel. If the decision is to stay, IP need to know their location in order to radio the information to rescue personnel. If the decision is to travel, IP must be able to use navigational tools to determine the best routes of travel, location of possible food and water, and hazardous areas which they should avoid. This chapter provides background information in the use of the maps, compasses, and other navigational tools.

18.1. Maps

A map is a pictorial representation of the Earth's surface drawn to scale and reproduced in two dimensions. Every map should have a title, legend, scale, north arrow, grid system, and contour lines. With these components, IP can determine the portion of the Earth's surface the map covers. IP should be able to understand all of the markings on the map and use them to their advantage. They should also be able to determine the distance between any two points on the map and be able to align the map with true north so it conforms to the actual features on the ground.

18.1.1. Description

A map is a conceptual picture of the Earth's surface as seen from above, simplified to bring out important details and lettered for added identification. A map represents what is known about the Earth rather than what can be seen by an observer. However, a map is selective in that only the information which is necessary for its intended use is included on any one map. Maps also include features which are not visible on Earth, such as parallels, meridians, and political boundaries.

18.1.2. Map Distortion

Since it is impossible to accurately portray a round object, such as the Earth, on a flat surface, all maps have some elements of distortion. Depending on the intended use, some maps sacrifice constant scale for accuracy in measurement of angles, while others sacrifice accurate measurement of angles for a constant scale. However, most maps used for ground navigation use a compromise projection in which a slight amount of distortion is introduced into the elements which a map portrays, but in which a fairly true picture is given.

18.1.2.1. Planimetric Map

A planimetric map presents only the horizontal positions for the features represented. It is distinguished from a topographic map by the omission of relief in a measurable form.

18.1.2.2. Topographic Map

A topographic map (Figure 18-1) portrays terrain and landforms in a measurable form and the horizontal positions of the features represented. The vertical positions, or relief, are normally represented by contours. On maps showing relief, the elevations and contours are measured from a specified vertical datum plane and usually mean sea level.

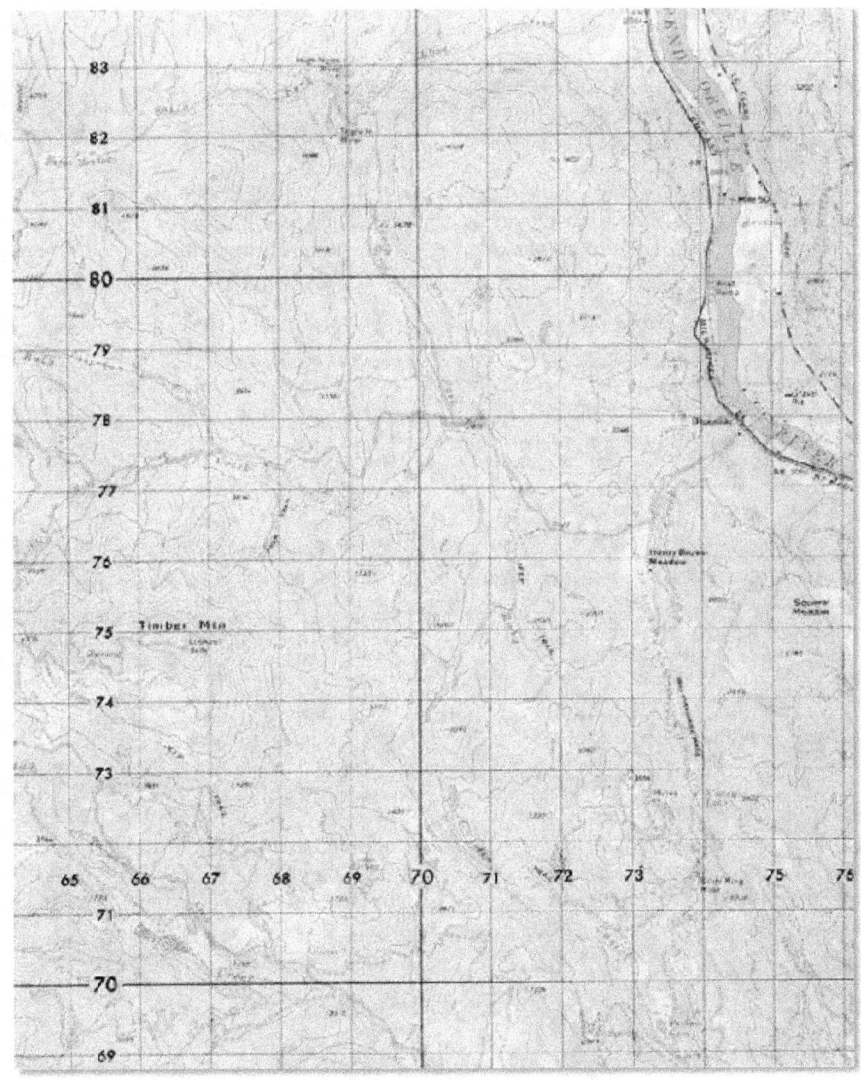

Figure 18-1 Topographical Map

18.1.2.3. Plastic Relief Map

A plastic relief map is a reproduction of an aerial photograph or a photo mosaic made from a series of aerial photographs upon which grid lines, marginal data, place names, route numbers, important elevations, boundaries, approximate scale, and approximate direction have been added.

18.1.2.4. PICTOMAP

A PICTOMAP is the acronym for photographic image conversion by tonal masking procedures. It is a map on which the photographic imagery of a standard photomap has been converted into interpretable colors and symbols.

18.1.2.5. Photo Mosaic

A photo mosaic is an assembly of aerial photographs and is commonly called a mosaic in topographic usage. Mosaics are useful when time does not permit the compilation of a more accurate map. The accuracy of a mosaic depends on the method used in its preparation and may vary from simply a good pictorial effect of the ground to that of a planimetric map.

18.1.2.6. Military City Map

A military city map is a topographic map, usually 1:12,500 scale, of a city, outlining streets and showing street names, important buildings, and other urban elements of military importance which are compatible with the scale of the map. The scales of military city maps can vary from 1:25,000 to 1:5,000, depending on the importance and size of the city, density of detail, and available intelligence information.

18.1.2.7. Special Maps

Special maps are for special purposes such as traffic ability, communications, and assault. These are usually overprinted maps of scales smaller than 1:100,000 but larger than 1:1,000,000. Other types of special maps are those made from organosol or materials other than paper to meet the requirements of special climatic conditions.

18.1.2.8. Terrain Model

A terrain model is a scale model of the terrain showing landforms, and in large scale models, industrial and cultural shapes. It is designed to provide a means for visualizing the terrain for planning or indoctrination purposes and for briefing on assault landings.

18.1.2.9. Special Purpose Map

A special purpose map is one that has been designed or modified to give information not covered on a standard map or to elaborate on standard map data. Special purpose maps are usually in the form of an overprint. Overprints may be in the form of individual sheets or combined and bound into a study of an area. A few of the items covered include:

- Landform
- Drainage characteristics
- Vegetation
- Climate
- Coast and landing beaches
- Railroads
- Airfields
- Urban areas

- Electric power
- Fuels
- Surface water resources
- Ground water resources
- Natural construction materials
- Cross-country movement
- Suitability for airfield construction
- Airborne operations

18.2. Aeronautical Charts

Air navigation and planning charts are used for flight planning. Each different series of charts is constructed at a different scale and format to meet the needs of a particular type of air navigation. The air navigation and planning charts are smaller in scale and less detailed than Army maps or air target materials. The control of positional error is less critical. The following list includes the charts most commonly used in intelligence operations. A description of each chart follows the listing:

CHART	SCALE	CODE
USAF Global Navigation and Planning Chart	1:5,000,000	GNC
USAF Jet Navigation Chart	1:3,000,000	JNC-A
USAF Operational Navigation Chart	1:1,000,000	ONC
USAF Tactical Pilotage Chart	1:500,000	TPC
USAF Jet Navigation Chart	1:2,000,000	JN
Joint Operations Graphic	1:250,000	JOG

18.2.1. Global Navigation and Planning Chart (GNC)

The GNC is designed for general planning purposes where large areas of interest and long-distance operations are involved. It serves as a navigation chart for long-range, high-altitude, and high-speed aircraft since sheet lines have been selected on the basis of primary areas of strategic interest. Some of these charts are produced on selected areas of strategic interest; others provide wide coverage. All general planning charts are produced at a small or very small scale which provides extensive area coverage on a single sheet.

18.2.2. USAF Jet Navigation Chart (JN/JNC-A)

The basic JNC is produced at a scale of 1:2,000,000. The JNC-A is produced on the north polar area and in the United States at a scale of 1:3,000,000. Both jet navigation charts are printed on 41 1/2- by 57 1/2-inch sheets.

18.2.2.1. JN Chart

The JN chart is used for preflight planning and en route navigation by long-range jet aircraft. The charts are designed so they can be joined to produce a strip chart which provides the necessary navigational information for any intended course. Relief is indicated through the use of contours, spot elevations, and gradient tints. Large, level terrain areas are indicated by a symbol that consists of narrow, parallel lines with the elevation annotated within the symbol.

- Principal cities and towns and principal roads and rail networks are shown on the JN chart. The transportation network is shown in the immediate area of populated places. Lakes and principal drainage patterns are also pictured. The elevations of major lakes are indicated so that the altitude may be determined by using the aircraft radar altimeter.

18.2.3. USAF Operational Navigation Chart (ONC)

The ONC was developed to meet military requirements for a chart adaptable to low-altitude navigation. The ONC is used for preflight planning and en route navigation. It is also used for operational planning, intelligence briefing and plotting, and flight planning displays.

18.2.3.1. ONC Description

This chart covers an area of $8°$ of latitude and $12°$ of longitude. ONC sheets are identified by combining a letter and a number. Letters identify $8°$ bands of latitude, starting at the North Pole and progressing southward. Numbers identify $12°$ sections of longitude from the prime meridian eastward. The successful execution of low-altitude missions depend upon visual and radar identification of ground features used as checkpoints and a rapid visual association of these features with their chart counterparts. The ONC portrays, by conventional signs and symbols, cultural features which have low-altitude checkpoint significance. Power lines are shown (except on cities) and are indicated by the usual line and pole symbol.

18.2.3.2. ONC Topographic Expression

The ONC portrays relief in perspective so that the user gets instantaneous appreciation of relative heights, slope gradients, and the forms of ground patterns. Topographic expression, illustrated basically with contours and spot elevations, is emphasized by the use of shaded relief and terrain characteristic tints defining the overall elevation levels. ONC contour intervals and terrain characteristic tints are selected regionally. This captures the relative significance of ground forms as a complete picture, and this feature aids preflight planning and in-flight identification.

18.2.4. USAF Tactical Pilotage Chart (TPC)

The TPC (Figure 18-2) is produced in a coordinated series at a scale of 1:500,000. Sheet sizes are the same dimensions as the ONC sheets; however, a TPC covers only one-fourth as much area as an ONC sheet. A TPC is identified by the ONC identification and the letter "A", "B", "C," or "D."

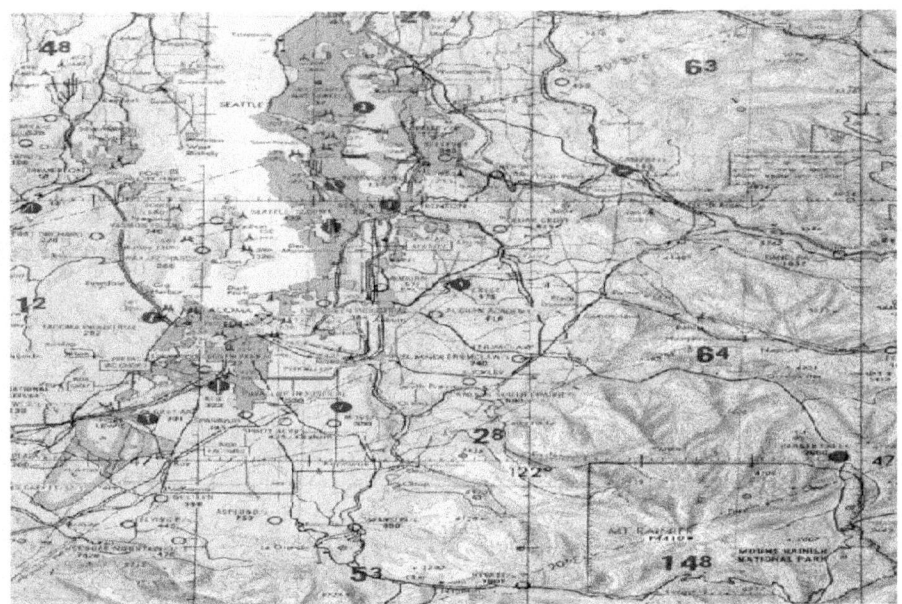

Figure 18-2 TPC Map

18.2.4.1. TPC Description

The TPC is used for detailed preflight planning and mission analysis. In designing the TPC, emphasis was placed on ground features which are significant for low-level, high-speed navigation, using visual and radar means. The selected ground features also permit immediate ground-chart orientation at predetermined checkpoints.

18.2.4.2. TPC Features

Relief on the TPC is displayed by contours (intervals may vary between 100 feet and 1,000 feet), spot elevations, relief shading, and terrain characteristic tints. Cultural features such as towns and cities, principal roads, railroads, power transmission lines, boundaries, and other features of value for low-altitude visual missions are included on the TPC. Pictorial symbols are used for features which provide the best checkpoints. Other features of the TPC which enhance its tactical air navigation qualities are as follows:

- UTM grid overprint.
- Vegetation color and symbol code.
- Enlarged vertical obstruction symbols.
- Enlarged road and railroad symbols.
- Emphasized radio aid to navigation symbols.
- Foreign place name glossary.

- Airdrome runway patterns to scale when information is available.
- Spot elevation, gradient tints, and shaded relief depicted for all elevations.
- The highest elevation for each 15-minute quadrangle is shown in thousands and hundreds of feet.

18.2.5. Joint Operations Graphic (JOG)

Joint Operations Graphics (JOGs) are a series of 1:250,000 scale military maps designed for joint ground and air operations. Both series emphasize the air-landing facilities but the air series has additional symbols to identify aids and obstructions to air navigation.

18.2.5.1. JOG Description

JOG was designed to provide a common-scale graphic for Army, Navy, and Air Force use. The Air Force makes use of it for tactical air operations, close air support, and interdiction by medium- and high-speed aircraft at low altitudes. The chart may also be used for dead reckoning and visual pilotage for short-range en route navigation. Due to its large scale, it is unsuitable for local area command planning for strategic and tactical operations.

18.2.5.2. JOG Topographic Expression

Relief on the JOG is indicated by contour lines (in feet). In some areas, the intervals may be in meters, with the approximate value in feet indicated in the margin of the chart. Spot elevations are used through all terrain levels. The ground series show elevations and contours in meters while the air series show the same elevations and contours in feet.

18.2.5.2.1. Shaded Relief

Relief is also shown through gradient tints, supplemented by shaded relief. The highest elevations in each 15-minute quadrangle are indicated in thousands and hundreds of feet.

18.2.5.2.2. Features

Cultural features, such as cities, towns, roads, trails, and railroads are illustrated in detail. The locations of boundaries and power transmission lines are also shown. Vegetation is shown by symbol. Detailed drainage patterns and water tint are used to illustrate water features, such as coastlines, oceans, lakes, rivers and streams, canals, swamps, and reefs. The JOG includes aeronautical information such as airfields, fixed radio navigation and communication facilities, and all known obstructions over 200 feet above ground. If the information is available, the airfield runway patterns are shown to scale by diagram.

18.2.5.2.3. JOG Numbering System

The basic numbering system of the JOG consists of two letters and a number which identifies an area 6° in longitude by 4° in latitude. If the chart covers an area north of the Equator, the first letter is "N;" a chart covering an area south of the Equator is identified with an initial "S." The second letter identifies the 4° bands of latitude lettered north and south from the Equator. The number identifies the 6° sections of longitude which are numbered from the 180° meridian eastward. The 6° x 4° areas identified by two letters and a number from one to 60 are further broken down to either 12 or 16 sheets. Charts produced in Canada use a slightly different sheet identification system. The DOD Aeronautical Chart Catalog contains an explanation of the system.

18.2.5.3. DOD Evasion Charts

DOD Evasion Charts (Figure 18-3) are specialized charts for evasion. The scale for these charts varies. The charts have both longitude and latitude and the UTM grid coordinate systems. The relief is duplicated by both contour lines and shading. Individual maps can have varying scale and elevation relief sectioned off. The magnetic variation is shown by a compass rose superimposed on the chart. The charts also indicate the direction of seasonal ocean currents. These charts include geographic environmental data consisting of a description of the people, climate, water, food, hazards, and vegetation. A conversion of elevation bar scale may aid in communicating with other forces. The star chart is provided to aid in night navigation.

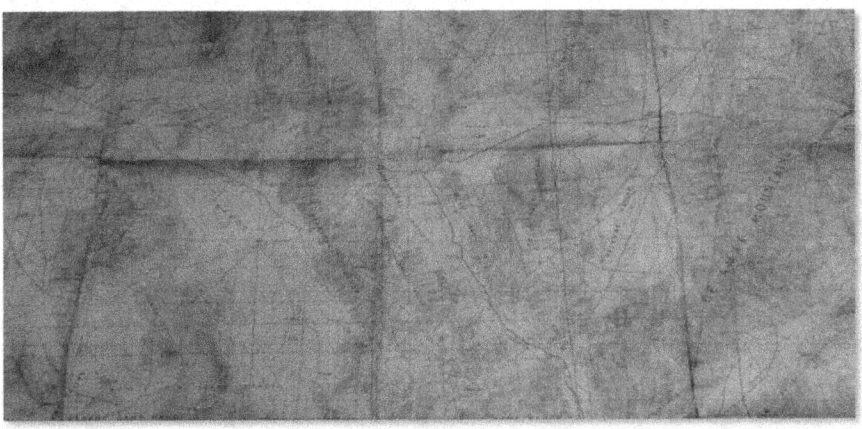

Figure 18-3 DOD Evasion Chart

18.3. Marginal Information

Map instructions are placed around the outer edges and are known as marginal information. All maps are not the same, so it becomes necessary each time a different map is used to carefully examine the marginal information.

18.3.1. Datum

A datum describes the model that was used to match the location of features on the ground to coordinates and locations on the map. Maps all start with some form of survey. Early maps and surveys were carried out by teams of surveyors on the ground using transits and distance measuring "chains". Surveyors start with a handful of locations in "known" positions and use them to locate other features. These methods did not span continents well. They also did not cross political borders. The "known points" and their positions are the information that the map datum is based. As space based surveying came into use, a standardized datum based on the center of the earth was developed. Every map that shows a geographic coordinate system such as UTM or Latitude and Longitude with any precision will also list the datum used on the map. Most USGS topographic maps are based on an earlier datum called the North American Datum (NAD) of 1927 or NAD 27. (Some Global Positioning System (GPS) units subdivide this datum into several datums spread over the continent. In the Continental United States use NAD27 CONUS.) In the Continental United States the difference between the World Geodetic System of 1984 (WGS 84) and NAD 27 can be as much as 200 meters. You should always set your GPS unit's datum to match the datum of the map you are using. It is critical that all navigational aids used by Personnel Recovery Command and Control (PR C2), support personnel, rescue forces, and the IP use the same datum usually found in PR Special Instructions.

18.3.2. Topographic Map Example

(Figure 18-4) is a large-scale (1:50,000) topographic map. The circled numbers indicate the marginal information with which the map user must be familiar. The location of the marginal information will vary with each different type of map. However, the following items are on most maps. The circled numbers correspond to the item numbers listed and described below.

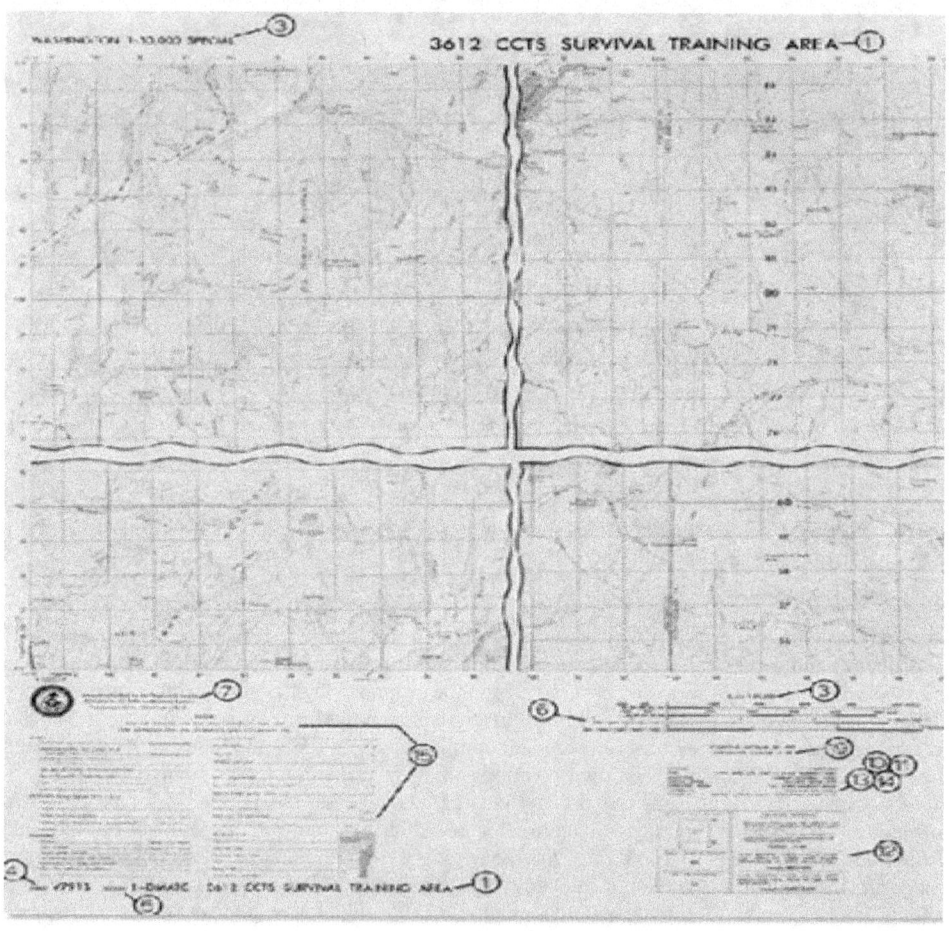

Figure 18-4 1:50,000 Topographic Map

18.3.2.1. Sheet Name

The sheet name (1 inFigure 18-4) is usually found in two places; the center of the upper margin and the right side of the lower margin. Generally, a map is named after its outstanding cultural or geographic feature. When possible, the name of the largest city on the map is used (not shown).

18.3.2.2. Sheet Number

The sheet number (2 inFigure 18-4) is normally in the upper right margin and is used as a reference number for that map sheet. For maps at 1: 100,000 scale and larger, sheet numbers are based on an arbitrary system which makes possible the ready orientation of maps at scales of 1:100,000, 1:50,000, and 1:25,000.

18.3.2.3. Series Name

The map series name and scale is usually in the upper left margin (3 in Figure 18-4). A map series usually comprises a group of similar maps at the same scale and on the same sheet lines or format designed to cover a particular geographic area. It may also be a group of maps which serve a common purpose, such as military city maps. The name given a series is of the most prominent area. The scale note is a representative fraction which gives the ratio of map distance to the corresponding distance on the Earth's surface. For example, the scale notes 1:50,000 indicate that one unit of measure on the map equals 50,000 units of the same measure on the ground.

18.3.2.4. Scale

The scale (3 in Figure 18-4) gives the ratio of map distance to ground distance. The terms "small scale," "medium scale," and "large scale" may be confusing when read with the numbers. However, if the number is viewed as a fraction, it quickly becomes apparent the 1:600,000 of something is smaller than 1:75,000 of the same thing. The larger the number (after the 1), the smaller the scale of the map.

18.3.2.4.1. Small Scale

Maps at scales of 1:600,000 and smaller are used for general planning and strategic studies at the high echelons. The standard small scale is 1:1,000,000.

18.3.2.4.2. Medium Scale

Maps at scales larger than 1:600,000 but smaller than 1:75,000 are used for planning operations, including the movement and concentration of troops and supplies. The standard medium scale is 1:250,000.

18.3.2.4.3. Large Scale

Maps at scales of 1:75,000 and larger are used to meet the tactical, technical, and administrative needs of field units. The standard large scale is 1:50,000.

18.3.2.5. Series Number

The series number (4 in Figure 18-4) appears in the upper right margin and the lower left margin. It is a comprehensive reference expressed either as a four-digit numeral (example, 1125) or as a letter, followed by a three- or four-digit numeral (example, V79 15).

18.3.2.6. Edition Number

The edition number (5 in Figure 18-4) is in the upper margin and lower left margin. It represents the age of the map in relation to other editions of the same map and the agency responsible for its production. Edition numbers run consecutively; a map bearing a higher edition number is assumed to contain more recent information than the same map bearing a lower edition number.

18.3.2.7. Bar Scales

The bar scales (6 in Figure 18-4) are generally located in the center of the lower margin. They are rulers used to convert map distance to ground distance. Maps normally have three or more bar scales, each a different unit of measure.

18.3.2.8. Credit Note

The credit note (7 in Figure 18-4) lists the producer, dates, and general methods of preparation or revision. This information is important to the map user in evaluating the reliability of the map as it indicates when and how the map information was obtained.

18.3.2.9. Adjoining Sheets Diagram

Maps at standard scales contain a diagram which illustrates the adjoining sheets (8 in Figure 18-4) (not shown). On maps at 1:100,000 and larger scales and at 1:1,000,000 scales, the diagram is called the Index to Adjoining Sheets, and consists of as many rectangles, representing adjoining sheets, as are necessary to surround the rectangle which represents the sheet under consideration. The diagram usually contains nine rectangles, but the number or names may vary depending on the location of the adjoining sheets. All represented sheets are identified by their sheet numbers. Sheets of an adjoining series, whether published or planned, using the same scales are represented by dashed lines. The series number of the adjoining series is indicated along the appropriate side of the division line between the series. On 1:50,000 scale maps, the sheet number and series number of the 1:250,000 scale map of the area are shown below the Index to Adjoining Sheets. On maps at 1:250,000 scale, the adjoining sheets are shown in the location diagram. Usually, the diagram consists of 25 rectangles, but the number may vary with the locations of the adjoining sheets.

18.3.2.10. Index to Boundaries

The index to boundaries diagram (9 in Figure 18-4) appears in the margin of all sheets 1: 100,000 scale or larger, and 1:1,000,000 scale. This diagram, which is a miniature of the map, shows the boundaries which occur within the map area, such as county lines and state boundaries. On 1:250,000 scale maps, the boundary information is included in the location diagram.

18.3.2.11. Projection

The projection system (10 in Figure 18-4) is the framework of the map. For maps, this framework is the conformal type; that is, small areas of the surface of the Earth retain their true shapes on the projection, measured angles closely approximate true values, and the scale factor is the same in all directions from a point. The projection is identified on the map by a note in the margin.

18.3.2.12. Grid Note

The grid note (11 in Figure 18-4) gives information pertaining to the grid system used, the interval of grid lines, and the number of digits omitted from the grid values. Notes pertaining to overlapping or secondary grids are also included when appropriate.

18.3.2.13. Grid Reference Box

The grid reference box (12 in Figure 18-4) has instructions for composing a grid reference.

18.3.2.14. Vertical Datum

The vertical datum note (13 in Figure 18-4) designates the basis for all vertical control stations, contours, and elevations appearing on the map.

18.3.2.15. Horizontal Datum Note

The horizontal datum note (14 in Figure 18-4) indicates the basis for all horizontal control stations appearing on the map. This network of stations controls the horizontal positions of all mapped features.

18.3.2.16. Legend

The legend illustrates and identifies the topographic symbols used to depict some of the more prominent features on the map. The symbols are not always the same on every map. To avoid error in the interpretation of symbols, the legend must always be referred to when a map is read.

18.3.2.17. Declination Diagram

The declination diagram indicates the angular relationships of true north, grid north, and magnetic north. On maps at 1:250,000 scale, this information is expressed as a note in the lower margin.

18.3.2.18. User's Note

A user's note requests cooperation in correcting errors or omissions on the map. Errors should be marked and the map forwarded to the agency identified in the note.

18.3.2.19. Unit Imprint

The unit imprint identifies the agency which printed the map and the printing date. The printing date should not be used to determine when the map information was obtained.

18.3.2.20. Contour Interval

The contour interval note states the vertical distance between adjacent contour lines on the map. When supplementary contours are used, the interval is indicated.

18.3.2.21. Special Notes and Scales

Under certain conditions, special notes or scales may be added to the margin information to aid the map user. The following are examples:

- A glossary is an explanation of technical terms or a translation of terms on maps of foreign areas where the native language is other than English.
- Certain maps require a note indicating the security classification. This is shown in the margin.
- A protractor scale is used to lay out the magnetic grid declination of the map which, in turn, is used to orient the map sheet with the aid of a magnetic compass.
- A coverage diagram may be used on maps at scales of 1:100,000 and larger. It indicates the methods by which the map was made, dates of photography, and reliability of the sources. On maps at 1:250,000 scale, the coverage diagram is replaced by a reliability diagram.
- On some maps at scales of 1:100,000 and larger, a miniature characterization of the terrain is shown by a diagram. The terrain is represented by bands of elevation, spot elevations, and

major drainage features. The elevation guide provides the map reader with a means of rapid recognition of major landforms.

- A special note is any statement of general information that relates specifically to the mapped area. For example, rice fields are generally subject to flooding; however, they may be seasonally dry.

18.3.2.22. Stock Number Identification

The stock number identification consists of the words "STOCK NO." followed by a unique designation which is composed of the series number, the sheet number of the individual map, and on recently printed sheets, and the edition number.

18.4. Topographic Map Symbols and Colors

The purpose of a map is to permit one to visualize an area of the Earth's surface with pertinent features properly positioned. Ideally, all the features within an area would appear on the map in their true proportion, position, and shape. However, this is not practical because many of the features would be unimportant and others would be unrecognizable because of size reduction. The mapmaker is required to use symbols to represent the natural and manmade features of the Earth's surface. These symbols resemble, as closely as possible, the actual features as viewed from above (Figure 18-5).

Figure 18-5 Area Viewed from Ground Position

18.4.1. Symbols and Colors

To facilitate identification of features on the map by providing more natural appearance and contrast, the topographic symbols are usually printed in different colors, with each color identifying a class of features. The colors vary with different types of maps, but on a standard large-scale topographic map, the colors used and the features represented are:

- Black - the majority of cultural or manmade features.

- Blue - water features such as lakes, rivers, and swamps.
- Green - vegetation such as woods, orchards, and vineyards.
- Brown - all relief features such as contours.
- Red - main roads, built-up areas, and special features.
- Occasionally, other colors may be used to show special information. (These, as a rule, are indicated in the marginal information. For example, aeronautical symbols and related information for air-ground operations are shown in purple on JOGS)

18.4.2. Symbol Location

In the process of making a map, everything must be reduced from its size on the ground to the size which appears on the map. For purposes of clarity, this requires some of the symbols to be exaggerated. They are positioned so that the center of the symbol remains in its true location. An exception to this would be the position of a feature adjacent to a major road. If the width of the road has been exaggerated, then the feature is moved from its true position to preserve its relation to the road.

18.4.3. Authorized Topographic Symbols and Abbreviations

Army Field Manual 3-25.26 gives a description of topographic symbols and abbreviations authorized for use on US military maps.

18.5. Coordinate Systems

The intersections of reference lines help to locate specific points on the Earth's surface. The primary reference line systems are the geographic coordinate system and the Universal Transverse Mercator grid system (UTM). Knowing how to use these plotting systems should help an IP to determine point locations.

18.5.1. Geographic Coordinate System

One of the oldest systematic methods of location is based upon the geographic coordinate system (latitude and longitude). By drawing a set of east-west rings around the globe (parallel to the equator), and a set of north-south rings crossing the equator at right angles and converging at the poles, a network of reference lines is formed from which any point on the earth's surface can be located.

18.5.2. Latitude

When looking at a map, latitude lines run horizontally. Latitude lines are also known as parallels since they are parallel and are an equal distant from each other. Each degree of latitude is approximately 69 miles (111 km) apart; there is a variation due to the fact that the earth is not a perfect sphere but an oblate ellipsoid (slightly egg-shaped). To remember latitude, imagine them as the horizontal rungs of a ladder. Degrees latitude are numbered from 0° to 90° north and south. Zero degrees are the equator, the imaginary line which divides our planet into the northern and southern hemispheres. 90° north is the North Pole and 90° south is the South Pole.

18.5.2.1. Longitude

The vertical longitude lines are also known as meridians. They converge at the poles and are widest at the equator (about 69 miles or 111 km apart). Zero degrees longitude is located at

Greenwich, England (0°). The degrees continue 180° east and 180° west where they meet and form the International Date Line in the Pacific Ocean. Greenwich, the site of the British Royal Greenwich Observatory, was established as the site of the prime meridian by an international conference in 1884.

18.5.2.2. How Latitude and Longitude Work Together

To precisely locate points on the earth's surface, degrees (°) longitude and latitude have been divided into minutes (') and seconds ("). There are 60 minutes in each degree. Each minute is divided into 60 seconds. Seconds can be further divided into tenths, hundredths, or even thousandths. For example, the U.S. Capitol is located at 38°53'23"N , 77°00'27"W (38 degrees, 53 minutes, and 23 seconds north of the equator and 77 degrees, zero (00) minutes and 27 seconds west of the meridian passing through Greenwich, England(Figure 18-6 and Figure 18-7).

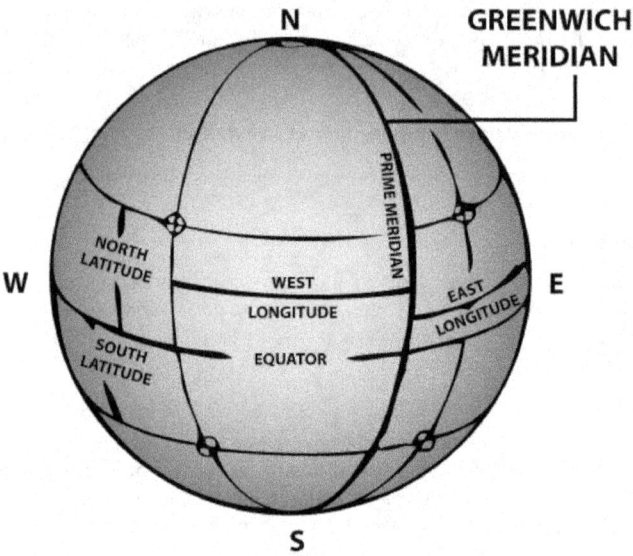

Figure 18-6 Prime Meridian and Equator

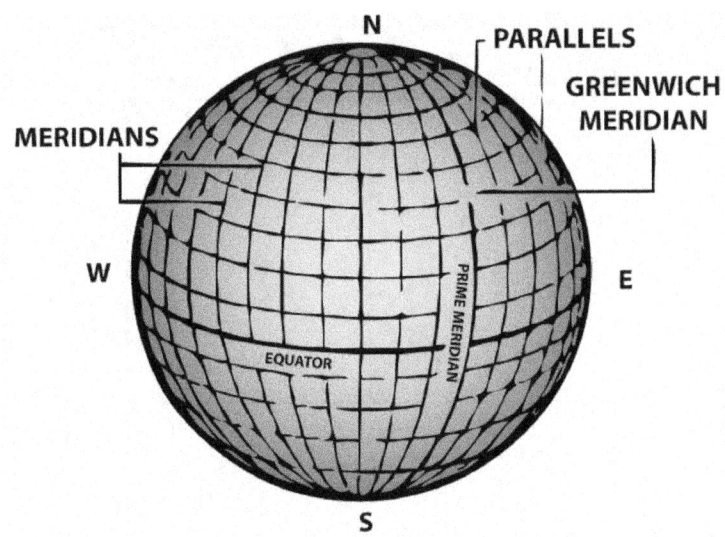

Figure 18-7 Reference Lines

18.5.2.3. Writing Geographic Coordinates

In general, there are general rules to follow in correctly writing geographic coordinates:

1. Write latitude first, followed by longitude.
2. Use two numbers of digits for latitude (00° -90°) and three numbers of digits for longitude (000° – 180°).
3. Do not use a dash or leave a space between latitude and longitude.
4. Use single upper case letter to indicate direction from the Equator (N or S) and prime meridians (E or W).

18.5.3. Plotting Geographic Coordinates

One can probably read the coordinates of point A and B in Figure 18-8 rather easily; however, plotting points on maps from given coordinates must also be done. To do this, first get acquainted with the map being used. Assume that (Figure 18-4) is the map being used. Note that it covers an area from 38N to 39N and from 104W to 105W. Also note that latitude and longitude are subdivided by 30' division lines and then with tick marks into five- and one-minute subdivisions.

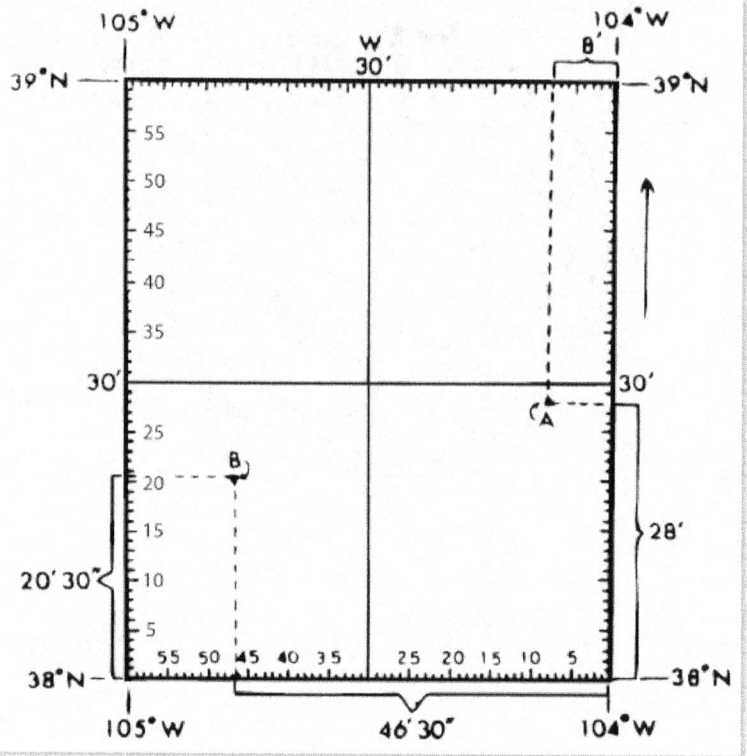

Figure 18-8 Plotting Geographic Coordinates

18.5.3.1. Plotting Points

Assume that the coordinates of the point which must be plotted are 38° 28'00"N 104° 08'00"W. Next, follow the general procedure listed below to plot the point on the map:

1. Locate the parallel of latitude for degrees (38° N).
2. Find the meridian of longitude for degrees (104° W).
3. Move to the meridian (usually a tick mark) for minutes (08° W).
4. Move to the parallel (usually a tick mark) for minutes (28'N).
5. Plot the point on the map (point A in Figure 18-9).

18.5.3.1.1. Points of Interest

Recovery points, rally points, and destination positions may be plotted or identified on a map or chart to enable rescue personnel, the survivors, and evaders to locate these positions. Seconds are not shown between the one-minute tick marks on maps and charts; they must be estimated. It is easy to estimate halfway tick marks (30 seconds); one-fourth (15 seconds) and three-fourths (45 seconds) are also reasonably easy to estimate. Then, as experience is gained, people will find that on large-scale maps they can estimate the sixths (10 seconds) and eights (about eight seconds). They cannot, however, accurately estimate to sixths or eights at the scale shown in Figure 18-9.

18.5.3.1.2. Writing Coordinates

To write geographic coordinates more precisely than minutes, merely carry the coordinates out to include seconds. In the previous example, the coordinates of a target located 30° 20' north of the Equator and 135° 06' east of the prime meridian were written as 30 ° 20'00"N135 ° 06'00"E. A more exact position of the target might be 30°20'05"N latitude and 135° 06'16"E longitude. This more precise position is correctly written as 30 ° 20'05"N135 ° 06'16"E.

18.5.4. Universal Transverse Mercator (UTM)

The Universal Transverse Mercator (UTM) geographic coordinate system is a grid-based method of specifying locations on the surface of the Earth that is a practical application of a two-dimensional Cartesian coordinate system. It is a horizontal position representation, i.e., it is used to identify locations on the earth independently of vertical position, but differs from the traditional method of latitude and longitude in several respects. The UTM system is not a single map projection. The system instead employs a series of sixty zones, each of which is based on a specifically defined secant transverse Mercator projection.

18.5.4.1. UTM Background

The UTM coordinate system was developed by the United States Army Corps of Engineers in the 1940s. The system was based on an ellipsoidal model of Earth. For areas within the conterminous United States, the Clarke 1866 ellipsoid was used. For the remaining areas of Earth, including Hawaii, the International Ellipsoid was used. Currently, the WGS84 ellipsoid is used as the underlying model of Earth in the UTM coordinate system.

18.5.4.2. UTM Zones

The UTM system divides the surface of Earth between 80°S and 84°N latitude into 60 zones, each 6° of longitude in width, and centered over a meridian of longitude. Zone 1 is bounded by longitude 180° to 174° W and is centered on the 177th West meridian. Zone numbering increases in an eastward direction. Each of the 60 longitude zones in the UTM system is based on a transverse Mercator projection, which is capable of mapping a region of large north-south extent with a low amount of distortion. There are special UTM zones between 0 degrees and 36 degrees longitude above 72 degrees latitude and a special zone 32 between 56 degrees and 64 degrees north latitude:

- UTM Zone 32 has been widened to 9° (at the expense of zone 31) between latitudes 56° and 64° (band V) to accommodate southwest Norway. Thus zone 32 extends westwards to 3°E in the North Sea.
- Similarly, between 72° and 84° (band X), zones 33 and 35 have been widened to 12° to accommodate Svalbard. To compensate for these 12° wide zones, zones 31 and 37 are widened to 9° and zones 32, 34, and 36 are eliminated. The W and E boundaries of zones are 31: 0 - 9 E, 33: 9 - 21 E, 35: 21 - 33 E and 37: 33 - 42 E.

18.5.4.3. Locating a Position Using UTM Coordinates

A position on the Earth is referenced in the UTM system by the UTM zone, and the easting and northing coordinate pair. The easting is the projected distance of the position eastward from the central meridian, while the northing is the projected distance of the point north from the equator (in the northern hemisphere). Easting and northing are measured in meters. The point of origin of each UTM zone is the intersection of the equator and the zone's central meridian. In order to avoid dealing with negative numbers, the central meridian of each zone is given a "false easting" value of 500,000 meters. Thus, anything west of the central meridian will have an easting less than 500,000 meters. For example, UTM easting's range from 167,000 meters to 833,000 meters at the equator (these ranges narrow towards the poles). In the northern hemisphere, positions are measured northward from the equator, which has an initial "northing" value of 0 meters and a maximum "northing" value of approximately 9,328,000 meters at the 84th parallel — the maximum northern extent of the UTM zones. In the southern hemisphere, northing's decreases as you go southward from the equator, which is given a "false northing" of 10,000,000 meters so that no point within the zone has a negative northing value. As an example, the destination point is located at the geographic position 43°38'33.24"N 79°23'13.7"W / 43.6425667°N 79.387139°W / 43.6425667; -79.387139 . This is in zone 17, and the grid position is 630084m east, 4833438m north. There are two points on the earth with these coordinates, one in the northern hemisphere and one in the southern. In order to define the position uniquely, one of two conventions is employed:

- Append a hemisphere designator to the zone number, "N" or "S", thus "17N 630084 4833438". This supplies the minimum additional information to define the position uniquely.

- Supply the grid zone, i.e., the latitude band designator appended to the zone number, thus "17T 630084 4833438". The provision of the latitude band along with northing supplies redundant information (which may, as a consequence, be contradictory).

- Because latitude band "S" is in the northern hemisphere, a designation such as "38S" is ambiguous. The "S" might refer to the latitude band (32°N – 40°N) or it might mean "South". It is therefore important to specify which convention is being used, e.g., by spelling out the hemisphere, "North" or "South", or using different symbols, such as - for south and + for north.

18.5.4.4. Military Grid Reference System (MGRS)

MGRS is an extension of the UTM system. UTM zone number and zone character are used to identify an area six degree in east-west extent and eight degrees in north-south extent. UTM zone number and designator are followed by 100 km square easting and northing identifiers. The system uses a set of alphabetic characters for the 100 km grid squares. Starting at the 180 degree meridian the characters A to Z (omitting I and O) are used for 18 degrees before starting over. From the equator north the characters A to V (omitting I and O) are used for 100 km squares, repeating every 2,000 km. Northing designators normally begin with 'A' at the equator for odd numbered UTM easting zones.

- For even numbered easting zones the northing designators are offset by five characters, starting at the equator with 'F'. South of the equator, the characters continue the pattern set north of the equator. Complicating the system, ellipsoid junctions (spheroid junctions in the terminology of MGRS) require a shift of 10 characters in the northing 100 km grid square designators. Different geodetic datums using different reference ellipsoids use different starting row offset numbers to accomplish this.

- If 10 numeric characters are used, a precision of one meter is assumed. two characters imply a precision of 10 km. From two to 10 numeric characters the precision changes from 10 km, 1 km, 100 m, 10 m, to 1 m. MGRS' 100,000-meter square identification:

- The 100,000-meter columns, including partial columns along zone, datum, and ellipsoid junctions, are lettered alphabetically, A through Z (with I and O omitted), north and south of the Equator, starting at the 180° meridian and proceeding easterly for 18°. The alphabetical sequence repeats at 18° intervals.
- To prevent ambiguity of identifications along ellipsoid junctions changes in the order of the row letters are necessary. The row alphabet (second letter) is shifted ten letters. This decreased the maximum distance in which the 100,000-meter square identification is repeated.
- The 100,000-meter row lettering is based on a 20-letter alphabetical sequence (A through V with I and O omitted). This alphabetical sequence is read from south to north, and repeated at 2,000,000-meter intervals from the Equator.
- The row letters in each odd numbered 6° grid zone are read in an A through V sequence from south to north.
- In each even-numbered 6° grid zone, the some lettering sequence is advanced five letters to F, continued sequentially through V and followed by A through V.
- The advancement or staggering of row letters for the even-numbered zones lengthens the distance between 100,000-meter squares of the same identification.
- Deviations from the preceding rules were mode in the past. These deviations were an attempt to provide unique grid references within a complicated and disparate world-wide mapping system.
- Determination of 100,000-meter grid square identification is further complicated by the use of different ellipsoids.
- The military grid reference. The MGRS coordinate for a position consists of a group of letters and numbers which include the following elements:
  - The Grid Zone Designation
  - The 100,000-meter square letter identification
  - The grid coordinates (also referred to as rectangular coordinates); the numerical portion of the reference expressed to a desired refinement
  - A reference is written as an entity without spaces, parentheses, dashes, or decimal points

18.5.4.4.1. MGRS Examples

- 18S (Locating a point within the Grid Zone Designation)
- 18SUU (Locating a point within a 100,000-meter square)
- 18SUU80 (Locating a point within a 10,000-meter square)
- 18SUU8401 (Locating a point within a 1,000-meter square)

- 18SUU836014 (Locating a point within a 100-meter square)
- 18SUU83630143 (Locating a point within a 10-meter square)
- 18SUU8362601432 (Locating a point within a 1-meter square)

18.6. Elevation and Relief

The map user must become proficient in recognizing various landforms and irregularities of the Earth's surface and be able to determine the elevation and differences in height of all terrain features.

18.6.1. Datum Plane

This is the reference used for vertical measurements. The datum plane for most maps is mean or average sea level.

18.6.2. Elevation

This is defined as the height (vertical distance) of an object above or below a datum plane.

18.6.3. Relief

Relief is the representation of the shape and height of landforms and characteristic of the Earth's surface.

18.6.4. Contour Lines

There are several ways of indicating elevation and relief on maps. The most common way is by contour lines. A contour line is an imaginary line connecting points of equal elevation. Contour lines indicate a vertical distance above or below a datum plane. Starting at sea level, each contour line represents an elevation above sea level. The vertical distance between adjacent contour lines is known as the contour interval. The amount of contour interval is given in the marginal information. On most maps, the contour lines are printed in brown. Starting at zero elevation, every fifth contour line is drawn with a heavier line. These are known as index contours and somewhere along each index contour, the line is broken and its elevation is given. The contour lines falling between index contours are called intermediate contours. They are drawn with a finer line than the index contours and usually do not have their elevations given.

1. Using the contour lines on a map, the elevation of any point may be determined by:
2. Finding the contour interval of the map from the marginal information, and noting the amount and unit of measure.
3. Finding the numbered contour line (or other given elevation) nearest the point for which elevation is being sought.
4. Determining the direction of slope from the numbered contour line to the desired point.
5. Counting the number of contour lines that must be crossed to go from the numbered line to the desired point and noting the direction-up or down. The number of lines crossed multiplied by the contour interval is the distance above or below the starting value. If the desired point is on a contour line, its elevation is that of the contour; for a point between contours, most military needs are satisfied by estimating the elevation to an accuracy of one-half the contour interval. All points less than one-fourth the distance between the lines are considered to be at the same elevation as the line. All points one-fourth to three-fourths

the distance from the lower line are considered to be at an elevation one-half the contour interval above the lower line (Figure 18-9).

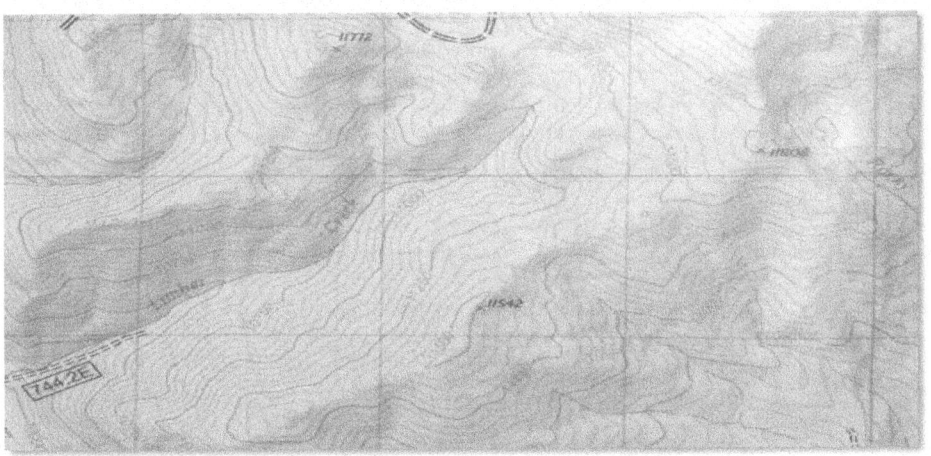

Figure 18-96 Estimating Elevation and Contour Lines

18.6.4.1.1. Terrain

Terrain may continue up to but less than the next contour interval; i.e., the elevation of the top of an unmarked hill may be up to one foot less than the contour interval of the highest contour line around the hill. This principle applies to estimating the elevation of the bottom of a depression.

18.6.4.1.2. Supplementary Contour

On maps where the index and intermediate contour lines do not show the elevation and relief in as much detail as may be needed, supplementary contour may be used. These contour lines are dashed brown lines, usually at one-half the contour interval for the map. A note in the marginal information indicates the interval used. They are used exactly as are the solid contour lines.

18.6.4.1.3. Approximate Contours

On some maps contour lines may not meet the standards of accuracy but are sufficiently accurate in both value and interval to be shown as contour rather than as form lines. In such cases, the contours are considered as approximate and are shown with a dashed symbol; elevation values are given at intervals along the heavier (index contour) dashed lines. The contour note in the map margin identifies them as approximate contours.

18.6.4.1.4. Benchmarks and Spot Elevations

In addition to the contour lines, bench marks and spot elevations are used to indicate points of known elevation on the map. Bench marks, the more accurate of the two, are symbolized by a black X, as in example X BM 124. The elevation value shown in black refers to the center of the X. Spot elevations are shown in brown and generally are located at road junctions, on hilltops, and other prominent landforms. The symbol designates an accurate horizontal control point. When a bench mark and a horizontal control point are located at the same point, the symbol BM is used.

18.6.4.1.5. Slope

The spacing of the contour lines indicates the nature of the slope. Contour lines evenly spaced and wide apart indicate a uniform, gentle slope (Figure 18-10). Contour lines evenly spaced and close together indicate a uniform, steep slope. The closer the contour lines to each other, the steeper the slope (Figure 18-11). Contour lines closely spaced at the top and widely spaced at the bottom indicate a concave slope (Figure 18-12). Contour lines widely spaced at the top and closely spaced at the bottom indicate a convex slope (Figure 18-13).

Figure 18-70 Uniform Gentle Slope

Figure 18-81 Uniform Steep Slope

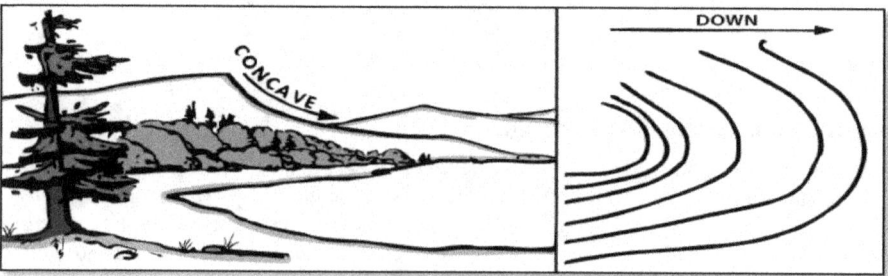

Figure 18-92 Concave Slope

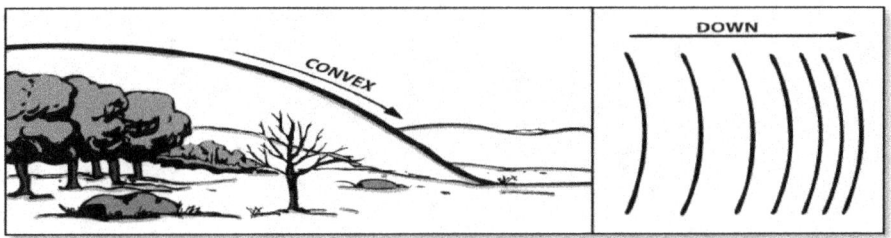

Figure 18-103 Convex Slope

18.6.4.1.6. Relief Features

To show the relationship of land formations to each other and how they are symbolized on a contour map, stylized panoramic sketches of the major relief formations were drawn and a contour map of each sketch developed. Each figure (Figure 18-14 through Figure 18-20) shows a sketch and a map with a different relief feature and its characteristic contour pattern.

- Hill. A point or small area of high ground (Figure 18-14). When one is located on a hilltop, the ground slopes down in all directions.

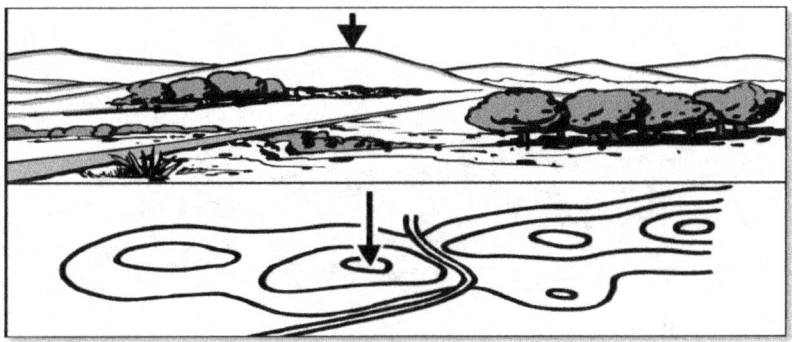

Figure 18-114 Hill

- Valley. Usually a stream course which has at least a limited extent of reasonably level ground bordered on the sides by higher ground (Figure 18-15). The valley generally has maneuvering room within its confines. Contours indicating a valley are U-shaped and tend to parallel a major stream before crossing it. The more gradual the fall of a stream, the farther each contour interval is apart. The curve of the contour crossing always points upstream.

- Drainage. A less-developed stream course in which there is essentially no level ground and, therefore, little or no maneuvering room within its confines (Figure 18-15). The ground slopes upward on each side and toward the head of the drainage. Drainages occur frequently along the sides of ridges, at right angles to the valleys between the ridges. Contours indicating drainage is V-shaped, with the point of the "V" toward the head of the drainage.

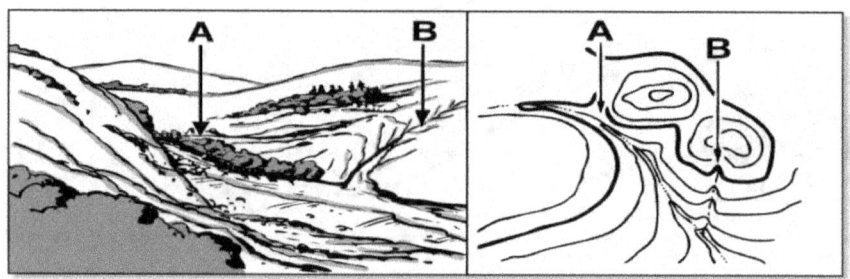

Figure 18-125 (A) Valley, (B) Drainage

- Ridge. A range of hills or mountains with normally minor variations along its crest (Figure 18-16). The ridge is not simply a line of hills; all points of the ridge crest are appreciably higher than the ground on both sides of the ridge.

- Finger Ridge. A ridgeline of elevation projecting from or subordinate to the main body of a mountain or mountain range (Figure 18-16). A finger ridge is often formed by two roughly parallel streams cutting drainages down the side of a ridge.

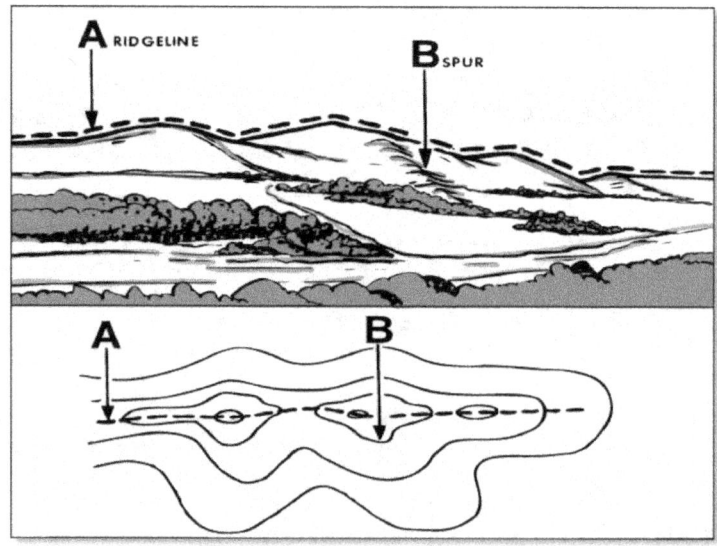

Figure 18-136 (A) Ridge Line, (B) Finger Ridge

- Saddle. A dip or low point along the crest of a ridge. A saddle is not necessarily the lower ground between two hilltops; it may simply be a dip or break along an otherwise level ridge crest (Figure 18-17).

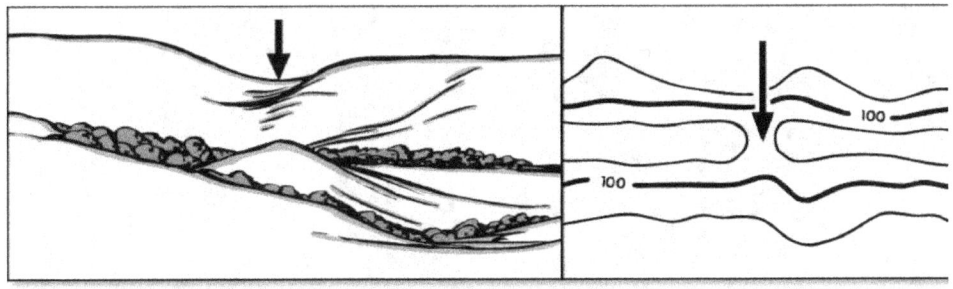

Figure 18-147 Saddle

- Depression. A low point or sinkhole surrounded on all sides by higher ground (Figure 18-18).

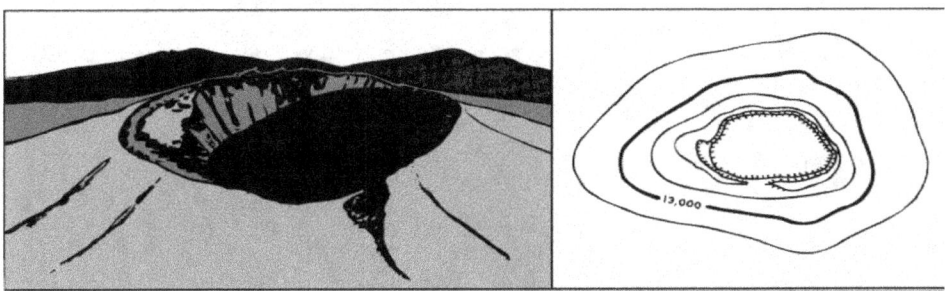

Figure 18-158 Depression

- Cuts and Fills. Manmade features like the bed of a road or railroad which is graded or leveled by cutting through high areas (Figure 18-19) and filling in low areas.

Figure 18-19 (A) Cut, (B) Fill

- Cliff. A vertical or near vertical slope (Figure 18-20). When a slope is so steep that it cannot be shown at the contour interval without the contours fusing, it is shown by a ticked "carrying" contour(s). The ticks always point toward lower ground.

Figure 18-160 Cliff

18.7. Representative Fraction (RF)

The numerical scale of a map expresses the ratio of horizontal distance on the map to the corresponding horizontal distance on the ground. It usually is written as a fraction, called the representative fraction (RF). The representative fraction is always written with the map distance as one (1). It is independent of any unit of measure. An RF of 1/50,000 or 1:50,000 means that one (1) unit of measure on the map is equal to 50,000 of the same units of measure on the ground.

18.7.1. Ground Distance

The ground distance between two points is determined by measuring between the points on the map and multiplying the map measurement by the denominator of the RF. Example: RF = 1:50,000 or 1 50,000 Map distance = 5 units (CM) 5 X 50,000 - 250,000 units (CM) of ground distance.

18.7.2. Accuracy

When determining ground distance from a map, the scale of the map affects the accuracy. As the scale becomes smaller, the accuracy of measurement decreases because some of the features on the map must be exaggerated so that they may be readily identified.

18.8. Graphic (Bar) Scales

On most maps, there is another method of determining ground distance. It is by means of the graphic (bar) scales. A graphic scale is a ruler printed on the map on which distances on the map may be measured as actual ground distances (Figure 18-21). To the right of the zero (0), the scale is marked in full units of measure and is called the primary scale. The part to the left of zero (0) is divided into tenths of a unit and is called the extension scale. Most maps have three or more graphic scales, each of which measures distance in a different unit of measure.

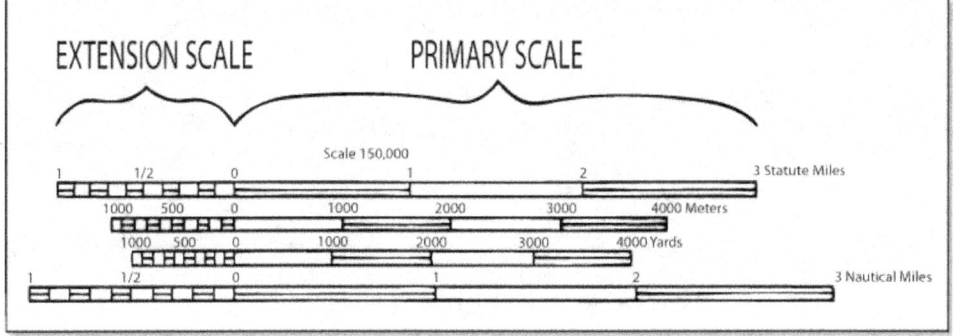

Figure 18-171 Graphic Bar Scale

18.8.1. Ground Distance – Straight-Line

To determine a straight-line ground distance between two points on a map, lay a straight-edged piece of paper on the map so that the edge of the paper touches both points. Mark the edge of the paper at each point. Move the paper down to the graphic scale and read the ground distance between the points. Be sure to use the scale that measures in the unit of measure desired (Figure 18-22).

Figure 18-182 Measuring Straight Line Map Distance

18.8.2. Ground Distance – Curved-Line

To measure distance along a winding road, stream, or any other curved line, the straightedge of a piece of paper is used again. Mark one end of the paper and place it at the point from which the curved line is to be measured. Align the edge of the paper along a straight portion and mark both the map and the paper at the end of the aligned portion. Keeping both marks together, place the point of the pencil on the mark on the paper to hold it in place. Pivot the paper until another approximately straight portion is aligned and again mark on the map and the paper. Continue in this manner until measurement is complete. Then place the paper on the graphic scale and read the ground distance (Figure 18-23).

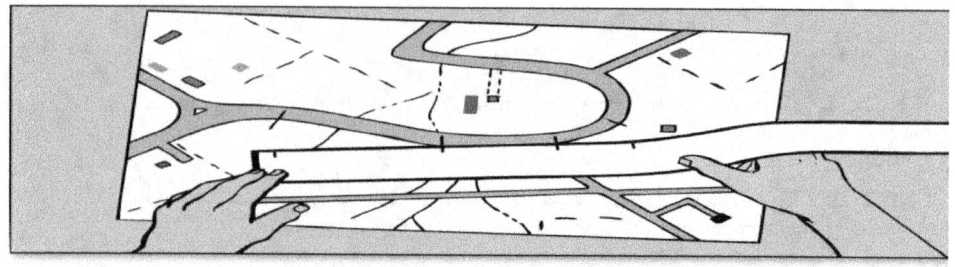

Figure 18-193 Measuring Curved Line Distances

18.8.3. Marginal Notes

Often, marginal notes give the road distance from the edge of the map to a town, highway, or junction of the map. If the road distance is desired from a point on the map to such a point off the map, measure the distance to the edge of the map and add the distance specified in the marginal note to that measurement.

18.9. Protractors

Protractors come in several forms - full circle, half circle, square, and rectangular (Figure 18-24). All of them divide a circle into units of angular measure, and regardless of their shape, consist of a scale around the outer edge and an index mark. The index mark is the center of the protractor circle from which all the direction lines radiate.

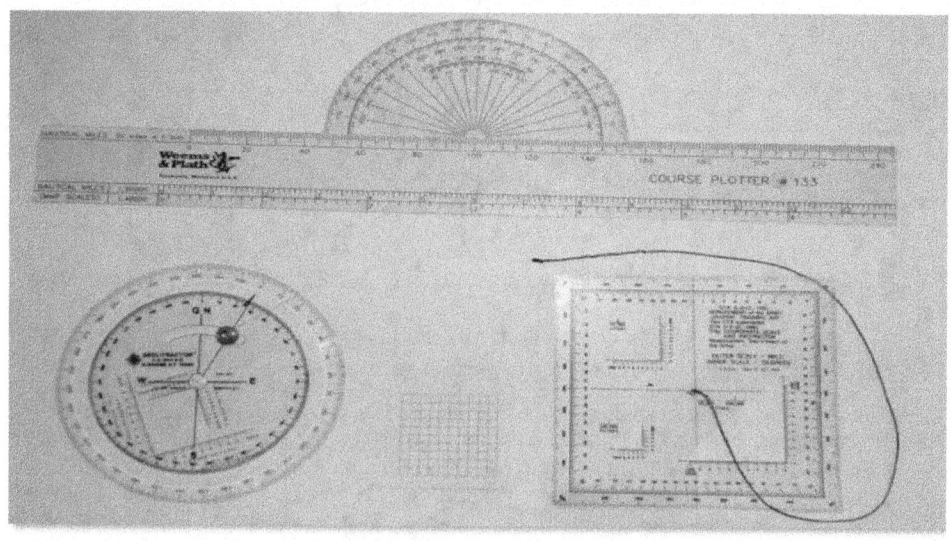

Figure 18-204 Types of Protractors

18.9.1. Measuring Azimuth

- To determine the grid azimuth of a line from one point to another on the map from (A to B or C to D) (Figure 18-25) draw a line connecting the two points.
- Place the index of the protractor at that point where the line crosses a vertical (north-south) grid line.
- Keeping the index at this point, align the 0° - 180° line of the protractor on the vertical grid line.
- Read the value of the angle from the scale; this is the grid azimuth to the point.

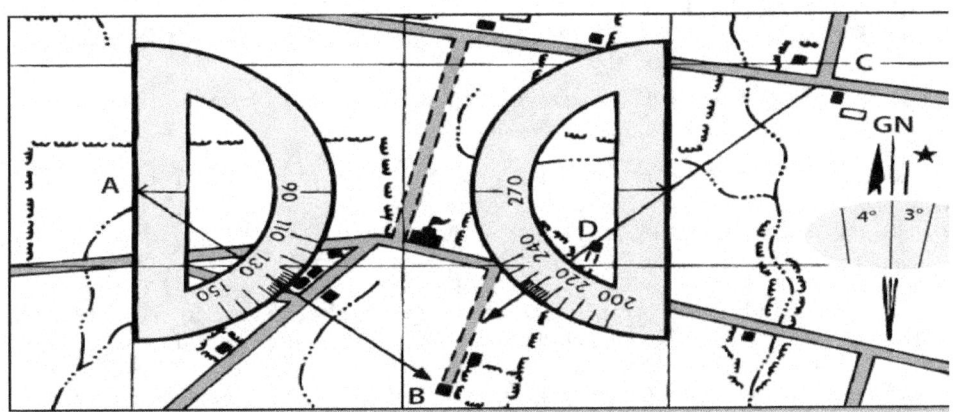

Figure 18-215 Measuring Azimuth on a Map using a Protractor

18.9.2. Plot a Direction Line

To plot a direction line from a known point on a map (Figure 18-26):

- Construct a north-south grid line through the known point:
- Generally, align the 0° - 180° line of the protractor in a north-south direction through the known point.
- Holding the 0° - 180° line of the protractor on the known point, slide the protractor in the north-south direction until the horizontal line of the protractor (connecting the protractor index and the 90° tick mark) is aligned on an east-west grid line.
- Then draw a line connecting 0°, the known point, and 180".
- Holding the 0° - 180° line on the north-south line, slide the protractor index to the known point.
- Make a mark on the map at the required angle. (Do not mark on the map.)
- Draw a line from the known point through the mark made on the map. This is the grid direction line.

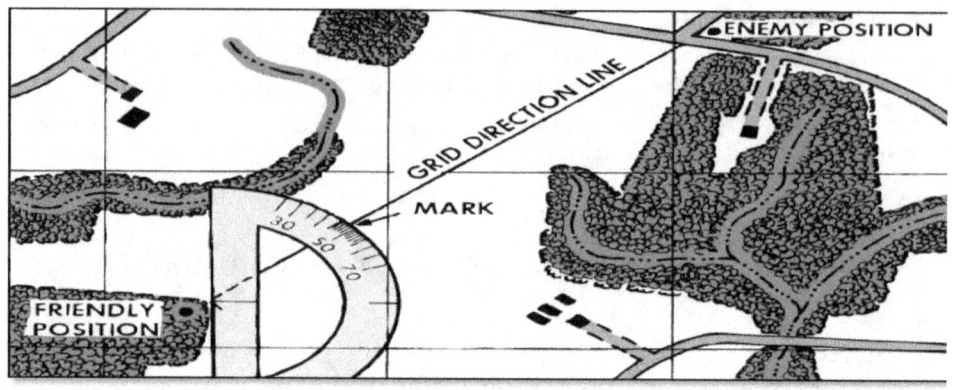

Figure 18-226 Plotting an Azimuth on a Map using a Protractor

18.10. Map Orientation

A map is oriented when it is in a horizontal position with its north and south corresponding to north and south on the ground. The best way to orient a map is with a compass. (NOTE: Caution should be used to ensure nothing (metal, mine ore, etc.) in the area will alter the compass reading.)

18.10.1. Lensatic Compass

With the map in a horizontal position, the lensatic compass is placed parallel to a north-south grid line with the cover side of the compass pointing toward the top of the map. This will place the black index line on the dial of the compass parallel to grid north. Since the needle on the compass points to magnetic north, a declination diagram is (on the face of the compass) formed by the index line and the compass needle.

- Rotate the map and compass until the directions of the declination diagram formed by the black index line and compass needle match the directions shown on the declination diagram printed in the margin of the map. If the variation is to the east of true north or the magnetic north arrow of the declination diagram is to the east (right) of the grid north line, subtract the degrees of variation from 360°. If it is to the left (west), add to 000°. East is least and west is best. The map is then oriented (grid north).

- If the magnetic north arrow on the map is to the left of grid north, the compass reading will equal the G-M angle (given in the declination diagram). If the magnetic north is to the right of grid north, the compass reading will equal 360° minus the G-M angle. In Figure 18-28, the declination diagram illustrates a magnetic north to the right of grid north and the compass reading will be 360° minus 21-1/2° or 338-1/2°. Remember to point the compass north arrow in the same direction as the magnetic north arrow, and the compass reading (equal to the G-M angle or the 360° minus G-M angle) will be quite apparent.

- If a grid line is not used, a true north-south line can be used. True north-south lines are longitudinal lines or lines formed by the vertical lines on a tick map (assuming the top of the map is north). The same procedure is used if magnetic variation is figured from true north-not grid north.

18.10.2. Floating Needle Compass

A floating needle compass (Figure 18-27) has a needle with a north direction marked on it. The degree and direction marks are stationary on the bottom inside of the compass. The button and wrist compasses may be floating dial or floating needle. To determine the heading, line up the north-seeking arrow over 360° by rotating the compass. Then read the desired heading. Orienting a map with a floating needle compass is similar to the method used with the floating dial. The only exception is with the adjustment for magnetic variation. If magnetic variation is to the east, turn the map and the compass to the left (the north axis of the compass should be aligned with the map north) so that the magnetic north-seeking arrow is pointing at the number of degrees on the compass which corresponds with the angle of declination.

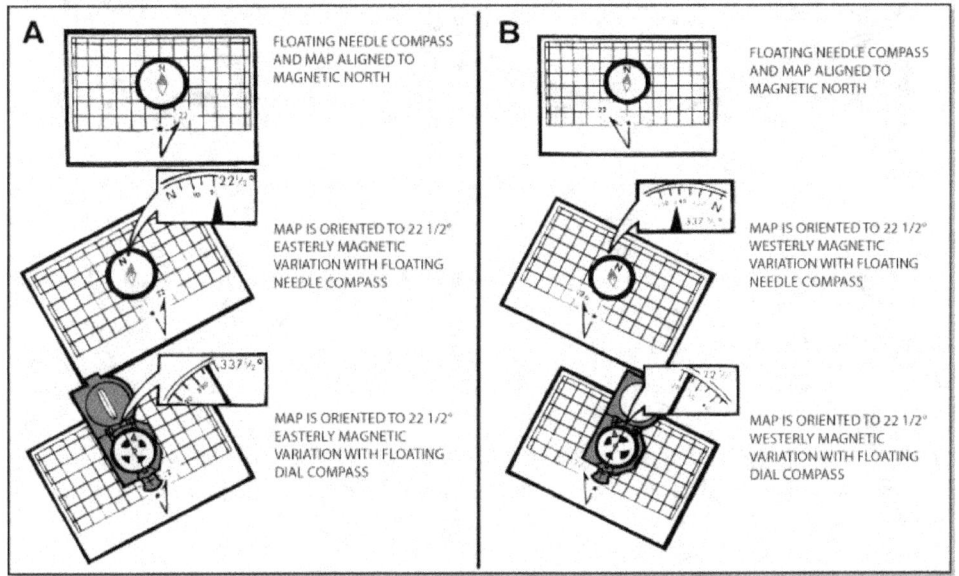

Figure 18-237 Floating Needle Compass

18.10.3. Linear Features

When a compass is not available, map orientation requires a careful examination of the map and the ground to find linear features common to both, such as roads, railroads, fence lines, power lines, etc. By aligning the feature on the map with the same feature on the ground (Figure 18-28) the map is oriented. Orientation by this method must be checked to prevent the reversal of directions which may occur if only one linear feature is used. This reversal may be prevented by aligning two or more map features (terrain or manmade). If no second linear feature is visible but the map user's position is known, a prominent object may be used. With the prominent object and the user's position connected with a straight line on the map, the map is rotated until the line points toward the feature.

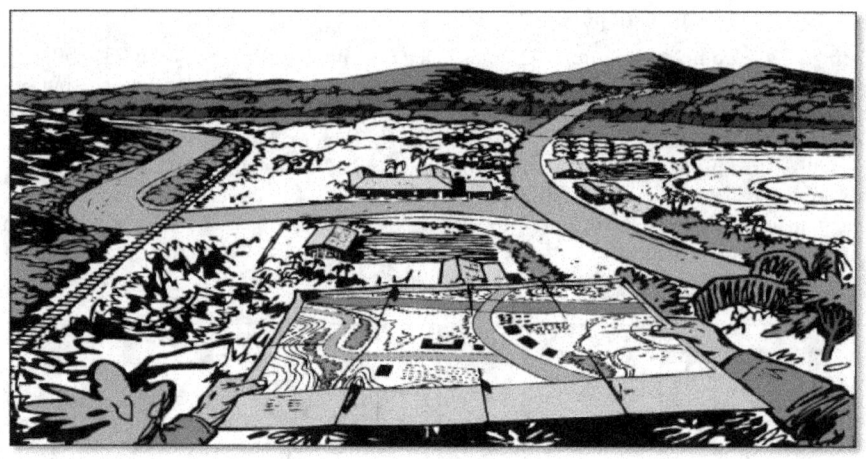

Figure 18-248 Map Orientation by Inspection

If two prominent objects are visible and plotted on the map and the position is not known, move to one of the plotted and known positions, place the straightedge or protractor on the line between the plotted positions, and turn the protractor and the map until the other plotted and visible point is seen along the edge. The map is then oriented.

18.10.4. Field Expedient Methods

When a compass is not available and there are no recognizable prominent landforms or other features, a map may be oriented by field expedient methods.

18.11. Determining Cardinal Directions Using Field Expedients

It is possible to determine the Cardinal Directions, or four basic points on a compass (north, east, south, west), by using field exponents.

18.11.1. Shadow Tip Method

The shadow tip method of determining direction and time is a simple method of finding direction by the Sun. It consists of only three basic steps (Figure 18-29).

1. Place a stick or branch into the ground at a fairly level spot where a distinct shadow will be cast. Mark the shadow tip with a stone, twig, or other means.

2. Wait until the shadow tip moves a few inches. If a four-foot stick is being used, about 10 minutes should be sufficient. Mark the new position of the shadow tip in the same way as the first.

3. Draw a straight line through the two marks to obtain an approximate east-west line. If uncertain which direction is east and which is west, observe this simple rule: The Sun "rises in the east and sets in the west" (but rarely DUE east and DUE west). The shadow tip moves in just the opposite direction. Therefore, the first shadow-tip mark is always in the west direction, and the second mark in the east direction, any place on Earth.

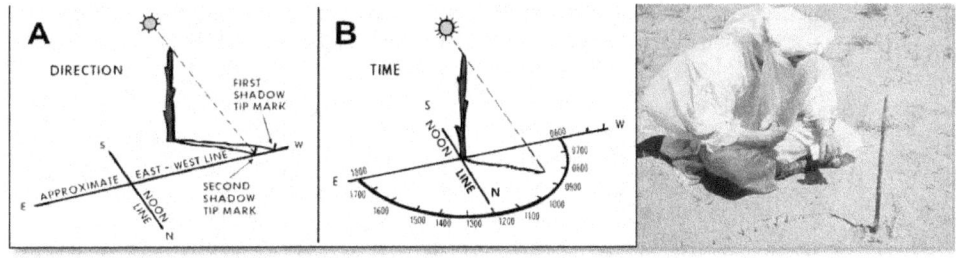

Figure 18-29 Determining Time and Direction by Shadow

18.11.1.1. Orientation

A line drawn at right angles to the east-west line at any point is the approximate north-south line, which will help orient a person to any desired direction of travel.

18.11.1.2. Conditions

Inclining the stick to obtain a more convenient shadow does not impair the accuracy of the shadow-tip method. Therefore, a traveler on sloping ground or in highly vegetated terrain need not waste valuable time looking for a large level area. A flat spot, the size of the hand, is all that is necessary for shadow-tip markings and the base of the stick can be either above, below, or to one side of it. Also, any stationary object (the end of a tree limb or the notch where branches are jointed) serves just as well as an implanted stick because only the shadow tip is marked.

18.11.1.3. Approximate Time of Day

The shadow-tip method can also be used to find the approximate time of day (Figure 18-30).

1. To find the time of day, move the stick to the intersection of the east-west line and the north-south line, and set it vertically in the ground. The west part of the east-west line indicates the time is 0600 and the east part is 1800.

2. The north-south line now becomes the noon line. The shadow of the stick is an hour hand in the shadow clock and with it the time can be estimated using the noon line and six o'clock line as the guides. Depending on the location and the season, the shadow may move either clockwise or counterclockwise, but this does not alter the manner of reading the shadow clock.

3. The shadow clock is not a timepiece in the ordinary sense. It always reads 0600 at sunrise and 1800 at sunset. However, it does provide a satisfactory means of telling time in the absence of properly set watches. Being able to establish the time of day is important for such purposes as keeping a rendezvous, prearranged concerted action by separated persons or groups, and estimating the remaining duration of daylight. Shadow-clock time is closest to conventional clock time at midday, but the spacing of the other hours, compared to conventional time, varies somewhat with the locality and date.

4. The shadow-tip system is ineffective for use beyond 66'/2° latitude in either hemisphere due to the position of the Sun above the horizon. Whether the Sun is north or south of a survivor at mid-day depends on the latitude. North of 23.4°N, the Sun is always due south at local noon and the shadow points north. South of 23.4°S, the Sun is always due north at

local noon and the shadow points south. In the tropics, the Sun can be either north or south at noon, depending on the date and location but the shadow progresses to the east regardless of the date.

18.11.2. Equal Shadow Method

The equal-shadow method of determining direction (Figure 18-30 and Figure 18-31) is a variation of the shadow-tip method. It is more accurate and may be used at all latitudes less than 66° at all times of the year. It consists of four steps:

1. Place a stick or branch into the ground vertically at a level spot where a shadow at least 12 inches long will be cast. Mark the shadow tip with a stone, twig, or other means. This must be done five to 10 minutes before noon (when the Sun is at its highest point (zenith)).
2. Trace an arc using the shadow as the radius and the base of the stick as the center. A piece of string, shoelace, or a second stick may be used to do this.
3. As noon is approached, the shadow becomes shorter. After noon, the shadow lengthens until it crosses the arc. Mark the spot as soon as the shadow tip touches the arc a second time.
4. Draw a straight line through the two marks to obtain an east-west line.

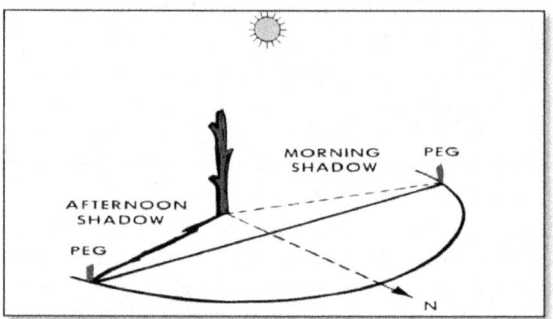

Figure 18-250 Equal Shadow Method of Determining Direction

18.11.2.1. Noon Requirement

Although this is the most accurate version of the shadow-tip method, it must be performed around noon. It requires the observer to watch the shadow and complete step three at the exact time the shadow tip touches the arc.

18.11.3. Direction by Using Stars

At night, the stars may be used to determine the north line in the northern hemisphere or the south line in the southern hemisphere. Figure 18-32 shows how this is done.

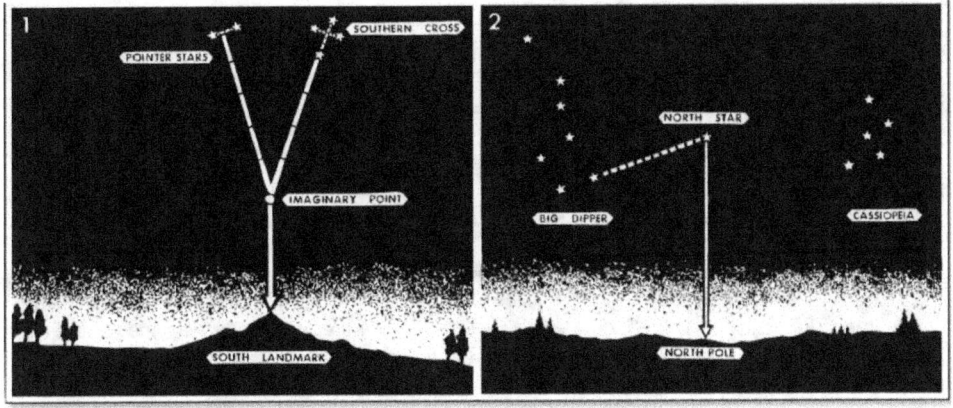

Figure 18-261 Determination of Direction by Using the Stars

18.11.4. Direction by Using a Watch

A watch can be used to determine the approximate true north or south (Figure 18-32). In the northern hemisphere, the hour hand is pointed toward the Sun. A south line can be found midway between the hour hand and 1200 standard time. During daylight savings time, the north-south line is midway between the hour hand and 1300. If there is any doubt as to which end of the line is north, remember that the Sun is in the east before noon and in the west in the afternoon.

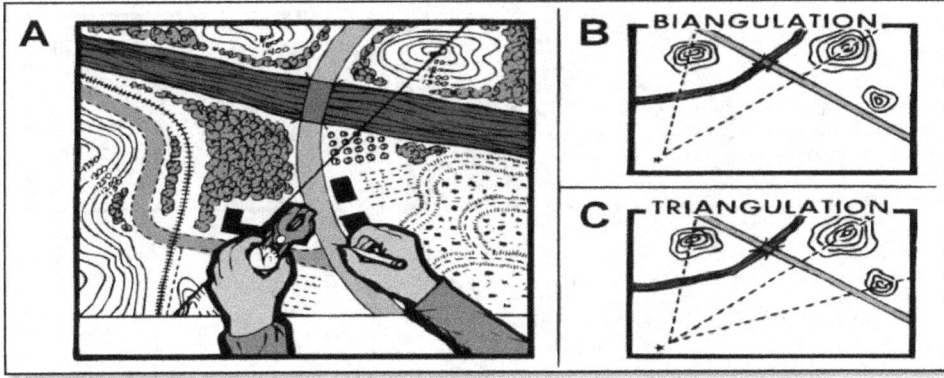

Figure 18-272 Directions Using a Watch

18.11.4.1. Southern Hemisphere

The watch may also be used to determine direction in the Southern Hemisphere; however, the method is different. The 1200-hour dial is pointed toward the Sun and halfway between 1200 and the hour hand will be a north line. During daylight savings time, the north line lies midway between the hour hand and 1300. The watch method can be in error, especially in the extreme latitudes.

18.12. Determining Specific Position

When using a map and compass, the map must be oriented using the method described earlier in this chapter. Next, locate two or three known positions on the ground and the map. Use the compass to shoot an azimuth to one of the known positions (Figure 18-33). Once the azimuth is determined, recheck the orientation of the map and plot the azimuth on the map. To plot the azimuth, place the front corner of the straightedge of the compass on the corresponding point on the map. Rotate the compass until the determined azimuth is directly beneath the stationary index line. Then create a line along the straightedge of the compass and extend the line past the estimated position on the map, creating a line of position. Lines can be created with the use of blades of grass, lanyard of the compass, twigs, and other nonmetallic straight materials. Repeat this procedure for the second point. If only two azimuths are used, the technique is referred to as biangulation. If a third azimuth is plotted to check the accuracy of the first two, the technique is called triangulation. Between procedures check orientation of the map. When using three lines, a triangle of error may be formed. If the triangle is large, the work should be checked. However, if a small triangle is formed, the user should evaluate the terrain to determine the actual position. One azimuth may be used with a linear land feature such as a river, road, railroad, etc., to determine specific position.

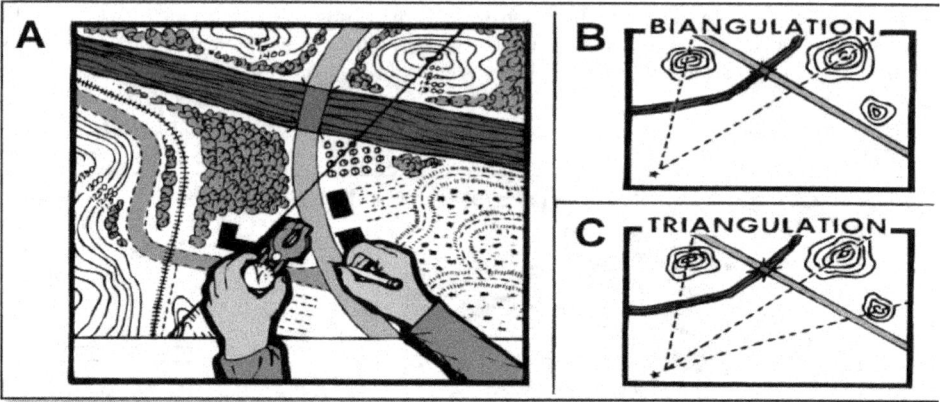

Figure 18-283 Azimuth, Biangulation, and Triangulation

18.13. Determining Specific Location without a Compass

A true north-south line determined by the stick and shadow, Sun and watch, or celestial constellation method may be used to orient the map without a compass. However, visible major land features can be used to orient the map to the lay of the land. Once the map is oriented, identify two or three landmarks and mark them on the map. Lay a straightedge on the map with the center of the straightedge at a known position as a pivot point and rotate the straightedge until the known position of the map is aligned with present position, and create a line. Repeat this for the second and third position. Each time a line of position is plotted, the map must still be aligned with true north and south. If three lines of position are plotted and form a small triangle, use terrain evaluation to determine present position. Recheck calculations for errors.

18.14. Dead Reckoning

Dead reckoning is the process of locating one's position by plotting the course and distance from the last known location. In areas where maps exist, even poor ones, travel is guided by them. It is a matter of knowing one's position at all times by associating the map features with the ground features and the distance travelled. The survivor may be required to travel in these areas without a map.

18.14.1. Movement on Land

Movement on land must be carefully planned. When the starting location and destination are known, and if a map is available, they are carefully plotted along with any known intermediate features along the route. These intermediate features, if clearly recognizable, serve as checkpoints while traveling. If a map is not available and the terrain or isolating event allows a clear view of the travel route, the plot is done on a blank sheet of paper (in a non-permissive environment IP should use a mental map). A scale is selected so the entire route will fit on one sheet. A north direction is clearly established. The starting point and destination are then plotted in relationship to each other. If the terrain and enemy situations permit, the ideal course is a straight line from starting point to destination. This is seldom possible or practicable. The route of travel usually consists of several courses, with an azimuth established at the starting point for the first course to be followed. Distance measurement begins with the departure and continues through the first course until a change in direction is made. A new azimuth is established for the second course and the distance is measured until a second change of direction is made, and so on. Records of all data are kept and all positions are plotted.

18.14.2. Measuring Distance by Paces

A pace is equal to the distance covered every time the same foot touches the ground. To measure distance, count the number of paces, estimate the distance travelled, and apply that to the map and route of travel. Usually, paces are counted in blocks of numbers (25, 50, etc.), and can be kept track of in many ways: making notes in a record book; counting individual fingers; placing small objects such as pebbles into an empty pocket; tying knots in a string; or using a mechanical hand counter. Distances measured this way are only approximate, but with practice can become very accurate. It is important prior to using dead reckoning navigation each person should establish the length of their average pace. This is done by pacing a measured course many times and computing the mean. In the field, an average pace must often be adjusted because of the following conditions:

- Slopes. The pace lengthens on a downgrade and shortens on an upgrade.
- Winds. A headwind shortens the pace while a tailwind increases it.
- Surfaces. Sand, gravel, mud and similar surface materials tend to shorten the pace.
- Elements. Snow, rain, or ice reduces the length of the pace.
- Clothing. Excess weight of clothing shortens the pace while the type of shoes affects traction and therefore the pace length.

18.14.3. Deliberate Offset

An offset is a planned magnetic deviation to the right or left of an azimuth to an objective. It is used when approaching a linear feature from the side and a point along the linear feature (such as a road junction) is the objective. One may reach the linear feature and not know whether the objective lies to the right or left. A deliberate offset by a known number of degrees in a known direction compensates for possible errors and ensures that, upon reaching the linear feature, the user knows whether to go right or left to reach the objective. Figure 18-34 shows an example of the use of offset to approach an objective. It should be remembered that the distance from "X" to the objective will vary directly with the distance to be traveled and the number of degrees offset. Each degree offset will move the course about 20 feet to the right or left for each 1,000 feet traveled. For example, in Figure 18-34, the number of degrees offset is 10 to the right. If the distance traveled to "X" is 1,000 feet, then "X" is located about 200 feet to the right of the objective.

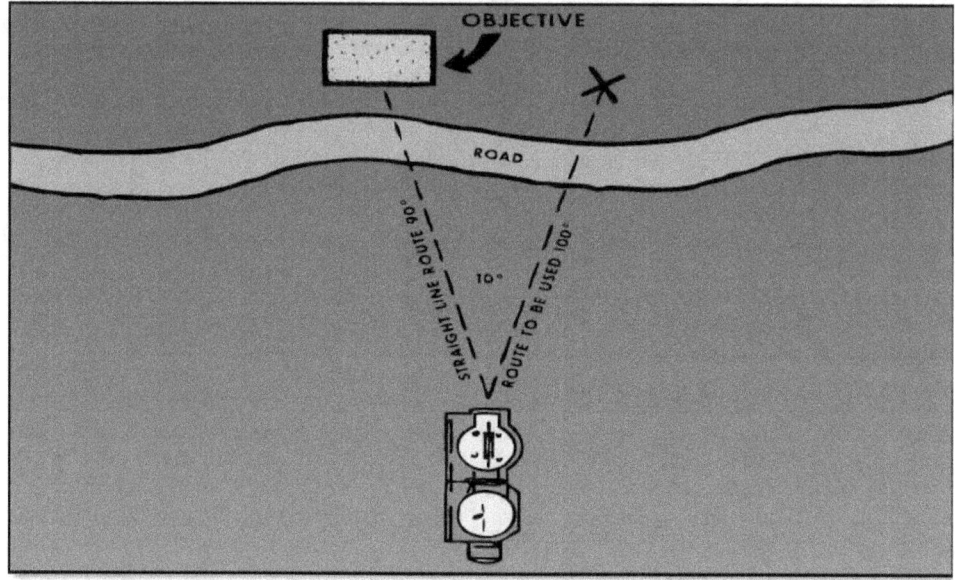

Figure 18-294 Deliberate Offset

18.14.4. Detour Around Enemy Position

Figure 18-35 shows an example of how to bypass enemy positions or obstacles by detouring around them and maintaining orientation by moving at right angles for specified distances; for example, moving on an azimuth 360° and wish to bypass an obstacle or position. Change direction to 90° and travel for 100 yards, change direction back to 360° and travel for 100 yards, change direction to 270° and travel for 100 yards, and then change direction to 360°, and back on the original azimuth. Bypassing an unexpected obstacle at night is done in the same way.

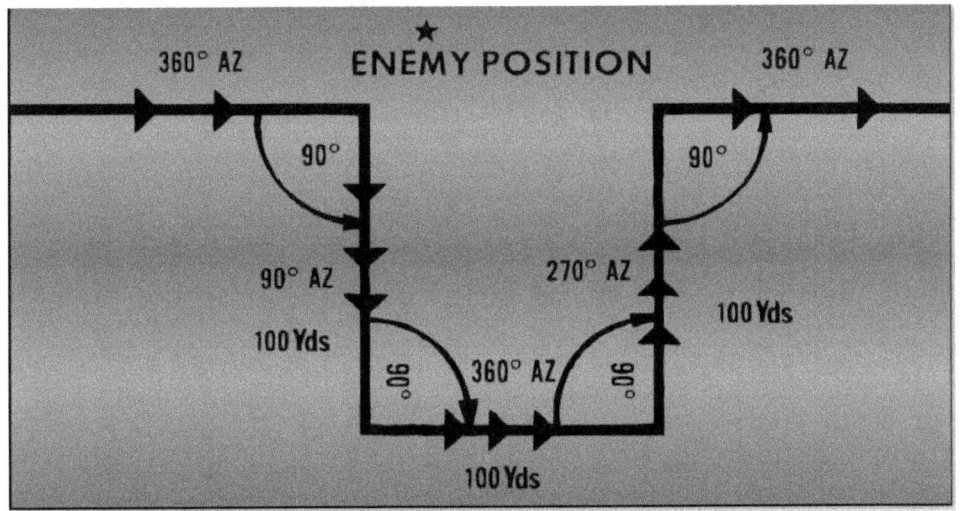

Figure 18-305 Detour Around Enemy Position

18.15. Position Determination

From the Sun at sunrise and sunset, Figure 18-36 shows the true azimuth of the rising Sun and the relative bearing of the setting Sun for all of the months in the year in the Northern and Southern Hemispheres (the table assumes a level horizon and is inaccurate in mountainous terrain).

18.15.1. Latitude by Compass

Latitude can be determined by using a compass to find the angle of the Sun at sunrise or sunset (subtracting or adding magnetic variation) and the date. According to the chart in figure 18-36, on January 26th, the azimuth of the rising Sun will be 120° to the left when facing the Sun in the Northern Hemisphere (it would be 120° to the right for setting Sun); therefore, the latitude would be 50°. If in the Southern Hemisphere, the direction of the Sun would be the opposite.

18.15.2. Latitude by Interpolation

The table does not list every day of the year, nor does it list every degree of latitude. If accuracy is desired within 1° of azimuth, interpolation may be necessary (split the difference) between the values given in the table. For example, between 45° latitude and 50° latitude is 5°. The difference in latitudes (5°) and the difference in azimuths (3°) split (5/3) is 1°/3'(1°40'), so the more accurate degree of latitude would be 46°40' latitude.

DATE		0°	5°	10°	15°	20°	25°	30°	35°	40°	45°	50°	55°	60°
JANUARY	1	113	113	113	114	115	116	117	118	121	124	127	133	141
	6	112	113	113	113	114	115	116	118	120	123	127	132	140
	11	112	112	112	113	113	114	115	117	119	122	125	130	138
	16	111	111	111	112	112	113	114	116	118	120	124	129	136
	21	110	110	110	111	111	112	113	115	117	119	122	127	133
	26	109	109	109	109	110	111	112	113	115	117	120	124	130
FEBRUARY	1	107	107	108	108	109	110	111	113	115	118	121	125
	6	106	106	106	106	107	107	108	109	111	113	115	118	123
	11	104	104	105	105	105	106	107	108	109	110	112	116	120
	16	103	103	103	103	103	104	105	106	107	108	110	112	116
	21	101	101	101	101	101	102	102	103	104	105	107	109	112
	26	99	99	99	99	100	100	100	101	102	103	104	106	108
MARCH	1	98	98	98	98	99	99	99	100	100	101	102	104	106
	6	96	96	96	96	96	97	97	97	98	98	99	100	102
	11	94	94	94	94	94	94	95	95	95	96	96	97	98
	16	92	92	92	92	92	92	92	92	93	93	93	93	94
	21	90	90	90	90	90	90	90	90	90	90	90	90	90
	26	88	88	88	88	88	88	88	88	87	87	87	87	86
APRIL	1	86	86	86	85	85	85	85	85	84	84	83	82	81
	6	84	84	84	83	83	83	83	83	82	82	81	80	79	77
	11	82	82	82	82	81	81	81	80	80	79	77	76	74
	16	80	80	80	80	79	79	78	78	77	76	74	72	70
	21	78	78	78	78	78	77	76	76	75	73	72	69	66
	26	77	77	76	76	76	75	75	74	72	71	69	66	63
MAY	1	75	75	75	74	74	73	73	72	70	69	66	63	59
	6	74	74	73	73	73	72	71	70	68	67	64	61	56
	11	72	72	72	72	71	70	69	68	67	64	62	58	52
	16	71	71	71	70	70	69	68	67	65	63	60	55	49
	21	70	70	70	69	69	68	67	65	63	61	58	53	47
	26	69	69	69	68	68	67	66	64	62	60	56	51	44
JUNE	1	68	68	68	67	66	66	64	63	61	58	54	49	41
	6	67	67	67	67	66	65	64	62	60	57	53	48	40
	11	67	67	67	67	66	65	63	62	59	56	53	47	39
	16	67	67	67	66	65	64	63	62	59	56	53	47	39
	21	67	67	67	66	65	64	63	62	59	56	53	47	39
	26	67	67	67	66	65	64	63	62	59	56	53	47	39
JULY	1	67	67	67	67	66	65	64	62	60	57	53	48	40
	6	67	67	67	67	66	65	64	62	60	57	53	48	40
	11	68	68	68	68	67	66	65	64	63	61	58	54	49	41
	16	69	69	69	68	68	67	66	65	63	62	59	55	50	43
	21	69	69	69	69	68	67	66	65	63	60	57	52	45
	26	70	70	70	69	69	68	67	66	64	62	59	54	48
AUGUST	1	72	72	72	71	71	70	69	68	66	64	61	57	51
	6	73	73	73	73	72	71	71	69	68	66	63	60	55
	11	75	75	74	74	74	73	72	71	70	68	66	63	58
	16	76	76	76	76	75	75	74	73	72	70	68	65	61
	21	78	78	77	77	77	76	75	74	73	72	71	68	65
	26	79	79	79	79	78	78	77	76	75	73	71	68
SEPTEMBER	1	82	82	82	81	81	81	80	80	79	78	77	75	73
	6	83	83	83	83	83	83	82	82	81	81	80	78	77
	11	85	85	85	85	85	85	85	84	84	83	83	82	81
	16	87	87	87	87	87	87	87	86	86	86	85	85	84
	21	89	89	89	89	89	89	89	89	89	89	88	88	88
	26	91	91	91	91	91	91	91	91	91	91	92	92	92
OCTOBER	1	93	93	93	93	93	93	93	94	94	94	95	95	96
	6	95	95	95	95	95	95	96	96	96	97	97	98	99	100
	11	97	97	97	97	97	98	98	99	99	100	101	102	104
	16	99	99	99	99	99	100	100	101	101	102	104	105	108
	21	101	101	101	101	101	102	102	103	104	105	107	109	112
	26	102	102	103	103	103	104	104	105	106	108	109	112	115
NOVEMBER	1	104	104	105	105	105	106	107	108	109	110	113	116	120
	6	106	106	106	107	107	108	109	110	111	113	115	119	123
	11	107	107	108	108	108	109	110	111	113	115	117	121	126
	16	109	109	109	109	110	110	111	112	113	115	117	120	124	130
	21	110	110	110	111	111	112	113	114	116	119	122	126	133
	26	111	111	111	112	112	113	114	116	118	120	124	128	135
DECEMBER	1	112	112	112	113	113	114	115	117	119	122	125	130	138
	6	112	112	113	113	114	115	116	118	120	123	126	132	140
	11	113	113	113	113	114	115	116	117	118	121	124	127	133	141
	16	113	113	113	114	114	115	116	117	118	121	124	127	133	141
	21	113	113	113	114	115	116	117	118	121	124	127	133	141
	26	113	113	113	114	115	116	117	118	121	124	127	133	141

NOTE: When the Sun is rising, the angle is reckoned from East to North. When the Sun is setting, the angle is reckoned from West to North.

Figure 18-316 Finding Direction from the Rising or Setting Sun

18.15.3. Latitude by Noon Altitude of the Sun

On any given day, there is only one latitude on Earth where the Sun passes directly overhead or through the zenith at noon. In all latitudes north of this, the Sun passes to the south of the zenith; and in those south of it, the Sun passes to the north. For each 1° change of latitude, the zenith distance also changes by 1 degree. Figure 18-37 gives the latitude for each day of the year where the Sun is in the zenith at noon. If a Weems plotter or other protractor is available, maximum altitude of the Sun should be used to find latitude by measuring the angular distance of the Sun from the zenith at noon. Local noon can be found using the methods described earlier. Stretch a string from the top of a stick to the point where the end of the noon shadow rested, place the plotter along the string and drop a plumb line from the center of the plotter. The intersection of the plumb line with the outer scale of the plotter shows the angular distance of the Sun from the zenith.

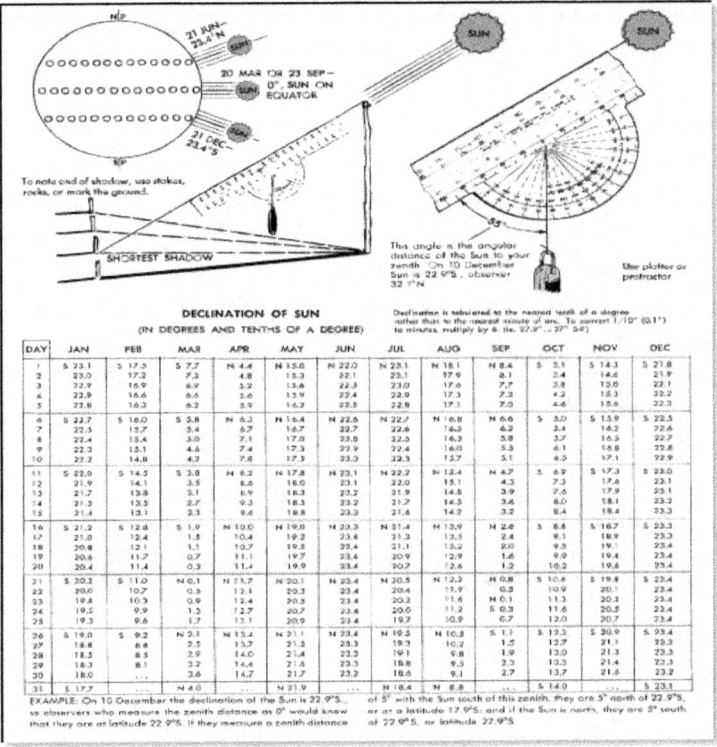

Figure 18-327 Determining Latitude by Noon Sun

18.15.4. Latitude by Length of Day

This method is used most effectively while on open seas. When in any latitude between 60ºN and 60ºS, the exact latitude within 30 nautical miles (1/2') can be determined if the length of the day within one minute is known. This is true throughout the year except for about 10 days before and 10 after the equinoxes -- approximately 11-31 March and 12 September thru 2 October. During these two periods, the day is about the same length at all latitudes. A level horizon is required to time sunrise and sunset accurately. Find the length of day from the instant the top of the Sun first appears above the ocean horizon to the instant it disappears below the horizon. This instant is often marked by a green flash. Write down the times of sunrise and sunset. Don't count on remembering them. Note that only the length of day counts in the determination of latitude; a watch may have an unknown error and yet serve to determine this factor. If only one water horizon is available, as on a sea coast, find local noon by the stick and shadow method. The length of day is twice the interval from sunrise to noon or from noon to sunset. Knowing the length of day, latitude can be found by using the nomogram shown in Figure 18-38.

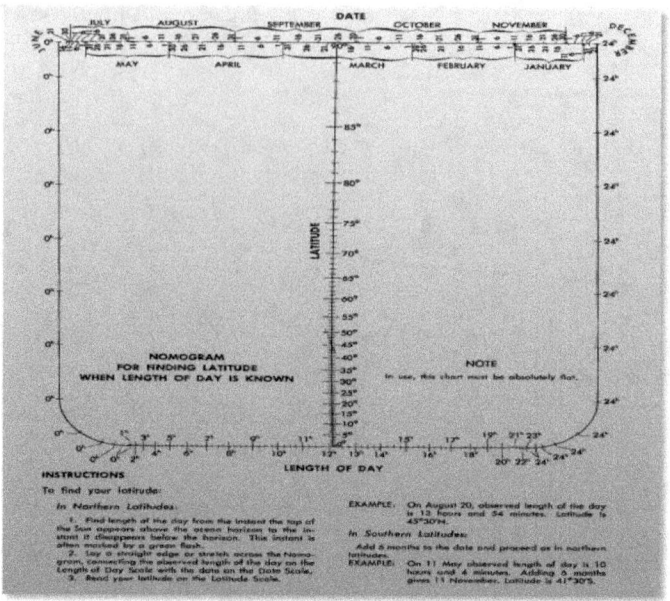

Figure 18-338 Nomogram

18.15.5. Longitude from Local Apparent Noon

To find longitude, an IP must know the correct time and the rate at which a watch gains or losses time. If this rate and the time the watch was last set is known, the correct time can be computed by adding or subtracting the gain or loss. Correct the zone time on the watch to Greenwich Time; for example, if the watch is on Eastern Standard Time, add 5 hours to get Greenwich Time. Longitude can be determined by timing the moment a celestial body passes the meridian. The easiest body to use is the Sun. Use one of the following methods:

18.15.5.1. Stick and Shadow

Put up a stick or rod (Figure 18-39) as nearly vertical as possible in a level place. Check the alignment of the stick by sighting along the line of a makeshift plumb bob. (To make a plumb bob, tie any heavy object to a string and let it hang free. The line of the string indicates the vertical.) Sometime before midday, begin marking the position of the end of the stick's shadow. Note the time for each mark. Continue marking until the shadow definitely lengthens. The time of the shortest shadow is the time when the Sun passed the local meridian or local apparent noon. A survivor will probably have to estimate the position of the shortest shadow by finding a line midway between two shadows of equal length, one before noon and one after. If the times of sunrise and sunset are accurately determined on a water horizon, local noon will be midway between these times.

Figure 18-39 Stick and Shadow

18.15.5.2. Double Plumb Bob

Erect two plumb bobs about one foot apart so that both strings line up on Polaris, much the same as a gun sight. Plumb bobs should be set up when Polaris is on the meridian and has no east-west correction. The next day, when the shadows of the two plumb bobs coincide, they will indicate local apparent noon.

18.15.5.3. Greenwich Time of Local Noon

The next step is to correct this observed time of meridian passage for the equation of time; that is, the number of minutes the real Sun is ahead of or behind the mean Sun. (The mean Sun was invented by astronomers to simplify the problems of measuring time. Mean Sun rolls along the Equator at a constant rate of 15° per hour. The real Sun is not so considerate; it changes its angular rate of travel around the Earth with the seasons.) Figure 18-40 gives the value in minutes of time to be added to or subtracted from mean (watch) time to get apparent (Sun) time.

Date		Eq. of Time*	Date		Eq. of Time*	Date		Eq. of Time*	Date		Eq. of Time*	Date		Eq. of Time*	Date		Eq. of Time*
Jan.	1	−3.5 min.	Mar.	4	−12.0	May	2	+3.0 min.	Aug.	4	−6.0	Oct.	1	+10.0 min.	Dec.	1	+11.0
	2	−4.0		8	−11.0		14	+3.8		12	−5.0		4	+11.0		4	+10.0
	4	−5.0		12	−10.0	May	28	+3.0		17	−4.0		7	+12.0		6	+9.0
	7	−6.0		16	−9.0					22	−3.0		11	+13.0		9	+8.0
	9	−7.0		19	−8.0	June	4	+2.0		26	−2.0		15	+14.0		11	+7.0
	12	−8.0		22	−7.0		9	+1.0	Aug.	29	−1.0		20	+15.0		13	+6.0
	14	−9.0		26	−6.0		14	0.0	Sept.	1	0.0	Oct.	27	+16.0		15	+5.0
	17	−10.0	Mar.	29	−5.0		19	−1.0		5	+1.0					17	+4.0
	20	−11.0					23	−2.0		8	+2.0					19	+3.0
	24	−12.0	Apr.	1	−4.0	June	28	−3.0		10	+3.0	Nov.	4	+16.4		21	+2.0
Jan.	28	−13.0		5	−3.0					13	+4.0		11	+16.0		23	+1.0
				8	−2.0	July	3	−4.0		16	+5.0		17	+15.0		25	0.0
Feb.	4	−14.0		12	−1.0		9	−5.0		19	+6.0		22	+14.0		27	−1.0
	13	−14.3		16	0.0		18	−6.0		22	+7.0		25	+13.0		29	−2.0
	19	−14.0		20	+1.0					25	+8.0						
Feb.	28	−13.0	Apr.	25	+2.0	July	27	−6.6	Sep.	28	+9.0	Nov.	28	+12.0	Dec.	31	−3.0

* Add plus values to mean time and subtract minus values from mean time to get apparent time.

Figure 18-340 Equation of Time

18.15.5.4. Convert Greenwich Time to Local Time

After computing the Greenwich Time of local noon, the difference of longitude between the survivor's position and Greenwich can be found by converting the interval between 1200 Greenwich and the local noon from time to arc. Remember that one hour equals 15° of longitude, four minutes equal 1° of longitude, and four seconds equal 1' of longitude. Example: The survivor's watch is on Eastern Standard Time, and it normally loses 30 seconds a day. It hasn't been set for four days. The local noon is timed at 15:08 on the watch on 4 February. Watch correction is 4 X 30 seconds, or plus 2 minutes. Zone time correction is plus 5 hours. Greenwich Time is 15:08 plus 2 minutes plus 5 hours or 20:10. The equation of time for 4 February is minus 14 minutes. Local noon is 20:10 minus 14 minutes or 19:56 Greenwich. The difference in time between Greenwich and present position is 19:56 minus 12:00 or 7:56. A time of 7:56 equals 119° of longitude. Since local noon is later than Greenwich noon, the location is west of Greenwich, longitude is 119° W.

18.15.6. Direction and Position Finding at Night

Direction and position can also be determined at nighttime.

18.15.6.1. Direction from Polaris

In the Northern Hemisphere, one star, Polaris (the Pole Star), is never more than approximately 1' from the North Celestial Pole. In other words, the line from any observer in the Northern Hemisphere to the Pole Star is never more than 1° away from true north. Find the Pole Star by locating the Big Dipper or Cassiopeia, two groups of stars which are very close to the North Celestial Pole. The two stars on the outer edge of the Big Dipper are called pointers because they point almost directly to Polaris. If the pointers are obscured by clouds, Polaris can be identified by its relationship to the constellation Cassiopeia. Figure 18-41 indicates the relation between the Big Dipper, Polaris, and Cassiopeia.

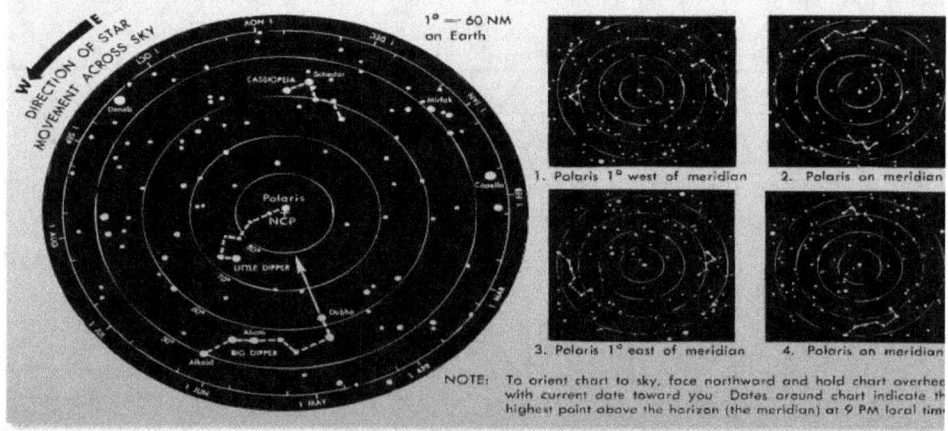

Figure 18-351 Finding Direction from Polaris

18.15.6.2. Direction from the Southern Cross

In the Southern Hemisphere, Polaris is not visible. There the Southern Cross is the most distinctive constellation. When flying south, the Southern Cross appears shortly before Polaris drops from sight astern. An imaginary line through the long axis of the Southern Cross, or True Cross, points toward the South Pole. The True Cross should not be confused with a larger cross nearby known as the False Cross, which is less bright and more widely spaced. Two of the four stars in the True Cross are among the brightest stars in the heavens; they are the stars on the southern and eastern arms. Those of the northern and western arms are not as conspicuous but are bright.

There is no conspicuous star above the South Pole to correspond to Polaris above the North Pole. In fact, the point where such a star would be, if one existed, lies in a region devoid of stars. This point is so dark in comparison with the rest of the sky that it is known as the Coal sack. Figure 18-42 shows the True Cross and-to the west of it-the False Cross.

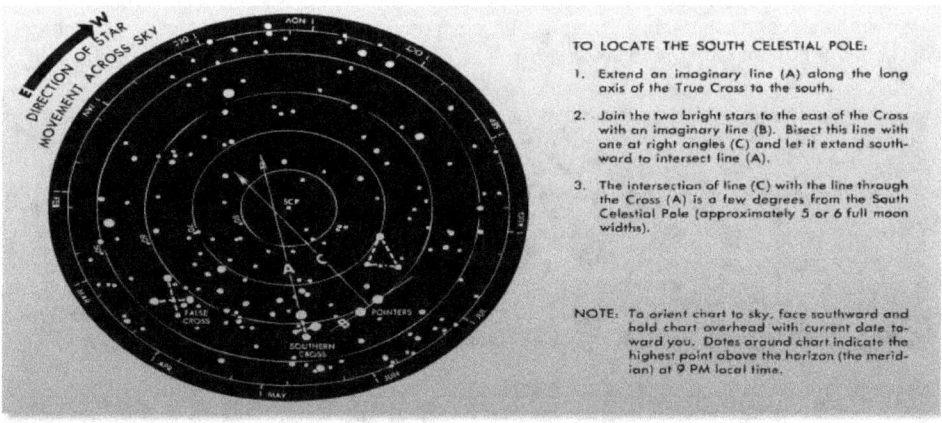

Figure 18-362 Finding Direction from Southern Cross

18.15.7. Finding Due East and West by Equatorial Stars

Due to the altitude of Polaris above the horizon, it may sometimes be difficult to use as a bearing. To use a point directly on the horizon may be more convenient.

18.15.7.1. Celestial Equator

The celestial equator, which is a projection of the Earth's equator onto the imaginary celestial sphere, always intersects the horizon line at the due east and west points of the compass. Therefore, any star on the celestial equator rises due east and sets due west (disallowing a small error because of atmospheric refraction). This holds true for all latitudes except those of the North and South Poles, where the celestial equator and the horizon have a common plane. However, if a survivor is at the North or South Pole, it will probably be known, so this technique can be assumed to be of universal use.

- Certain difficulties arise in the practical use of this technique. Unless a survivor is quite familiar with the constellations, it may be difficult to spot a specific rising star as it first appears above the eastern horizon. It will probably be simpler to depend upon the identification of an equatorial star before it sets in the west.

- Another problem is caused by atmospheric extinction. As stars near the horizon, they grow fainter in brightness because the line of sight between the observer's eyes and the star passes through a constantly thickening atmosphere. Therefore, faint stars disappear from view before they actually set. However, a fairly accurate estimate of the setting point of a star can be made some time before it actually sets. The atmospheric conditions of the area have a great effect on obstructing a star's light as it sets. Atmospheric haze, for example, is much less a problem on deserts than along temperate zone coastal strips.

18.15.7.1.1. Equatorial Stars

Figure 18-43 shows the brighter stars and some prominent star groups which lie along the celestial equator. There are few bright stars actually on the celestial equator. However, there are a number of stars which lie quite near it, so an approximation within a degree or so can be made. Also, a rough knowledge of the more conspicuous equatorial constellations will give a continuing checkpoint for maintaining orientation.

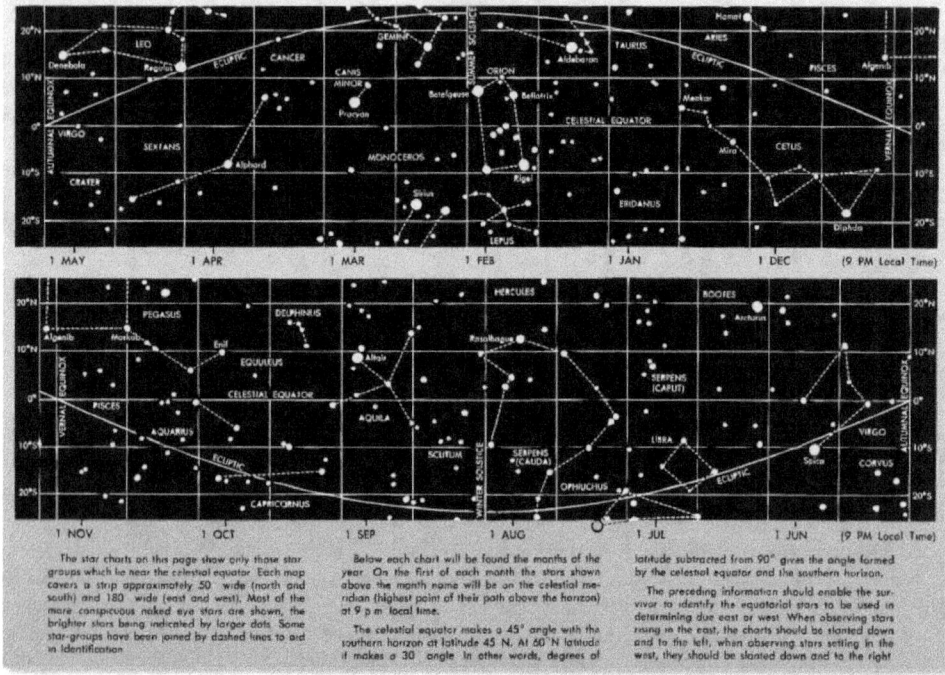

Figure 18-373 Charts of Equatorial Stars

18.15.8. Finding Latitude from Polaris

The latitude in the Northern Hemisphere north of 10°N can be found by measuring the angular altitude of Polaris above the horizon, as shown in Figure 18-44.

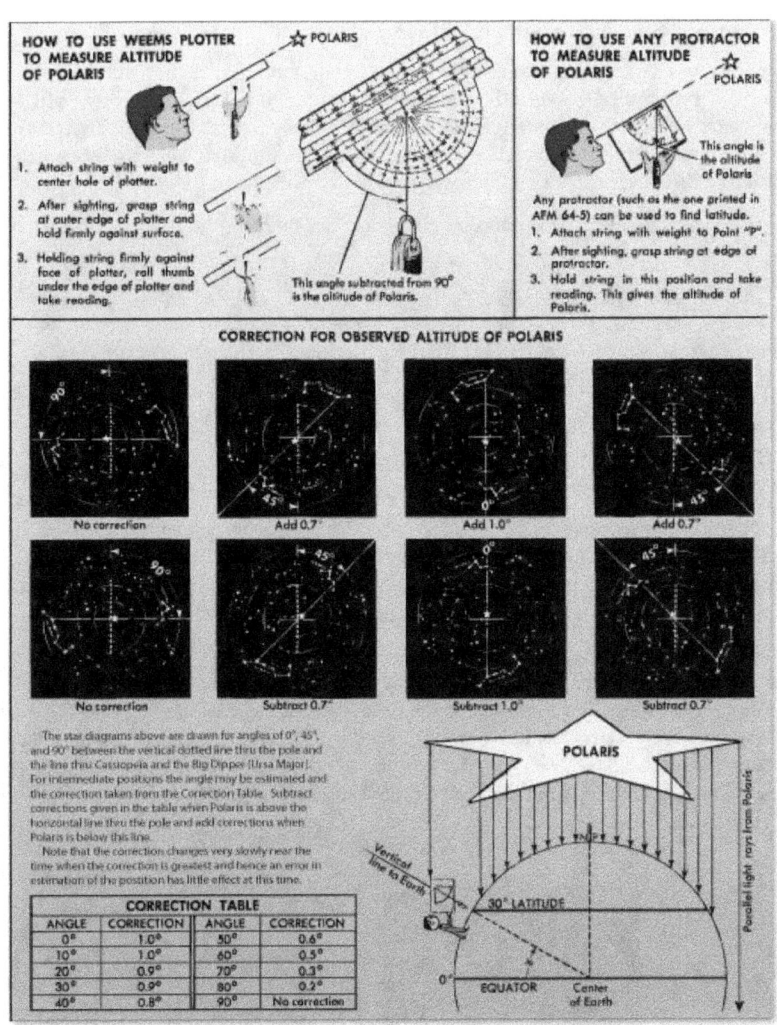

Figure 18-384 Finding Latitude by Polaris

18.15.9. Direction by Overhead Stars

Direction (North) can be determined from overhead stars that are not in the general location of the celestial poles. At times, may not be able to locate Polaris (the North Star) may not be located due to partial cloud cover, or its position below the observer's horizon. In this situation, it would seem that one would be unable to locate direction. Fortunately, those who wish to initially find direction or who wish to check a course of travel during the night need not worry about being lost or unable to travel if Polaris cannot be identified.

18.15.9.1. Modified Stick and Shadow Method

The following is an adaptation of the stick and shadow method of direction finding. This method is based on the principles that all the heavenly bodies (Sun, Moon, planets, and stars) rise (generally) in the east and set (generally) in the west. This technique can be used anywhere on Earth with any stars except those which are circumpolar. Circumpolar stars are those which appear to travel around Polaris instead of apparently "moving" from east to west.

- To use this technique, keep in mind that any star other than a circumpolar one may be used.
- Prepare a device to aid in knowing general direction. This can be done by placing a stick (about five feet in length) at a slight angle in the ground in an open area. Thin material (suspension line, string, vine, braided cloth, etc.) is then attached to the tip of the stick. This material should be longer than what is required to reach the ground (Figure 18-45).
- Lie on the back with the head next to this hanging line. Pull the cord up to the temple area and hold it tautly.
- Next, IP move around on the ground until the taut line is pointing directly at the selected, bright, non-circumpolar star (or planet).
- The taut line is now in position to simulate the star's (or planet's) shadow. Remember that this method of finding direction is an adaptation of the Sun, stick, and shadow approach. Here the more distant stars and (or) planets take the place of our Sun. Since these objects are too distant from the Earth to create shadow, the string represents the shadow.
- With the taut line simulating the star's shadow, mark the point on the ground where the line touches with a stick, stone, etc. Repeat this sighting on the same star (or planet) after about 15 to 20 minutes (marking the spot at which the line "shadow" touches the ground). A line scribed on the ground which connects these two points will run west-east (as the stars and planets move from east to west, the "shadow" will move in the opposite direction). The first mark will be in the west. Drawing a line perpendicular to the west-east line will give a north-south line to aid in travel.

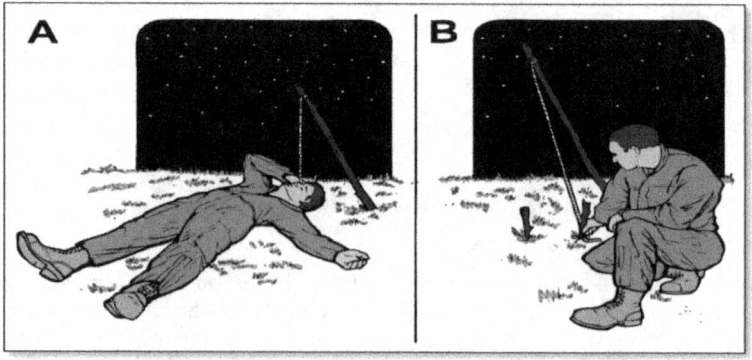

Figure 18-395 Stick and String Direction Finding

18.16. The Compass and Its Uses

The magnetic compass is the most commonly used and simplest instrument for measuring directions and angles in the field. The lensatic compass (Figure 18-46) is the standard magnetic compass for military use today.

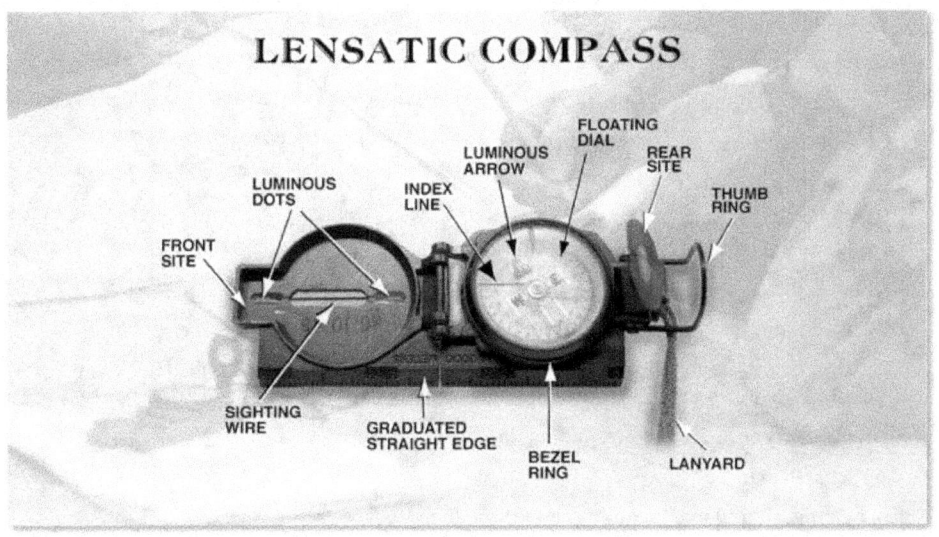

Figure 18-406 Lensatic Compass

18.16.1. Sighting

The lensatic compass must always be held level and firm when sighting on an object and reading an azimuth. The compass is comprised of three major components: the cover, the base and the lens. The cover serves the purpose of protecting the compass rose dial. It contains sighting wires and dots utilized in nighttime navigation.

18.16.2. Base

The base is where all movable parts of the compass reside. A floating dial rotates indicating direction each time the compass maintains a level position. Incandescent figures depict the directions east (E) and west (W). In the center lies the directional arrow. This always points in the direction of north (N). East falls at 90° and west at 270°. There are also two scales. The outer scale denotes miles and the inner scale denotes degrees. The inner scale indicator is in red color.

18.16.3. Bezel Ring

A bezel ring is also inside the compass base. This ratchet device turns 120 clicks on a full rotation. Each individual click represents 3°. A short incandescent line works with the north-directional arrow in navigation. This line lies in the glass face of the bezel ring. The floating dial contains a fixed black index line as well. The final component of the base is a thumb loop. This simply attaches to the base as a handling mechanism.

18.16.4. Lens

The last component of the lensatic compass is the lens. This is what reads the floating dial. A rear-sight navigational slot works with the front sight wires in the cover to locate objects. In addition, this slot protects the compass when in the closed position. This works by a lock and clamp system. The rear-sight mechanism must remain open at a minimum of 45° in order for the compass floating dial to work.

18.16.5. Bearing

To take a bearing (Figure 18-47):

1. Unlock the cover half way so that the compass card forms a 90° angle.
2. Raise the lens arm to a 45° angle.
3. Stabilize the compass by placing thumb inside the thumb hook. Ensure the hook is all the way towards the bottom before doing so.
4. Locate the target object.
5. Adjust the sighting wire so that it lies in the center of the target object.
6. Read fine degree marking on the compass card. Do this by moving the lens up and down until the degree mark reads without taking your eye off the target.
7. Read the bearing in degrees (red numbers).

18.16.6. Set a Bearing

To set a bearing, follow the directions above, and then:

1. Bring the marking on the bezel in line with the north direction arrow. Indication of this comes off the compass card.
2. Once the bezel marking and directional arrow align, orientation is set.
3. Proceed on course in direction indicated by sighting wire.

18.16.7. Follow a Bearing

To follow a bearing, follow the directions above, and then:

1. Pinpoint landmark in the distance to serve as a reference point.
2. When reference point disappears, due to weather or trees, use compass to stay the course.
3. Occasionally set a new bearing for your selected reference point.

Figure 18-417 Holding the Compass

18.16.8. Night Use of the Lensatic Compass

For night use, special features of the compass include the luminous markings, the bezel ring, and two luminous sighting dots. Turning the bezel ring counterclockwise causes an increase in azimuth, while turning it clockwise causes a decrease. The bezel ring has a stop and spring which allows turns at 3° intervals per click and holds it at any desired position. One accepted method for determining compass directions at night is:

1. Rotate the bezel ring until the luminous line is over the black index line.
2. Hold the compass with one hand and rotate the bezel ring in a counterclockwise direction with the other hand to the number of clicks required. The number of clicks is determined by dividing the value of the required azimuth by three. For example, for an azimuth of 51°, the bezel ring would be rotated 17 clicks counterclockwise (Figure 18-48).
3. Turn the compass until the north arrow is directly under the luminous line on the bezel.
4. Hold the compass open and level in the palm of the left hand with the thumb along the side of the compass. In this manner, the compass can be held consistently in the same position. Position the compass approximately halfway between the chin and the belt, pointing to the direct front. (Practice in daylight will make a person proficient in pointing the compass the same way every time.) Looking directly down into the compass, turn the body until the north arrow is under the luminous line. Then proceed forward in the direction of the luminous sighting dots (Figure 18-49). When the compass is to be used in darkness, an initial azimuth should be set while light is still available. With this initial azimuth as a base, any other azimuth which is a multiple of 3 can be established through use of the clicking feature of the bezel ring.

SETTING THE COMPASS FOR NIGHT TRAVEL

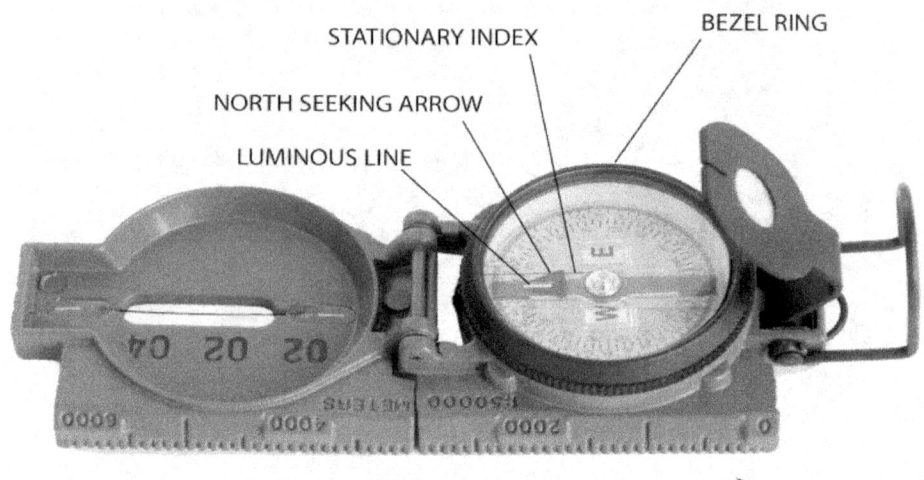

Figure 18-428 Night Travel

18.16.8.1. Floating Needle Compass

IP may find themselves with a floating needle compass (i.e., button compass or improvised compass). A floating needle compass uses a magnetic needle mounted or suspended and free to pivot until aligned with the earth's magnetic field. Improvised floating needle compasses have been created by rubbing a metal needle in one direction against silk material, sliding the needle in one direction against a magnet, and running the needle along the both charge points of a low volt battery, such as a 9 volt. Then suspend the needle in water or tie a piece of thread in the middle of the needle. The needle will seek north, but will not stay charged for a long time.

18.16.8.2. Note on Readings

The magnetic compass is a delicate instrument, especially the dial balance. The IP should take care in its use. Compass readings should never be taken near visible masses of iron or electrical circuits.

18.17. Using a Map and Compass, and Expressing Direction

To use a map, the map must correspond to the lay of the land, and the user must have knowledge of direction and how the map relates to the cardinal directions. In essence, to use a map for land navigation, the map must be "oriented" to the lay of the land. This is usually done with a compass. On most maps, a declination diagram, compass rose, and lines of map magnetic variations are provided to inform the user of the difference between magnetic north, grid north, and/or True north.

18.17.1. Directions

Directions are expressed in everyday life as right, left, straight ahead, etc. but the question arises, "to the right of what?" Military personnel require a method of expressing direction which is accurate, adaptable for use in any area of the world, and has a common unit of measure. Directions are expressed as units of angular measure. The most commonly used unit of angular measure is the degree with its subdivisions of minutes and seconds.

18.17.1.1. Baselines

To measure anything, there must always be a starting point or zero measurement. To express a direction as a unit of angular measure, there must be a starting point or zero measure and a point of reference. These two points designate the base or reference line. There are three baselines-true north, magnetic north, and grid north. Those most commonly used are magnetic and grid-the magnetic when working with a compass, and the grid when working with a military map.

18.17.1.1.1. True North

True North is a line from any position on the Earth's surface to the North Pole. All lines of longitude are true north lines. True north is usually symbolized by a star (Figure 18-49).

18.17.1.1.2. Magnetic North

Magnetic North is the direction to the north magnetic pole, as indicated by the north-seeking arrow of a magnetic instrument (compass). Magnetic North is usually symbolized by a half arrowhead (Figure 18-49).

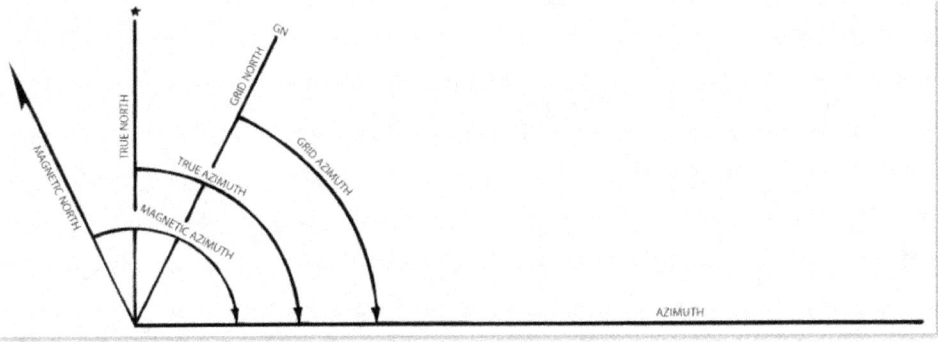

Figure 18-49 True, Grid, and Magnetic Azimuths

18.17.1.1.3. Grid North

Grid North is the North established by the vertical grid lines on the map. Grid North may be symbolized by the letters GN or the letter Y.

18.17.1.2. Azimuth

The most common method used by the military for expressing a direction is azimuths. An azimuth is defined as a horizontal angle, measured in a clockwise manner from a north baseline. When the azimuth between two points on a map is desired, the points are joined by a straight line and a protractor is used to measure the angle between north and the drawn line. This measured angle is the azimuth of the drawn line (Figure 18-50). When using an azimuth, the point from which the azimuth originates is imagined to be the center of the azimuth circle. Azimuths take their name from the baseline from which they are measured; true azimuths from true north, magnetic azimuths from magnetic north, and grid azimuths from grid north (Figure 18-49). Therefore, any given direction can be expressed in three different ways: a grid azimuth if measured on a military map, a magnetic azimuth if measured by a compass, or a true azimuth if measured from a meridian of longitude.

Figure 18-430 Azimuth Angle

18.17.1.2.1. Back Azimuth

A back azimuth is the reverse direction of an azimuth. It is comparable to doing an "about face." To obtain a back azimuth from an azimuth, add 180° if the azimuth is 180° or less, or subtract 180° if the azimuth is 181° or more. The back azimuth of 180° may be stated as either 000° or 360°.

18.17.1.3. Declination Diagram

A declination diagram is placed on most large-scale maps to enable the user to properly orient the map. The diagram shows the interrelationship of magnetic north, grid north, and true north (Figure 18-51). There are two declinations, a magnetic declination (Figure 18-51) and a grid declination.

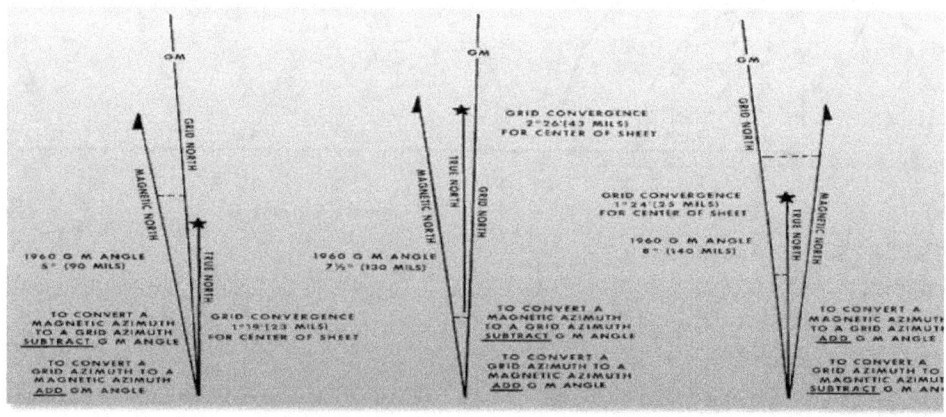

Figure 18-441 Declination Diagram (East and West)

18.17.1.3.1. Grid-Magnetic (G-M) Angle

Grid-Magnetic (G-M) Angle is an arc indicated by a dashed line, which connects the grid north and the magnetic north prongs. The value of this arc (G-M ANGLE) states the size of the angle between grid north and magnetic north and the year it was prepared. This value is expressed to the nearest 1/2', with mil equivalents shown to the nearest 10 mils.

18.17.1.3.2. Grid Convergence

Grid Convergence is an arc, indicated by a dashed line, which connects the prongs for true north and grid north. The value of the angle for the center of the sheet is given to the nearest full minute with its equivalent to the nearest mil. These data are shown in the form of a grid convergence note.

18.17.1.3.3. Convergence Notes

Conversion notes may also appear with the diagram explaining the use of the G-M angle. One note provides instructions for converting magnetic azimuth to grid azimuth; the other note provides for converting grid azimuth to magnetic azimuth. The conversion (add or subtract) is governed by the direction of the magnetic north prong relative to the grid north prong. The expression "East is least and West is best" represents that when True North is East of Magnetic North you subtract (least) variation and when True North is West of Magnetic North you add (best) variation (Figure 18-52).

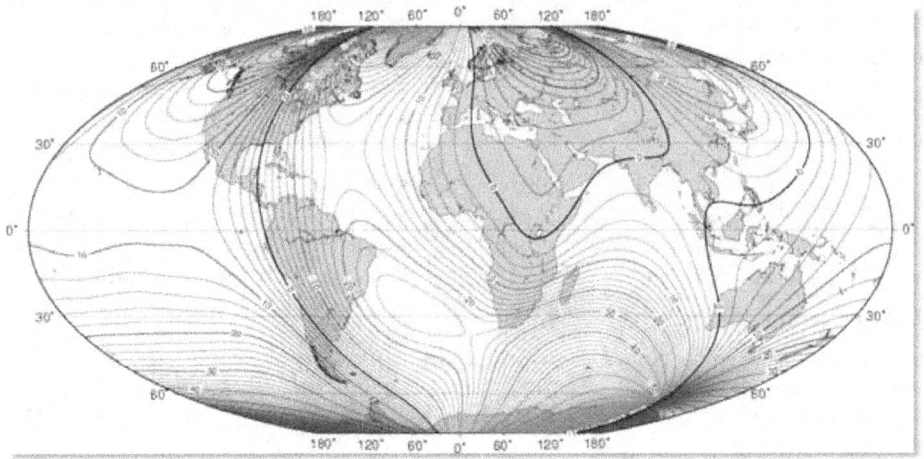

Figure 18-452 Lines of Magnetic Variation

18.17.1.3.4. Relative Position

The relative position of the directions is obtained from the diagram, but the numerical value should not be measured from it. For example, if the amount of declination from grid north to magnetic north is 1°, the arc shown in the diagram may be exaggerated if measured, having an actual value of 5'. The position of the three prongs in relation to each other varies according to the declination data for each map.

Chapter 19

LAND TRAVEL AND EVACUATION TECHNIQUES

19. Land Travel

In any isolating situation following an aircraft or vehicular emergency, a decision must be made to either move or remain as close as possible to the aircraft, parachute, or vehicle crash site. In this chapter, land travel will be discussed and the various considerations that IP should address before determining if travel is or is not a necessity. As an IP, the ability to walk effectively is important in conserving energy and safety. Additionally, in rough terrain, travel may need to be done with the aid of a rope. The techniques of ascending and descending steep terrain are fundamental to understanding and performing rescue from rough terrain. These techniques, as well as techniques for snow travel, are covered. Travel may not be easy, but knowledgeable IP can travel safely and effectively while saving time and energy.

19.1. Decision to Stay or Travel

In non-permissive areas, the decision to travel should be planned prior to the mission with modifications based on the situation. To stay in the vicinity of the crash or parachute landing site may lead to capture. In permissive areas, a choice exists. Staying near the initial incident site has assisted in the recovery of most IP. IP should only leave the area when they are certain of their location and know that water, shelter, food, and help can be reached, or after having waited several days, they are convinced that rescue is not coming and they are equipped to travel.

19.1.1. Physical Condition

Note that IP may need to carry supplies and equipment while traveling to sustain life. Before making any decision, IP should consider their personal physical condition and the condition of others in the party when estimating their ability to sustain travel. If people are injured, try to get help. If travel for help is required, send the people who are in the best physical and mental condition. Send two people if possible. Travelling alone is dangerous. Before any decision is made, IP should consider all of the facts.

19.1.2. Vehicle Wreckage

From the air, it is easier to spot vehicle wreckage than it is to spot people traveling on the ground. Someone may have seen the crash and investigate. The wreck or parts from it can provide shelter, signaling aids, and other equipment (cowling for reflecting signals, tubing for shelter framework, gasoline and oil for fires, etc.).

19.1.3. Hazards

Avoiding the hazards and difficulties of travel is another reason to stay at the incident site. Rescue chances are good if IP made radio contact, the isolating event occurred on course or near a traveled route, and weather conditions are fair.

19.1.4. Present Location

Present location must be known to decide intelligently whether to wait for rescue or to determine a destination and route of travel. The IP should try to locate their position by studying maps, landmarks, GPS, and flight data, or by taking celestial observations. IP should try to determine rescue destination, the distance to it, the possible difficulties and hazards of travel, and the probable facilities and supplies en route and at the destination.

19.1.5. Decision to Stay

If the decision is to stay, in addition to all the primary needs (except travel) these problems should be considered:

- Environmental conditions
- Health and body care
- Rest and shelter
- Water supplies
- Food

19.1.6. Decision is Travel

If the decision is to travel: In addition to all primary survival needs, the following must be considered:

- Direction of travel and why
- Travel plan
- Equipment required
- Environmental conditions
- Potential of recovery

19.1.6.1. Permissive Situations

In permissive situations, before departing the incident site, IP should leave information at the site stating departure time, destination, route of travel, personal condition, and available supplies.

19.1.6.2. Other Factors to Consider

In addition, there are a number of other factors that should be considered when deciding to travel.

- The equipment and materials required for travel should be analyzed. Travel is extremely risky unless the necessities of survival are available to provide support during travel. IP should have sufficient water to reach the next probable water source indicated on a map or chart and enough food to last until they can procure additional food. To leave a safe shelter to travel in adverse weather conditions is foolhardy unless in an escape or evasion situation.
- In addition to the basic requirements, the physical condition of the IP must be considered in any decision to travel. If in good condition, IP should be able to move an appreciable distance, but if an IP is not in good condition or is injured, the ability to travel extended distances may be reduced. Analyze all injuries received during the isolating event. For example, if a leg or ankle injury occurred during the isolating event, this must be considered before traveling.
- If possible, IP should avoid making any decision immediately after the isolating event. They should wait a period of time to allow for recovery from the mental-if not the physical-shock resulting from the event. When shock has subsided IP can then evaluate the situation, analyze the factors involved, and make valid decisions.

19.2. Travel

Once an IP decides to travel, there are several considerations that apply regardless of the circumstances.

19.2.1. Leadership

The ranking person must assume leadership, and the IP must work as a team to ensure that all tasks are done in an equitable manner. Full use should be made of any experience or knowledge possessed by members of the group, such as that of members of the Guardian Angel weapon system, U.S. Army Rangers, physician/medics, etc. The leader is responsible for ensuring that the talents of all IP are used.

19.2.2. Energy Output

IP should keep the body's energy output at a steady rate to reduce the effects of unaccustomed physical demands.

19.2.2.1. Pace

A realistic pace should be maintained to save energy. It increases durability and keeps body temperature stable because it reduces the practice of quick starts and lengthy rests. Travel speed should provide for each IP's physical condition and daily needs, and the group pace should be governed by the pace of the slowest group member. Additionally, rhythmic breathing should be practiced to prevent headache, nausea, lack of appetite, and irritability. In high altitudes, a slow and steady pace is essential in avoiding the risks of lapse of judgment and hallucinations due to lack of oxygen (hypoxia).

19.2.2.2. Rest Stops

Rest stops should be short since it requires added energy to begin again after cooling off. IP should wear their clothing in layers (layer system) and make adjustments to provide for climate, temperature, and precipitation. Start with fewer layers while cool, when beginning to travel. Stop and shed a layer when beginning to warm up.

19.2.2.3. Clothing

Wearing loose clothing provides for air circulation, allows body moisture to evaporate, and retains body heat. Loose clothing also allows freedom of movement.

19.2.2.4. Travel Time

IP should keep in mind when planning travel time and distance that the larger the group, the slower the progress will be. Time must be added for those IP who must acclimate themselves to the climate, altitudes, and the task of travel. IP should also allow time for unexpected obstacles and problems which could occur.

19.2.2.5. Food and Water

Proper nutrition and water are essential to building and preserving energy and strength. Several small meals a day are preferred to a couple of large ones so that calories and fluids are constantly available to keep the body and mind in the best possible condition. IP should try to have water and a snack available while trekking, and should eat and drink often to restore energy and prevent chills in cold temperatures. This also applies at night.

19.3. Land Travel Techniques

Land travel techniques are based largely on experience, which is acquired through performance. However, experience can be partially replaced by the intelligent application of specialized practices that can be learned through instruction and observation. For example, travel routes may be established by observing the direction of a bird's flight, the actions of wild animals, the way a tree grows, or even the shape of a snowdrift. Bearings read from a compass, GPS, the Sun, or stars will improve on these observations and confirm original headings. All observations are influenced by the location and physical characteristics of the area where they are made and by the season of the year.

19.3.1. Route Finding

The novice should follow a compass heading or GPS, whereas a more experienced person may follows lines of least resistance, such as selecting a faster or easier curved route under certain circumstances, and supports it with navigational aids. Use game trails when they follow a projected course only. On scree or rockslides, game trails may be critical. Game trails offer varying prospects, in addition to an easier route of travel such as the chance of securing game or locating waterholes. Successful land travel requires knowledge beyond mere travel techniques. IP should have at least a general idea of direction of travel and their ultimate destination. They should also have knowledge of the people and terrain through which they will travel. Whether the population is friendly, hostile, or unknown, they must adapt their entire method of travel and mode of living to these conditions.

19.3.2. Wilderness

Wilderness travel requires constant awareness. Survey the surrounding countryside, and plan travel only after carefully surveying the terrain. A distant blur may be mist or smoke, a faint, winding line on a far-off hill may be manmade or an animal trail; a blur in the lowlands may be a herd of caribou, fainting goats, pigs, or cattle. Study distant landmarks for characteristics that can be recognized from other locations or angles. Careful and intelligent observation will help IP to correctly interpret the things they see, distant landmarks, or a broken twig at their feet. Before leaving a place, travelers should study their backtrail carefully. IP should know the route forward and backward. An error in route planning may make it necessary to backtrack in order to take a new course. For this reason, all trails should be marked (Figure 19-1).

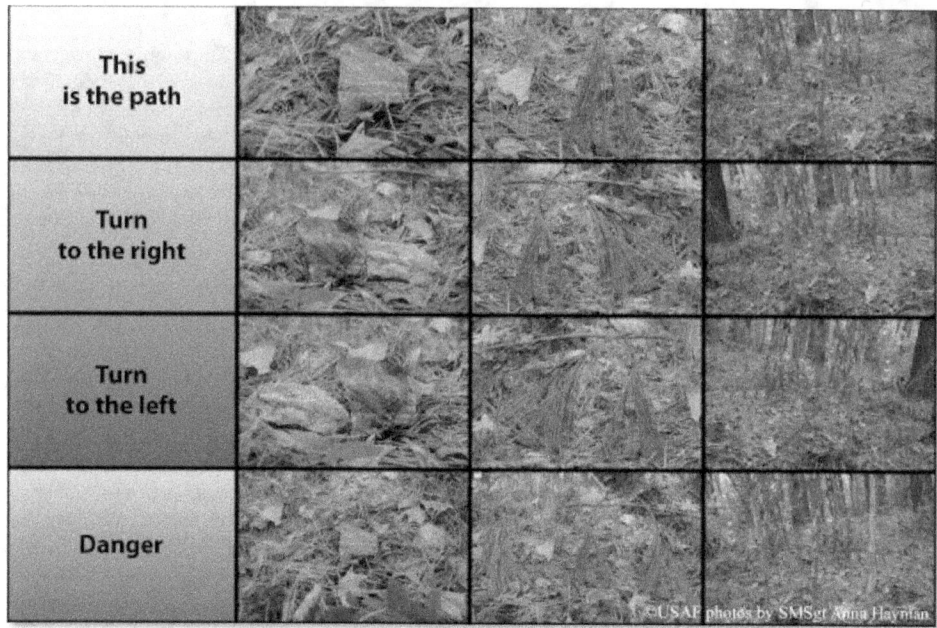

Figure 19-1 Examples for Marking a Trail

19.3.3. Mountain Ranges

Mountain ranges frequently affect the climate of a region and the climate in turn influences the vegetation, wildlife, and the character and number of people living in the region. For example, the ocean side of mountains has more fog, rain, and snow than the inland side of a range. Forests may grow on the ocean side, while inland, it may be semi-dry. Therefore, a complete change of survival techniques may be necessary when crossing a mountain range.

- Travel in mountainous country is simplified by conspicuous drainage landmarks, but it is complicated by the roughness of the terrain. A mountain traveler can readily determine the direction in which rivers or streams flow; however, surveying is necessary to determine if a river is safe for rafting, or if a snowfield or mountainside can be traversed safely.

- Mountain travel differs from travel through rolling or level country, and certain cardinal rules govern climbing methods. A group, descending into a valley, where descent becomes increasingly steep and walls progressively more perpendicular may be obliged to climb up again in order to follow a ridge until an easier descent is possible.

- In mountains, travelers must avoid possible avalanches of earth, rock, and snow, as well as deep crevices in ice fields.

- In mountainous country, it may be better to travel on the wind packed side of ridges, where the snow surface is typically firmer and there is a better view of the route from above. IP should watch for snow and ice overhanging steep slopes. Avalanches are a hazard on steep snow-covered slopes, especially on warm days and after heavy snowfalls.

19.3.3.1. Avalanches

Snow avalanches occur most commonly and frequently in mountainous country during wintertime, but they also occur with the warm temperatures and rainfalls of springtime. Both small and large avalanches are a serious threat to IP traveling during winter as they have tremendous force. The natural phenomena of snow avalanches are complex. It is difficult to definitely predict impending avalanches, but knowing general behaviors of avalanches and how to identify them can help to avoid avalanche hazard areas.

19.3.3.1.1. Loose Snow (Sluff) or Slab Avalanches

The loose snow avalanche is one kind of avalanche that starts over a small area or in one specific spot. It begins small and builds up in size as it descends. As the quantity of snow increases, the avalanche moves downward as a shapeless mass with little cohesion. Slab avalanches are when a large area, possibly hundreds of yards wide and several yards thick, starts moving almost instantaneously. These avalanches account for approximately 90 percent of avalanche-related fatalities.

19.3.3.1.2. Terrain Factors Affecting Avalanches

- Steepness. Most commonly, avalanches occur on slopes ranging from 30 to 45 degrees (60 to 100 percent grade), but large avalanches do occur on slopes ranging from 25 to 60 degrees (40 to 173 percent grade). (See Figure 19-2.)

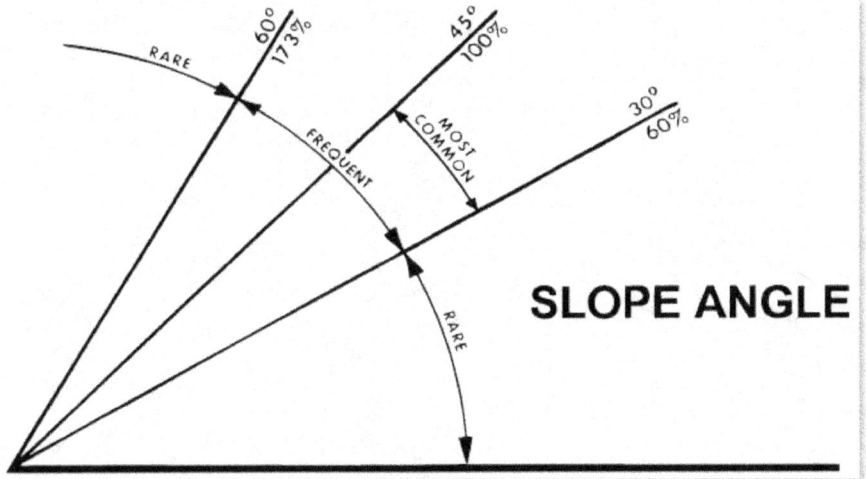

Figure 19-2 Slope Angle

- Profile. The dangerous slab avalanches have more chance of occurring on convex slopes because of the angle and the gravitational pull. Concave slopes cause a danger from slides that originate at the upper, steep part of the slope (Figure 19-3).

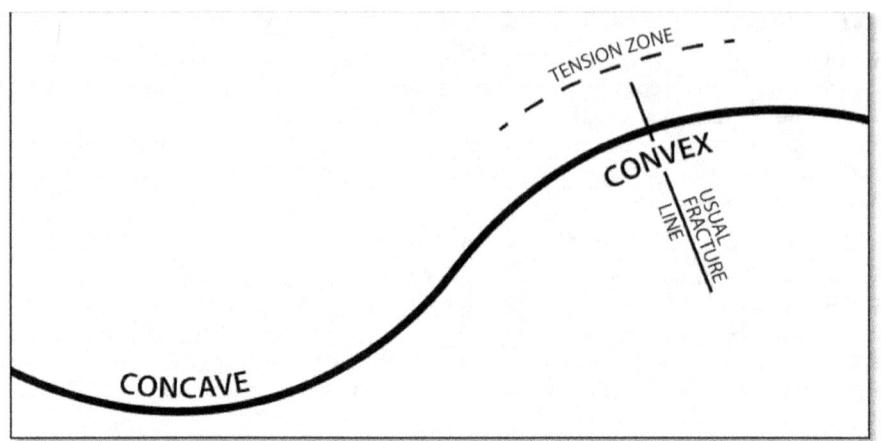

Figure 19-3 Profile of Slope

- Slopes. Midwinter snow slides usually occur on north-facing slopes. This is because the north slopes do not receive the required sunlight which would heat and stabilize the snowpack. South-facing slope slides occur on sunny, spring days when sufficient warmth melts the snow crystals causing them to change into wet, watery, slippery slides. Leeward slopes are dangerous because the wind blows the snow into well packed drifts just below the crest. If the drifts have not adhered to the snow underneath, a slab avalanche can occur. Windward slopes generally have less snow and are compact. It is usually strong enough to resist movement, but avalanches may still occur with warm temperature and moisture.
- Surface Features. Most avalanches are common on smooth, grassy slopes that offer no resistance. Brush, trees, and large rocks bind and anchor the snow, but avalanches can still occur in areas with trees (Figure 19-4).

Figure 19-4 Snow Slides

19.3.3.1.3. Weather Factors

Old snow depth covers up natural anchors (rocks, brush, and fallen trees) so that the new snowfall slides easier. The type of old snow surface is important. Sun crust is a type of snow crust formed by refreezing after surface snow crystals have been melted by the sun. Sun crusts and smooth surface snows are unstable; whereas a rough, jagged surface would offers more stability and anchorage. A loose snow layer underneath is far more hazardous than a compacted one as the upper layer of snow will slide more easily with no rough texture to restrain it. Travelers should check the underlying snow by using a rod or stick.

- Winds, 15 miles per hour or more, cause the danger of avalanches to develop rapidly. Leeward slopes will collect snow that has been blown from the windward sides, forming slabs or sluffs, depending upon the temperature and moisture. Snow plumes or cornices indicate this condition (Figure 19-5).

    – A high percentage of all avalanches occur during or shortly after storms. Layers of different types of snow from different storms will cause unstable snow because the bond between layers will vary in strength. The rate of snowfall also has a significant effect on stability. A heavy snowfall spread out over several days is not as dangerous as a heavy snowfall in a few hours because slow accumulation allows time for the snow to settle and stabilize. A large amount of snow over a short period of time results in the snow constantly changing and building up, giving it little time to settle and stabilize. If the snow is light and dry, little settling or cohesion occurs, resulting in instability.

    – Under extremely cold temperatures, snow is unstable. In temperatures around freezing or just above, the snow tends to settle and stabilize quickly. Storms, starting at low temperature with light, dry snow which are followed by rising temperature, cause the top layer of snow to be moist and heavy, providing opportune conditions for avalanching. The light, dry snow underneath lacks the strength and elastic bondage necessary to hold the heavier, wetter snow deposited on top; therefore, the upper layer slides off. Also, extreme temperature differences between night and day cause the same problems. Rapid changes in weather conditions cause adjustments and movement within the snowpack. IP should be alert to rapid changes in winds, temperatures, and snowfall which may affect snow stability.

    – Avalanches of wet snow are more likely to occur on south slopes. Sun, rainstorms, or warmer temperatures brought by spring weather are absorbed by the snow causing it to become less stable.

Figure 19-5 Forming Slides

19.3.3.1.4. Warning Signs

Avalanches generally occur in the same area. After a path has been smoothed, it's easier for another avalanche to occur. Steep, open gullies and slopes, pushed over small trees, trees with limbs broken off, and tumbled rocks are signs of slide paths (Figure 19-6). Snowballs tumbling downhill or sliding snow is an indication of an avalanche area on leeward slopes. If the snow echoes or sounds hollow, conditions are dangerous. If the snow cracks and the cracks persist or run, the danger of a slab avalanche is imminent. The deeper the snow, the more the terrain features will be obscured. Knowledge of common terrain features can help IP visualize what they may be up against, what to avoid, and the safest areas to travel. Knowing the general weather pattern for the area is helpful. IP should try to determine what kind of weather will be coming by observing and knowing the signs that indicate certain weather conditions.

Figure 19-6 Avalanche Warning Signs

19.3.3.1.5. Route Selection

If avoiding the mountains and avalanche danger areas is impossible, there are precautions IP can take when confronting dangerous slopes. They should decide which slopes will be the safest by analyzing the factors that determine what makes one slope safe and another deadly. Study the slope terrain and keep in mind why avalanches occur. Take into consideration the time of day; the best times are early morning and later in the evening due to cooler temperatures creating more stability in the snow.

- When IP decide to cross a slope, one person at a time should cross. If all go together, they should not tie together since there is no way one person can hold another against an avalanche. Instead, they should tie a line about 100 feet long to each person. If they should get caught in an avalanche, the line will help identify their position if it is exposed and contrasts the snow. IP should select escape routes before and throughout the climb and keep these routes in mind at all times. They should also stay to the fall line (natural path an object will follow when falling or sliding down the slope). When climbing do not zigzag or climb a different route because it seems easier. Staying to the fall line will prevent making fractures and at the same time, compact the snow, making it more stable for others who follow. If traversing, they should travel above the danger area. IP should travel quickly and quietly to avoid extended exposure to the probable danger of avalanches.
    - If caught in an avalanche, any equipment that weighs IP down should be dropped. The pack snowshoes, and any other articles should be jettisoned.
    - The standard rule is to use swimming motions to try to move towards the snow surface. Further, IP should go for the sides and not try to out swim the avalanche. If near the surface, they should try to keep one arm or hand above the surface to mark their position. If buried, a person should inhale deeply (nose down) before the snow stops moving to make room for their chest. Trapped persons should try to make breathing space around their faces. They shouldn't struggle, but should relax and conserve their energy and oxygen. Only when fellow IP or rescuers are nearby should the trapped individual shout. Rescue should be done as quickly as possible. Avalanche victims generally have a 50 percent chance of surviving after 30 minutes have passed and the snow hardens.
    - Whenever possible glacier travel should be avoided. Glacier crossing demands special knowledge, techniques, and equipment such as the use of a lifeline and poles for locating crevasses. There are many places where mountain ranges can be negotiated on foot in a single day by following glaciers. IP must be especially careful on glaciers and watch for crevasses or deep crevices covered by snow. If traveling in groups of three persons or more, they should be roped together at intervals of 40 to 45 feet. The rope should be kept tight enough that it does not drag on the ground, but also not overly taunt (smiley face shape of rope between travelers). Every step should be probed with a pole. Snow-bridged crevasses should be crossed at right angles to their course. The strongest part of the bridge can be found by probing. When crossing a bridged crevasse, weight should be distributed over a large area by crawling or by wearing snowshoes.

19.4. Glaciers and Glacial Travel

Glaciers have many hidden dangers. Glacial streams may run just under the surface of the snow or ice, creating weak spots, or they may run on the surface and cause slick ice. Crevasses which run across the glacier can be a few feet to several hundred feet deep. Quite often crevasses are covered over with a thin layer of snow, making them practically invisible. IP could fall into crevasses and sustain severe injuries or death. If glacier travel is required, it is best to use a probe pole to test the footing ahead.

19.4.1. Features

To cope with the problems which can arise in using glaciers as avenues of travel, it is important to understand something of the nature and composition of glaciers.

19.4.1.1. Parts

A valley glacier is essentially a river of ice and it flows at a rate of speed that depends largely on its mass and the slope of its bed. A glacier consists of two parts:

1. The lower glacier, which has an ice surface void of snow during the summer.
2. The upper glacier, where the ice is covered, even in summer with layers of accumulated snow that changes into glacier ice.

19.4.1.1.1. Adjacent Features

To these two integral parts of a glacier may be added two others which, although not a part of the glacier proper, are generally adjacent to it and are of similar composition. These adjacent features, the ice and snow slopes, are immobile since they are anchored to underlying rock. A large crevasse separates such slopes from the glacier proper and defines the boundary between moving and anchored ice.

19.4.1.2. Surface and Crevasses

Ice has a plastic-like surface, but is not smooth enough to prevent cracking as it moves forward over irregularities in the ice bed. Fractures in a glacier surface, called crevasses, vary in width and depth from only a few inches to many feet. Crevasses form at right angles to the direction of greatest tension and due to a limited area, tension is usually in the same direction. Crevasses in any given area tend to be roughly parallel to each other. Generally, crevasses develop across a slope. Therefore, when traveling up the middle of a glacier, people usually encounter only transverse crevasses (crossing at right angles to the main direction of the glacier). Near the margins or edges of a glacier, the ice moves more slowly than it does in midstream. This speed differential causes the formation of crevasses diagonally upstream away from the margins or sides. While crevasses are almost certain to be encountered along the margins of a glacier and in areas where a steepening in gradient occurs, the gentlest slopes may also contain crevasses.

19.4.1.3. Icefall

An icefall forms where an abrupt steeping of slope occurs in the course of a glacier. These stresses are set up in many directions. As a result, the icefall consists of a varied mass of ice blocks and troughs with no well-defined trend to the many crevasses.

19.4.1.4. Debris

As a glacier moves forward, debris from the valley slopes on either side is deposited on its surface. Shrinkage of the glacier from subsequent melting causes this debris to be deposited along the receding margins of the glacier. Such ridges are called lateral (side) moraines. Where two glaciers join and flow as a single river of ice, the debris on the adjoining lateral margins of the glaciers also unites and flows with the major ice stream, forming a medial (middle) moraine. (By examining the lower part of a glacier, it is often possible to tell how many tributaries have joined to form the lower trunk of the glacier.) Terminal (end) moraine is usually found where the frontage of the glacier has pushed forward as far as it can go; that is, to the point at which the rate of melting equals the speed of advance of the ice mass. This moraine may be formed of debris pushed forward by the advancing edge or it may be formed by a combination of this and other processes (Figure 19-7).

19.4.1.4.1. Moraines

Lateral and medial moraines may provide excellent avenues of travel. When the glacier is heavily crevassed, moraines may be the only practical routes. Ease of progress along moraines depends upon the stability of the debris composition. If the material consists of small rocks, pebbles, and earth, the moraine is usually loose and unstable and the crest may break away at each footstep. If large blocks compose the moraine, they have probably settled into a compact mass and progress may be easy.

19.4.1.4.2. Moraine Travel

On moraine travel, it is best either to proceed along the crest or, in the case of lateral moraines, to follow the trough which separates it from the mountainside. Since the slopes of moraines are usually unstable, there is a great risk of spraining an ankle on them. Medial moraines are usually less pronounced than lateral moraines because a large part of their material is transported within the ice. Travel on them is usually easy but should not be relied upon as routes for long distances since they may disappear beneath the glacier surface. Only rarely is it necessary for a party traveling along or across moraines to be roped together.

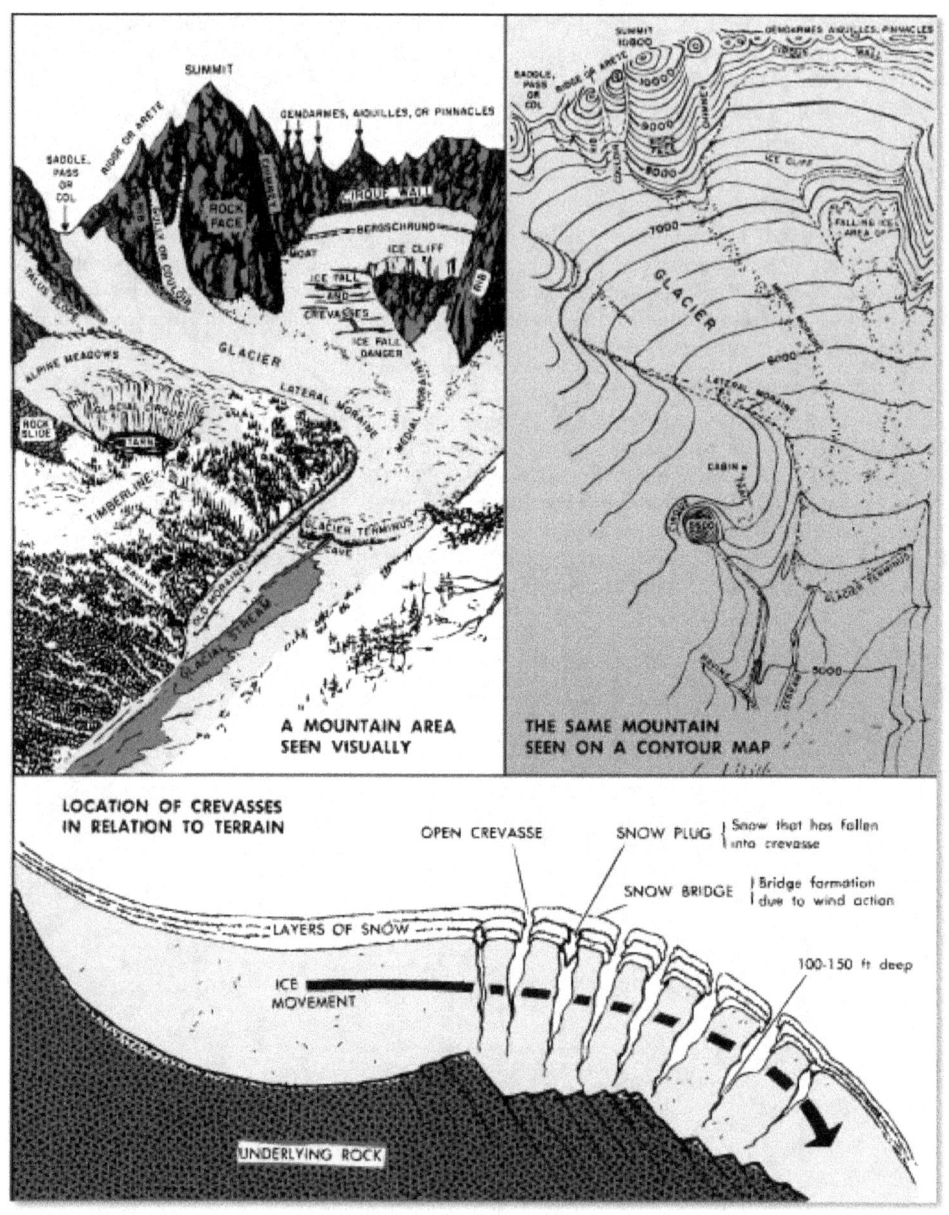

Figure 19-7 Typical Glacier Construction

19.4.1.4.3. Glacial Rivers

Glacial rivers are varied in type and present numerous problems to those who must cross or navigate them. Wherever mountains and highlands exist in the arctic regions, melting snows produce concentrations of water pouring downward in a series of falls and swift chutes. Rivers flowing from icecaps, hanging piedmonts (lake-like), or serpentine (winding or valley) glaciers are all notoriously treacherous. Northern glaciers may be vast in size and the heat of the summer sun can release vast quantities of water from them. Glacier ice is extremely unpredictable. An ice field may look innocent from above, but countless sub-glacial streams and water reservoirs may be under its smooth surface. These reservoirs are either draining or temporarily blocked. Mile-long lakes may lie under the upper snowfield, waiting only for a slight movement in the glacier to liberate them sending their waters into the valleys below. Because of variations in the amounts of water released by the Sun's heat, all glacial rivers fluctuate in water level. The peak of the flood water usually occurs in the afternoon as a result of the noonday heat of the Sun on the ice. For some time after the peak has passed, rivers which drain glaciers may not be fordable or even navigable. However, by midnight or the following morning, the water may recede so fording is both safe and easy. When following a glacial river broken up into many shifting channels, choose routes next to the bank rather than taking a chance on getting caught between two dangerous channels.

19.4.1.4.4. Flooding Glaciers

Glaciers from which torrents of water descend are called flooding glaciers. Two basic causes of such glaciers are the violent release of water which the glacier carried on its surface as lakes, or the violent release of large lakes which have been dammed up in tributary glaciers because of the blocking of the tributary valley by the main glacier. This release is caused by a crevasse or a break in the moving glacial dam; the water then roars down in an all-enveloping flood. Flooding glaciers can be recognized from above by the flood-swept character of the lower valleys. The influence of such glaciers is sometimes felt for many miles below. Prospectors have lost their lives while rafting otherwise safe rivers because a sudden flood entered by a side tributary and descended as a wall of white, rushing water.

19.4.1.4.5. Surface Streams

On those portions of a glacier where melting occurs, runoff water cuts deep channels in the ice surface and forms surface streams. Many such channels exceed 20 feet in depth and width. They usually have smooth sides and undercut banks. Many of these streams terminate at the margins of the glacier where in summer they contribute to the torrent that constantly flows between the ice and the lateral moraine. Size increases greatly as the heat of the day moves to an end. The greatest caution must be taken in crossing a glacial surface stream since the bed and undercut banks are usually hard, smooth ice which offers no secure footing.

19.4.1.4.6. Glacial Mills

Some streams disappear into crevasses or into round holes known as glacial mills, and then flow as sub-glacial streams. Glacial mills are cut into the ice by the churning action of water. They vary in diameter. Glacial mills differ from crevasses, not only in shape but also in origin, since they do not develop as a result of moving ice. In places, the depth of a glacial mill may equal the thickness of the glacier.

19.4.2. Glacier Operations

The principal dangers and obstacles to operations in glacier areas are crevasses and icefalls. Hidden crevasses present unique problems and situations since their presence is often difficult to detect. When one is detected, often it is due to a team member having fallen through the unstable surface cover. The following techniques and procedures should be followed when performing glacier operations.

19.4.2.1. Snow Bridges

Any snow bridge (over a crevasse or running water) should be closely and completely examined before use. If overhanging snow obscures the bridge, explore at closer range by probing the depth and smashing at the sides while walking delicately, ready for an arrest or sudden drop. When there is doubt about the integrity of a bridge find a different route. When it is the only possible route, the lightest climber in the team should be the first across, with the following climbers walking with light steps and taking care to step exactly in the same tracks.

- Bridges vary in strength with changes in temperatures. In the cold of winter or early morning, the thinnest and most fragile of bridges may have incredible structural strength. However, when the ice crystals melt in the afternoon temperature, even the largest bridge may suddenly collapse. Each bridge must be tested with care, being neither abandoned nor trusted until its worth is determined.

19.5. Snow and Ice Areas

Travel in snow and ice areas is not recommended except to move from an unsafe area to a safe area, or from an area that has few natural resources to an area of greater resources (shelter material, food, and signaling area).

19.5.1. Hazards

The greatest hazards in snow and ice areas are the intense cold, high winds, and deep snow. Judging distance is difficult due to the lack of landmarks and the clear arctic air. Image distortion is a common phenomenon. "White-out" conditions exist and IP should not travel during this time. A white-out condition occurs when there is complete snow cover and the clouds are so thick and uniform that light reflected by the snow is about the same intensity as that from the sky. If traveling during bad weather, great care must be taken to avoid becoming disoriented or falling into crevasses, going over cliffs or high snow ridges, or walking into open leads. A walking stick is very useful to probe the area in the line of travel.

19.5.2. Whiteouts

Strong winds often sweep unchecked across tundra areas (due to the lack of vegetation) causing whiteout conditions. Because of blowing snow, fog, and lack of landmarks, a compass is a must for travel, yet it is still difficult to navigate a true course since the magnetic variation in the high latitudes (polar areas) is often extreme.

19.5.3. Summer Months

During the summer months, the area is a mass of bogs, swamps, and standing water. Crossing these areas will be difficult at best. Rain and fog are common. Insects such as mosquitoes, midges, and black flies can and will cause the IP physical discomfort and may cause travel problems. If the body is not completely covered with clothing, or if IP do not use a head net or insect repellant, insect bites may be severe and infection can set in.

19.5.4. Mountainous Areas

In mountainous country, it is often best to travel along ridge lines because it provides a firmer walking surface and there is usually less vegetation to contend with. High winds make travel impractical if not impossible at times.

19.5.5. Timbered Areas

Summer travel in timbered areas should not present any major problems; however, travel on ridges is preferred since the terrain is drier and there are usually fewer insects. During the cold months, snow may be deep and travel will be difficult without some type of snowshoes or skis. Travel is generally easier on frozen rivers, streams, and lakes since there is less snow or wind-packed snow and they are easier to walk on.

19.5.6. Rivers

Rivers that are comparatively straight are that way because of the volume of water flow and extremely fast currents. These rivers tend to have very thin ice in the winter (cold climates), especially where snow banks extend out over the water. If an object protrudes through the ice, the immediate area will be weak and should be avoided if possible. Where two rivers and streams come together, the current is swift and the ice will be weaker than the ice on the rest of the river. Very often after freeze up, the source of the river or stream dries up so rapidly that air pockets are formed under the ice and can be dangerous if fallen into.

- During the runoff months (spring and summer), rivers and streams usually have a large volume of water which is very cold and can cause cold injuries. Wading across or down rivers and streams should be done with proper footwear and exposure protection due to the depth, swiftness, unsure footing, and coldness of the water. Generally, streams are too small and shallow for rafting. Streams are often bordered by high cliffs or banks at the headwaters. As a stream progresses, its banks are often choked with alder, devil's club, and other thick vegetation making traveling very slow and difficult. Many smaller streams will simply lead the traveler to a bog or swamp where they end, causing more problems for the survivor.

19.5.7. Sea Ice

Sea ice conditions vary greatly from place to place and season to season. During the winter months, there is generally little open water except between the edges of floes. Crossing from one floe to another can be done by jumping across the open-water area, but footing may be dangerous. When large floes are touching each other, the ice between is usually ground into brash ice by the action of the floes against each other and this ground-up ice will not support a person's weight. Pressure ridges are long ridges in sea ice caused by the horizontal pressure of two ice floes coming together. Pressure ridges may be 100 feet high and several miles long; they may occur in a gulf or bay, or on polar seas. They must be crossed with great care because of the ruggedness of ice formations, weak ice in the area, and possibility of open water covered with a thin layer of snow or ground-up ice. During summer months, the ice surface becomes very rough and covered with water. The ice also becomes soft and honeycombed (candlestick ice) even though the air temperature may be below freezing. Traveling over sea ice in the summer months is very dangerous.

19.5.8. Icebergs

Icebergs are great masses of ice and are driven by currents and winds. Approximately two-thirds of an iceberg is below the surface of the sea. Icebergs in open seas are always dangerous because the ice under the water will melt faster than the surface exposed to the air, upsetting the equilibrium and toppling them over. The resulting waves can throw small pieces of ice in all directions. Avoid pinnacle-shaped icebergs; low, flat-topped icebergs are safer.

19.6. Dry Climates

Before traveling in the desert, the decision to travel must be weighed against the environmental factors of terrain and climate, condition of IP, possibility of rescue, and the amount of water and food required.

19.6.1. Day or Night Travel

The time of day for traveling is greatly dependent on two significant factors: the first and most apparent is temperature, and the second factor is type of terrain. For example, in rocky or mountainous deserts, the eroded drainages and canyons may not be seen at night and could result in a serious fall. Additionally, manmade features such as mining shafts or pits and irrigation channels could cause similar problems. IP have reported that the surface became so hot their feet became blistered through their shoes. If the temperature is not conducive to day travel, IP should travel during the cooler parts of the day (in early morning or late evening). Traveling on moonlit nights is another possibility; however, IP must be aware that moonlight can cast deceiving shadows. This problem can be decreased by scanning the ground to allow the eye time to pick up the slight differences in lighting. In hot desert areas where these hazards do not occur, traveling at night is a very practical solution. During the winter in the mid-latitude deserts, the cold temperatures make day travel most sensible.

19.6.2. Desert Types

There are three types of deserts: mountain, rocky plateau, and sandy or dune deserts. Each type of desert can present difficult travel problems.

19.6.2.1. Mountain Deserts

Mountain deserts are characterized by scattered ranges or areas of barren hills or mountains, separated by dry, flat basins. High ground may rise gradually or abruptly from flat areas to a height of several thousand feet above sea level. Most desert rainfall occurs at high elevations and the rapid runoff causes flash floods, eroding deep gullies or ravines and depositing sand and gravel around the edges of the basins. These floods are a problem on high and low grounds for travel by the IP. The flood waters rapidly evaporate, leaving the land as barren as before, except for plush vegetation which rapidly becomes dormant. Basins without shallow lakes will have alkaline flats which can cause problems with chemical burns and can destroy clothing and equipment. Whenever possible, IP should avoid crossing these areas.

19.6.2.2. Rocky Plateau Deserts

Rocky plateau deserts have relief interspersed by extensive flat areas solid or broken rock at or near the surface. They may be cut by dry, steep-walled, eroded valleys, known as wadis in the Middle East and as arroyos or canyons in the United States and Mexico. The narrower of these valleys can be extremely dangerous to humans and material due to flash flooding. Travel in these valleys may present another problem: an IP can lose sight of reference points, travel farther than intended, and get lost. The Syrian Golan Heights in Israel is an example of a rocky plateau desert.

19.6.2.3. Sandy (Dune) Deserts

Sandy (dune) deserts are extensive flat areas covered with sand or gravel. They are the product of ancient and modern wind erosion. "Flat" is relative in this case, as some sand dunes are over 1,000 feet high and 10 to 15 miles long. These dunes can help the IP determine general direction. Longitudinal or seif dunes are continuous banks of sand at even heights that lie parallel with the dominant wind. Other areas, however, may be totally flat for distances in excess of two miles. Plant life varies from none to scrub reaching over six feet. Examples of this desert include the ergs of the Sahara Desert, Empty Quarter of the Arabian Desert, areas of California and New Mexico, and the Kalahari Desert in South Africa. Horseshoe-shaped crescent dunes have a hollow portion that faces downward. Ripples caused by wind in the sand may indicate the direction of the prevailing winds. These ripples generally lie perpendicular to the prevailing winds. In deserts, it is easier to travel on the windward side of the tops of dunes. Even though these ridges may not continue in a straight line and may wander, they offer a better route of travel than traveling in straight lines. A great deal of energy and time can be expended walking up and down dunes, especially in the loose sand on the leeward side of dunes.

19.6.3. Night Travel

IP should travel the desert at night orienting themselves by compass, GPS, and/or celestial aids such as the stars and Moon. IP should use all directional aids during travel and each aid should be frequently crosschecked against each other.

19.6.3.1. Landmarks

Without any navigational aids, landmarks must be used for navigation. This can lead to difficulties. Mirages can cause considerable trouble. Ground haze throughout the day may obscure vision. Distances are deceptive in the deserts and IP have reported difficulty in estimating distances and the size of objects. In southern Egypt, one IP reported large boulders always appeared smaller than they were and in other cases small obstacles appeared insurmountable. IP in Saudi Arabia and in Tunisia warned that it is difficult to maintain a single landmark in navigation. Several groups reported they found it necessary to take turns keeping an eye on a specific mountain, peak, or object which was their goal. Objects have a way of vanishing in some cases when the eye is moved for an instant, and in other cases, many peaks or hills looked alike and cause difficulties in determining the original object. In Tunisia, twin peaks are not reliable landmarks because of their frequent occurrence. IP have found after a short time of traveling they may have up to a dozen twin peaks for reference in the same vicinity. The Great Sand Sea (Egypt, Libya, and Sudan) was the emergency landing site of several groups of IP and posed navigational difficulties. In these rolling sand hills, it is impossible to keep one object in view, and even footmarks fail to provide a reliable back trail for determining travel directions. The extreme flatness of other stretches of desert terrain in North Africa also makes navigation difficult. With no landmarks to follow, and no objectives to sight, IP may walk in circles or large arcs before realizing their difficulties.

- A Marine pilot who made an emergency landing in the Arizona desert took the precaution of immediately spreading his parachute on the ground and putting rocks on the edges to ensure maximum visibility from the air. Then he decided to walk to his crashed plane, a distance he estimated to be 500 yards from his landing spot. He reached the plane and found it gutted by fire, and spent 5 days wandering the flat desert trying to find his parachute.

19.6.3.2. Navigational Difficulties

Navigational difficulties of a different type may be experienced in Ethiopia, Kenya, and Somalia. Here the density of the thorn brush, primarily acacia with small leaves, makes it extremely hard to navigate from one point to another. In this area, IP should follow animal trails and hope they lead to rivers or waterholes. Elephant trails seem to offer the best and clearest route.

19.6.3.3. Travel Routes

In the Sinai Desert area and in portions of Egypt, travel routes may be used and IP can use the trails. One IP, who made it to a trail, encountered a camel caravan almost immediately; although he reported it bothered him that he had not seen them approach, as they suddenly appeared out of a mirage. Another commented it was awfully hard to be alone in his section of the desert, for in every direction, he saw wandering tribes, camel herds, or people watching him. Two IP independently suggested that IP pay attention to the wind as an aid in navigation. One IP, on the Arabian Peninsula, noted the wind blew consistently from the same direction. The other, in the Libyan Desert, was able to judge direction of travel by the angle at which the wind blew his clothes or struck his body. IP in certain areas may orient themselves to the prevailing winds once it is established that these are consistent.

19.6.4. Sandstorms

Extreme winds can blow into sandstorms. Generally, IP reported they could see the approach of such storms and were able to take proper precautions; however, sandstorms completely surprised a few groups and they had difficulty navigating. IP should not underestimate the power and danger of such storms. Protection from the storms should be a priority in an IP's mind. Rock cairns, natural ledges, boulders, depressions, or wells can be used for shelter. If time permits, IP should attempt to dig depressions and rig a shelter from blankets, parachutes, tarps, or any available material. If time does not permit for construction of an adequate shelter, IP can wrap themselves in an EVC (evasion chart).

19.6.4.1. Orientation

Nearly all IP have made some comment on orientation before, during, and after a sandstorm. They warned specifically that it is necessary to adequately mark the direction of travel before the storm. A few IP have said when the storm was over they had no idea which way they had been traveling and all their landmarks were forgotten, obliterated, or indistinguishable. The general plan for marking travel routes before a sandstorm is to place a stick to indicate direction.

- One IP oriented himself with one rock a few feet in front of his position. He commented after the storm, that one point was not adequate and recommended using a row of stones, sticks, or heavy gear about 10 yards in length to give adequate direction following such an event.

19.6.5. Mirages

Mirages are common in desert areas. They are optical phenomena due to refraction of light by uneven density distribution in the lower layer of the air. The most common desert mirage occurs during the heat of the day when the air close to the ground is much warmer than the air aloft. Under this condition, atmospheric refraction is less than normal and the image of the distant low sky appears on the ground looking like a sheet of water. Distant objects may appear to be reflected in the "water." When the air close to the ground is much colder than the air aloft, as in the early morning under a clear sky, atmospheric refraction is greater than normal. When this condition occurs, distant objects appear larger and closer than they are and objects below the normal horizon are visible. Unless the density distribution in the lower layers is such that the light rays from an object reach the observer along two or more paths, they will see a distorted image or multiple images of the object.

19.6.6. Manmade Characteristics

The following are manmade characteristics of the desert:

- Wells, pipelines, refineries, quarries, and crushing plants may identify a route of travel. Additionally, pipelines are often elevated, making them visible from a distance.
- Many desert areas are irrigated for agricultural and habitation purposes. Agriculture and irrigation canals are signs which can lead an IP to people.

19.7. Tropical Climates

The inexperienced person's view of jungle travel may range from difficult to nearly impossible. However, with patience and good planning, the best and least difficult route can be selected. In some cases, the easiest routes are rivers, trails, and ridge lines. However, there may be hazards associated with these routes.

19.7.1. Waterways

Rivers and streams may be overgrown making them difficult to reach and impossible to raft. These waterways may also be infested with leeches. Trails may have traps or animal pits set on them. Trails can also lead to a dead end or into thick brush or swamps. Ridges may end abruptly at a cliff. The vegetation along a ridge may also conceal crevices or extend out past cliffs creating a false floor, making the cliff unnoticed until it's too late.

19.7.2. Machete Use

The machete is an aid to survival in the jungle. However, IP should not use it unless there is no other way. They should part the brush rather than cut it if possible. If the machete must be used, cut at a down-and-out angle, instead of flat and level, as this method requires less effort.

19.7.3. Surroundings

IP should take their time and not hurry. This allows them to observe their surroundings and gives better insight as to the best route of travel. Watch the ground for the best footing as some areas may be slippery or give way easily. Avoid grabbing bushes or plants when traveling. Falling may be a painful experience as many plants have sharp edges, thorns, or hooks. Wear gloves and fully button clothes for personal protection.

19.7.4. Quicksand

Quicksand can be a problem. In appearance, quicksand looks just like the surrounding area with an absence of vegetation. It is usually located near the mouths of large rivers and on flat shores. The simplest description of quicksand is a natural water tank filled with sand and supplied with water. The bottom consists of clay or other substances capable of holding water. The sand grains are rounded, as opposed to normal sharper-edged sand. This is caused by water movement which also prevents it from settling and stabilizing. The density of this sand-water solution will support a person's body weight. The potential danger if an IP panics may be drowning. In quicksand, the

IP should use the spread-eagle position to help disperse the body weight to keep from sinking and a swimming technique to return to solid ground. Remember to avoid panicking and struggling, and spread out and swim or pull along the surface.

19.8. Forested Areas

When forested areas are dense, river travel and ridges usually afford the easiest travel routes. In open forests, land travel is easier and offers a better selection of travel routes.

19.8.1. Second-growth Timber

After a fire, windstorm, or logging operation, second-growth timber usually grows thick. After it grows about 20 feet high, any space between the trees is filled in by branches as the overhead timber isn't thick enough to cause the lower branches to die from lack of sunlight.

19.8.1.1. Brush

Deciduous brush also contributes to the overgrowth. Blowdowns, avalanche fans, and logging slash are difficult to negotiate. Such obstructions, even of a few hundred feet, may require major changes to the original travel route. Do not travel through dense brush if it can be avoided.

19.8.1.2. Dense Vegetation

Dense forests are hard to penetrate. IP can use fallen trees as walkways to provide a route of travel through dense vegetation to a clear area. Gloves should be worn when penetrating thorny vegetation. Overlaying bushes can be separated to allow passage. When land is steep, brush can be used to provide handholds if it is strong and anchored well.

19.8.1.3. Dangers

Brush can be dangerous. IP should be aware of the possibility of slipping while going downhill. Therefore, they should ensure each step is firmly placed. IP should be aware of travel difficulties presented by cliffs, boulders, and ravines which are covered by brush.

19.8.1.4. Travel

Travel on trails rather than taking shortcuts through the brush. Brush is frequently easier to travel through (over) during the winter season when it is covered by snow and when snowshoes are available or improvised. The heaviest timber areas are best for travel because little or no brush will be growing on the forest floor.

19.8.1.5. Streams

Try to avoid areas near streams and valley floors because they have more brush and trees than the valley walls and ridges. However, traveling in the stream channels may be preferable when the area is covered with dense brush and vegetation. IP may have to wade, but the stream may offer the best route through the brush.

19.8.1.6. Timberline

Traveling high above the brush at the timberline may be worthwhile if the bottom and sides of the valley look dense.

19.9. Mountain Walking Techniques

Depending upon the terrain formation, mountain walking can be divided into four different techniques-walking on hard ground, walking on grassy slopes, walking on scree slopes, and walking on talus slopes. Included in all of these techniques are two fundamental rules which must be mastered in order to expend a minimum of energy and time: the weight of the body must be kept directly over the feet and the sole of the boot must be placed flat on the ground (Figure 19-8). These fundamentals are most easily accomplished by taking small steps at a slow steady pace. An angle of descent which is too steep should be avoided and any indentations or protrusions of the ground, however small, should be used to advantage.

Figure 19-8 Weight of Body Over Feet

19.9.1.1. Hard Ground

Hard ground is usually considered to be firmly packed dirt which will not give way under the weight of a person's step. When ascending, the above fundamentals should be applied with the addition of locking the knees on every step in order to rest the leg muscles (Figure 19-9). When steep slopes are encountered, they can be traversed easier than climbed straight up. Turning at the end of each traverse should be done by stepping off in the new direction with the uphill foot. This prevents crossing the feet and possible loss of balance. In traversing, the full sole principle is done by rolling the ankle away from the hill on each step. For narrow stretches, the herringbone step may be used; that is, ascending straight up a slope with toes pointed out and using the principles stated above. Descending is usually easiest by coming straight down a slope without traversing. The back must be kept straight and the knees bent in such a manner that they take up the slack of each step. Again, remember the weight must be directly over the feet, and the full sole must be placed on the ground with every step.

Figure 19-9 Locking Knees, Traversing, Walking Uphill, Walking Downhill

19.9.2. Grassy Slopes

Grassy slopes are usually made up of small tussocks of growth rather than one continuous field. In ascending, the techniques previously mentioned are applicable; however, it is better to step on the upper side of each tussock (Figure 19-10) where the ground is more level than on the lower side. Descending is best done by traversing.

Figure 19-10 Walking on Grassy Slopes

19.9.3. Scree Slopes

Scree slopes consist of small rocks and gravel which have collected below rock ridges and cliffs. The size of the scree varies from small particles to the size of a fist. Occasionally, it is a mixture of all size rocks, but normally scree slopes will be made up of rocks the same size. Ascending scree slopes is difficult, tiring, and should be avoided when possible. All principles of ascending hard ground apply, but each step must be picked carefully so the foot does not slide down when weight is placed on it. This is best done by kicking in with the toe of the upper foot so a step is formed in the scree (Figure 19-11). After determining the step is stable, carefully transfer weight from the lower foot to upper and repeat the process. The best method for descending scree is to come straight down the slope with feet in a slight pigeon-toed position using a short shuffling step with the knees bent and back straight (Figure 19-12). When several climbers descend a scree slope together, they should be as close together as possible, one behind the other, to prevent injury from dislodged rocks. Since there is a tendency to run down scree slopes, care must be taken to ensure that this is avoided and control is not lost. By leaning forward, one can obtain greater control. When a scree slope must be traversed with no gain or loss of altitude, use the hop-skip method. This is a hopping motion in which the lower foot takes all the weight and the upper foot is used for balance.

Figure 19-11 Walking on Scree Slopes

Figure 19-12 Walking Down Scree Slopes

19.9.4. Talus Slopes

Talus slopes are similar in makeup to the scree slopes, except the rock pieces are larger. The technique of walking on talus is to step on top of, and on the uphill side of, the rocks (Figure 19-13). This prevents them from tilting and rolling downhill. All other previously mentioned fundamentals are applicable. Usually, talus is easier to ascend and traverse, while scree is a more desirable avenue of descent.

Figure 19-13 Walking on Talus Slopes

19.10. Burden Carrying

Using approved packing and carrying techniques can eliminate unnecessary hardships and help in transporting the load with greater safety and comfort. Carrying a burden initially creates mental irritation and fatigue, either of which can lower morale. IP should keep their minds occupied with other thoughts when packing a heavy load. Adjustments should be made during each rest stop to improve the fit and comfort of the pack. Additionally, the rate of travel should be adjusted to the weight of the pack and the environmental characteristics of the terrain being crossed.

19.10.1. Packs

Often IP must quickly gather their equipment and move out without the assistance of a good pack. The gear may have to be carried in the arms while rapidly leaving the area. In such an instance, it would be better to fashion a roll of the gear and wear it over the shoulder, time permitting. When time is not a factor, it may be desirable to make a semi-rigid pack such as a square pack. The convenience of being able to keep track of equipment, particularly small items, can be critical in survival situations.

19.10.1.1. Packsack

A packsack can be fashioned from available survival kit containers, several layers of the fabric from the parachute canopy, shirt, or materials from the vehicle. The sack can be taped, glued, tied, or sewed.

19.10.1.2. Square Pack

To improvise a square pack (Figure 19-14), lay a rectangle of material, waterproof if available, (five feet by five feet minimum size) flat on the ground (examples are plastic, tarp, poncho, or suitable material). Visualize the material divided into squares like a tic-tat-toe board. The largest piece of soft, bulky equipment (sleeping bag, parachute canopy, etc.) is placed in the center square in an "S" fold. This places the softest item in the pack against the tarp which rests on the back while traveling. If using a poncho, place the sleeping bag just below the hood opening. Place hard, heavy objects between the top layer and the middle layer of the "S" fold near the top of the pack. Soft items can be placed between the middle and bottom layer.

- After all desired items are inside the folds; tie the inner pack in the fashion shown in Figure 19-14. Start with a one-inch diameter loop in the end of a long piece of parachute suspension line or other suitable line and loop it around the "S" fold laterally. Standing at the bottom of the pack, divide it into thirds and secure the running end of the line to the loop with a trucker's hitch. Both of these hitches should be at the intersection of the thirds so as to divide the pack vertically into thirds. Wrap the running ends around the pack at 90 degrees (working toward the center) to the line and when it crosses another line, use a jam (postal) hitch to secure it and pull both ways to ensure tightness in all directions. When returning to the original starting position, use the loop of the tied trucker's hitch to secure another trucker's hitch and the inner part is complete. The waterproof materials are then folded around the inner pack as shown. Tie the "outer" pack in the same manner ensuring that it is waterproof with all edges folded in securely. If a poncho is used, the head portion may be used to get into the pack if necessary. However, it must be properly secured to ensure that the inner items are protected. With a square pack constructed in this manner, the equipment should not get wet. (NOTE: With practice, an excellent pack can be constructed by tying the inner pack and outer cover simultaneously).

Figure 19-14 Square Pack

19.10.1.3. Horseshoe Pack

A horseshow pack is simple to make and use and relatively comfortable to carry over one shoulder. To construct, (Figure 19-15): lay available square-shaped (preferably five feet by five feet) material (waterproof if available) flat on the ground and place all gear on the long edge of the material, leaving about six inches at each end. All hard items should be padded. Roll the material with the gear to the opposite edge of the square. Tie each end securely. Place at least two or three evenly spaced ties around the roll. Bring both tied ends together and secure. This pack is compact and comfortable if all hard, heavy items are packed well inside the padding of the soft gear. If one's shoulder is injured, the pack can be carried on the other shoulder. It is easy to put on and remove.

Figure 19-15 Horseshoe Pack

19.10.1.4. Alaskan Packstrap

The most widely used improvised packstrap is called an Alaskan packstrap (Figure 19-16). This type of packstrap can be fashioned out of any pliable and strong material. Some suitable materials for constructing the packstrap are animal skins, canvas, and parachute harness webbing. The pack should be worn so it can be released from the strap with a single pull of the cord in the event of an emergency, such as falling into water. The knot securing the pack should be made with an end readily available which can be pulled to drop the pack quickly; for example, a trucker's hitch with safety for normal terrain travel and with the safety removed when in areas of danger, such as water or rough terrain.

19.10.1.4.1. Advantages

Some advantages of the Alaskan packstrap include:

- Small in bulk and light in weight.
- Easily carried in a pocket while traveling.
- Quickly released in an emergency.
- Can be adjusted to efficiently pack items of a variety of shapes and sizes.

19.10.1.4.2. Disadvantages

Some disadvantages of the Alaskan packstrap include:

- Difficult to put on (without practice).
- Experience and ingenuity are necessary to use it with maximum efficiency.

Figure 19-16 Alaskan Packstrap

19.10.2. Pack Principles

The following principles should be considered when packing and carrying a pack:

1. The pack or burden-carrying device should be adequate for the intended job.
2. The pack or burden may be adaptable to a pack frame. The pack frame could have a belly band to distribute the weight between the shoulders and hips and prevent undue swaying of the pack. Pack frames are also used to carry other burdens such as meat, brush, and firewood.
3. Proper weight distribution is achieved by ensuring that the weight is equally apportioned on each side of the pack and as close as possible to the body's center of gravity. This enhances balance and the ability to walk in an upright position. If heavy objects are attached to the outside of the pack, the body will be forced to lean forward. A pack bundle without a frame or packboard should be carried high on the back or shoulders. For travel on level terrain, weight should be carried high. When traveling on rough terrain, weight should be carried low or midway on the back to help maintain balance and footing.
4. Emergency and other essential items (extra and/or protective clothing, first-aid kit, radio, flashlight, etc.) should be readily available by being placed in the top of the pack.
5. Fragile items should be protected by padding them with extra clothing or soft material and placing them in the pack where they won't shift or bounce around. Hard and/or sharp objects cannot damage the pack or other items if cutting edges are properly sheathed, padded, and not pointed toward the bearer. Items outside the pack should be firmly secured but not protruding where they could snag on branches and rocks.
6. Adjust and carry the pack so that overloading or straining of muscles or muscle groups is avoided. When using a pack, the straps should be adjusted so they ride comfortably on the trapezium muscles and avoid movement when walking. Back support should be tight and placed to ensure good ventilation and support. During breaks on the trail, rest using the proper position to ease the weight of the pack and take the strain off muscles (see Figure 19-17 for methods of resting). A comfortable pack is adjustable to the physique of the

person. A waistband will support 80 to 90 percent of the weight and is fitted relatively tight. The waistband should be cinched down around the pelvic girdle/crest area to avoid constricting circulation or restricting muscle movement.

Figure 19-17 Methods of Resting

19.11. Rough Land Travel and Evacuation Techniques

There are isolating situations where traveling over rough terrain is required. However, if it is necessary to travel in such areas, specialized skills, knowledge, and equipment are required.

19.11.1. Safety Rope

A safety rope must be used when there is danger of the rescue team or climber falling. Environmental factors may require a rapid retreat. During these circumstances, when speed is critical, it may be desirable to unrope.

19.12. Specialized Knots for Climbing and Evacuation

Each of the following knots has a specific purpose. These knots have survived the test of time and are used in maintaining operations. They are designed to have the least effect on the fiber of a rope lock without slipping, and they are easy to untie when wet and icy. All knots reduce the strength of ropes; however, these knots reduce the strength of the rope as little as possible. Most knots should be safe tied with an overhand knot or two half hitches. A knot does not have to be safe tied if the knot is designed for the middle of a line or if it is 15 feet or more from the end of the rope.

19.12.1. Water Knot

The water knot or right bend is used for joining nylon webbing (Figure 19-18).

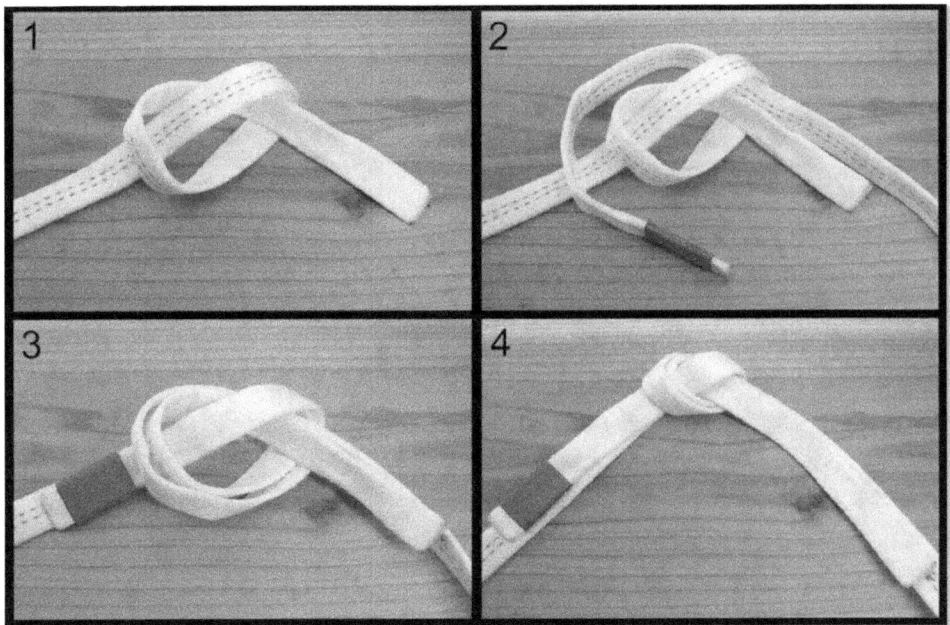

Figure 19-18 The Water Knot

19.12.2. Double Fisherman's Bend

The double fisherman's bend is used to securely join two lines of unequal; diameter or hard lay lines (Figure 19-19).

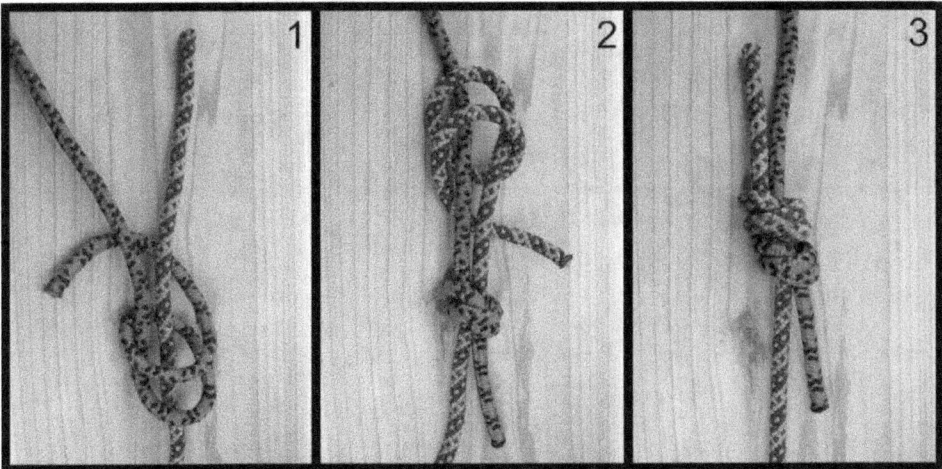

Figure 19-19 Double Fisherman's Bend

19.12.3. Double Figure Eight Knot

The double figure-eight knot is used for temporarily joining two rope or hard lay ropes (rappels, Tyrolean traverses) (Figure 19-20).

Figure 19-20 Figure-Eight Loop

Figure 19-21 Figure-Eight with Two Loops

19.12.4. Butterfly Knot

The butterfly is used to make a fixed loop in the middle of a line where the direction is between 120- to 180-degree angles (Figure 19-22). For angles less than 120 degrees, a figure eight in a loop will suffice. For an angle of pull greater than 120 degrees, the figure eight will become weakened and begin to split.

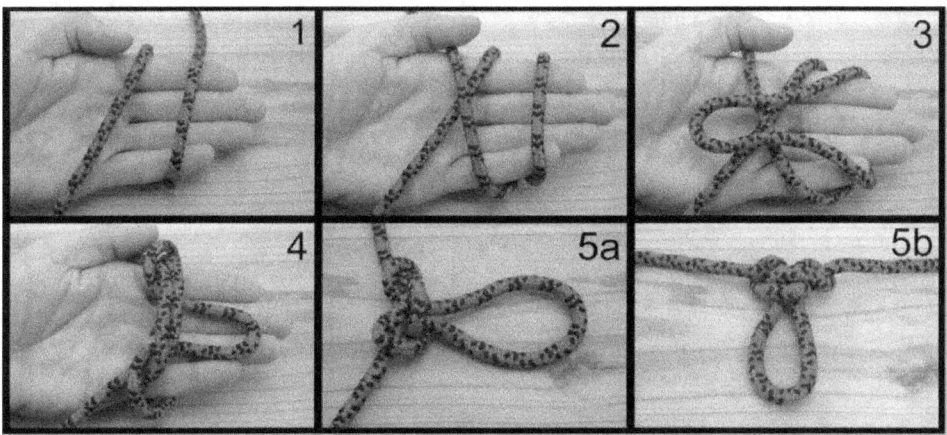

Figure 19-22 Butterfly Knot

19.12.5. Prusik Knot

The prusik knot may be used to ascend a fixed line, tying into a safety line, or safety a rappel (Figure 19-23).

Figure 19-23 Prusik Knot

19.12.6. Three-Loop Bowline

A three-loop bowline is a variation of the bowline on a bight. It is used for three anchor points or as an improvised harness (Figure 19-24).

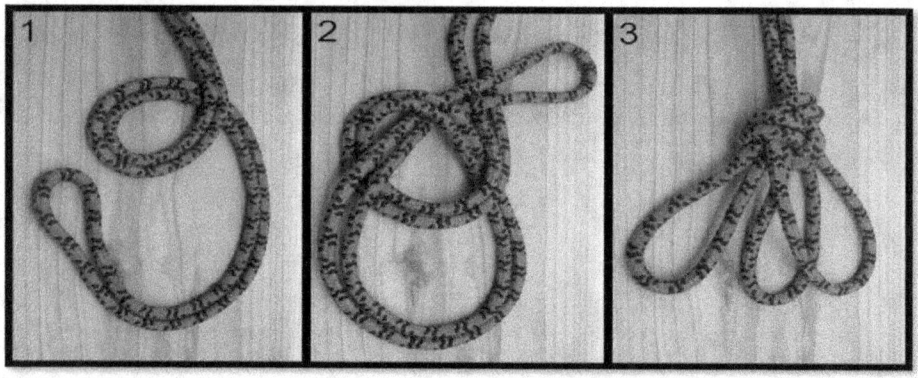

Figure 19-24 Three-Loop Bowline

19.13. Seat Harness

The seat harness is a safety sling which is used to attach the rope to the IP. It must be tied correctly for safety and comfort reasons. An improvised seat harness can be made of one-inch tubular nylon tape or rope (Figure 19-25). The tape is placed across the back so the midpoint (center) is on the hip opposite the hand that will be used for braking during belaying or rappelling. Keep the midpoint on the appropriate hip, cross the ends of the tape in front of the body, and tie half of a surgeon's knot (three or four overhand wraps) where the tapes cross. The ends of the tape are brought between the legs (front to rear), around the legs, and then secured with a jam hitch to tape around the waist on both sides. The tapes are tightened by pulling down on the running ends of the tape. This must be done to prevent the tape from crossing between the legs (Figure 19-25). Bring both ends around to the front and across the tape again. Then bring the tape to the opposite side of the intended brake hand and tie a square knot with an overhand knot or two half-hitch safety knots on either side of the square knot. The safety knots should be passed around as much of the tape as possible. Once the seat harness has been properly tied, attach a single locking carabineer to the harness by clipping all of the web around the waist and the web of the half surgeon knot together. The gate of the carabineer should open on top and away from the climber.

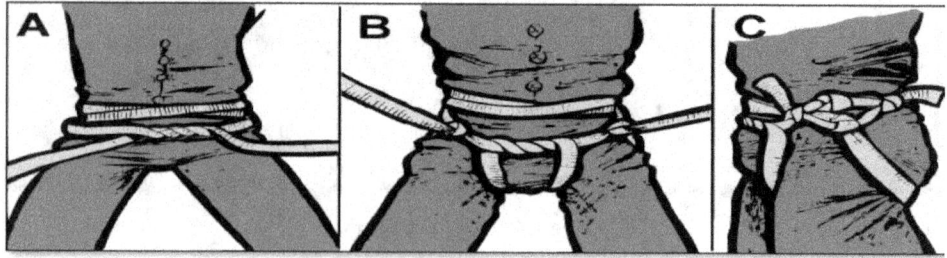

Figure 19-25 Improvised Seat Harness

19.13.1. Carabineers

Carabineers should not be dropped or used for other than the designed purpose since small fracture lines may develop and weaken the structure. Carabineers should not be used as a hammer nor loaded (stressed) beyond their maximum breaking strength. The moving parts, hinge and sleeve, of locking carabineers should be kept clean for free movement. If a carabineer "binds," do not oil it - discard it! Carabineers should not be filed, stamped, or marked with an engraving tool. All moving parts (gate, locking sleeve) should operate freely and the locking pin must properly align with the locking notch. Obvious fractures, regardless of size, are cause for condemning a carabineer.

19.14. Route Selection

Route selection can be the deciding factor in planning a climb whether it is on the side of a mountain or down the side of a building. A direct line is seldom the proper route from a given point to the area of the IP. Time spent at the beginning of the climbing operation in proper route selection may save a large amount of time once the operation has started. The entire route must be planned before it is carried out, with the safest route selected. Natural hazards present, retreat routes available, time involved to perform the climb, and logistics will be major influencing factors in selecting the route.

- Terrain must be analyzed to find an efficient route of travel. The IP must make a detailed reconnaissance, noting each obstacle, the best approach, height, angle, type of rock, difficulty, distance between positions, and equipment available to accomplish the mission.
- At least two vantage points should be used so a three-dimensional understanding of the climb can be attained. Use of early morning or late afternoon light, with its longer shadows, is helpful in this respect.

19.15. Dangers to Avoid

There are several dangers to avoid while traveling.

19.15.1. Weather

On long routes, changing weather will be an important consideration. Wet or icy surfaces can make an otherwise easy route almost impassable; cold may reduce climbing efficiency; snow may cover holds. A weather forecast should be obtained if possible. Smooth rock slabs are treacherous, especially when wet or iced after freezing rain. Ledges should then be sought. Rocks overgrown with moss, lichens, or grass become treacherous when wet. Under these conditions, cleated boots are by far better than composition soles.

19.15.2. Grass and Bushes

Tufts of grass and small bushes that appear firm may be growing from loosely packed and unanchored soil, all of which may give way if the grass or bush is pulled upon. Grass and bushes should be used only for balance by touch or as push holds-never as pull holds. Gently inclined but smooth slopes of rock may be covered with pebbles that may roll treacherously underfoot.

19.15.3. Ridges

Ridges can be free of loose rock, but topped with unstable blocks. A route along the side of a ridge just below the top is usually best. Gullies provide the best protection and often the easiest routes, but are more subject to rockfalls. The side of the gully is relatively free from this danger. Climbing slopes of talus, moraines, or other loose rock are not only tiring to the individual but dangerous because of the hazards of rolling rocks to others in the party. Rescuers should close up intervals when climbing simultaneously. In electrical storms, lightning can endanger the climber. Peaks, ridges, pinnacles, and lone trees should be avoided.

19.15.4. Rockfalls

Rockfalls are the most common mountaineering danger. The most frequent causes of rockfalls are other climbers, heavy rain and extreme temperature changes in high mountains, and resultant splitting action caused by intermittent freezing and thawing. Warning of a rockfall may be the cry "ROCK," a whistling sound, a grating sound, a thunderous crashing, or sparks where the rocks strike at night. A rockfall can be a single rock or a rockslide covering a relatively large area. Rockfalls occur on all steep slopes, particularly in gullies and chutes. Areas of frequent rockfalls may be indicated by fresh scars on the rock walls, fine dust on the talus piles, or lines, grooves, and rock-strewn areas on snow beneath cliffs. Immediate action is to seek cover, if possible. If there is not enough time to avoid the rockfall, the climber should lean into the slope to minimize exposure. Danger from falling rocks can be minimized by careful climbing and route selection. The route selected must be commensurate with the ability of the least experienced team member. (Note: In a permissive environment yell "ROCK" when equipment is dropped.)

19.16. Climbing

Balance climbing is a type of movement used to climb multiple surfaces. During the process of route selection, the climber should mentally climb the route to know what is expected. Climbers should not wear gloves when balance climbing.

19.16.1. Body Position

The climber must keep good balance when climbing (the weight placed over the feet during movement) (Figure 19-26). The feet, not the hands, should carry the weight (except on the steepest cliffs). The hands are for balance. The feet do not provide proper traction when the climber leans in toward the rock. With the body in balance, the climber moves with a slow, rhythmic motion. Three points of support, such as two feet and one hand are used when possible. The preferred handholds are waist to shoulder high. Resting is necessary when climbing because tense muscles tire quickly. When resting, the arms should be kept low where circulation is not impaired. Use of small intermediate holds is preferable to stretching and clinging to widely-separated big holds. A spread-eagle position, where a climber stretches too far (and cannot let go), should be avoided.

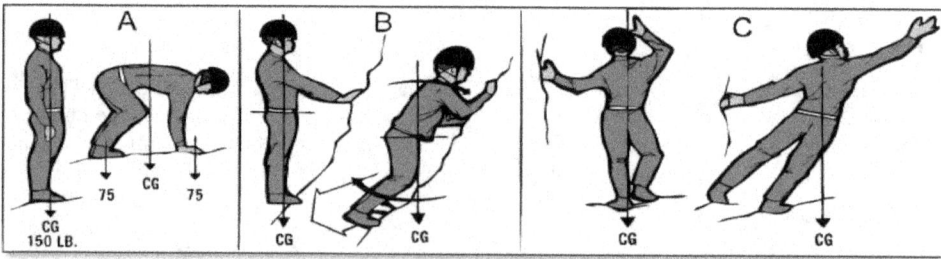

Figure 19-26 Body Position

19.16.2. Types of Holds

There are various types of climbing holds.

19.16.2.1. Push Holds

Push holds (Figure 19-27) are desirable because they help the climber keep the arms low; however, they are more difficult to hold onto in case of a slip. A push hold is often used to advantage in combination with a pull hold.

Figure 19-27 Push Holds

19.16.2.2. Pull Holds

Pull holds (Figure 19-28) are those that are pulled down upon and are the easiest holds to use. They are also the most likely to break out.

Figure 19-28 Pull Holds

19.16.2.3. Jam Holds

Jam holds (Figure 19-29) involve jamming any part of the body or extremity into a crack. This is done by putting the hand into the crack and clenching it into a fist or by placing the arm into the crack and twisting the elbow against one side and the hand against the other side. When using the foot in a jam hold, care should be taken to ensure the boot is placed so it can be removed easily when climbing is continued.

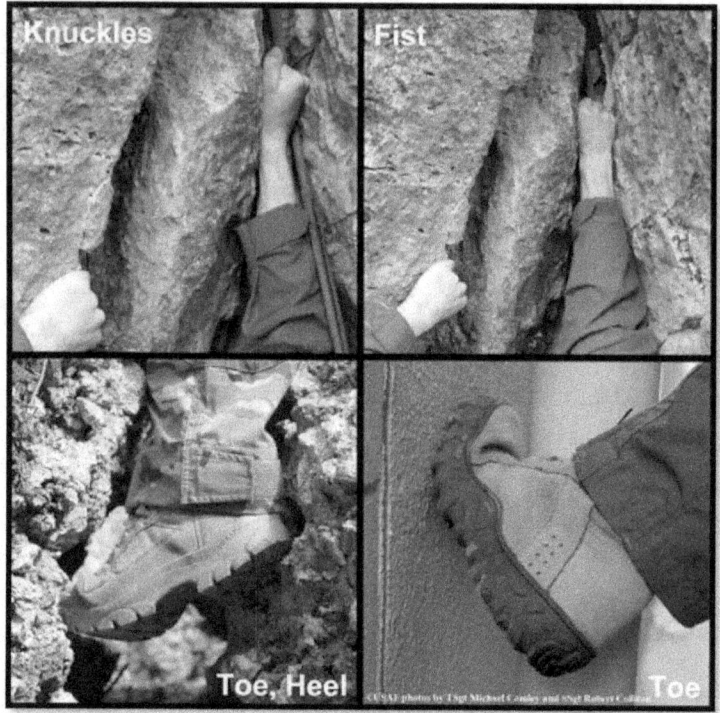

Figure 19-29 Jam Holds

19.16.2.4. Combination Holds

The holds previously mentioned are considered basic and from these any number of combinations and variations can be used. The number of these variations depends only on the limit of the individual's imagination. The following are a few of the more common ones:

- The counterforce (Figure 19-30) is attained by pinching a protruding part between the thumb and fingers and pulling outward or pressing inward with the arms.

Figure 19-30 Combination Holds

- The lay-back (Figure 19-31) is done by leaning to one side of an offset crack with the hands pulling and the feet pushing against the offset side. Lay-backing is a classic form of force or counterforce where the hands and feet pull and push in opposite directions enabling the climber to move up in a series of shifting moves. It is very strenuous.

Figure 19-31 Lay-Back

- Underclings (Figure 19-32) permit cross pressure between hands and feet.

Figure 19-32 Underclings

- Mantleshelving, or mantling, takes advantage of down pressure exerted by one or both hands on a slab or shelf. By straightening and locking the arm, the body is raised, allowing a leg to be placed on a higher hold (Figure 19-33).

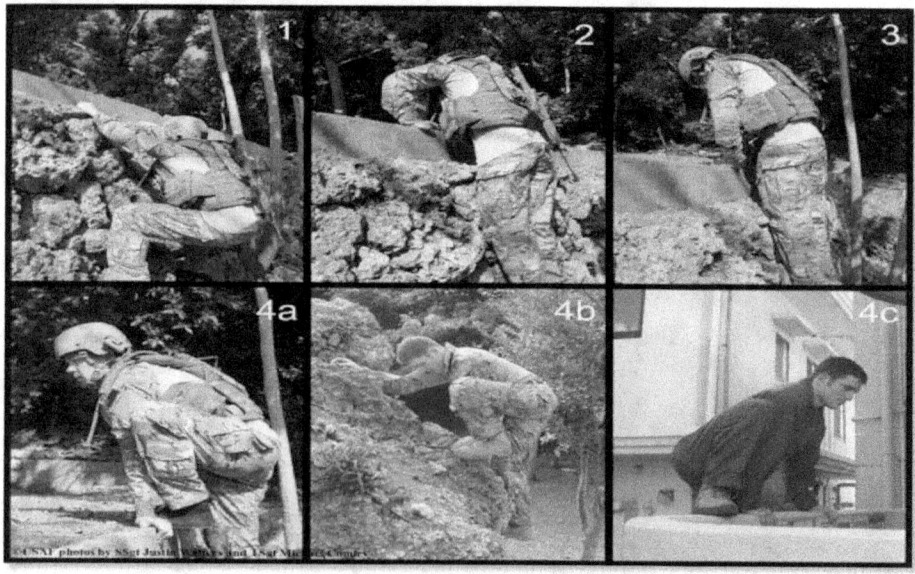

Figure 19-33 Mantleshelving

19.16.3. Chimney Climb

The chimney climb is a body-jam hold used in very wide cracks (Figure 19-34). The arms and legs are used to apply pressure against the opposite faces of the rock in a counterforce move. The outstretched hands hold the body while the legs are drawn as high as possible. The legs are flexed forcing the body up. This procedure is continued as necessary. Another method is to place the back against one wall and the legs and arms against the other and "worm" upward (Figure 19-34).

Figure 19-34 Chimney Climbing

19.16.4. Friction Climbing

A slab is a relatively smooth portion of rock lying at an angle. When traversing a slab, the lower foot is pointed slightly downhill to increase balance and friction of the foot. All irregularities in the slope should be used for additional friction. On steep slabs, it may be necessary to squat with the body weight well over the feet with hands used alongside for added friction. This position may be used for ascending, traversing, or descending. A slip may result if the climber leans back or lets the buttocks down. Wet, icy, mossy, or scree-covered slabs are the most dangerous.

- Friction holds (Figure 19-35) depend solely on the friction of hands or feet against a relatively smooth surface with a shallow hold. They are difficult to use because they give a feeling of insecurity which the inexperienced climber tries to correct by leaning close to the rock, thereby increasing the insecurity. They often serve well as intermediate holds, giving needed support while the climber moves over them; however, they would not hold if the climber decided to stop.

Figure 19-35 Friction Holds

19.17. Overland Snow Travel

Cold-weather operations conducted in snow-covered regions magnify the difficulties of reaching IP and effecting an extraction. Routes of travel surveyed from the air may not be possible from the ground. A straight line from the initial point to the objective is the most desired route; however, inherent dangers (avalanches, collapsing cornices, etc.) may necessitate an alternate route entailing a longer trek. During all operations, safety, and not ease of travel, will be the primary concern.

19.17.1. Snow Conditions

Travel time varies from hour to hour. Certain indicators may assist in the direction of travel; that is, the best snow condition is one which supports a person on or near the surface when wearing boots and the second best is calf-deep snow conditions. If possible, avoid traveling in thigh- or waist-deep snow. Snowshoes should be worn when conditions dictate.

19.17.1.1. Travel Considerations

South and west slopes offer hard surfaces late in the day after exposure to the Sun and the surface is refrozen. East and north slopes tend to remain soft and unstable. Walking on one side of a ridge, gully, clump of trees, or large boulders is often more solid than the other side. Dirty snow absorbs more heat than clean snow. Slopes darkened by rocks, dust, or uprooted vegetation usually provide more solid footing. Travel should be done in the early morning after a cold night to take advantage of stable snow conditions. Since sunlight affects the stability of snow, travel should be concentrated in shaded areas where footing should remain stable.

19.17.1.2. Seasonal Snowfall

In areas covered by early seasonal snowfall, travel between deep snow, and clear ground must be done cautiously. Snow on slopes tends to slip away from rocks on the downhill side, forming openings. These openings, called moats, are filled by subsequent snowfalls. During the snow season, moats below large rocks or cliffs may become extremely wide and deep, presenting a hazard to the rescue team.

19.17.2. Travel Speed

An over-zealous drive to reach an objective may be too fast for the endurance of the IP. Fast starts at the point of insertion usually result in frequent stops for recuperation. The best way to reach an objective is to start with a steady pace and continue that pace throughout. Movement at reasonable speeds, with rest stops as required, will help prevent team "burnout." To further ensure steady advancement with minimal degradation to the IP, maintain a steady pace to help maintain an even rate of breathing, and, after the initial period of travel (one half hour), initiate a shakedown rest to adjust boots, snowshoes, packs, or to remove or add layers of clothing, etc.

19.17.3. Snowshoe Technique

A striding technique is used for movement with snowshoes. In taking a stride, the toe of the snowshoe is lifted upward to clear the snow and thrust forward. Energy is conserved by lifting the snowshoe no higher than is necessary to clear the snow. If the front of the snowshoe catches, the foot is pulled back to free it and then lifted before proceeding with the stride. The best and least exertive method of travel is a loose-kneed rocking gait in a normal rhythmic stride. Care should be taken not to step on or catch the other snowshoe.

19.17.3.1. Ascent

On gentle slopes, ascent is made by climbing straight upward. (Traction is generally very poor on hard-packed or crusty snow.) Steeper terrain is ascended by traversing and packing a trail similar to a shelf. When climbing, the snowshoe is placed horizontally in the snow. On hard snow, the snowshoe is placed flat on the surface with the toe of the upper one diagonally uphill to get more traction. If the snow will support the weight of a person, it is better to remove the snowshoes and temporarily proceed on foot. In turning, the best method is to swing the leg up and turn the snowshoe in the new direction of travel.

19.17.3.2. Obstacles

Obstacles such as logs, tree stumps, ditches, and small streams should be stepped over. Care must be taken not to place too much strain on the snowshoe ends by bridging a gap, since the frame may break. In shallow snow, there is danger of catching and tearing the webbing on tree stumps or snags. Wet snow will frequently ball up under the feet, making walking uncomfortable. This snow should be knocked off with a stick or pole.

19.17.3.3. Poles and Bindings

Generally, ski poles are not used in snowshoeing; however, one or two improvised poles are desirable when carrying heavy loads, especially in mountainous terrain. The bindings must not be fastened too tightly or circulation will be impaired and frostbite can occur. During stops, bindings should be checked for fit and possible readjustment.

19.17.4. Uphill Travel

Maximum altitude may be obtained with less effort by traversing a slope. A zigzag or switch-back route used to traverse steep slopes places body weight over the entire foot as opposed to the balls of the feet as in a straight line uphill climb. An additional advantage to zigzagging or switch backing is alternating the stress and strain placed on the feet, ankles, legs, and arms when a change in direction is made.

19.17.4.1. Turning

When a change in direction is made, the body is temporarily out of balance. The proper method for turning on the steep slope is to pivot on the outside foot (the one away from the slope). With the upper slope on the right side, the left foot (pivot foot) is kicked directly into the slope. The body weight is transferred onto the left foot while pivoting toward the slope. The slope is then positioned on the left side and the right foot is on the outside.

19.17.4.2. Snow Types

In soft snow on steep slopes, pit steps must be stamped in for solid footing. On hard snow, the surface is solid but slippery, and level pit steps must be made. In both cases, the steps are made by swinging the entire leg in toward the slope, not by merely pushing the boot into the snow. In hard snow, when one or two blows do not suffice, crampons should be used. Space steps evenly and close together to facilitate ease of travel and balance. Additionally, the lead climber must consider the other team members, especially those who have a shorter stride.

19.17.4.3. Lead Function

The IP should travel in single file when ascending, permitting the leader to establish the route. The physical exertion of the climbing leader is greater than that of any other team member. The climbing leader must remain alert to safeguard other IP while choosing the best route of travel. The lead function should be changed frequently to prevent exhaustion of any one individual. IP following the leader should use the same leg swing technique to establish foot positions, improving each step as they climb. Each foot must be firmly kicked into place, securely positioning the boot in the step. In compact snow, the kick should be somewhat low, shaving off snow during each step, thus enlarging the hole by deepening. In very soft snow, it is usually easier to bring the boot down from above, dragging a layer of snow into the step to strengthen and decrease the depth of it.

19.17.4.4. Level Elevation

When it is necessary to traverse a slope without an increase in elevation, the heels rather than the toes form the step. During the stride, the climber twists the leading leg so that the boot heel strikes the slope first, carrying most of the weight into the step. The toe is pointed up and out. Similar to the plunge step, the heel makes the platform secure by compacting the snow more effectively than the toe.

19.17.5. Descending

The route down a slope may be different from the route up a slope. Route variations may be required for descending different sides of a mountain or moving just a few feet from icy shadows onto sun-softened slopes. A good surface snow condition is ideal for descending rapidly since it yields comfortably underfoot. The primary techniques for descending snow-covered slopes are plunge stepping and descending step by step.

19.17.5.1. Plunge Step

The plunge step makes extensive use of the heels of the feet and is applicable on scree as well as snow. Ideally, the plunging route should be at an angle, one that is within the capabilities of all the IP and affords a safe descent. The angle at which the heel should enter the surface varies with the surface hardness. On soft snow slopes, almost any angle suffices; however, if the person leans too far forward, there is a risk of lodging the foot in a rut and inflicting injuries. On hard snow, the heel will not penetrate the surface unless it has sufficient force behind it. Failure to firmly drive the heel into the snow can cause a slip and subsequent slide. The quickest way to check a slip is to shift the weight onto the other heel, making several short, stiff-legged stomps. If roped, plunging requires coordination and awareness of all team members' progress. Speed of the IP's travel must be limited to the slowest member.

19.17.5.2. Step-by-Step

The technique of step-by-step descending is used when the terrain is extremely steep, snow is significantly deep, or circumstances dictate a slower pace. On near-vertical walls, it is necessary to face the slope and cautiously lower oneself step by step, thrusting the toe of the boot into the snow while maintaining an anchor or handhold. Once the new foothold withstands the body's full weight, the technique is repeated. On moderately angled terrain, the IP should face away from the slope and descend by step-kicking with the heels.

19.18. Snow and Ice Climbing Procedures and Techniques

Traveling in snow and ice may include the need for specialized and/or improvised equipment such as an ice axe.

19.18.1. Ice Axe Techniques

The axe can be used for braking assistance when a climber begins sliding down a steep snow-covered incline. Improvised ice axes can be made from available parts of the IP's vehicle, poles made from branches/limbs, repurposed equipment items (i.e. an entrenching tool), and debris. (Figure 19-36)

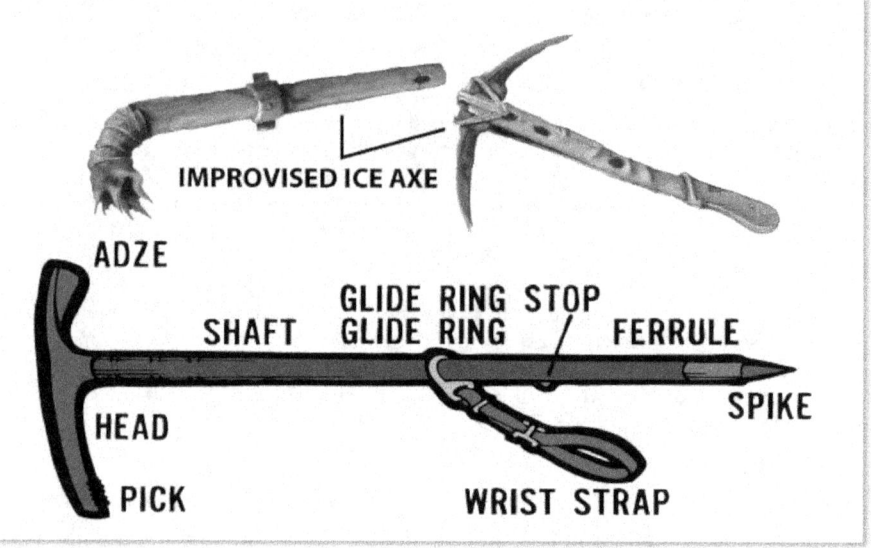

Figure 19-36 Ice Axe

19.18.1.1. Self-Arrest Technique

IP should practice the self-arrest technique before venturing onto steep grades. The improvised ice axe, whether sharp or not, is a lethal weapon when flying about on the attached cord. Physically, IP rig for arrest by rolling down shirt sleeves, putting on mittens, securing loose gear, and most important of all, making certain the axe is held correctly. Mentally prepare by recognizing the importance of instantaneous application. A quick arrest, before the fall picks up speed, has a better chance of success than a slow arrest. Preparation for an ice axe arrest should be taken when traveling on terrain which could result in a fall. The proper method of self-arrest (Figure 19-37) is to press strongly on the shaft and jam the spike into the ice or ground. The legs should be stiff and spread apart, toes digging in and hang on to the axe.

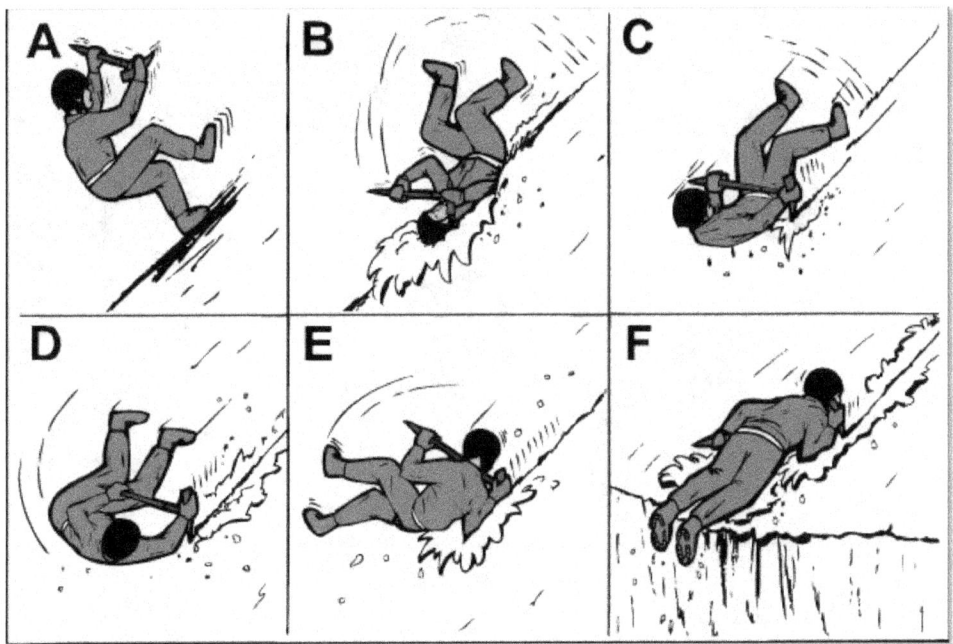

Figure 19-37 Ice Axe Self-Arrest

19.18.2. Glissading

Glissading is a means of rapidly descending a slope. Consisting of two basic positions, glissading offers a speedy means of travel with less energy exerted than using the descending step-by-step or plunging techniques (Figure 19-38).

19.18.2.1. Sitting Glissade

When snow conditions permit, the sitting glissade position is the easiest way to descend. The IP simply sits in the snow and slides down the slope while holding the improvised ice axe in an arrest position. Any tendency of the body to pivot head downwards may be checked by running the spike of the axe rudder-like along the surface of the snow. Speed is increased by lying on the back to spread the body weight over a greater area and by lifting the feet in the air. Sitting back up and returning the feet to the snow surface reduces speed. On crusted or firmly packed snow, the IP should sit fairly erect with the heels drawn up against the buttocks and the boot soles skimming along the surface. Turns are nearly impossible in a sitting glissade. However, the spike, dragged as a rudder and assisted by body contortions, can effect a change in direction of several degrees. Obstructions on the slope are best avoided by rising into a standing glissade (Figure 19-38) for the turn, and then returning to the sitting position. Speed is decreased by dragging the spike and increasing pressure on it. After the momentum has been checked by the spike, the heels are dug in for the final halt, but not while sliding at a fast rate as the result is likely to be a somersault. Emergency stops at high speeds are made by arresting.

Figure 19-38 Glissading

19.18.2.2. Standing Glissade

The standing glissade is similar to skiing. Positioned in a semi-crouch stance with the knees bent as if sitting in a chair, the legs are spread laterally for stability, and one foot is advanced slightly to anticipate bumps and ruts. For additional stability, the spike of the axe can be skimmed along the surface, the shaft held alongside the knee in the arrest grasp, with the pick pointing down or to the outside away from the body. Stability is increased by widening the spread of the legs, deepening the crouch, and putting more weight on the spike. A decrease in speed increases muscular strain and the technique becomes awkward and trying, although safe. Speed is increased by bringing the feet close together, reducing weight on the spike, and leaning forward until the boot soles are running flat along the surface like short skis. If the slide is too shallow, a long skating stride helps.

19.18.2.3. Usage

A glissade should be made only when there is a safe runout. Unless a view of the entire run can be obtained beforehand, the first person down the run must use extreme caution, stopping frequently to study the terrain ahead. Equipment must be adjusted before beginning the descent. Mittens or gloves are worn to protect the hands and to maintain control of the axe. Heavy waterproof pants provide protection to the buttocks. Gaiters are also helpful for all glissading. Glissades should never be attempted in terrain where the axe safety cord is required. The hazards of a flailing ice axe should never be risked during a glissade.

19.19. Evacuation Principles and Techniques

Evacuation of an IP from the position in a downhill direction is an easier task than establishing a mechanical leverage for pulling an IP to the top of a hill. The IP's medical condition will dictate the method of evacuation and equipment needed. The primary litter used is the Stokes litter (tubular frame litter with a wire basket), but it is unlikely that an IP will have one available. Therefore, improvised liters must be made using available resources such as clothing, cloth materials, poles made of wood or vehicle wreckage, etc.

19.19.1. Preparing the Patient for Transport

The litter must be secured to prevent its loss or further injury to the patient. Additionally, the litter may be padded or insulated (blankets or foam pads) for protection. The ties for securing the feet and pelvis should be attached to the litter. Before evacuating, all emergency medical treatment appropriate to the situation should be performed (splinting fractures, maintaining an open airway, etc.). The IP should be insulated from environmental conditions such as cold, wind, or rain. The person in charge of the IP's medical condition should ensure that the IP's condition is stable enough for transporting. In mountainous terrain, the IP should be protected from further injury due to rock fall by wearing a helmet at all times. A litter team generally consists of four to six people. Fewer than six cannot withstand the fatigue of frequent or long trips while carrying an injured person.

19.19.2. Three or Four-Man Lift

Three bearers take up positions on one side of the IP, one at the shoulder, one at the hip, and one at the knees. If one side is injured, the three bearers should be on the uninjured side. A fourth bearer, if available, takes a position on the opposite side, at the IP's hip.

19.19.2.1. Bearers

The bearers should kneel next to the IP. Then, simultaneously, the bearer at the IP's shoulder puts one arm under the IP's head, neck, and shoulder, and the other under the upper part of the victim's back. Each bearer at the IP's hips places one arm under the IP's back and the other under the IP's thighs. The bearer at the IP's knees places one arm under the IP's knees and the other under the ankles (Figure 19-39).

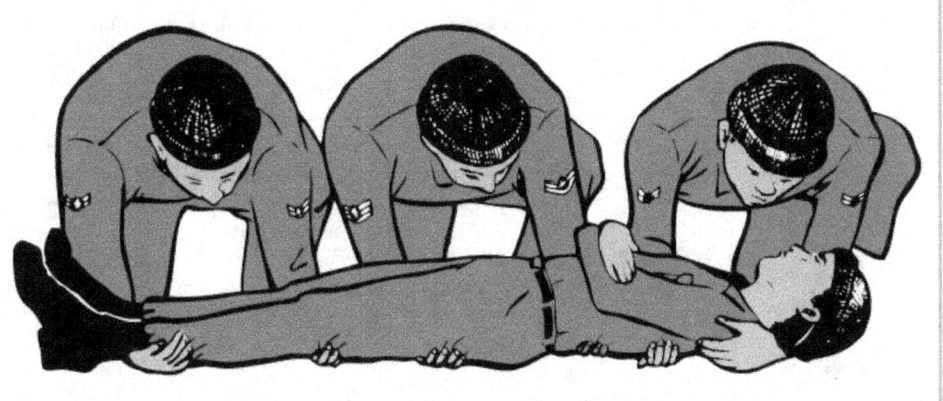

Figure 19-39 Lifting the Patient

19.19.2.2. Commands

The person at the IP's head gives all the commands. The command "prepare to lift!" is followed by the command "lift." Immediately, all the bearers lift simultaneously and place the IP in line on their knees. If the IP needs to be moved any distance to the litter, move as shown in Figure 19-40. The fourth bearer, if available, places a stretcher under the IP and against the toes of the three kneeling bearers. The command "Prepare to lower!" is followed by the command "Lower!" and the IP is gently lowered to the litter.

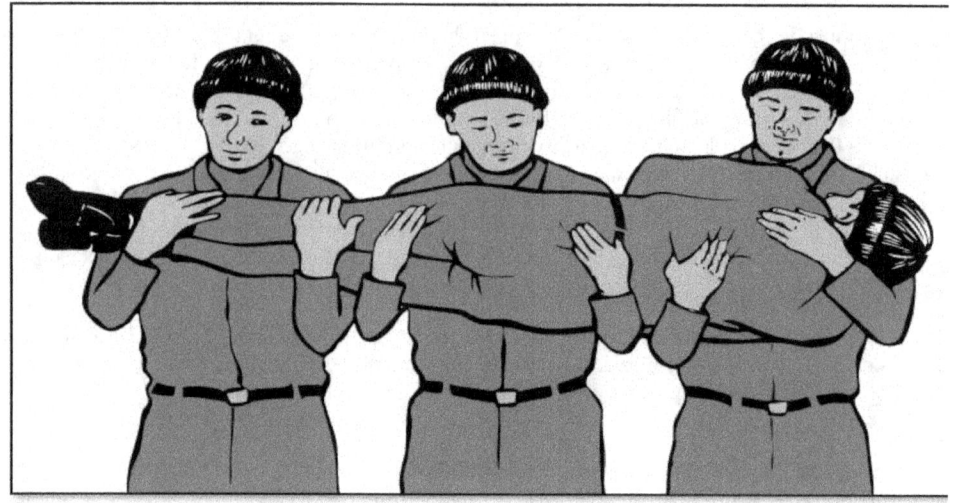

Figure 19-40 Moving the Patient

19.19.3. Securing Patient in the Litter

Once properly positioned in the litter, the IP must be secured in a manner to prevent further injury. The IP may be secured to the litter in a variety of ways depending upon the evacuation route and the IP's condition.

- The tape sling used to secure the feet is tied to the framework of the litter which separates the legs near the groin area. The tape sling should be tied with a clove hitch in the middle of the tape in a manner to prevent the tape from sliding down to the feet when pulled tight. The feet are secured by running the tape across the legs to the window on the outside of the litter then across the patient's legs to the feet. An overhand knot is made in each tape which can be passed over the corresponding foot. When the feet are secured, there should be ample room to apply tension to the head if needed. The tape is then tied at the foot of the litter to a major support bar on the inside of the litter frame. The ties on a litter should never be made on any outside rail as they are subject to abrasion. The tape slings should be tied off with a two-round turn and two half hitches. If the two-round turn does not hold tension, then a clove hitch can be used in its place.

- The tape used to secure the pelvis should be tied just above the tape used to secure the feet, and secured in the same fashion. Each end of the rope is passed over the leg to the larger upright cross-member of the Stokes litter between the outer rail and inner basket rail. This cross-member corresponds with the side of the hip. The tape is secured with a two-round turn and two half hitches or a clove hitch and two half hitches. The ends of the tapes are then tied together at the middle of the patient's waistline with a square knot and two half hitches on either side of the knot.

- The upper torso is secured by placing the middle of the tape in the center of the patient's chest and the two ends of the tape are secured to the large upright cross-member. The running ends of the tape are then passed diagonally across the IP to the cross-member which is next to the abdomen. The tape is secured again and the ends are tied at the midline of the body. The head is secured by running a tape sling over the helmet and securing the tape at the corresponding cross-members. The helmet can be used with a tape sling to provide traction; however, it is not a substitute for the neck collar (Figure 19-41).

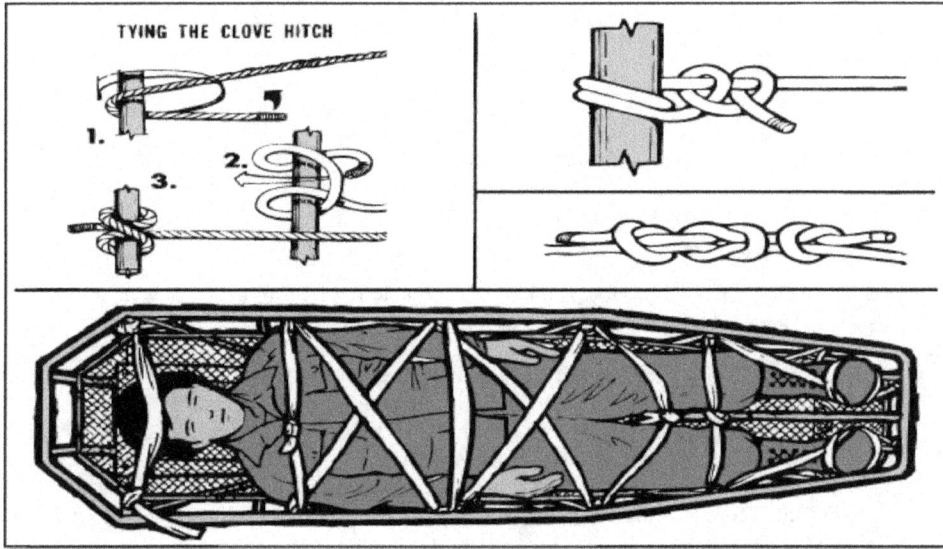

Figure 19-41 Tying the Patient into a Stokes Litter

19.19.4. Evacuation

Once the patient and the system are ready for the low-angle evacuation, the entire system must be double-checked. Once the litter and patient are prepared as described, ascent or descent is made through a team of litter bearers

19.19.4.1. Descent

In descending, the most direct, practical passage should be used. Communication is made through a series of commands. As litter bearing is rapidly exhausting, team members should alternate roles. Additionally, a sling attached with a girth hitch to the litter may be used to transfer some of the weight from the arm to the skeletal system via the shoulder.

19.19.4.2. Scout Usage

A scout may precede the team to pick a trail, make the passage more negotiable, or make a reconnaissance so the team need not retrace its course if an impasse is encountered.

Chapter 20

RIVER TRAVEL

20. River Travel

The techniques of river travel may be adapted to other water features such as swamps and lakes. Information provided in the open sea chapter may be adapted to enhance an IP's survival during river travel. A thorough knowledge of water travel techniques will greatly increase an IP's chances of success.

20.1. River Travel

Rivers have been used as a safe means of travel and are the reason many of the cities of the world are located on rivers. It is not uncommon for a river to flow at four or five knots per hour. An IP could travel 20 to 25 miles in five hours. This may contrast greatly with the rate of travel on land. Finally, the amount of energy required to carry an IP's equipment and supplies across 20 to 25 miles of land is much greater than if transported by river.

20.1.1. Rivers as Hazards

Each major continent has thousands of miles of navigable rivers. Some rivers such as the Nile, Amazon, Mississippi, Lena, and Mackenzie have hundreds of miles of navigable water with seldom a ripple. These navigable sections are generally found flowing through the flatlands, plains, tundra, and basins of the world. In these areas, only the temperature of the water and the plant and animal life may present hazards. In contrast, the headwaters of rivers, like the Mackenzie, Yangtze, and Ganges, are so rough that they would best be categorized as a threat to life. This would also be true of the Snake, Salmon, and Rogue Rivers of the Northwestern United States. These rivers, although traveled by white water rafters, pose an unreasonable hazard to IP. IP must take into account individual or group skills, injuries, type and severity of rapids, the temperature of the water, and direction the river flows in making the decision to travel. Even if a portage of several miles is required, the energy saved by floating on a river might warrant river travel. However, once the energy expended for portage exceeds the energy conserved by floating a section of a river, the river as a mode of travel should be abandoned.

20.1.2. Rivers to Transition Zones

In a permissive environment, rivers will most likely carry IP to indigenous people who could aid the IP in meeting the basic needs for sustaining life and effecting rescue. Even if some form of civilization is not encountered, IP would most likely reach a lakeshore or even the seacoast. These environments, particularly the seacoast, provide transition zones from land to water which are rich in food and other survival resources. In these areas, the resources could improve the chances of survival. It is much easier to spot signs of IP along a shore versus the interior of a landmass.

20.2. Using Safe Judgment and Rules for River Travel

There are certain safety rules and guidelines that must be followed to reduce the dangers associated with river travel. Respect for these rules and guidelines are necessary to reduce the potential dangers.

20.2.1. Personal Preparation

The most important safety rule is personal preparation. Preparation should begin by thoroughly scouting the river. The conditions of the river will determine the intermittent stops. High ridges along river edges provide needed visibility to plan each leg of travel. If there are numerous bends and poor lookout points to view the river, stops are frequent. Sound judgment must be used when planning routes. Patience in planning each leg of travel helps prevent disaster. All IP must know the plans and be able to handle the route safely, considering their skills and strength. IP should be aware of and avoid river hazards and have alternate routes and communication signals in case flow conditions suddenly change, making the run more difficult. All rapids which cannot be seen clearly from the river should be scouted. The route should be discussed by all IP. The skills, knowledge, and abilities of the IP must be considered, including swimming abilities and physical condition. Areas of high risk should not be attempted. Before reaching an area of suspected great difficulty, rafts should be beached and carried to the next point of travel. This is called portaging.

20.2.2. Donning Suitable Clothing for Adequate Protection

Before entering the raft, IP should don life preservers and suitable clothing for adequate protection. The equipment should be tested to ensure it is serviceable. Bulkiness is not advisable due to the possibility of the raft capsizing and being weighted down with water. The anti-exposure suit (if available) should be worn. Items that might absorb water should be packed in a waterproof container. The IP should ensure:

- All equipment is secured in the raft.
- Additional equipment and issued survival kits are checked and inventoried.
- Extra efforts should be made to keep supplies and equipment in good condition.
- All items are secured to the raft to prevent loss and (or) injuries.
- The raft is checked for leaks and necessary repairs are made.

20.2.3. Operating One-Man Rafts

When using a one-man raft for river travel (Figure 20-1), it may be advisable to tie or cut off the ballast bucket, fasten the spray shield in the opened position, and remove the sea anchor to prevent problems with swamping or entanglement with subsurface obstacles. Without the ballast bucket, the raft can be easily maneuvered by paddling with the backstroke or for slight adjustment, with a front stroke. When using either the backstroke or the front stroke, the IP will find it easier if the two underarm cells of the life preserver underarms (LPU) are disconnected in front and the cells placed behind the back (Figure 20-2). This gives the IP a full range of motion. When rough water is encountered, the IP should fasten the LPU and face downstream, feet first.

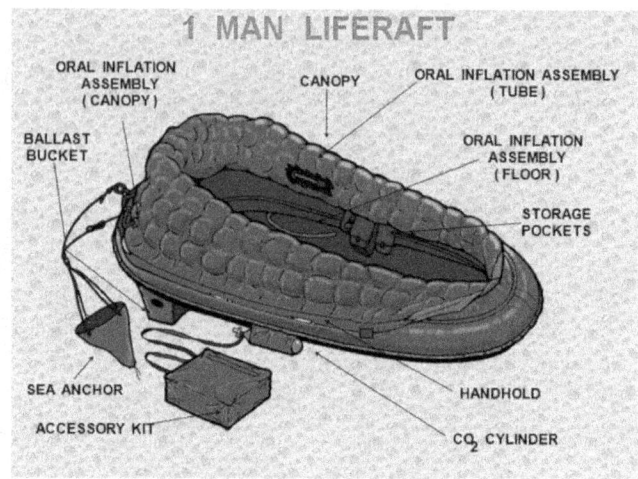

Figure 20-1 One-Man Raft

20.2.4. Avoiding Obstacles in a One-Man Raft

One of the primary methods of avoiding hazards on the river is to slow the speed of the raft and move across the river, avoiding a collision with the obstacle. This ferry position should be initiated early to avoid large rocks and reversals in the river. If the collision obstacles are to be avoided, the ferry position is to point the bow of the raft toward them and backstroke against the current to slow the speed of the raft's downstream progress and move it across the river. Usually, the best angle is about 45 degrees to the current. The greater the angle, the quicker the movement across the river, but this also increases the downstream speed of the raft. Decreasing the angle will slow downstream speed, but movement across the width of the river will also be decreased (Figure 20-2). The raft will be more maneuverable if it is well inflated. If the raft should pass over a rock, arch the back up to prevent injury to the buttocks or back.

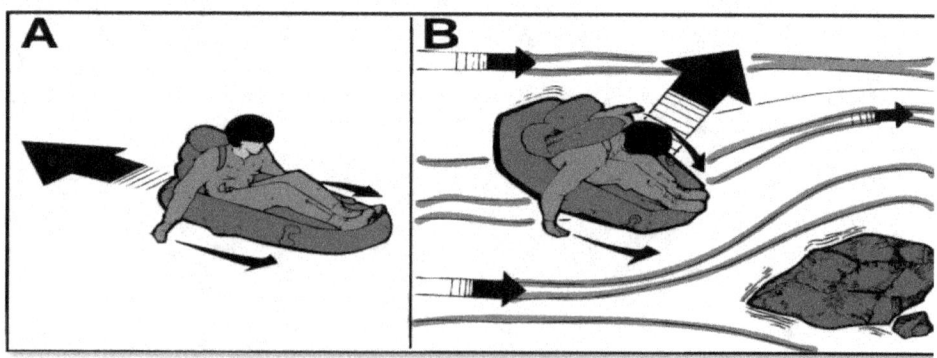

Figure 20-2 One-Man Raft (Paddling)

20.2.5. Operating Multi-Place Rafts

When using multi-place rafts, the boarding line and sea anchor should be removed to prevent entanglement (Figure 20-3). If available, about 50 feet of line should be tied on the bow and stern of the raft to be used for tie-offs. An additional 200 feet of line should be coiled and tucked away for emergency and rescue work; one end is secured to the raft while the other end has a fixed loop. An improvised suspension line rope may be used for this; for example, three-strand braid or a two-strand twist.

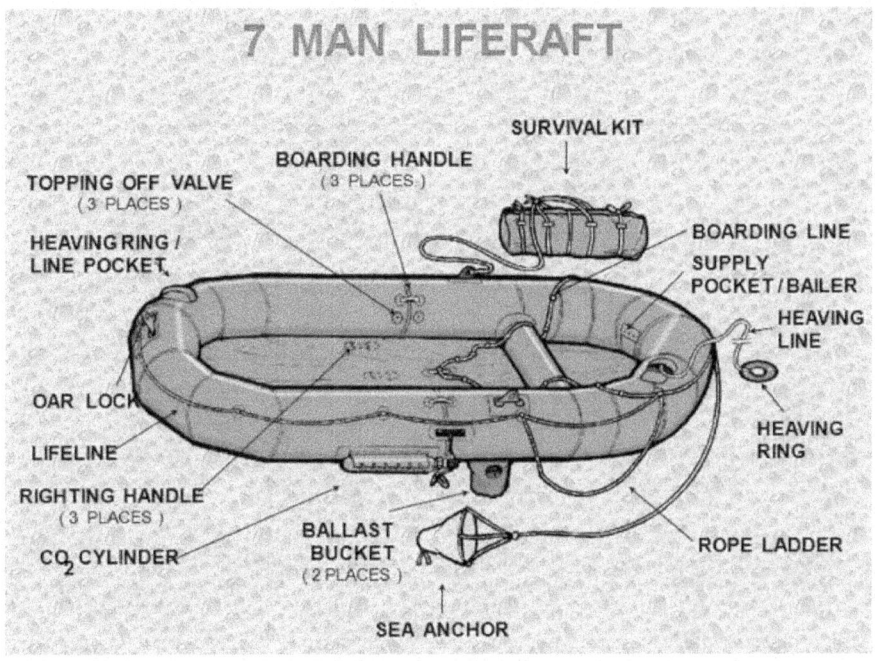

Figure 20-3 Seven-Man Raft

20.2.6. Equipment and Personnel in Multi-Placed Rafts

Proper placement of equipment and personnel should equalize weight distribution to ensure stable control. Overloading should be avoided. Assign personnel crew positions and responsibilities in the raft; captain (person in charge), stern paddler (maneuvers raft), and side paddlers. Twilight and night rafting should be avoided (permissive) as poor visibility increases danger.

20.2.7. Steering a Multi-Placed Raft

Steer a raft by using paddles and poles. A pole is more efficient in fairly shallow water, but a paddle is preferable in deep water. Poles and paddles from both ends of the raft are used. The person in the bow (front) can see any obstructions ahead, and the one in the stern (rear) can follow directions for steering. Poles are also useful for pushing a raft in quiet water.

20.2.7.1. Paddle Techniques

Paddle techniques are used to maneuver the raft. When paddling, there are three possible body positions on a raft. The best way is to sit on the upper buoyancy tube with both legs angled to the inside of the raft. The body should be perpendicular to the sides of the raft, enabling the rafter to paddle. Another way is to sit cowboy style, straddling the upper buoyancy tube of the raft with one leg on either side, and folding at the knee with each leg back. However, the outside knee may collide with obstacles and cause injury. The third way is normally used in calmer waters because it consists of partially straddling the upper tube, with legs comfortably extended. In a smaller raft, the IP may be able to sit down inside the raft and reach over the buoyancy tube. The following strokes can be done from these positions. Knowing the parts of paddles will help in explaining the different paddling strokes (Figure 20-4).

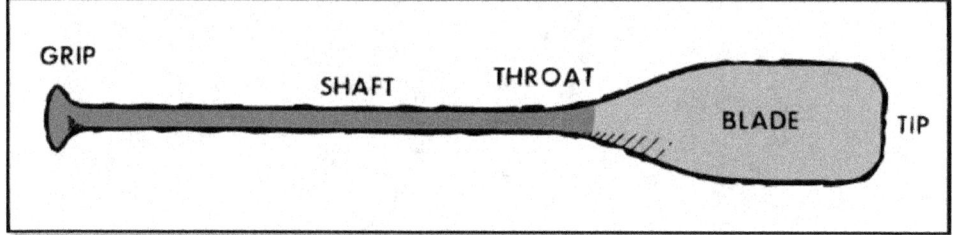

Figure 20-4 Paddle

20.2.7.1.1. The Forward Stroke

One of the easiest is the forward stroke which is done in smooth continuous movements using these techniques:

- Thrust the blade of the paddle forward using the outboard arm, and then momentarily keeping the outboard arm stiff and away from the raft, push the grip. The inboard hand is then moved forward to cut the blade deeply into the water. Continue the stroke by pushing on the grip and pulling on the shaft keeping the blade at a 90-degree angle to the raft. Stop the motion as the blade comes slightly past the hip, because a full follow-through provides little forward power and wastes valuable energy. Slide the blade out of the water by pushing down on the grip and swinging it toward the inboard hip and turning the blade at a parallel angle to the water once it has cleared the water. By paralleling the blade, it cuts wind and wave resistance and saves time and energy. This cycle is repeated until the strokes are changed.

- In mild water, there is no need to over reach or excessively twist the upper trunk of the body. When extra speed is needed, lean deeply into the strokes which brings the entire body into play. Position the inboard hand across the tip of the paddle grip and the outboard hand halfway to three-fourths of the way down the shaft.

20.2.7.1.2. The Backward Stroke

The opposite of the forward stroke is the backward stroke. The blade is thrust into the water just behind the hip, and pressure applied by simultaneously pushing forward on the shaft and pulling back on the grip. End the stroke where the forward stroke would begin, and again angle the blade out of the water back to the beginning of the backward stroke.

20.2.7.1.3. The Draw and Pry Strokes

The draw and pry strokes are opposite sideways strokes. These strokes are good for small sideways maneuvers and for turning the raft when used from the front or rear of the raft.

- Draw stroke. Reach out from the raft, dip the blade in parallel to the raft, and pull on the shaft while pushing on the grip. Pull the blade flat to the side of the raft. Pull the paddle out and repeat.
- Pry stroke. Dip the blade in close to the raft, and push out on the shaft while pulling in on the grip.

20.2.7.1.4. The Calm Water Crawl

The calm water crawl is used alternately with the forward stroke when paddling through long calms. Sit cowboy fashion while facing the stem and hold the paddle diagonally in front with the shaft which is held by the outboard hand against the outboard hip and the grip held by the inboard hand in front of the inboard shoulder. Extend the inboard arm to swing the blade behind, dip the blade in the water, and pull back on the grip, prying it forcefully, using the hip as a fulcrum (the point of support on which a lever works). Using the shoulder, hip, and hand as assisters, the crawl is easy yet powerful (Figure 20-5).

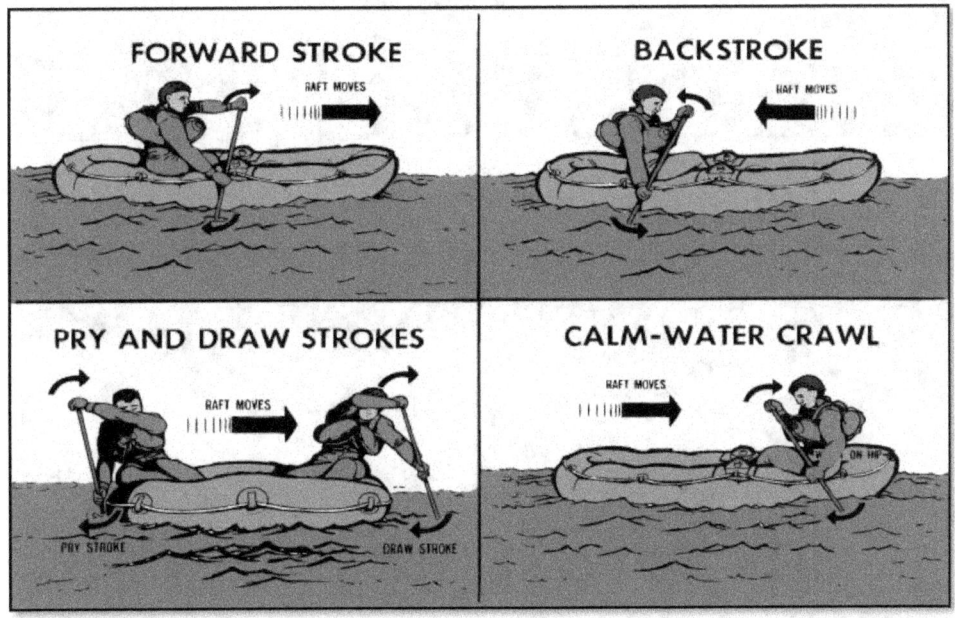

Figure 20-5 The Paddle Strokes

20.2.7.1.5. The Ferry

The ferry is a maneuvering to navigate bends and to sidestep obstacles in swift currents. The ferry is essentially paddling upstream at an angle to move the raft sideways in the current. Paddle rafts can ferry either with the bow (front) angled upstream or downstream. The bow-upstream ferry is stronger because it uses the more powerful and easier forward stroke. It is carried out by placing the raft at a 45-degree angle to the current with the bow angled upstream and the side toward the desired direction. The bow-downstream ferry is weaker because it uses the less powerful backstroke, but it does offer certain advantages. It enables paddlers to look ahead and makes it easy to put the bow into waves (Figure 20-6 and Figure 20-7). It is carried out by backstroking with the stern (back) angled upstream at a 45-degree angle and the side facing the desired direction.

Figure 20-6 Ferry

Figure 20-7 Bow Downstream Ferry

20.2.7.1.6. Entering a Small or Violent Eddy

There may be times when the only way for a heavy raft to enter a small or violent eddy is with a reverse ferry (Figure 20-8). Eddies are caused by currents reaction to obstacles in its path. The following steps may be used for an oar or paddle raft, except the paddle raft approaches the eddy bow first and finishes in a bow-upstream position:

- Raft approaches sideways.
- Raft turns around to angle its bow downstream.
- With careful timing, the captain should have the crew begin to pull powerfully on the paddle. The angle of the raft to the current can be close to 90 degrees, but is best at about 45 degrees
- While aiming for the eddy, the crew should continue with the front stroke and gain momentum.
- With the crew still using the front stroke, the raft breaks through the eddy fence.
- With the bow in the upstream eddy current and the stern still in the downstream current, the raft is spun into a normal ferry angle. The crew continues with the front stroke while making the necessary turn to bring the boat entirely into the eddy.
- The raft rides easily in the eddy. The reverse ferry and eddy turns are not only used to enter eddies, but can also be used to dodge through tight places. The reverse ferry (or sometimes an extreme ferry) scoots the raft sideways, the eddy turn snaps the bow into a bow-downstream position, and the raft, rather than entering the eddy, rides the eddy fence past a major obstruction or hole.

Figure 20-8 Entering an Eddy with a Reverse Ferry

20.2.7.1.7. The Straight Forward and Back Paddle

The straight forward paddle is used in calm and moderate waters where there is ample maneuvering time. Simply point the bow in the desired direction and follow the forward stroke method of paddling. The back paddle is performed the exact opposite of the forward paddle. Point the stern (back) in the desired direction and follow the backstroke method of paddling.

20.2.7.1.8. Making Left and Right Hand Turns

To make a left turn, the left side of the raft will back paddle, while the right side paddles forward. It is just the opposite to make the raft turn right. The right side on the raft back paddles while the left side paddles forward-both performs the paddling maneuvers at the same time (Figure 20-9).

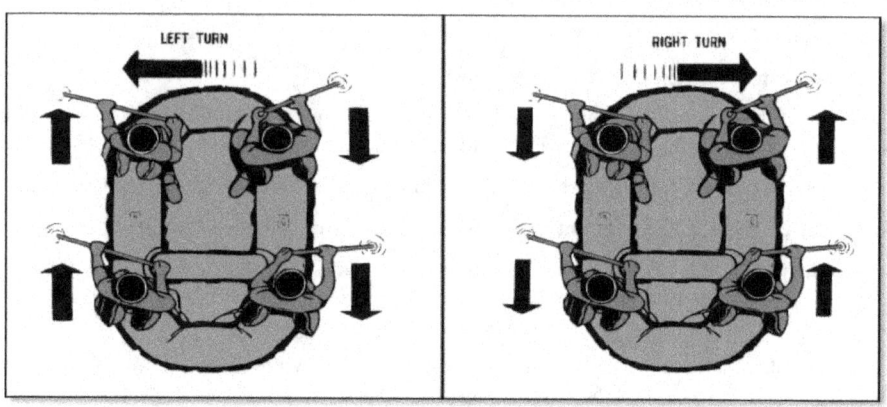

Figure 20-9 Turning the Raft to the Right or Left

20.2.7.1.9. The Pry and Draw Strokes

The pry and draw strokes are used to move the raft sideways (Figure 20-10).

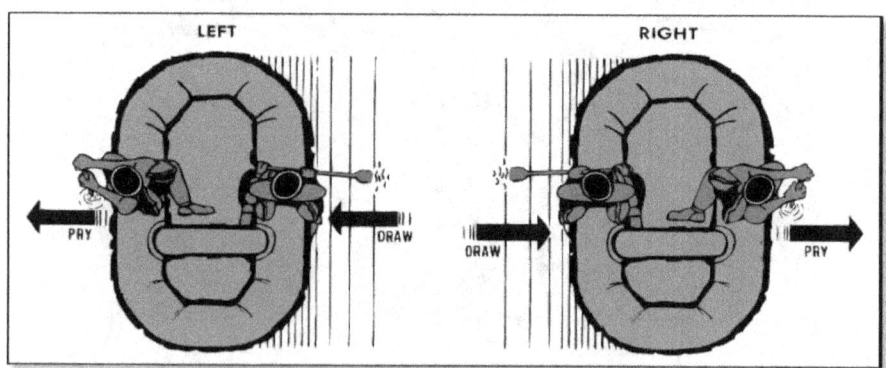

Figure 20-10 Making the Raft Move Left and Right with Pry and Draw Stroke

20.2.7.1.10. Stern Maneuvers

Stern maneuvers are used to increase the maneuverability of the raft. The paddle used at the stern of the boat is basically a rudder which controls direction. To turn right, the paddle blade is held to the right, square against the direction of the current. To turn left, the paddle is held to the left.

20.2.7.1.11. Other Strokes

Other strokes, such as a forward stroke or draw stroke, used at the stern of the raft, will cause it to turn or move faster. If stroking is done slightly to the side (either right or left) of the raft, it will help move the raft in the opposite direction (Figure 20-11).

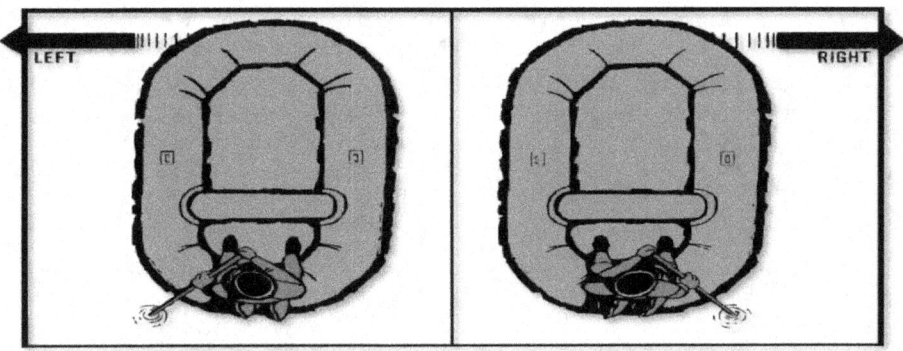

Figure 20-11 Stern Maneuvers

20.2.8. The Captain and Crew

River travel requires fast, decisive action. Therefore, a paddle raft needs a captain to coordinate the crew's actions by the use of commands or signals. Communications between captain and crew are crucial; all members must agree on a set of short, clear commands. The following are suggested commands:

CAPTAIN'S COMMANDS:	CREW RESPONSE:
Forward	Crew paddles forward.
Back paddle	Crew does backstroke.
Turn right	Left side paddles forward, right side does the backstroke.
Turn left	Right side paddles forward, left side does the backstroke.
Draw right	Right side uses draw stroke, left side uses pry stroke.
Draw left	Left side uses draw stroke, right side uses pry stroke
Stop	Paddlers relax.

20.2.8.1. Commands

Commands must be carried out immediately, so the crews should practice until they can snap through all the commands without hesitation. The captain controls both the direction and speed of the raft with a specific tone of voice and commands. This control, and the captain's ability to anticipate how the water ahead will affect the raft, will help avoid under compensation and overcompensation of maneuvers through obstacles. Good captains think well ahead and move with the river, issuing commands precisely and sparingly, working their crews as little as possible.

- When using commands to maneuver the boat in harmony with the river's currents, paddling can be easy and effective, even fun. When instant action is necessary, the captain may say, "Paddle at will." When time permits, the captain should introduce commands with a preparatory statement such as, "we're going to ferry to the right of that big rock. OK-(gives command)." This gives the crew time to prepare for the next response. If a command is not heard or understood, it should be repeated with zest until it is understood.

- If a raft member spots a better way through a rapid or channel, a fully extended arm is used to point it out. This signal, like the others agreed upon, should be repeated until it is understood.

20.3. River Hydraulics

An understanding of river hydraulics is important to the IP. Knowledge of the types of obstacles and why they should be avoided or overcome is necessary for a safe river journey.

20.3.1. Laminar Flow

The drag produced when moving water flows over or past various types of objects and surfaces is called a laminar flow (Figure 20-12). The laminar flow principle is that various layers or channels of water move at different speeds. The lower layer of the river moves more slowly than the top layer. This is due to the friction on the bottom and sides of the river which is caused by soil, vegetation, or contours of the riverbank. The layers next to the bottom and sides are the slowest; each subsequent layer will increase in speed. The top layer of the river is only affected by the air. The fastest part of the flow on smooth straight stretches of water will be between five and 15 percent of the river depth below the surface. Even straight running riverbeds are not smooth; they have jutting and receding banks on the sides which affect the laminar flow. The friction caused by the banks causes the sides of the flow to be slower than the midstream. The areas near the banks are shallower and have fewer layers. When the river travels at four to five knots, turbulence begins to develop which interferes with the regular flow of the current. When this rate of river flow is achieved, the friction between the layers of water will cause whirling and spinning actions which agitates the smooth flow of water, creating more resistance.

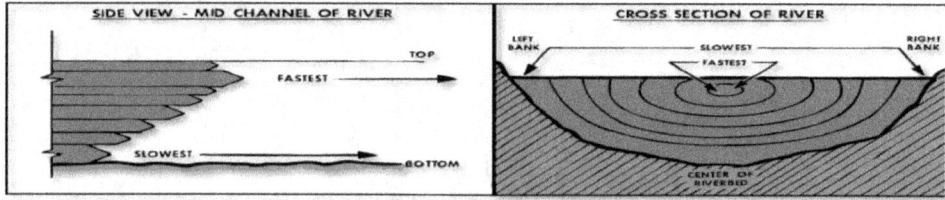

Figure 20-12 Laminar Flow

20.3.2. River Currents

River currents describe the movement of water. The current varies depending upon the volume of water, the channel's gradient, and various other factors.

20.3.2.1. Reflex Currents

When a current of a river is deflected by obstructions, the overall downward flow of a river will respond. These responses vary from mild to radical deflection, creating direction and speed changes of water-flow. These changes are called reflex current. The reflex current responds to an obstruction such as bends or submerged rocks.

20.3.2.2. Helical Currents

One response to the laminar flow is a spiraling, coil-springing flow called a helical current (Figure 20-13) which corkscrews as a result of the friction with the riverbank. Going downriver, on the left side of the supposed straight-line river, the helical flow turns clockwise to the main current, and on the right side the helical flow is counterclockwise. This results from friction and drag caused by shallow banks combined with the strong force of the main current flowing down. The helical flow and the mainstream create a circular, whirling secondary current which travels down along a line near the point of maximum flow. Helical current flow starts along the bottom of the river going out toward the riverbank, surfacing, and then spiraling back into the mainstream at a downward angle. This flow causes floating objects around the edges to be pulled into the mainstream and held there. By understanding where the fast water is and how to observe the characteristics which show the current, an IP can maneuver the raft to take advantage of the faster water to increase the rate of travel. Even at the quietest edge of a flow, particles are still drawn into the strongest part of the current. Laminar and helical flows are always present in fast-flowing rivers.

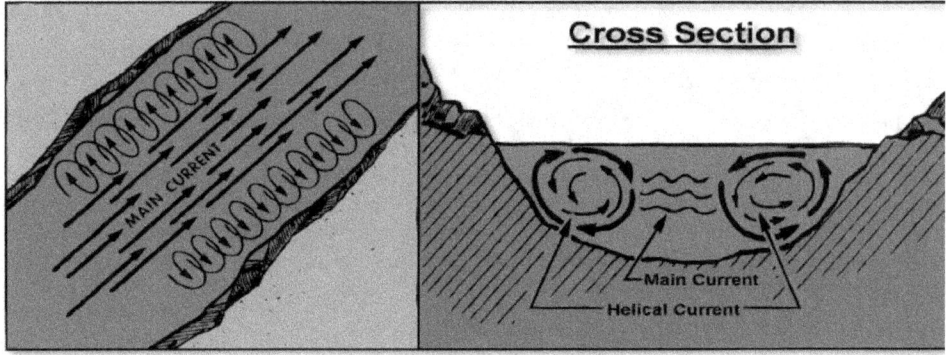

Figure 20-13 Helical Flow

20.3.2.3. The Main Channel

The main channel is the deepest part of the river and can wander from bank to bank. The turbulence caused by the wandering main current erodes wide curves into sharper, more defined bends, creating indirect courses.

20.3.2.4. Backwater

When the river makes sharp turns, the current is affected by centrifugal force swinging it wide into the outside bank. The helical current diminishes, being smothered by the laminar flow, thereby increasing the corkscrew effect on the inside of the curve. The surface water is being whirled hard in the direction of the outside curve of the bank. The faster the water flow, the stronger the push. Floating objects are forced, with the surface water, to the outside of the curve and into the banks, usually getting lodged against and onto the shore.

20.3.2.4.1. Point Bars

A powerful helical flow not only pushes the surface outward, but as it swirls up from the bottom it carries sediment up with it. The sediment and other debris is deposited at the highest point of the inside bank of the bend. The sediment is then dropped during high water, and when the water recedes, a point bar (made of the sediment and debris) is revealed. The point bar generally sticks out far enough to funnel floating objects into the swiftest part of the river during high waters, avoiding the sandbar.

20.3.2.4.2. Super-Elevation

Super-elevation is a feature where the water is being increased in volume, intensity, and height. When both the stream volume and movement are high, centrifugal force exerts another type of influence on flow characteristics. The river surface water tends to curve in a dish shape towards the outer bend, like a banked turn on a racetrack. The dished inside curve is the easiest and safest route to travel through. If maneuvered correctly, the slight rise of the water and the force of the current around the curve will cause the raft to slip gently off the wave and into the quiet pools of water below. But if the raft was maneuvered across the line of the currents, the raft may either be sucked under by the dominant helical flow, or the power and force of the river on the outside of the dish on the curve could smash and pin any floating device against the outside bank. During river travel, personnel floating in their one man life rafts have been sucked down entirely into the dominant helical flow.

20.3.2.5. Macroturbulence

Macroturbulence is any extreme, unpredictable turbulence (Figure 20-14). It is an especially dangerous phenomenon caused by a drop or decline in the river bottom. The phenomenon also occurs when the water comes in contact with the river bend or rocks. The extreme amounts of froth created from the turbulence and the gravitational pull cause rafts to spin. Raft control is extremely limited because the lack of water viscosity causes resistance against paddles and a lack of buoyancy, making it difficult to float or maneuver. This type of white water can be impassable, depending upon how extreme the dip and amount of water flow.

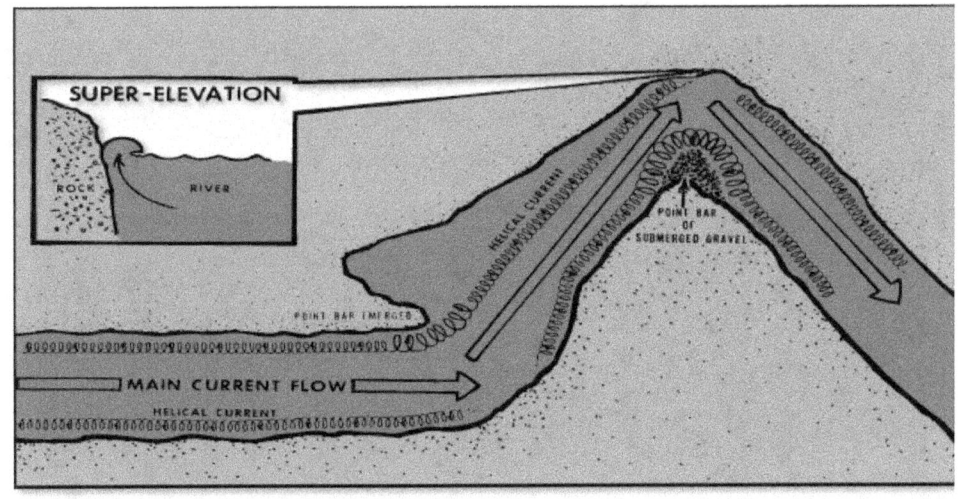

Figure 20-14 Macroturbulence

20.3.2.6. Mixed Currents in River Bends

Coming out of a sharp bend, the river currents are mixed, but the dangerous movement is still pulling. The result of the laminar flow shooting into a bank creates a helical flow effect immediately below the turn, where the river is still trying to assume a natural "straight" flow. Being a liquid, water cannot resist stress and it responds to a variety of obstacles (most common are submerged boulders) (Figure 20-15).

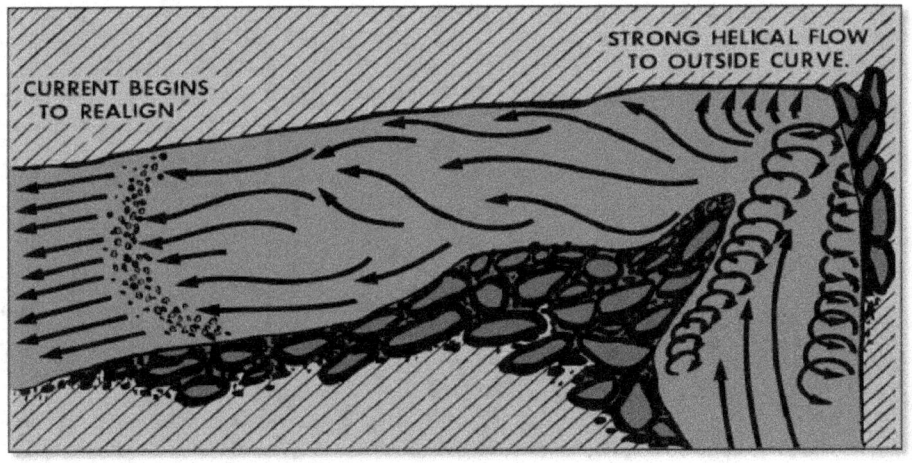

Figure 20-15 Anatomy of a River Bend

20.3.2.7. Increased Laminar Flow

When water flows over obstructions, such as submerged boulders, the character of the laminar flow is changed. As the water flows over the top of the rock, the layers of the laminar flow increase in speed. This is known as a venturi effect. The hydraulic area is a type of "vacuum" formed as water flows around the rock. Created directly below the obstruction are confused and disordered currents which accelerate the layers of the laminar flow (Figure 20-16).

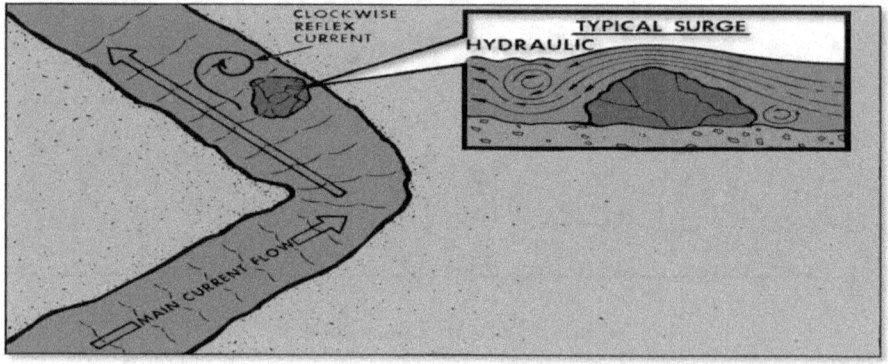

Figure 20-16 Response to a Submerged Rock

20.3.2.8. Surges

One type of hydraulic is the surge, which usually occurs when the current is slow and the water is deep. This hydraulic is formed downstream from an obstruction with a surge in the water volume. When obstacles no longer have the ability to hold the water back, the pressure is released. Surges present few problems if the boulder or obstruction is covered with enough water-flow to prevent contact when floating over the top. IP should be aware of obstructions (known as "sleepers") if the water does not sufficiently cover them (Figure 20-17). Failing to recognize a sleeper can result in raft destruction and severe bodily injuries. With large sleepers, the water flows over the top creating a powerful current. This powerful, secondary current is trying to fill the vacuum created by the hydraulic downstream action.

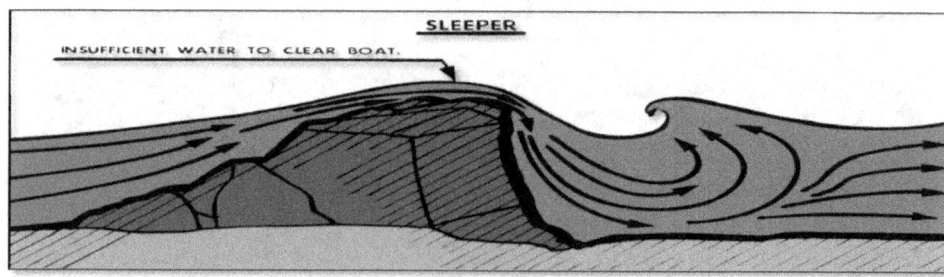

Figure 20-17 Sleeper

20.3.2.9. Dribbling Fall

Another form of large sleeper is referred to as a dribbling fall. These are caused by minimal water flowing over submerged obstacles with considerable drop below. This type of sleeper causes a bumpy ride, reducing speed, and can capsize the raft.

20.3.2.10. Breaking Holes

Breaking holes occur where a large quantity of water flows over a sleeper and the drop is not steep enough to create a suction hole (Figure 20-18). A wave of standing water, much like an ocean breaker, is found downstream. This wave is stationary and can vary from one to 10 feet high, and even though it lacks the strong upriver flow of a suction hole, it can be a trap for rafts too small to climb up and over the crest. The size of the breaking hole and the IP's seamanship must be considered before tackling this obstacle.

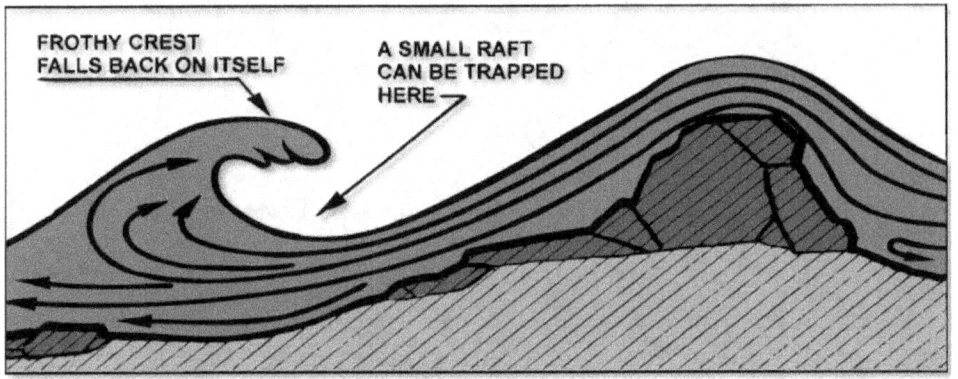

Figure 20-18 Breaking Hole

20.3.3. Suction Holes

The vacuum created by a suction hole is strong enough to pull an IP wearing a LPU beneath the surface. If pulled down into a suction hole, the IP will normally be whirled to the surface downriver and returned to the suction hole by the upriver flow, to be pushed under once again. Objects too buoyant to sink normally remain trapped. It's usually difficult to identify a suction hole because there is no frothing, no obvious curling water, and little noise. Extreme caution should be used when a large bulge appears in the water (Figure 20-19). There are three possible ways of surviving and avoiding serious injury in suction holes. One way is to find the layer of water below the surface which is moving in the same desired direction. The second way is to reach down with a paddle or hand and feel for a current which is moving out of the hole. However, in a large suction hole the downstream flow will be too deep to reach. The IP should attempt to cut across through the side of the eddy into the water rushing by. The final and best solution is to scout ahead and try to identify the location of suction holes and avoid them.

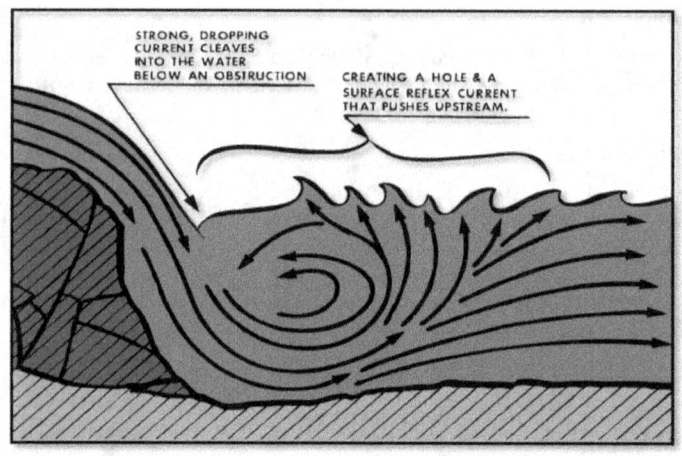

Figure 20-19 Suction Hole

20.3.4. Eddies

An eddy (Figure 20-20) is a reaction to an obstruction. The type of eddy which occurs next to the bank is caused by portions of the main current being deflected and forced to flow back upriver where it again joins the mainstream. These areas are usually associated with quiet and slow-flowing water. They are also associated with areas where the river widens or just above or below a bend in the river. An eddy has two distinct currents: the upstream current, and the downstream current. The dividing line between the two is called an eddy fence. It is a line of small whirlpools spun off the upriver current by the power of the downstream current.

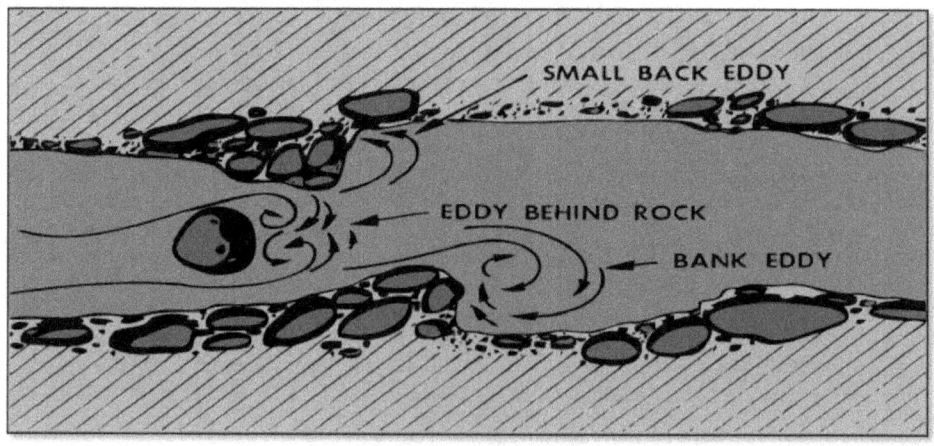

Figure 20-20 Eddies

20.3.4.1. Two-Dimensional Eddy

A two-dimensional eddy is when the tip of an obstruction is slightly above the water and causes a two-dimensional flow around the obstacle. Because of the speed and power of the current, the water is super-elevated and is significantly higher than the level of water directly behind the obstacle. This creates a hole which is filled by the flow of water around the obstacle (Figure 20-21). Two-dimensional eddies which occur in midstream will create two eddy fences, one on each side of the obstacle. The water will enter the depression from both sides and will travel in a circular motion, clockwise from the right bank, counterclockwise from the left bank, and back upstream directly behind the obstacle. If the projection is large enough and a strong circular motion is created, it becomes a whirlpool. The outer reaches of the swirling water are super-elevated by centrifugal force and a suction is created, similar to a drain in a bathtub. These vortexes are very rare and usually occur on huge rivers. IP may stop and rest where eddies occur since there should be no strong swift currents in the eddy. If the obstacle is huge, it may be impossible to paddle fast enough to cross the eddy fence without being spun around in a pin wheeling manner.

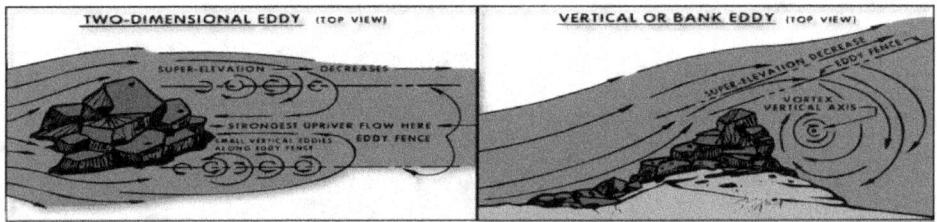

Figure 20-21 Two-Dimensional Eddies

20.3.5. Falls

In most falls, there are two reflex currents or suction holes forming, both whirling on crosscurrent axes into the falls (Figure 20-22). One current falls behind the crashing main stream of water while the other falls in front. It is not too dangerous on a two- or three-foot drop, but when heights of six or more feet are present, it could be a death trap. The suction holes formed below these falls are inherently inescapable because of their power. The foam and froth formed at the bottom of falls can be dangerous because it takes away the buoyancy. Jagged boulders and other hazards may be hidden.

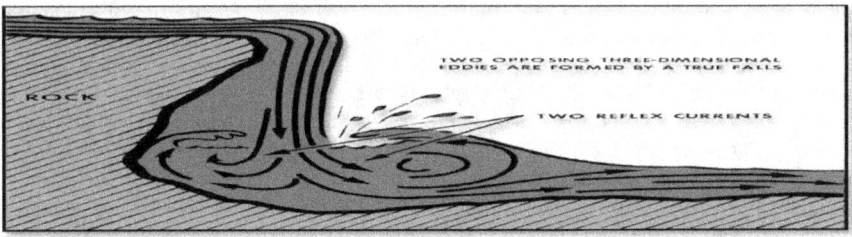

Figure 20-22 Falls

20.3.6. Boils

A boil may occur below a fall or sleeper, downstream of the curling or reverse-current suction. This appears as a dome or mound-shaped water formation. Boils are the result of layers of flows hitting bottom, aimed upward, and reaching the surface parting into a flowerlike flow. The water billowing out in boils is super-oxygenated, taking away the resistance needed to push with the paddles or to suspend an IP in a life preserver.

20.3.7. Rollers

Rollers are another difficulty found when traveling on fast rivers. Rollers are large, cresting waves caused by a variety of situations. A wave seen below a breaking hole is one type of roller. Velocity waves are another type which occur on straight stretches of fast dropping waters and caused by the drag of sandy banks and submerged sandbars. They may be large enough to overturn a raft, but are easily recognized, and are usually regular and easy to navigate by keeping the raft direction of travel in line with the crest of the wave (Figure 20-23).

Figure 20-23 Tall Waves and Rollers

20.3.8. Tail Waves

Tail waves are quiet waves which are a reflex from the current hitting small rocks along the bed of the river and deflecting the current toward the surface. They are usually so calm that going over them is not noticeable.

20.3.9. Bank Rollers

Bank rollers are similar in appearance to crested tail waves and occur when there is a sharp bend in the riverbed which turns so sharply the water can't readily turn in the bend. The water slams into the outside bank and super-elevates, falling back upon itself. Small bank rollers cause few problems, but large ones cresting five feet or higher can capsize a raft (Figure 20-24). There are three terms used to specify the severity or height of rollers; one being washboard. Washboard rollers are a series of swells which gently ripple and are safe and easy to ride. The next stage of rollers is called standing water (cresting rollers), where the speed of contours on the bottom are such that the tops of the swells fall back onto themselves. The most dangerous and insurmountable rollers are referred to as haystacks or roosters (Figure 20-25). They resemble haystacks of water coming down from every direction and appear as giant frothing bulges. They may be so high a raft cannot go over them. It would be easy to become trapped in a deep trough (depression) and be buried under tons of frothing water. These troughs also hide dangers such as sharp rocks which could tear a raft or break through the bottom of a boat.

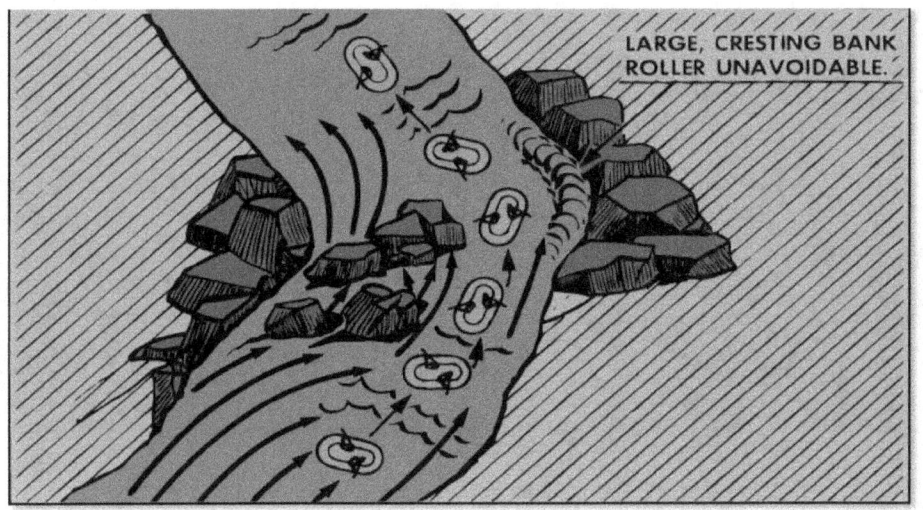

Figure 20-24 Bank Roller

Figure 20-25 Haystack

20.3.10. Chutes

River chutes are good, logical travel routes, but they may harbor dangers. A chute (or tongue) is a swift-running narrow passage between river obstructions caused by a damming effect. An example would be water forced between two large boulders. Because the water flow is restricted, it accelerates and a powerful current rushes through. Because of the water velocity, there may be a suction hole on either side of the chute (Figure 20-26).

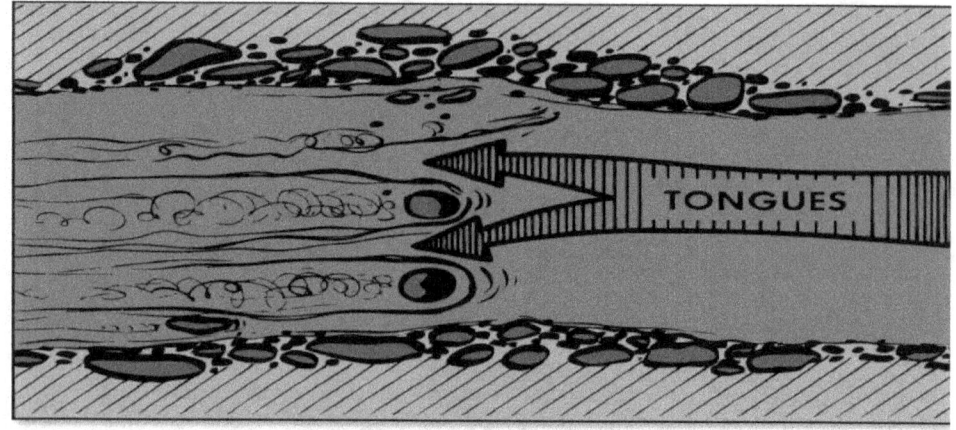

Figure 20-26 Tongue of the Rapids

20.3.11. Log Jams (Tongue of the Rapids)

Log jams are extremely dangerous. They consist of logs, brush, and debris collected from high waters that become lodged across the current. They remain stationary while the river flows through them. If a craft should be swept up against the stationary logs, it will be pinned in place by the current. Should the craft be tipped and swamped, it could be swept under the log jam. If the current is strong enough to do this, the occupants may also be pinned underneath the jam.

20.3.12. Sweepers

Sweepers can be the most dangerous obstructions in rivers. A sweeper is a large tree growing on a riverbank which has fallen over and is resting at or near the surface of the water. It may bounce up and down with the current. IP may be suddenly confronted with a sweeper which blocks the channel while rounding a bend in a river. The IP are relatively helpless when it encounters sweepers in swift water. The only precautionary measure is to land above a bend in order to study the river ahead. Many people have met disaster by hitting sweepers.

20.3.13. Glacial Rivers Sediment

While any river may have obstacles, glacial rivers are known for their large amount of sediment and silt which may make debris and submerged obstacles difficult to see. Rivers with large amounts of sediment and silt can create a sandblasting effect on anything attached to the outside of the life raft.

20.4. Emergency Situations

Emergency situations may be encountered.

20.4.1. Rock Collisions

Collisions with rocks above the surface of the water are common occurrences on a river. If the collision is unavoidable, IP should spin the raft powerfully just before contact, or hit the rock bow-on. If the IP is able to spin the raft, it will usually turn the raft off and around the rock. If they hit the rock bow-on, it will stop the raft momentarily, giving time to manipulate a spin-off with a few turn strokes. When the bow-on method is used, occupants in the stern (aft) should move to the center of the raft before impact. This allows the stern to raise and the rushing current to slide under the raft. If the stern is low, the water will pile against it causing the raft to be swamped. If IP are colliding broadside with a rock, the entire crew must immediately jump to the side of the raft nearest the rock-always being the raft's downstream side (Figure 20-27). This should be done before contact. If not, the river will flow over and suck down the raft's upstream tube. The raft will be flooded and the powerful force of the current will wrap it around the rock, possibly trapping some or all of the crew between the rock and the raft. Once it is un-swamped, the broached (broadsided) raft on a rock is usually freed easily (Figure 20-28). If two people push with both feet on the rock in the direction of the current, the raft will swing or slide into the pull of the current. The rest of the crew should shift to the end which is swinging into the current. These methods rarely fail; however, if the raft refuses to budge, it can be freed using the enormous power of the current. Large gear bags or sea anchors are securely tied to a long rope and secured to the end of the raft expected to swing downstream. The sea anchors (gear bags) are then tossed downstream into the current. A safety line should be used.

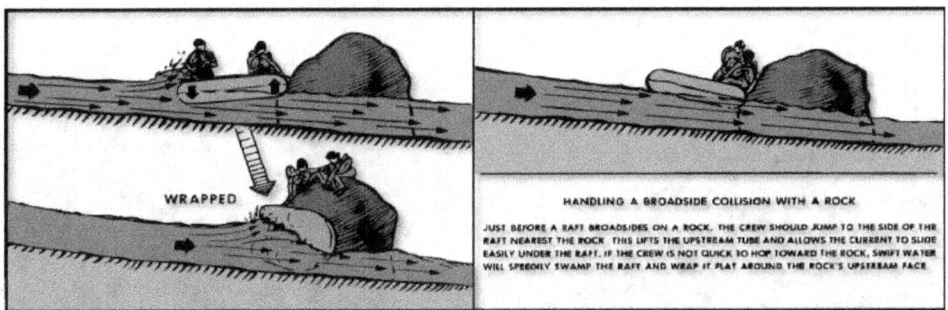

Figure 20-27 Handling Broadside Collision with Rock

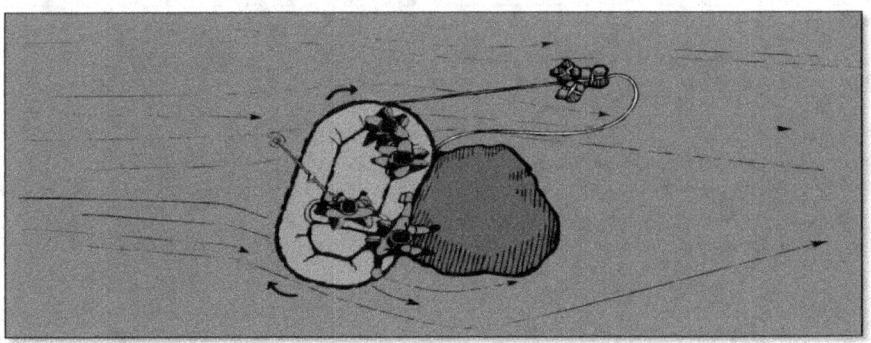

Figure 20-28 Freeing Un-swamped Raft

20.4.2. Freeing-a Wrapped Raft

Sometimes the powerful force of swift water may pin and wrap a raft around a rock. It is unusual for a raft to be equally balanced around the rock, so it will move more easily one way than it will the other. The part of the raft with more weight and bulk should be moved toward the flow of the current. Lines are attached to at least two points on the raft so, when pulled, the force is equally distributed. One of these points is on the far end of the raft, around the tube and is called the hauling line. The second tie-off point can be a cross tube. One person should hold the line attached to the stern of the raft. The raft should be moved over the rock into the pull of the current. Once it is freed, the person holding the stern line can move the raft to a safe position.

20.4.3. Raft Flips

When a raft is about to flip, there is little time to react to the situation. If the raft is diving into a big hole, the primary danger is being violently thrown forward into a solid object in the raft. IP should protect themselves by dropping low and flattening themselves against the backside of the baggage or cross tube. If the raft is being upset by a rock, fallen tree, or other obstacles, members should jump clear of the raft to prevent being crushed against the obstruction or struck by the falling raft. If the raft is pinned flat against an obstruction, the members should stay with the raft and try to safely climb up the obstruction (Figure 20-29).

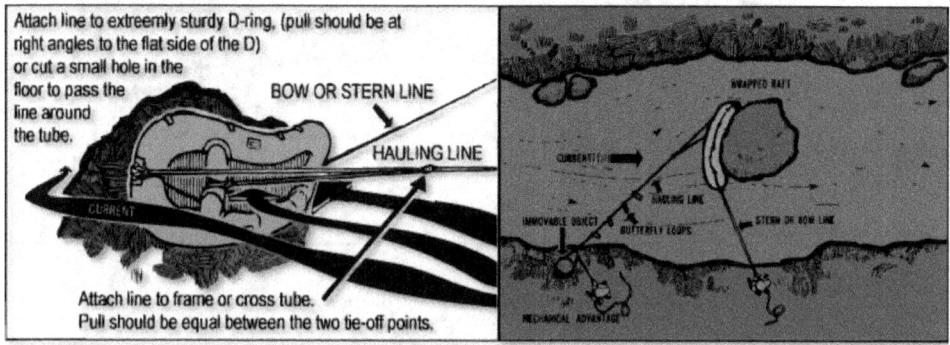

Figure 20-29 Freeing a Wrapped Raft

20.4.4. Lining Unrunnable Rafts

Lining a raft through rapids is basically letting it run through rapids with a crew on shore controlling it by attached lines (Figure 20-30). The raft should be moved slowly by maintaining tight control of the lines attached to the bow and stern. If strong eddies or steep narrow chutes are not present, one member should walk along the shore and control the raft with a strong, long line. The lines running to shore should be long enough to allow the raft full travel through the rapids.

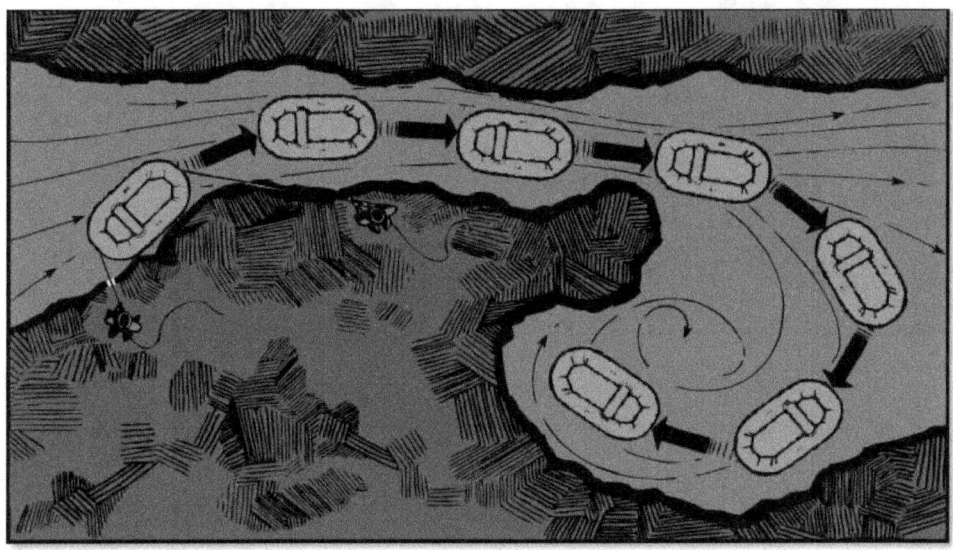

Figure 20-30 Lining a Chute

20.4.5. Rescuing a Swimmer from Shore

The rescuer should carefully choose the right spot where the coil of rope thrown to the swimmer will not cross hazardous areas or obstacles, and yet be near a rock or tree which can be used for belaying (securing without being tied) the end of the line. The person throwing the line should make sure the line has a flotation device (life preserver) at the end before it is thrown to the swimmer. The rope should be coiled in a manner which will allow it to flow smoothly, without entanglement, to full extension. One hand holds one-half to one-third of the coil while the other throws the remainder of the coil out. The weighted coil should be thrown to a spot where the swimmer will drift, which is usually downstream, in front of the swimmer. As the rope travels out, all of the line except the last 10 or 12 feet should be uncoiled. The member pulling the IP should be braced or have the line around a rock or tree to hold the swimmer once the line has been reached by the IP.

20.4.6. The Swimmer's Responsibility

The swimmer should be aware of the rescuer's location and face downstream when waiting for the hauling line. When the line is thrown, the swimmer will normally be required to swim to reach it. Once holding the line, the swimmer should be prepared for a very strong pull from the current and line. The line should be held tightly, but not wrapped around a wrist or hand! Entanglement must be avoided. The swimmer should pull, hand-over-hand, until reaching shallow water, and then use the rope for steadiness while walking to shore (Figure 20-31).

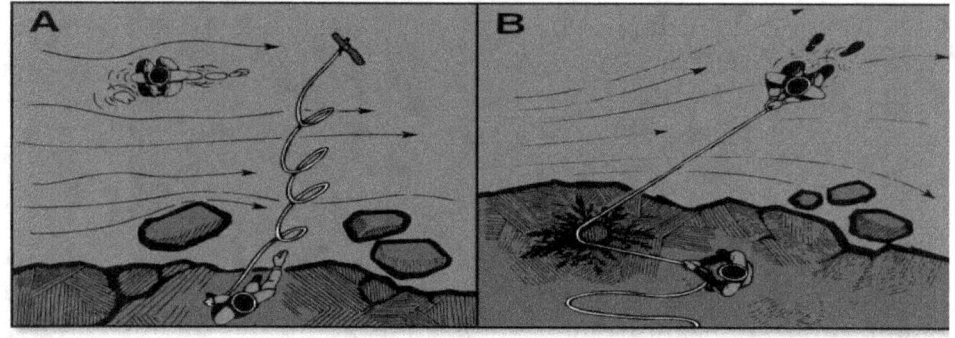

Figure 20-31 Making a Rescue

20.5. Improvised Rafts

There are various types of flotation devices which may be improvised and used as rafts for equipment and personnel.

20.5.1. The Donut Raft

The donut raft (Figure 20-32) can be used for transporting equipment, but is not a good vehicle for people. The raft is constructed by using saplings or pliable willows and a waterproof cover (poncho or other material). A hoop-shaped framework of saplings or pliable willow is constructed within a circle of stakes. The hoop is tied with cordage or suspension line and removed from the circle of stakes and placed on the waterproof cover to which it will be attached. Clothing and (or) equipment is then placed in the raft and the IP swims, pushing the raft.

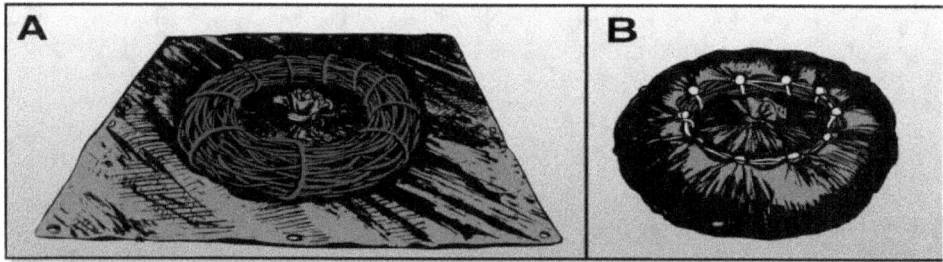

Figure 20-32 Donut Raft

20.5.2. The Bull Boat

The bull boat (Figure 20-33) is a shallow-draft skin boat shaped like a tub and formerly used by Native Americans in the Great Plains area. IP should construct an oval frame, similar to a canoe, of willow or other pliable materials and cover the framework with waterproof material or skins. This makes a craft which is suitable for transporting equipment across a river with the IP propelling it from behind.

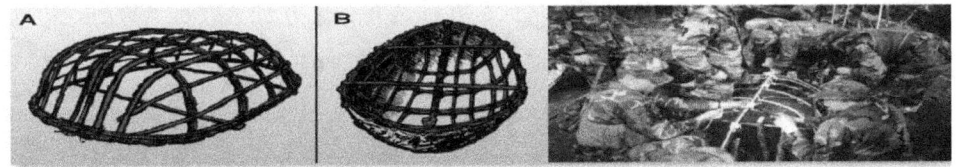

Figure 20-33 Bull Boat

20.5.3. The Emergency Boat

An emergency boat can be made by stretching a tarpaulin or light canvas cover over a skillfully shaped framework of willows and adding a well-framed keel of green wood, such as slender pieces of spruce. Gunwales (sides) of slender saplings are attached at both ends and the spreaders or thwarts are attached as in a canoe. Ribs of strong willows are tied to the keel. The ends of the ribs are bent upward and tied to the gunwales. The inside of the frame is closely covered with willows to form a deck upon which to stand. Such a boat is easy to handle and is buoyant, but lacks the strength necessary for long journeys. This boat is entirely satisfactory for ferrying a group across a broad, quiet stretch of river. When such a boat has served its purpose, the cover should be removed for later use.

20.5.4. The Vegetation Raft

The vegetation raft is built of small vegetation which will float and is placed within clothing or parachute to form a raft for an IP and (or) equipment. Plants such as water hyacinth or cattail may be used (Figure 20-34).

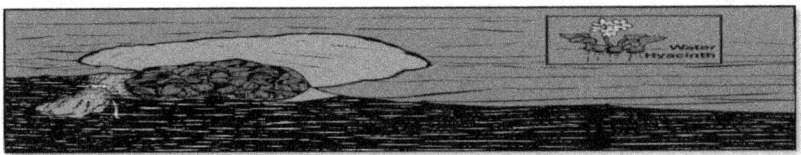

Figure 20-34 Vegetation Raft

20.5.5. Balsa Log Floating Device

A good floating device for the single IP can be fabricated by using two balsa logs or other lightweight wood. The logs should be placed about 2 feet apart and tied together. The IP sits on the lines and travels with the current (Figure 20-35).

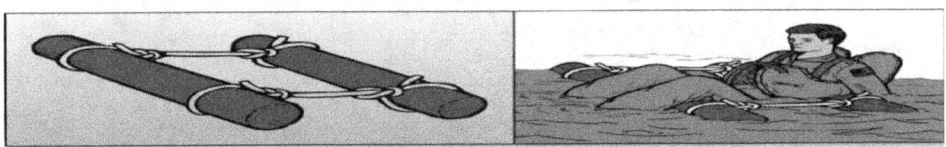

Figure 20-35 Log Flotation

20.5.6. Building a Raft

The greatest problem in raft construction (Figure 20-36) is being able to construct a craft strong enough to withstand the buffeting it may have to take from rocks and swift water. Rafts have been built by using notches, spikes, or lashing (Figure 20-37). Each has its advantages and disadvantages. Notches tend to be very durable, but take a great deal of time to construct. Spikes, even 6"-8" ones, tend to hold and are faster than notching, but may pull/twist out. Lashing is the fastest of all three methods, but may wear out due to constant chaffing and contact with obstacles.

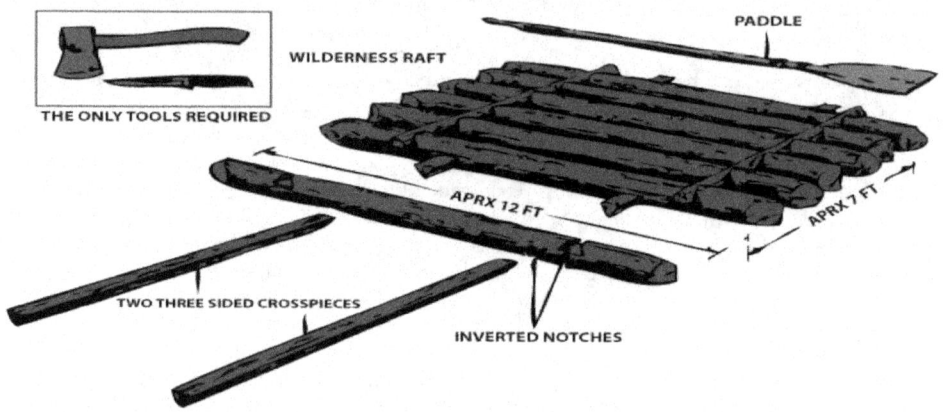

RAFT CONSTRUCTION

A raft for three persons should be about 12 feet long and 6 feet wide, depending on the size of the logs used. The logs should be 12 to 14 inches in diameter and so well matched in size that notches you make in them are level when crosspieces are driven into place.

Build the raft on two skid logs placed so that they slope downward to the bank. Smooth the logs with an ax so that the raft logs lie evenly on them. Cut two sets of slightly offset inverted notches, one in the top and bottom of both ends of each log. Make the notches broader at the base than at the outer edge of the log, as shown in the illustration. Use small poles with straight edges or a string pulled taut to make the notches. A three-sided wooden crosspiece about a foot longer than the total width of the raft is to be driven through each end of the four sets of notches.

Complete the notches on all logs at the top of the logs. Turn the logs over and drive a 3-sided crosspiece through both sets of notches on the underside of the raft. Then complete the top set of notches and drive through the two additional sets of crosspieces.

You can lash together the overhanging ends of the two crosspieces at each end of the raft to give it added strength; however, when the crosspieces are immersed in water they swell and tightly bind the raft logs together.

If the crosspieces fit too loosely, wedge them with thin, boardlike pieces of wood split from a dead log. When the raft is in water, the wood swells, and the crosspieces become very tight and strong.

Make a deck of light poles on top of the raft to keep packs and other gear dry.

Figure 20-36 Raft Construction

Figure 20-37 Spikes and Lashed Log Raft

20.6. Fording Streams

- IP traveling on foot through wilderness areas may have to ford some streams. These can range from small, ankle-deep brooks to large rivers. Rivers are often so swift an IP can hear boulders on the bottom being crashed together by the current. If these streams are of glacial origin, the IP should wait until early morning to ford due to thawing which causes increased water flow later in the day.

- Careful study is required to find a place to safely ford a stream; IP should use any high vantage point for the study. Finding a safe crossing area may be easy if the river breaks into a number of small channels. The area on the opposite bank should be surveyed to make sure travel will be easier after crossing. When selecting a fording site, the IP should:

- When possible, select a travel course which leads across the current at about a 45degree angle downstream. An example would be crossing the Euphrates in Iraq, IP should survey exit sites because the clay banks could be too steep and slick to cross.

- Never attempt to ford a stream directly above, or close to, a deep or rapid waterfall or a deep channel. The stream should be crossed where the opposite side is comprised of shallow banks or sandbars.

- Avoid rocky places, since a fall may cause serious injury. However, an occasional rock which breaks the current may be of some assistance. The depth of the water is not necessarily a deterrent. Deep water may run more slowly and be safer than shallow water.

- Before entering the water, IP should have a plan of action for making the crossing. Use all possible precautions, and if the stream appears treacherous, take the steps shown in Figure 20-38.

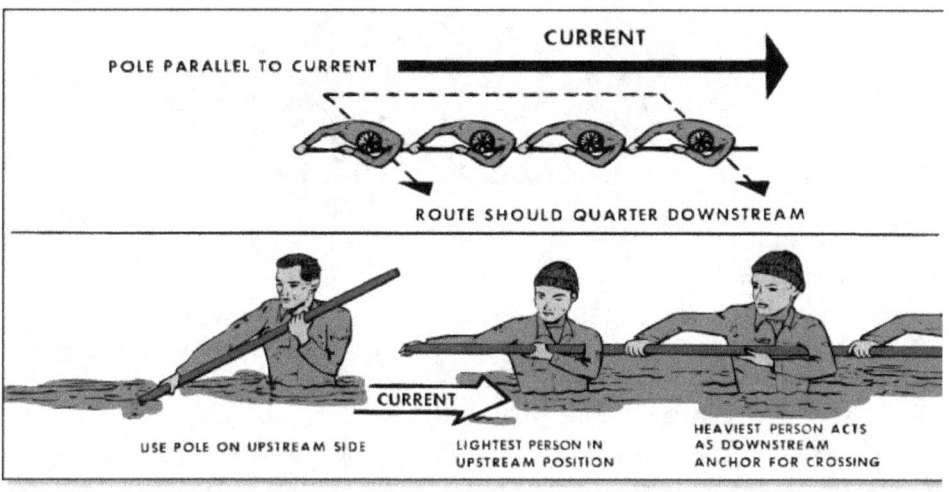

Figure 20-38 Fording a Treacherous Stream

Chapter 21

SIGNALING AND COMMUNICATION

21. Signaling and Communication

Most successful recoveries have resulted primarily because IP were able to assist in their own recovery. Many rescue efforts failed because IP lacked the knowledge and ability necessary to assist. When needed, this knowledge and ability could have made the difference between life or death, freedom or captivity (Figure 21-1).

Figure 21-1 Signaling and Recovery

21.1. Communications Equipment

IP need to know how to operate the communications equipment in the survival kit and when to put each item into use. IP should also be able to improvise signals to improve their chances of being sighted and to supplement the issued equipment.

21.2. Emergency Equipment

It is not easy to spot one isolated person, a group of IP, or even a crashed aircraft or ground vehicle from the air. Emergency signaling equipment is designed to assist in the location of IP by rescue forces. Emergency equipment may be used to provide rescue personnel with information about IP's condition, plans, position, or the availability of a rescue site (Figure 21-2).

Figure 21-2 Signaling

21.3. Visualizing Emergency Development

A part of pre-mission planning should be to visualize how emergencies might develop, recognize them when they occur, and, at the appropriate time, let friendly forces know about the problem. The length of time before IP are rescued often depends on effectively reporting the isolating event and providing an accurate location of the IP. Signal sites should be carefully selected to enhance the signal and have natural or manufactured materials readily available for immediate location identification when rescue forces arrive. IP should only use pyrotechnic signals when directed by or within visual range of rescue forces. Signals used correctly can hasten recovery and eliminate the possibility of a long, hard isolating event. IP should:

- Know how to safely use their emergency signals.
- Know when to use their emergency signals.
- Be able to use their signals on short notice.

21.4. Furnishing Information in Permissive Environments

The situation on the ground governs the type of information which IP can furnish the rescue team, and will govern the type of signaling they should use. In permissive environments, there are no limitations on the ways and means IP may use to furnish information.

21.5. Furnishing Information in Non-Permissive Environments

In non-permissive areas, limitations on the use of signals should be expected. The use of some signaling devices will pinpoint IP to the enemy as well as to friendly personnel. Remember the signal enhances the visibility of the IP.

21.5.1. Manufactured Signals and Communication Devices

Maintain care and use of all electronic devices. Operate devices in accordance with Special Instructions (SPINS) and directions. It is critical that potential IP are familiar with issued equipment and how to operate it, (i.e., how to operate the radio they hand carry compared to a radio packed in their accessory kit).

21.5.1.1.1. Radios

Radios are equipped for transmitting and receiving tone, voice, or text. Radio range can be line of sight and/or over the horizon. The ranges vary depending on the altitude of the receiving aircraft, terrain factors, forest density, weather, battery strength, type(s) of radio and interference. Some radios are equipped with GPS, beacons, distance measuring equipment (DME) capability, and multiple frequencies.

21.5.1.1.2. Beacons

Beacons are equipped for transmitting tone and data. The range of beacons can be line of sight and/or over the horizon. The range varies depending on the altitude of the receiving aircraft, terrain factors, forest density, weather, battery strength, type(s) of beacon and interference. Some beacons have multiple frequencies, GPS, and can transmit text.

21.5.1.1.3. Precautions Before Using Radios and Beacons

Before using radios and beacons, a few basic precautions should be observed. These will help obtain maximum performance.

- The best transmission range will be obtained when operating in clear, unobstructed terrain.
- Extending from the top and bottom of the radio antenna is an area referred to as the "cone of silence." To avoid the "cone of silence" problem, keep the radio/beacon antenna at a right angle to the path of the rescue aircraft.
- Never allow the antenna to ground out on clothing, body, foliage or the ground. This will severely decrease the effective range of the signals.
- To conserve battery power use in accordance with SPINS.
- If another radio or beacon is broadcasting, it may interfere with the radio or beacon transmissions.

21.5.1.2. Satellite Assisted Recovery Systems

These satellite systems consists of a constellation of satellites and ground receiving stations which process digital emergency messages and coordinate rescue activities.

21.5.1.2.1. COSPAS-SARSAT

The COSPAS-SARSAT system is an internationally supported search and rescue alert and information distribution system that provides global coverage. The COSPAS-SARSAT System is composed of:

- The IP's distress radio/beacons.
- Transponders on board satellites in geostationary and Low Earth Orbits (LEO) which detect the signals transmitted by distress radio beacons.
- Ground receiving stations, referred to as Local Users Terminals (LUTs), which receive and process the satellite downlink signal to generate distress alerts.
- Mission Control Centers (MCCs) which receive alerts produced by LUTs and forward them to Rescue Coordination Centers (RCCs), Joint Personnel Recovery Centers (JPRCs), and/or Personnel Recovery Coordination Cells (PRCCs).

- Search and Rescue Points of Contact (SPOCs) or other MCCs.

21.5.1.2.2. UHF SATCOM

A two-way, DOD controlled, geosynchronous satellite communications system providing worldwide coverage between 70° North latitude and 70° South latitude. The Combat Survivor/Evader Locator (CSEL) system uses this constellation of satellites to provide the primary means of communications with a radio directly to the JPRCs and/or PRCCs.

21.5.1.2.3. Low Probability of Exploitation (LPE) Transmissions

The system of satellites provides LPE of radio signals by hostile forces using a number of transmission techniques to significantly reduce user vulnerability to detection and interception. When LPE is available, a rescue center can send a message that automatically places the radio in a LPE only transmit mode. The LPE system uses a waveform and a variety of national assets to provide the primary communications from the IP's radio data directly to the JPRCs and/or PRCCs.

21.5.1.2.4. HOOK2® GPS CSAR System

Comprised of the software-defined, upgradeable AN/PRC-112G® transceiver or the AN/PRC-112B1 transceiver, plus a handheld GPS Quickdraw2® Interrogator, it delivers encrypted two-way messaging convenience plus GPS positioning data for precise, accurate location, and turns virtually any aircraft into a CSAR platform simply by plugging the Quickdraw2 into the intercom. The HOOK2 GPS CSAR System uses a variety of national assets that include, but are not limited to COSPAS-SARSAT, SATCOM, LPE too provide the primary communications from the IP's radio data directly to the JPRCs and/or PRCCs.

21.5.2. Pyrotechnics

IP may be required to use a variety of flares. They must know the types of flares in their survival kits, aircraft, and/or vehicle. IP should learn how to use each type of flare before they face an emergency. Flares are designed to be used during the day or night. Day flares produce a unique bright-colored smoke which stands out very clearly against most backgrounds. Night flares are extremely bright and may be seen for miles by air, ground, or naval recovery forces.

21.5.2.1. Hand-Launched Flares

Hand-launched flares fall in the pyrotechnic category. They were designed to overcome the problems of terrain masking and climatic conditions. For example, a person may be faced with multi-layer vegetation or atmospheric conditions known as an inversion which keeps the smoke next to the ground.

- Flares must be fired at the right time to be of maximum value. Smoke flares, for example, take a second or two after activation before they produce a full volume of smoke. Therefore, the flare should be ignited just before the time it can be seen by rescue personnel. Signal flares should not be used in tactical environments unless directed to do so.
- Because of the rapid changing technology in pyrotechnic signaling devices, potential IP should check regularly for new and improved models, making special note of the firing procedures and safety precautions necessary for their operations.

21.5.2.2. Pyrotechnic Devices Around Flammable Materials

Care should be used when operating any pyrotechnic device around flammable materials (grass, fuel, etc) or life rafts.

21.5.3. Sea Dye Marker:

- Of the many dyes and metallic powders tested at various times for marking the sea, the most successful is the fluorescent, water-soluble, orange powder. When released in the water, a highly visible, light green, fluorescent color is produced. Sea dye marker has rapid dispersion power; a packet spreads into a slick about 150 feet in diameter and lasts approximately for an hour in calm weather. Rough seas will stream it into a long streak, which may disperse rapidly (Figure 21-3).

Figure 21-3 Sea Dye Marker

- Under ideal weather conditions, the sea dye marker can be sighted at five miles with the aircraft operating at 1,000 feet. The dye has also been spotted at seven miles away from an aircraft operating at 2,000 feet.

- Sea dye marker should be used in friendly areas during daytime and only when there is a chance of being sighted (aircraft seen or heard in the immediate area). It is not as effective in heavy fog, solid overcast and storms with high winds and waves. The release tab on the packet of dye is pulled to open for use. In calm water, the dye can be dispersed more rapidly by stirring the water.

- If left open in the raft, the escaping powder penetrates clothing, stains hands, face, and hair, and eventually may contaminate food and water. After using the dye, it should be rewrapped or placed back into its plastic bag to conserve the remainder of the packet.

21.5.4. Paulin Signals

The paulin is a signaling device packed with some multi-place life raft accessory kits. The paulin is constructed of rubberized nylon material and is blue on one side and yellow on the other. These colors contrast against each other so when one side is folded over the other, the designs are easily distinguished. The size is seven feet by 11 feet which is a disadvantage when folded because it makes a small signal. The paulin has numerous uses such as a camouflage cloth, sunshade, tent, or sail, or it can be used to catch drinking water.

21.5.5. Space Blankets

A space blanket can be used in the same manner as the signal paulin because it is highly reflective (silver on one side and various colors on the other side). The space blanket has multiple uses beyond signaling (similar to the signal paulin).

21.5.6. Audio Signals

Sounds carry far over water under ideal conditions; however, they are easily distorted and deadened by the wind, rain, or snow. On land, heavy foliage cuts down on the distance sound will travel. Shouting and whistling signals have been effective at short ranges for summoning rescue forces. Most contacts using these means were made at less than 200 yards, although a few reports claim success at ranges of up to a mile. A weapon can be used to attract attention by firing shots in a series of three. The number of available rounds and the situation determine whether this is practical or advisable. IP have used a multitude of devices to produce sound. Some examples are: striking two poles together, striking one pole against a hollow tree or log, and improvising whistles out of wood, metal, and grass.

21.5.7. Light Signals

When tested away from other manufactured lights, aircraft lights have been seen up to 85 miles. At night, an IP should use any type of light to attract attention. A signal with a flashlight, vehicle light, or fire can be seen from a long distance. A flashing light or strobes light are also very effective.

21.5.8. Signal Mirror

It is the most valuable daytime means of visual signaling. A mirror flash has been visible up to 100 miles under ideal conditions, but its value is significantly decreased unless it is used correctly. It also works on overcast days. Practice is the key to effective use of the signal mirror. Whether the mirror is factory manufactured or improvised, aim it so the beam of light reflected from its surface is directed to hit rescue forces.

21.5.8.1. Aiming Manufactured Mirrors

Instructions are printed on the back of the mirror. IP should:

- Reflect sunlight from the mirror onto a nearby surface-raft, hand, etc.
- Slowly bring the mirror up to eye-level and look through the sighting hole where a bright spot of light will be seen. This is the aim indicator.
- Hold mirror near the eye and slowly turn and manipulate it so the bright spot of light is on the target.

21.5.8.2. Signal Mirror in a Hostile Environment

In a hostile environment, the exact location of the flash is extremely important. The signal mirror should be covered when not in use. IP should understand that even if the mirror flash is directly on the rescue forces that same flash may be visible to others (possibly the enemy) located at the proper angle in regard to the IP's position.

21.5.8.3. Signal Mirror in a Friendly Environment

In friendly areas, where rescue by friendly forces is anticipated, free use of the mirror is recommended. IP should continue to sweep the horizon even though no aircraft or ships are in sight (Figure 21-4).

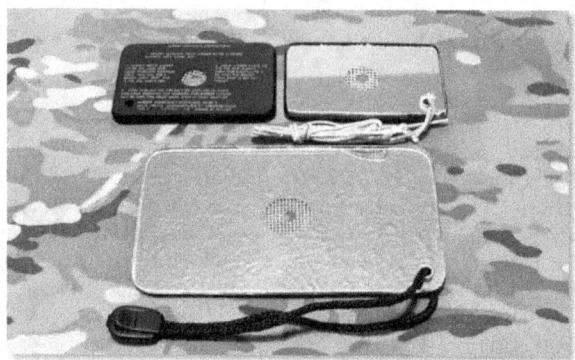

Figure 21-4 Signal Mirrors

21.5.9. Improvised Signal Mirrors

Improvised signal mirrors can be made from cans, parts from a vehicle, polished aluminum, glass, reflective ID card, and the foil from rations or cigarette packs. However, the mirror must be accurately aimed if the reflection of the Sun in the mirror is to be seen by the pilot of a passing aircraft or the crew of a ship.

21.5.9.1. Aiming Improvised Mirrors

The simple way to aim an improvised mirror is to place one hand out in front of the mirror at arm's length and form a "V" with two fingers. With the target in the "V" the mirror can be manipulated so that the majority of light reflected passes through the "V" (Figure 21-5). This method can be used with all mirrors. In evasion, some IP use the space between their hand and thumb for aiming; using their hand to hide the reflected light until they "sight" the "target" in the space between their hand and thumb.

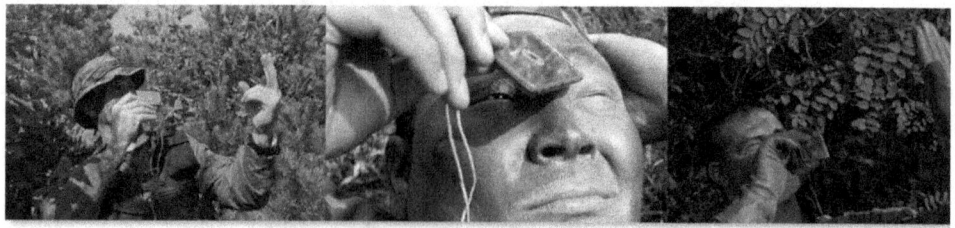

Figure 21-5 Aiming Signal Mirror

21.5.9.1.1. Aiming Stake

A second method is to use an aiming stake. Any object four to five feet high can serve as the point of reference. IP should hold the mirror so they can sight along its upper edge. Changing their position until the top of the stick and target line up, they should adjust the angle of the mirror until the beam of reflected light hits the top of the stick. If stick and target are then kept in the sighting line, the reflection will be visible to the rescue vehicle (Figure 21-6). A method used when the angle is greater than 90 degrees is to lie on the ground in a large clearing and aim the mirror using one of the methods previously discussed (Figure 21-7).

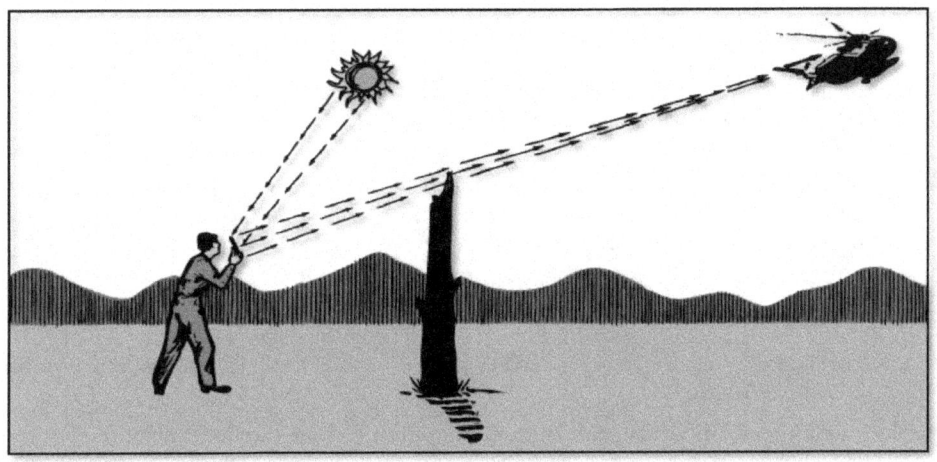

Figure 21-6 Aiming Signal Mirror – Stationary Object

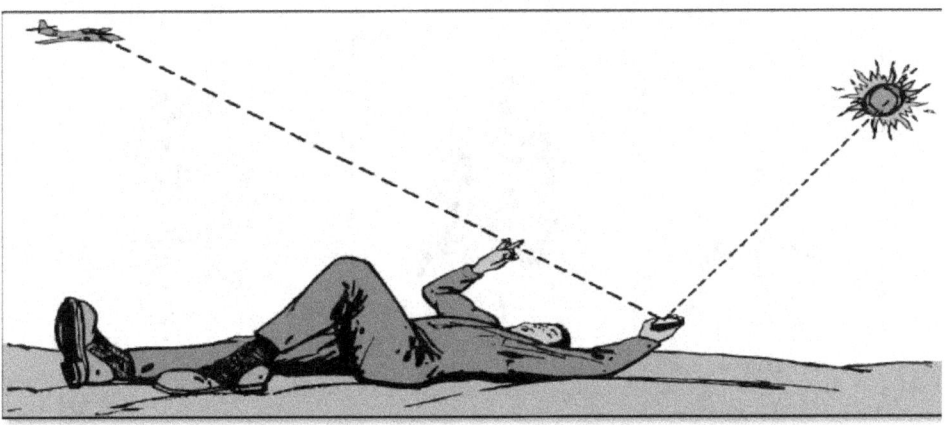

Figure 21-7 Aiming Signal Mirror – By Lying on the Ground

21.5.10. Fire and Smoke Signals

Fire and smoke can be used to attract the attention of recovery forces. Three evenly spaced fires, 100 feet apart, arranged in a triangle or in a straight line, serve as an international distress signal. One signal fire will usually work for an IP. During the night, the flames should be as bright as possible, and during the day, as much smoke as possible should be produced.

- Smoke signals are most effective on clear and calm days. They have been sighted from up to 50 miles away. Heavily wooded areas, high winds, rain, or snow tend to disperse the smoke and lessen the chances of it being seen.

- The smoke produced should contrast with its background. Against snow or a light background, dark smoke is most effective. Smoke can be darkened by burning rags soaked in oil, pieces of rubber, matting, insulation, padding from vehicle seats, or plastic being added to the fire. Against a dark background, white smoke is best. Green leaves, moss, ferns, or wet vegetation produces white smoke.

- Signal fires must be prepared before the recovery vehicle enters the area. The fires used by IP for heat and cooking may be used as a signal fire as long as the necessary materials are prepared and in the immediate vicinity. IP should supplement the fire to provide the desired signal.

21.5.10.1. Raised Platform Smoke Generator

To build a raised platform smoke generator (Figure 21-8), IP should:

- Build a raised platform above wet ground or snow, if lashing the limbs and platform together the IP should use wire.

- Place highly combustible materials in the center of the platform then build out; adding layers of flammable material protecting the center, so when the highly combustible material in the center is lit it quickly ignites.

- Then place smoke-producing materials over the platform and light when search aircraft is in the immediate vicinity.

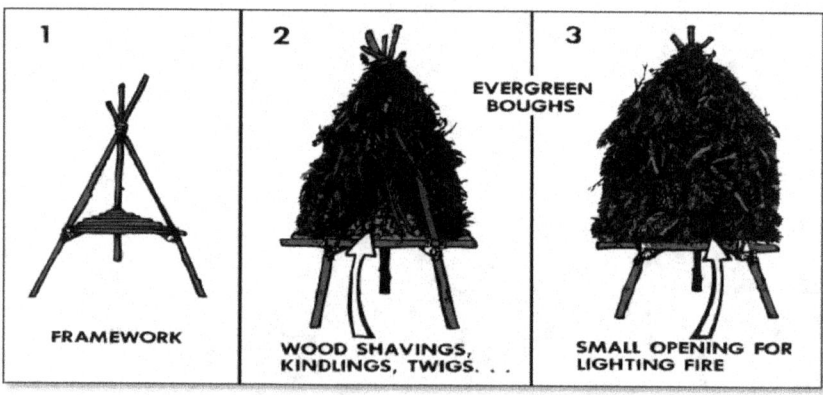

Figure 21-8 Smoke Generator – Platform

21.5.10.2. Ground Smoke Generator

To build a ground smoke generator (Figure 21-9), IP should:

- Construct a base platform using green logs, and then build a large log cabin fire configuration on top of this. This provides good ventilation and supports the green boughs used for producing smoke.
- Place smoke-producing materials over the fire lay; ignite when a search aircraft is in the immediate vicinity.

Figure 21-9 Smoke Generator – Ground

21.5.10.3. Tree Torch Smoke Generator

To build device tree torch smoke generator (Figure 21-10), IP should:

- Locate a tree in a clearing and clear area under tree to bare earth to prevent a forest fire hazard.
- Add additional smoke-producing materials; weaving them into the branches of the tree.
- Create a "nesting area" to add a large amount of tinder and other highly combustible material to be ignited.
- Ignite when a search aircraft is in the immediate vicinity.

Figure 21-10 Tree Torch

21.5.10.4. Fuel Smoke Generator

If IP are with the aircraft or vehicle, they can improvise a generator by burning fuels, lubricating oil, or a mixture of both. One to two inches of sand or fine gravel should be placed in the bottom of a container and saturated with fuel. Care should be used when lighting the fuel as an explosion may occur initially. If there is no container available, a hole can be dug in the ground, filled with sand or gravel, saturated with fuel, and ignited. Care should be taken to protect the hands and face.

21.5.11. Ground-to-Air Signals (GTAS)

The construction and use of pattern signals must take many factors into account. Size, ratio, angularity, contrast, location, and meaning are each important if the IP's signals are to be effective. The type of signal constructed will depend on the material available to an IP. Ingenuity plays an important role in the construction of the signal. IP should remember to judge their signals from the standpoint of aircrew members who are flying over their location searching for them.

- Size. The signal should be as large as possible, but IP should try to make the GTAS at least 3 feet wide and 18 feet long (Figure 21-11).

- Ratio. When creating specific symbols such as letters, arrows, or an "X" for a GTAS, the IP must remember to keep the ratio of the symbol in proper proportion. For example, if the baseline (bottom) of an "L" is 12 feet long, then the vertical line of the "L" must be longer (18 feet). In most cases, a two to three ratio is needed to maintain the GTAS symbol in proper proportion.

- Angularity. Straight lines and square corners are not found in nature, so they tend to attract the attention of aircrew flying over. IP should make all GTAS with straight lines and square corners.

- Contrast. The signal should stand out sharply against the background. Contrast can be improved by outlining the signal with green boughs, piling brush and rocks to produce shadows, or raising the panel on sticks to cast its own shadow. The idea is to make the signal look highlighted by the difference so it stands out (Figure 21-12). On snow, the fluorescent sea dye marker can be used to add contrast around the signal. In grass and scrubland, the grass should be stamped down or turned over to allow the signal to be easily seen from the air. A burned grass pattern is also effective. When in snow, a trampled out signal is very effective. IP should use only one path to and from the signal to avoid disrupting the signal pattern. Avoid using orange parachute material on a green or brown background as it has a tendency to blend in.
- Location. The signal should be located so it can be seen from all directions. IP should make sure the signal is located away from shadows and overhangs. A large high open area is preferable. It can serve a dual function- for signaling and for the rescue aircraft to land.Meaning. If possible, the signal should tell the rescue forces something pertaining to the situation. For example: "require medical assistance" or a coded symbol used during evasion, etc. Figure 21-13 shows accepted symbols.

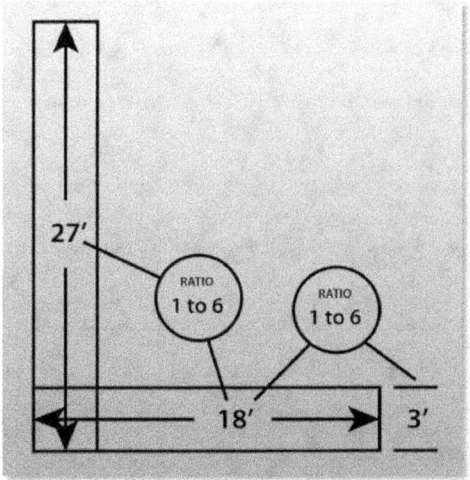

Figure 21-11 GTAS Sizes

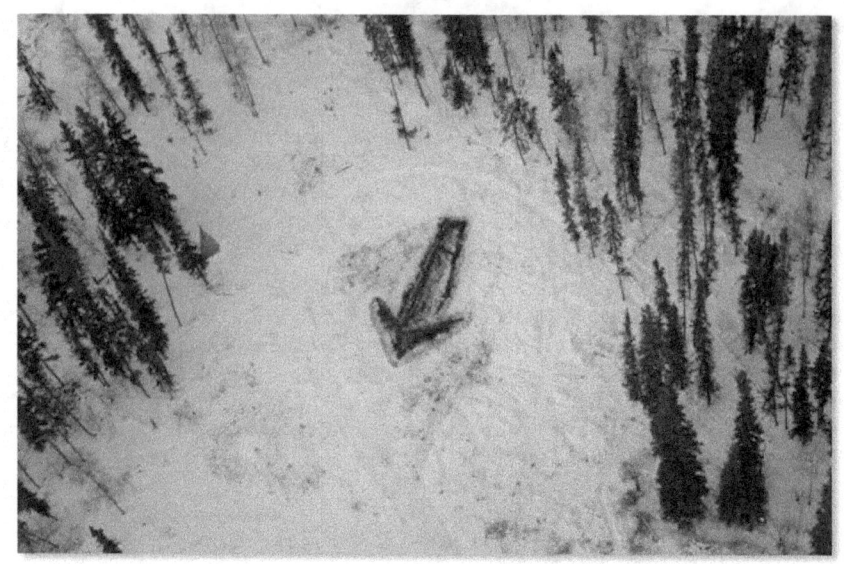

Figure 21-12 Contrast (Snow and Boughs)

NO.	MESSAGE	CODE SYMBOL
1	REQUIRE ASSISTANCE	V
2	REQUIRE MEDICAL ASSISTANCE	X
3	NO or NEGATIVE	N
4	YES or AFFIRMATIVE	Y
5	PROCEEDING IN THIS DIRECTION	↑

Figure 21-13 Signal Key

21.5.11.1. Parachute Signals

Parachute material can be used effectively to construct pattern signals. A rectangular section of parachute material can be formed as shown in Figure 21-14. When making a pattern signal, IP should ensure the edges are staked down so the wind will not blow the panels away.

- A parachute caught in a tree will also serve as a signal. IP should try to spread the material over the tree to provide the maximum amount of signal.

- When open areas are not available, IP should stretch the chute over low trees and brush or across small streams.

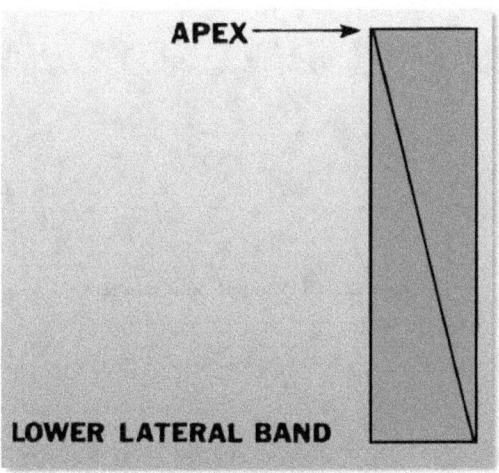

Figure 21-14 Parachute Strips

21.5.11.2. Shadow Signals

If other means are not available, construct mounds that will use the Sun to cast shadows. These mounds should be constructed in one of the international distress patterns. Brush, foliage, rocks, or snowblocks may be used to cast shadows. To be effective, these shadow signals must be oriented to the Sun to produce the best shadow. In areas close to the Equator, a north-south line gives a shadow at any time except noon. Areas farther north or south require the use of an east-west line or some point of the compass in between to give the best results.

21.5.11.3. Flags, Pennants, and Banners

Attach contrasting material to a pole, branch, the top of a tree (cleared of top layer of vegetation), or anywhere the material can be easily seen. By allowing the material to move with the wind you are attracting attention to the signal by the movement and contrast, as well as acting as a ground wind indicator to rescue forces (Figure 21-15).

Figure 21-15 Pennants and Banners

21.5.12. Aircraft Acknowledgements

Rescue personnel will normally inform IP they have been sighted by:

- Using their radio.
- Flying low with landing lights on (Figure 21-16) and (or) rocking the wings.

Figure 21-16 Standard Aircraft Acknowledgements

Chapter 22

RECOVERY

22. Recovery

The success of a rescue effort depends on many factors, such as the availability of rescue forces, the proximity of enemy forces, and current weather conditions. Above all, an IP's knowledge of what to do in the rescue effort may make the difference between success and failure (Figure 22-1).

Figure 22-1 Recovery

22.1. IP's Role

The role of an IP in effecting their rescue changes continuously as recovery vehicles and rescue equipment become more sophisticated. There are several independent organizations engaged in search and rescue (SAR) operations or influencing the SAR system. The organizations may be international, federal, state, county, local governmental, commercial, or private organizations. IP are responsible for being familiar with procedures used by international SAR systems in order to assist in rescue efforts. Some international organizations include:

- International Civil Aviation Organization (ICAO).
- Intergovernmental Maritime Consultative Organization (IMCO).
- Automated Mutual-Assistance Vessel Rescue (AMVER) System.

22.2. National Search and Rescue (SAR) Plan

The National SAR Plan is implemented the instant an aircraft is known to be down. There are three primary SAR regions; they are the Inland Region, the Maritime Region, and the Overseas Region.

- United States Air Force: Recognized SAR Coordinator for the United States aeronautical SRR corresponding to the continental United States other than Alaska.

- United States Pacific Command: Recognized SAR Coordinator for the United States aeronautical SRR corresponding to Alaska.
- United States Coast Guard: Recognized SAR Coordinator for all other United States aeronautical and maritime SRRs. This includes the State of Hawaii as well as waters over which the United States has jurisdiction, such as navigable waters of the United States.

22.2.1. The National SAR Supplement

The *U.S. National SAR Supplement* (NSS) plans and addendums provide a long-range rescue plan which personnel should study for additional information.

22.3. IP Responsibilities

The IP have various responsibilities pre-mission and during the PR cycle.

22.3.1. Pre-Mission Preparation

Pre-mission preparation is a vital element in the PR cycle. It entails activities that at risk personnel accomplish in order to prepare for an isolating event to include their role in the recovery. It may include self-study, intelligence briefs, SERE training, country studies, etc. The potential IP has recovery responsibilities to themselves, to the rescue forces which may recover them, and to the USAF. These responsibilities include, but are not limited to: correctly filling out isolated personnel report (ISOPREP), knowing how to maintain and operate signaling/communication equipment, and preparing themselves with a working knowledge of rescue forces and likely recovery devices.

22.3.2. Reporting IP Event

Once the emergency occurs, IP are responsible for an immediate radio message or signal which identifies the isolating event. This is the "report" task in execution phase of personnel recovery. The 'Save Chain' is an illustration of the basic PR tasks that take place immediately following an isolating event (Figure 22-2). If possible, the message should include position, course, altitude, and actions planned. This information is essential for initiating efficient recovery operations.

Figure 22-2 Save Chain

22.3.3. Providing Recovery Forces with Required Information

Once recovery operations have been initiated, IP have a continuing responsibility to work with recovery forces providing information and signaling as required or directed. If a group of IP should become separated, each group member should, when contacted by rescue forces, provide as much information about themselves and the entire group as possible. IP should follow all instructions from rescue forces. If instructions are not followed, IP could be responsible for causing their own death and/or the death of rescue personnel.

22.4. Recovery Site

- Consideration must be given to a recovery site. Other major considerations of IP are the type of recovery vehicle and the effects of the weather and terrain on it.
- IP should try to pick high, clear terrain in the immediate area for pickup. When locating this recovery site, IP should watch for obstacles such as trees, cliffs, overhangs, or sides of steep slopes etc., which could limit the aircraft's ability to maneuver.
- IP should try to pick terrain that is safe and accessible to expected ground rescue vehicles in the immediate area.
- Even though IP should select a recovery site, it is ultimately the responsibility of rescue personnel to decide if the selected site is suitable.

22.5. Recovery Procedures

Since procedures involving recovery vary with changes in equipment and rescue capability, IP must always know the current procedures and techniques. This is particularly true of the procedures used for wartime recovery, which are covered in the Evasion Chapter 23 of this handbook. Specifics on combat recovery are found in an AOR's Combat Search and Rescue (CSAR) Special Instructions (SPINS).

22.5.1. Supplies for Delayed Recoveries

If a delay in recovery is expected based on location of the isolating event, distance from rescue forces, or environmental conditions, supplies can be dropped to IP to help sustain and protect them while they await rescue.

22.5.2. Rescue by Helicopter

Helicopters make rescues by landing or hoisting. Helicopter landings are made for all rescues when a suitable landing site is available, and danger from enemy forces is not an issue. Hovering the helicopters and hoisting IP aboard requires more helicopter power than landing and presents a hazard to both the aircraft and the IP. Altitude and temperature can further affect the helicopter's power and ability to hover. Rotor wash from helicopters and vertical lift aircraft can move equipment and debris so IP should stow equipment such as parachutes, rafts, or other survival equipment, to avoid danger to the recovery vehicle. Other debris affected by rotor wash such as vegetation or water may potentially blind IP to lowered rescue devices or instructions indicated by rescue forces.

22.5.2.1. Rescue by Landing Helicopter

After landing, crewmembers will usually depart the aircraft to make contact with the IP. If for some reason this cannot be done, IP should approach the helicopter from either the three o'clock or nine o'clock position relative to the nose of the helicopter, or as directed by aircrew (Figure 22-3).

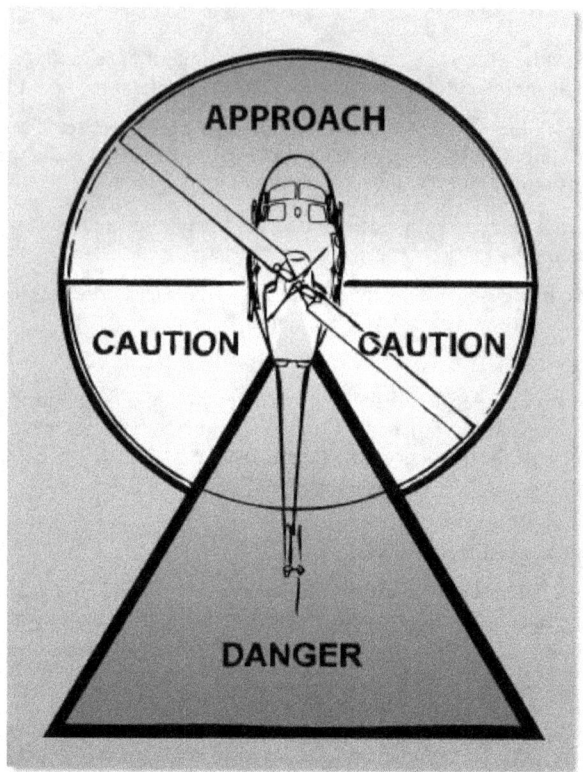

Figure 22-3 Approaching Helicopter

22.5.2.2. Pickup Devices

When rescue forces are in the immediate area of IP, they will, if conditions permit, deploy personnel to assist IP. Unfortunately, conditions may not always permit this, so IP should know how to use different types of pickup devices.

22.5.2.2.1. Common Factors for All Pickup Devices

Some common factors concerning all pickup devices include:

- Helicopter recovery devices should be allowed to ground to discharge static electricity before donning.

- To ensure stability, IP should sit or kneel when donning a pickup device. Do not straddle the device.

- IP must be careful to not get their fingers between the hoist's hook and any part of the pickup device. Also care must be taken to not wrap any access hoist cable around the IP's body or appendages.

- If no audio is available, IP should visually signal the hoist operator when ready for lift-off with a "thumbs up" or vigorously shake the cable from side to side.
- Most devices can be used as a sling by creating or finding a fixed loop.
- IP must remember to follow all instructions provided by the recovery crew. When lifted to the door of the helicopter, IP should not attempt to grab the door or assist the hoist operator in any way. They must not try to get out of the pickup device. The crew will remove the device after the IP is well inside the aircraft.

22.5.2.2.2. Specific Pickup Devices

- Rescue Sling. Before donning the rescue sling, IP should face the drop cable and make sure the cable has touched the water or ground and lost its charge of static electricity.
    - On land, the most commonly accepted method for lowering the rescue sling is as a fixed loop. Grasp the sling with both hands lifting it over the head and bring it down under the arms and around the body. Regardless of the method used, the IP should remember the webbing and metal hardware of the device should be directly in front of the face preventing IP from slipping out of the device while being hoisted. The webbing under the metal ring can be held until tension is put on the cable. The IP's hands may then be interlocked and rested on the chest. This tends to lock the IP into the sling as upward pressure is applied (Figure 22-4).

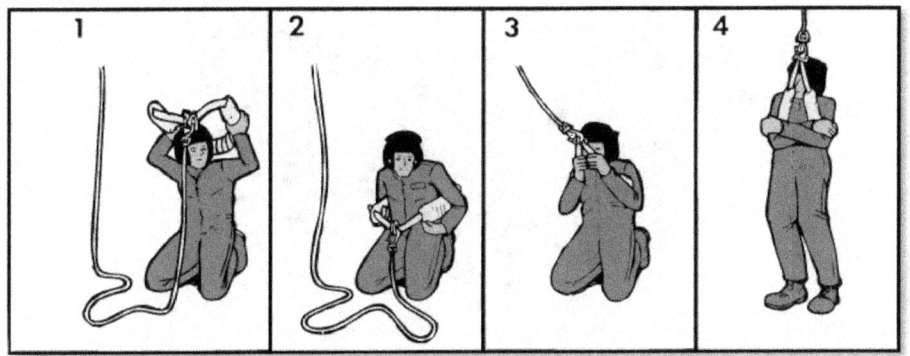

Figure 22-4 Rescue Sling

- In water, the most commonly accepted method for lowering the rescue sling is with only one end attached to the rescue hook (not in a fixed loop). Grasp the free end of the sling, pass it hand to hand behind the back, and then clip free end of the sling into the rescue hook forming a fixed loop. Again IP should remember the webbing and metal hardware of the device should be directly in front of the face preventing the IP from slipping out of the device while being hoisted. The webbing under the metal ring can be held until tension is put on the cable. The IP's hands may then be interlocked and rested on the chest. This tends to lock the IP into the sling as upward pressure is applied.

- Rescue Basket. There are two types of rescue baskets: a basket (which is similar to a shopping cart) and a liter.
    - The rescue basket (basket-style) is commonly used by the US Coast Guard and US Navy. Some rescue forces will lower it a distance away, allowing it to ground out, and then drag it through the water to the IP. Others will drop the device next to the IP or directly into a multi-person liferaft. Either way, the IP should allow the device to ground out and be brought to them. Once it reaches the IP's location, the IP should climb in; sit in the center of the basket similar to sitting in a chair, wrap arms around legs, and lock hands together taking care to avoid grabbing the edges or the floor of the basket which can lead to injury (Figure 22-5).
    - If rescue basket (litter-style) is used, it will be accompanied to the water or ground by a member of the helicopter crew. The crewmember will assist IP into the litter.

Figure 22-5 Basket

- Forest Penetrator
    - The forest penetrator rescue seat is designed to make its way through interlacing tree branches and dense jungle growth. It can also be used in open terrain or over water. The device is equipped with three seats which are spring-loaded in a folded position against the body or main shaft and must be pulled down to the locked position for use. On the main shaft of the tube, above the seats, there is a zippered fabric storage pouch for the safety (body) straps which are stowed when lowered to the IP for a land pickup. If the forest penetration is used for water pickup, it will be equipped with the flotation collar which enables the device to float with the upper one-third (approximately) of the device protruding above the water. Additionally, for water pickups, one strap will

usually be removed from the stowed position with one seat locked in the down position to assist the IP in using the penetrator.

- The safety strap is pulled from the storage pouch and placed around the body to hold the person on the penetrator seat. The strap should not be unhooked unless there is no other way to fasten it around the body such in the case of an IP wearing an LPU. The IP must make certain the safety strap does not become fouled in the hoist cable. After the strap is in place, the seat should be pulled down sharply to engage the hook which holds it in the extended position. The IP can then place the seat between the legs. Then the IP should pull the safety strap as tight as possible ensuring the device fits snugly against the body. The IP must always keep the arms down, elbows locked against the body, and not attempt to grab the cable or weighted snap link above the device. After making certain the body is not entangled in the hoist cable, the signal to be lifted can be given (Figure 22-6).

Figure 22-6 Forest Penetrator.

- In a combat area, under fire, IP may be lifted out of the area with the cable suspended before being brought into the helicopter. It is important to be correctly and securely positioned on the pickup device. The seat should always be held tightly against the crotch to prevent injury when slack in the cable is taken up. The hands should be kept below and away from the swivel on the cable with the arms around the body of the penetrator. IP should keep their head close to the body of the penetrator so that tree branches or other obstructions will not come between the body and the hoist cable.

- When IP reach a position level with the helicopter door, the hoist operator will turn them so they face away from the helicopter and then pull them inside. The crewmember will disconnect the IP from the penetrator once the device is safely inside the helicopter.

- The forest penetrator is designed to lift as many as three persons. When two or three IP are picked up, heads should be kept tucked in on opposite sides. The penetrator can be used to lower a paramedic or crewmember to assist injured personnel, and both (IP and paramedic) can be hoisted to the helicopter. If the forest penetrator seat blades have been lowered in a tree area, and if for any reason the pickup cannot be made, the blades should be returned to the folded position to prevent possible hang-up on tree limbs or other objects while the device is being retracted.

- With all types of devices, it is necessary to watch the device as it is lowered. The devices weigh about 23 pounds. If the device were to hit an IP, it could cause a serious injury or death. The weight and design also dictates that the IP maintain control of the device once it has made contact with the ground or water.

• Rope Ladder. This device is used primarily by the Army and special ground forces. If this device is used, it should be approached from the side and not the front. The IP should climb up a few rungs, sit down on a rung, and intertwine the body with rungs (Figure 22-7). The IP should not try to climb up the ladder and into the helicopter.

Figure 22-7 Rope Ladder

22.5.3. Rescue by Fixed-Wing Aircraft on Land

The most significant role played by fixed-wing aircraft in rescue operations is providing immediate assistance to IP and serving as the "eyes" of approaching rescue units.

22.5.3.1. Fixed-Wing Performing Rescue

The role of fixed-wing aircraft in actually performing a rescue is limited to instances where there is a suitable runway or surface near the IP where the aircraft is designed to operate. Fixed-wing aircraft rescues have often been made in extremely cold climates where the aircraft have either used frozen lakes or rivers as runways or, when fitted with skis, have operated from snow-covered surfaces and glaciers. However, landing in unknown terrain under what appears to be ideal conditions is extremely hazardous.

22.5.4. Rescue by Ship

When IP are a considerable distance from shore, rescue will normally be performed by ship. The rescue methods used by these ships vary considerably according to their displacement and whether the rescue is made in mid-ocean or close to land. Weather, tides, currents, sea conditions, shallow water, reefs, daylight, or darkness may be important factors.

22.5.4.1. Pickup of IP by Ship

Removal of IP from water, liferafts, lifeboats, or other vessels to the safety of the rescue vessel deck may be the most difficult phase of a maritime search and rescue mission. In most cases, IP will have to be assisted aboard. The most commonly used methods are to recover personnel directly from the water or directly from the distressed vessel/liferaft.

22.5.4.2. Methods for Rescuing IP from Water

When rescuing people from the water, the following methods are generally used:

- Ship arrives and deploys a rescue swimmer to assist IP.
- Ship arrives and deploys a line to the IP. IP uses a fixed loop around themselves and are pulled to vessel by vessel crew.
- Ship arrives and launches small boat to recover IP and return them to vessel.
- Ship arrives, deploys a long trail line and circles IP. Once IP has line, they use a fixed loop around themselves and are pulled to vessel by vessel crew.
- The most commonly used methods for rescuing personnel aboard distressed vessels/liferafts include:
    - Ship arrives and drops a ladder or hoist directly to IP in distressed vessel or liferaft. When a ladder is used the IP should grab the ladder at the high point of any wave and be prepared to climb.
    - Ship arrives and deploys a small boat to recover IP from distressed vessel/liferaft and return to larger vessel.
    - Ship arrives and deploys line to distressed vessel/liferaft. IP ties line to distressed vessel/liferaft to be hauled in.
    - Rescue boats are usually small and may not be able to take all IP on board at one time;

therefore, a sufficient number of boats to offset the rescue may be dispatched to the distress scene.

22.5.5. Boat and Helicopter Rescues

Occasionally, boats and helicopters will be dispatched for a rescue operation. Generally the first rescue unit to arrive in the vicinity of the IP will attempt the first rescue. If the boat arrives first and makes the rescue, it may transfer the IP to the helicopter to affect a rapid delivery to medical facilities.

22.6. Preparations for Open Seas Recovery

If helicopter recovery is unassisted, IP will be expected to do the following before pickup:

- Secure all loose equipment in raft, accessory bag, or in pockets. IP must clear any lines/lanyards or other gear which could cause entanglement during recovery. After all items are secure, IP should put on their helmet (if available).
- Canopies and sails should be taken down to ensure a safer pickup. Deploy sea anchor, ballast bags, and accessory kit(s).
- Partially deflate raft and fill with water.
- With a one man liferaft unsnap survival kit container from parachute harness.
- Grasp raft handhold and roll out of raft when directed.
- Allow pickup device and (or) cable to ground out on water surface.
- Maintain handhold with liferaft until pickup device is in the other hand.
- Mount recovery device (avoid raft lanyard entanglement).
- Signal hoist operator for pickup.

22.7. Reintegration

The fifth task of the five critical tasks of recovery is reintegration. For a successful reintegration the IP must understand general reintegration objectives and processes, as well as their own responsibilities.

22.7.1. Reintegration Process

Reintegration is a critical task that allows the DOD to gather necessary intelligence and SERE information while coordinating multiple activities and protecting the health and well-being of returned IP. IP are returned by escape, release, rescue, or evasion. Returned IP include US military personnel, DOD civilian employees, DOD contractors or other persons of interest as determined by the POTUS or Secretary of Defense. Reintegration team chiefs, combat rescue officers (CROs), SERE specialists, intelligence debriefers, SERE psychologists, and others who assist the recovered isolated personnel decompress and reintegrate to their unit, family, and society are key to accomplishing reintegration. Reintegration normally consists of three distinct phases. The number of phases a returned IP goes through is based on the isolating event, the medical attention they need, and the decision process of the COCOM Commander (Figure 22-8).

Figure 22-8 Reintegration

22.7.1.1. Reintegration Phases Are:

- Phase I, Release and/or Recovery, Reception, and Returnee Assessment.
- Phase II, Theater reintegration, usually conducted at a predetermined transition location.
- Phase III, Air Force reintegration. In-depth medical support, decompression, and debrief.

22.7.1.2. Gathering Intel, SERE Lessons Learned, and IP Health

All three phases focus on gathering intelligence, collecting operational and SERE lessons learned, and ensuring the physical and mental health of the returnee. When more than one person is returned, all returnees should be moved together to the theater transition point. Experience indicates that returnees benefit greatly from the opportunity to achieve closure with one another. This improves their ability to overcome the isolation experience and reintegrate with their unit and family.

22.7.2. IP Reintegration Responsibilities

IP reintegration is the final piece of the IP's mission to "return to friendly control without giving aid or comfort to the enemy, and to return early and in good physical and mental condition". It is the hot-wash and conclusion to their isolating event. The returned IP is the critical member of the reintegration process. The IP's responsibilities to the reintegration process include:

- Understanding of the importance of the information they provide. It may save lives, provide time critical intelligence, and lead to successful execution of future missions.
- Honesty - when telling their story, giving the best detail and insight they can provide.
- Communicate openly with the reintegration team. Let them know problems, concerns, and any possible barriers to communication.

Chapter 23

EVASION

23. Evasion

IP who encounter traumatic circumstances must be prepared to survive and evade the enemy to return to friendly control. An active commitment to solving problems and to individual survival (the "will to survive") is essential. IP must be prepared to exert extreme effort, both mental and physical, to successfully evade capture.

23.1. Convert Survival

This section addresses covert survival; that is, evasion. Areas to be covered include:

- Evader's moral obligations and legal status.
- Principles and techniques of evasion, camouflage, and travel (assisted or unassisted).
- The special aspects of food and water procurement.
- Combat signaling and recovery.

23.2. Definitions

- Evader: An "evader" is "any person isolated in hostile or unfriendly territory who eludes capture" (JP 1-02).
- Evasion: "Evasion" is "the process whereby isolated personnel avoid capture with the goal of successfully returning to areas under friendly control" (JP 1-02).

23.3. Five Phases of Evasion

While each isolating event is as unique as the IP who experiences it, when an IP is under evasion conditions, it is likely they will find themselves coping by using the five phases of evasion. The Air Land Sea Application (ALSA) Center, in accordance with the memorandum of agreement between the Headquarters of the Army, Navy, Air Force, and Marine Corps doctrine commanders has developed a publication to meet the immediate needs of the IP. This multi-Service TTP publication (FM 3-50.3/NTTP 3-50.3/AFTTP (I) 3-2.26) for *Survival, Evasion, and Recovery,* has prepared a checklist for the FIVE PHASES OF EVASION. A potential IP can refer back to this checklist taken from the multi-service publication and the support material within the handbook:

1. Immediate Actions – REMAIN CALM. THINK BEFORE YOU ACT!
    - Assess immediate situation.
    - Assess medical condition; treat as necessary.
    - Protect from chemical, biological, radiological, and nuclear hazards.
    - Gather equipment; determine the need to travel.
    - Establish initial hide site.
    - Make initial radio contact in accordance with communication plan/special instructions (SPINS).

- Sanitize self of compromising information.
- Sanitize area; hide equipment you decide to leave.
- Retain personal protection equipment (i.e., body armor).
- Apply initial personal camouflage.

2. Initial Movement
   - Move in the direction of your evasion plan of action (EPA)/contingency plan.
   - Break line of sight from your initial isolation area and move uphill if possible.
   - Move out of area (chapter II). Zigzag pattern recommended.
   - Use terrain and concealment to your advantage.
   - Move to hide site.

3. Hide Site
   - Select hide site that provides:
     - Concealment from ground and air searches.
     - Safe distance from enemy/high traffic areas/natural lines of drift.
     - Listening and observation points.
     - Multiple avenues of escape.
     - Protection from the environment.
     - Communications/signaling.
   - Be prepared to authenticate in accordance with communications plan/SPINS.
   - Establish radio contact in accordance with communication plan/SPINS.
     - Note: Communications/signaling devices may compromise position.
   - Drink water, treat injuries.
   - Reevaluate the tactical situation.
   - Inventory equipment.
   - Review and execute your EPA/contingency plan.
   - Determine location.
   - Improve camouflage.
   - Stay alert, maintain security, and be flexible (combat mindset).

4. Evasion Movement (Chapters I and II)
   - Travel slowly and deliberately. Intermittently stop, look, listen, and smell.
   - Do not leave evidence of travel.

- Maintain noise and light discipline.
- Stay away from heavily trafficked areas/natural lines of drift.
- Move from one point of concealment to another point of concealment.
- Use evasion movement techniques.
  - Note: You are more at risk of detection during movement.
5. Recovery (Chapters III and IV)
   - Select suitable area for recovery.
   - Prepare for use of communications and signaling devices.
   - Prepare to transmit position (distance and direction).
   - Select site(s) in accordance with criteria in theater recovery plans.
   - Observe/report enemy activity and hazards.
   - Secure equipment.
   - Stay concealed until recovery is imminent.
   - Be prepared to authenticate via isolated personnel report (ISOPREP).
   - During recovery:
     - Follow recovery force instructions.
     - Secure weapon.
     - Assume non-threatening posture.
     - Beware of rotors/propellers from aircraft.

23.4. Value of Evasion

If the opportunity to evade exists, IP must be motivated to take advantage of it. The motivation to make a total effort to adhere to every evasion principle, 24 hours a day, may be personal or gained through training. It may be motivated by a desire to return to a family or loved ones. This strong central drive will give IP the necessary push to make these efforts. Motivation may involve one or all of the following reasons applicable to all military personnel:

- To return and fight again.
- To deny the enemy a source of military information.
- To deny the enemy a source of propaganda.
- To deny the enemy a source of forced labor.
- To tie up enemy forces, transportation, and communications that otherwise might be committed to the war effort.
- To return with intelligence information.

23.4.1. Personal Motivation

Additionally, it is suggested that other personal reasons for being motivated to evade include: fear of death, pain, suffering, humiliation, degradation, disease, illness, torture, uncertainty, and fear of the unknown.

23.4.2. Value

Even if evasion is unsuccessful and the evader is captured, every hour spent eluding the enemy ties up enemy forces and lessens the evader's intelligence value: "Americans, indeed all free men, remember that in the final choice a soldier's pack is not so heavy a burden as a prisoner's chains."

23.4.3. IP Obligations

In addition to the above reasons for evading capture, IP also have moral and legal obligations to fulfill.

23.4.3.1. Moral Obligation

Moral obligation is implied throughout the Articles of the Code of Conduct, specifically Article II, which states: "I will never surrender of my own free will. If in command, I will never surrender the members of my command while they still have the means to resist." This Article of the Code should guide an evader's behavior during evasion just as it does in any other combat situation. Surrender is always dishonorable and never allowed. When there is no chance for meaningful resistance, evasion is impossible, and further fighting would lead to death with no significant loss to the enemy, members of Armed Forces should view themselves as "captured" against their will versus a circumstance that is seen as voluntarily "surrendering." They must remember that the capture was dictated by the futility of the situation and overwhelming enemy strengths. In this case, capture is not dishonorable.

23.4.3.2. Legal

The UCMJ continues to apply to evader's conduct during evasion or captivity. Particularly applicable are Article 99, Misbehavior before the Enemy, Article 104, Aiding the Enemy, and Article 92, Failure to Obey a Lawful Order. Thus, one can be tried for misconduct as a combatant or as a noncombatant. The Geneva Conventions characterize lawful combatants and noncombatants:

- "Lawful Combatants. A lawful combatant is an individual authorized by governmental authority or the Laws of Armed Conflict (LOAC) to engage in hostilities. A lawful combatant may be a member of a regular armed force or an irregular force. In either case, the lawful combatant must be commanded by a person responsible for subordinates; have fixed distinctive emblems recognizable at a distance, such as uniforms; carry arms openly; and conduct his or her combat operations according to the LOAC. The LOAC applies to lawful combatants who engage in the hostilities of armed conflict and provides combatant immunity for their lawful warlike acts during conflict, except for LOAC violations.

- Noncombatants. These individuals are not authorized by governmental authority or the LOAC to engage in hostilities. In fact, they do not engage in hostilities. This category includes combatants who are out of combat, such as POWs, the wounded, sick, shipwrecked, medical personnel and chaplains. Noncombatants may not be made the object of direct attack.

23.4.3.2.1. Laws of War

Various countries around the world have developed written and unwritten laws of war. Four

Geneva treaties were entered into by the United States and 60 other countries in 1949 and these treaties, as since amended, are in AFP 110-20. AFP 110-31 provides additional guidance for the conduct of armed conflict and air operations.

23.4.3.3. Combatant Status

An evader is a combatant and retains this status as a fighting person under arms according to international law until captured. Evaders are considered instruments of their government, under orders to evade capture, and never to surrender of their own free will. The evader is still militarily-effective and may take such steps as necessary, within the rules of warfare, to accomplish the mission, which includes returning after striking an enemy target. While in combatant status, the evader may continue to strike legitimate military targets and enemy troops without being held liable to prosecution after capture for violation of the local criminal law. To do so while evading capture is the legal function of a combatant.

23.4.3.4. Prisoner of War (POW)

During wartime, once captured, an evader becomes a prisoner of war (POW). International law provides certain rights to a POW and requires issuance of an identity card showing name, rank, serial number, and date of birth. When questioned, a POW is bound only to provide this information. An individual is a POW until: he or she rejoins the armed forces of his or her country or the armed forces of a friendly power (including ships); or he or she has left the territory under the control of the Detaining Power or of an ally. Once this occurs, combatant status is regained and no punishment may be imposed by the Detaining Power in the event of subsequent recapture. A POW who attempts to escape and is recaptured before rejoining his or her armed forces is liable under the Geneva Convention only for disciplinary punishment in respect of this act, even if it is a repeated offense. A POW who kills or wounds enemy personnel of the Detaining Power in an attempt to escape or evade may be tried and punished for the offense. However, special surveillance may be imposed upon a POW who fails in an escape attempt. Special surveillance must be undergone in the POW camp and cannot adversely affect the POW's health or his or her rights under the Geneva Convention.

23.4.3.4.1. Guidance

In other operations besides war, specific guidance regarding peacetime governmental detention, hostage detention, evasion, and combatant status will be provided.

23.4.3.5. Disguise

Most often, the use of disguise raises issues of retention of legal status during combat operations. The law of armed conflict makes a distinction between IP evading initial capture, and escaping after capture. While there are certain restrictions, there is also some latitude to what is considered acceptable conduct. This is particularly true in the case of disguise. During an escape or an evasion, the adoption of varying degrees of disguise may be logical, appropriate, and legal. For instance, if the population density is such that movement in uniform is not possible, IP may be required to adopt some sort of disguise to transit the area. Likewise, if contact with an indigenous assistance group has been established, isolated personnel may disguise themselves to facilitate movement. In this instance, the judgment of the assistance group regarding adopting a disguise should be respected. However, IP need to understand that in the event of capture, they will likely be treated exactly like members of the assistance group, unless they can convince their captors that they are lawful combatants. If the disguise is essentially civilian clothing, the individual should retain at least some of their uniform or personal identification (e.g., ID tags, US Armed Forces/Geneva Conventions ID card, blood chit) to use as proof of their status in the event of capture. Captured US military personnel (other than escaping POWs) wearing civilian apparel without a fixed distinctive sign and without visible weapons may be considered spies by the captors. However, unless they commit an independent law of war violation, history indicates their actions will not be regarded illegal.

23.4.3.6. Assisted Verses Unassisted Evasion

Evasion can be classified as either assisted or unassisted. Assistance can be defined as any help which is offered to an IP by any person. This help may include food, clothing, medicine, shelter, money, and even such a small item as a shoelace. Evaders should, in fact, consider they have been assisted even if, while evading through hostile territory, their presence has been ignored by indigenous personnel. The term "unassisted evasion" relates to situations where the IP, as an evader, must rely solely on their own knowledge and abilities to successfully emerge from an enemy-held or hostile area and return to areas under friendly control. The process of emerging may include aerial recovery, water recovery, or assisted evasion, but it is primarily an unassisted individual effort. Certain principles apply to all evasion situations, certain procedures should be followed, and certain techniques have widespread application. The beginning of an evasion experience is often the most critical phase. This is particularly true for a person who bails out over enemy territory during daylight hours and in sight of enemy personnel. The IP can count on the enemy to make a determined effort to capture them if seen. Evading the enemy is also a matter of effort and luck. Luck plays its part initially in establishing where an isolating event starts in relation to the location of any enemy personnel. An IP who lands in a heavily populated area (city, military installation, or combat area) may be taken prisoner immediately. The problem becomes one of early escape rather than one of evasion. For example, on February 2, 1991, during the first Iraq war, Capt. R. Dale Storr was captured by Iraqi soldiers after his A-10 Thunderbolt was shot down near Kuwait. Capt Storr ejected from his A-10 almost immediately over the military target he had been attacking. "The next thing I see is this truck coming from the truck park that I had just strafed coming after me. I landed and started to hide behind this sand dune. The Iraqis were there within a minute." The 29-year-old Air Force pilot from Spokane was a prisoner of war for 33 days, spending a portion of that time in Baghdad's Abu Ghraib prison, while his friends and family believed he had died in the plane crash. He was regularly beaten and interrogated by the secret police, but used techniques taught to him at the USAF survival school to get through it.

23.4.3.7. Evasion Planning

All IP should be prepared to evade until rescued, no matter how long the evasion experience might last. An evasion situation should not be categorized in terms of length. History has proven that predicting the length of any specific evasion situation is practically impossible. Emphasizing the advantages of "short-term" evasion over "long-term" may cause an overly optimistic, possibly even foolhardy, attitude toward evasion planning. Or, evaders may decide, if they are not rescued in a "short" period of time, it is no longer worth the effort; thereby, taking on a defeatist attitude.

23.5. Preparation

The actions one takes before an evasion situation can make the difference between being able to evade or being captured. In a hostile area, IP should remember evasion is an integral part of their mission and plan accordingly. The enemy may make mistakes of every conceivable form and not suffer more than indignation, anger, and fatigue. The evader, on the other hand, must constantly guard against mistakes of any sort. Being seen is the greatest mistake an evader can make. The evader must prepare for this task. Physical condition is the responsibility of the potential evader and has a great effect on the evader's ability to survive. Once on the ground, it is too late to get in shape. One more aspect is an evader's personal habits. Personal grooming habits might not be considered an important pre-mission briefing item, but using aftershave lotion, hair dressing, or cologne could add to the problems of an evader. The odor can carry for great distances and give away the evader's presence. Other preparation which will enhance the chance of success for the potential IP include:

- Study the physical features of the land, noting the location of mountains, swamps, plains, deserts, or forests, type of vegetation, and availability of water.
- Know the climatic characteristics and typical weather conditions of the area which they will be operating in/over and prepare accordingly.
- Study Isolated Personnel Guides (IPGs), CIA country studies, State Department information, US Marine Corp Intelligence Center documents, and other resources before a mission and learn some of the customs and habits of the local people. Such knowledge will aid in mission planning and development of evasion plans of action. For example, it may give the evader the ability to avoid hostile people or to identify and deal with "friendlies." This knowledge may also allow for blending into the local populace.
- Know their equipment - the location of each item, its operation, and its value. An evader must preplan which equipment should be retained and which should be left behind in different isolating situations.

23.5.1. Basic Problems

Three basic problems during evasion include:

1. Evading the enemy.
2. Surviving.
3. Returning to friendly control.

23.5.2. Successful Evasion

Chances for a successful evasion are improved if evaders:

- Observe the elementary rules of movement, camouflage, and concealment.
- Have a definite plan of action.
- Be patient, especially while traveling. Hurrying increases fatigue and decreases alertness. Patience, preparation, and determination are key words in evasion.
- Conserve resources.
- Conserve as much strength as possible for critical periods.
- Rest and sleep as much as possible.
- Maintain a highly developed "will to survive" and "can do" attitude. Evasion may require living off the land for extended periods of time and traveling on foot over difficult terrain, often during inclement weather.

- Constantly assess and evaluate land features to improve or take advantage of survival needs while evading.
- Use SA and available resources to mitigate the effects of weather conditions.

23.5.3. Planning of Travel

Once in the evasion situation, planning for travel may be a consideration for evaders. If planning to travel, a definite objective with a confident approach and ability to achieve it are a must. There will normally be several options with variations to choose from in selecting a plan of action or destination. The enemy force deployment, search procedures, terrain, population distribution, climate, distance, and environment will influence destination selection. Examples of options and destinations include:

- Await recovery forces.
- Evade to a hold-up area.
- Evade to a neutral country. (Note: Border areas not disrupted by combat may have a security system intact.)

23.5.4. Forward Edge of the Battle Area (FEBA)

If evaders are in the Forward Edge of the Battle Area (FEBA) and feel sure that friendly forces are moving in their direction, they should seek concealment and allow the FEBA to overrun their position. Evaders' attempts at penetrating the FEBA should be avoided. Evaders face stiff opposition from both sides.

23.5.5. Choosing Destination and Direction

A consideration in choosing destination and direction of evasion is whether there are suitable areas for pickup or possible contact with friendly forces.

23.5.6. Opportunity

Potential evaders must take advantage of any and all opportunities to evade, following current, approved emergency procedures for the theater of operations. This starts when an emergency is declared. In cases where ejection, bailout, or ditching appears imminent, the aircraft commander will attempt to establish radio contact. When communication is established, the SPINS will determine what will be transmitted. Whenever possible, the IP must remain alert and travel away from potential threats (populated areas, gun emplacements, troop concentrations, etc.) or out to sea (feet wet). The IP must be proficient in the use of the survival/evasion equipment to facilitate evasion (for example, use of the compass in conjunction with the survival radio). In addition to the opportunity to evade, motivation is essential to the evader's success.

23.6. Evasion Principles

Evaders should remain flexible, avoid detection, sanitize evidence of their presence, and orient themselves to their location.

23.6.1. Flexibility

Flexibility is one of the most important keys to successful evasion. The best thing an evader can do is to stay open to new ideas, suggestions, and changes of events. Having several backup plans of action can give the evader organized flexibility. If one plan of action is upset by enemy activity, the evader can then rapidly switch to a backup plan without panic.

23.6.2. Avoiding Detection

The evader is primarily interested in avoiding detection. Each evader should remember that people catch people. If the evader avoids detection, success is almost assured. Evaders should:

- Observe and listen for sounds of enemy fire and vehicle activity during parachute descent or vehicle ditching, and move away from those enemy positions once on the ground. IP who become isolated during daylight should assume they were seen and expect a search to center on their location (Figure 23-1).
- Be patient and determined while traveling.
- Use weather conditions and terrain as an aid while evading.
- Select times, routes, and methods of travel to avoid detection, as circumstances permit.
- Avoid lines of communication (waterways, roads, power lines, cell towers, etc.).

Figure 23-1 Avoiding Detection

23.6.3. Sanitize Area

One of the evaders' objectives is immediate recovery. In hostile areas or situations, IP must sanitize all evidence of presence and direction of travel (Figure 23-2). IP may never be certain rescue is imminent.

Figure 23-2 Sanitize Area

23.6.4. Initial Concealment

Although evaders would not normally move too far if rescue is imminent, in many situations they will have to leave the area quickly and travel as far as practical before selecting a hiding place. Evaders should leave no sign which indicates the direction or presence of travel. All hiding places should be chosen with extreme care. The time evaders will remain in the first location is governed by enemy activity in the area, their physical condition, availability of water and food, rescue capabilities, and patience. It is in this place of initial concealment that the evaders should regain strength, examine the current situation, and plan for the evasion situation ahead.

23.6.5. Alternate Destinations

Once in a place of concealment, evaders should make use of all available navigation aids to orient themselves. After determining their location, evaders should also select an ultimate destination and any necessary alternate destinations. The best possible route of travel should then be decided upon. When the time comes to move, they should have a primary plan and alternate plans for travel that address eventualities they may encounter.

23.6.5.1. Location of Assets/Forces

Evaders need to be aware of the location of friendly assets/forces. Assistance may be close at hand. These friendly assets/forces may come from several forms, each of which, under particular circumstances, may prove to be the most effective.

23.6.5.2. Emergency Rescue

The situation at the time of the emergency will determine the evader's best course of action. High ground is often the best position from which to await rescue; evaders may expect the best results from signaling devices, may observe the surrounding terrain, and may be kept under observation by friendly air cover. Whatever position is chosen, it should be clear of obstacles that would prevent a successful rescue.

23.6.5.3. Evading to a New Position

If not rescued immediately, the situation may compel evaders to move. Evaders must plan a course of action before leaving their position. When the evaders are certain their position is known to friendly elements, they might expect ground forces to attempt a rescue. They should remember their position may be detected as the enemy search parties approach. They must be prepared to evade to a new position.

23.6.5.4. Safest Routes

Evaders should remember that when traveling they are probably more vulnerable to capture. Once past the danger of an immediate search, evaders must avoid people. Inhabited areas should be bypassed rather than penetrated. As a rule, the safest route avoids major roads and populated areas, even if it takes more time and energy. Many evaders have been captured because they followed the easiest and shortest route, or failed to employ simple techniques such as observing, evading, camouflage, and concealment.

23.7. Camouflage

Camouflage consists of those measures evaders use to conceal their presence from the enemy. As a tool for evasion, it enables evaders to carry out life supporting activities and to travel unseen, undetected, and free to return to friendly control. They should try to blend in with the surrounding environment. Evaders should remember that camouflage is a continuous, never-ending process if they want to protect themselves from enemy observation and capture (Figure 23-3).

Figure 23-3 Camouflaging

23.8. Types of Observation

23.8.1. Five Senses

Of the five senses, sight is by far the most useful to the enemy. Hearing is second, while smell is of only occasional importance. These same senses can be of equal value to the evader.

23.8.2. Camouflaging

How useful these senses are depends primarily on range. For this reason, basic camouflage stresses visual concealment which is relatively long range. Most people are accustomed to looking from one position on the ground to another position on the ground.

23.8.3. Direct and Indirect Observation

The evaders must also have an understanding of the types of observation used by the enemy. There are two categories of observation: direct and indirect.

23.8.3.1. Direct Observation

Direct observation refers to the process whereby the observer looks directly at the object itself without the use of any type of augmentation such as real-time satellite and unmanned aerial vehicles (UAV) feeds, field glasses, and/or sniper-scopes. Direct observation may be made from the ground or from the air. Reported direct observations (such as an IP) can be immediately fired upon or troops can be sent in to investigate (Figure 23-4).

Figure 23-4 Direct Observation

23.8.3.2. Physical Signs

The enemy may also use dogs, foot patrols, and mechanized units to patrol a given area. Such teams could also physically search an area for signs of the passage of strangers, such as footprints, old campfires, discarded or lost equipment, and other "telltale" signs which would indicate that someone had been in the area.

23.8.3.3. Local Populace

Observation by the local populace is also a possibility. Upon seeing an evader or "telltale" signs an evader left behind, they may contact the local authorities, who initiate organized searches.

23.8.3.4. Indirect Observation

Indirect observation refers to the study of a photograph or an image of the subject via satellites, UAV, and reconnaissance photography, radar, or surveillance feeds. This form of observation is becoming increasingly more varied and widespread, and may be used from either manned or unmanned positions.

23.8.3.5. Aerial View

Views from the ground are familiar, but views from the air are usually quite unfamiliar. In modern warfare, the enemy may put emphasis on aerial photographs for information. It is important to become familiar with the "bird's-eye view" of the terrain as well as the ground view in order to learn how to guard against both kinds of observation.

23.9. Preventing Detection

Camouflage disciplines are those actions which contribute to an evader's ability to remain undetected. Proper use of camouflage discipline avoids any activity that changes an area or reveals objects to an enemy. Examples of common breaches of camouflage discipline include reflections from brightly shining objects (watches, glasses, rings, etc. Figure 23-5), over camouflaging, or using camouflage materials which are foreign to the area presently occupied by an evader. Evaders must also watch for signs that may reveal enemy camouflage efforts. Inadvertently walking into a camouflaged enemy position may result in capture.

Figure 23-5 Reflections

23.10. Factors of Recognition

The eight factors of recognition are movement, position, shape, shadow, texture, color, tone, and shine. These factors must be considered in camouflage to ensure that one or more of these factors do not reveal the location of the evaders.

23.10.1. Movement

Of the eight factors of recognition, movement is the quickest and easiest to detect. The eye is very quick to notice any movement in an otherwise still scene. Observation of movement can be the IP moving and/or movement of objects the IP has disturbed (Figure 23-6).

Figure 23-6 Movement

23.10.2. Position

Position is the relation of an object or person to its background. When choosing a position for concealment, a background should be chosen which will virtually absorb the evader (Figure 23-7).

Figure 23-7 Position

23.10.3. Shape

Shape is the outward or visible form of an object or person as distinguished from its surface characteristics and color. Shape refers to outline or form. Color or texture is not considered. At a distance, the forms or outlines of objects can be recognized before the observer can make out details in their appearance. For this reason, camouflage should disrupt the normal shape of an object or person (Figure 23-8).

Figure 23-8 Shape

23.10.4. Shadow

A shadow may be more revealing than the object itself, especially when seen from the air. Conversely, shadows may sometimes assist in concealment. Objects in the shadow of another object are more likely to be overlooked. As with shape, it is more important to disrupt the shadow pattern than to totally conceal the object or person. The identifiable shadows can be broken up by the addition of natural vegetation at various points on the body. Wearing "shapeless" garments will also disrupt the outline. For example, a soft and shapeless field cap can be used instead of a helmet or flight cap (Figure 23-9).

Figure 23-9 Shadow / Shadow Breakup

23.10.5. Texture

Texture is a term used to describe the relative characteristics of a surface, whether that surface is a part of an object or an area of terrain. Texture affects the tone and apparent coloration of things because of its absorption and scattering of light. Highly textured surfaces tend to appear dark and remain constant in tone regardless of the direction of view and lighting, whereas relatively smooth surfaces change from dark to light with a change in direction of viewing or lighting. The application of texture to an object often has the added quality of disrupting its shape and the shape of its shadow, making it more difficult to detect and identify as something foreign to the surroundings in which it exists. Looking straight down, the aerial observer sees all of the shadows, whereas a person on the ground may not. The textured surface may look light at ground level, but to the aerial observer the same surface produces an effect of relative darkness. The material used to conceal a person or an object must approximate the texture of the terrain in order to blend in with the terrain. Personnel walking or vehicles moving across the terrain will change the texture by mashing down the growth. Therefore, this will show up clearly from the air as foot paths or vehicle tracks.

23.10.6. Color

Pronounced color differences at close range distinguish one object from another. Thermal imagery works by sensing and contrasting the temperature and energy emitted or reflected from objects. The contrast between the color of the object and the color of its background can be an aid to enemy observers. The greater the contrast in color, the more visible the object appears (Figure 23-10).

Figure 23-10 Color

23.10.6.1. Hues

Color differences, such as green versus blue, or differences in hue, such as light green versus dark green, become increasingly difficult to distinguish as the viewing range is increased. The principal contrast is in their dark and light qualities. Therefore, as a first general principle, camouflage should match the darker and lighter qualities of the background and be increasingly concerned with the colors involved as the viewing range is decreased or the size of the object or installation becomes larger. A second general rule to follow is to avoid contrasts of hues. This is especially true in areas with heavy vegetation. Light-toned colors, such as leaf bottoms, should be avoided as they tend to attract attention.

23.10.7. Tone

Tone is the amount of contrast between variations of the same color. It is the effect achieved by the combination of light, shade, and color. The principal contrast is the dark and light relationship existing between an object and its background. Camouflage blending is the process of eliminating or reducing these contrasts. The two principal means available for reducing tone contrast are the application of matching or neutral coloration and the use of texturing to form disruptive patterns (Figure 23-11).

Figure 23-11 Tone

23.10.8. Shine

Shine is a particularly revealing signal to an observer. In undisturbed, natural surroundings, there are comparatively few objects which cause a reflected shine. Skin, clean clothing, metallic insignia, rings, glasses, watches, buckles, identification bracelets, and similar items produce shine. Such items must be neutralized by staining, covering, or removing to prevent their shine from revealing an evader's location. This is especially true at night (Figure 23-12).

Figure 23-22 Reflected Shine

23.11. Principles and Methods of Camouflage

Camouflage uses three basic methods of deceiving the enemy: hiding, disguising, and blending.

- Hiding. The complete concealment of a person or object by physical screening.

- Disguising. Changing the physical characteristics of an object or person in such a manner as to fool the enemy.
- Blending. The arrangement of camouflage on or about an object in such a manner as to make the object appear to be part of the background.

23.11.1. Basic Principles of Camouflage

The basic principles of camouflage are location, camouflage discipline, and camouflage construction.

23.11.1.1. Location

Evaders should look for an existing location which can be used almost as is, such as a cave or thicket if there are many like it in the area. The location should be changed as little as possible. Isolated locations such as individual trees, haystacks, or houses should be avoided. They tend to attract attention and are likely to be searched first because they are so obvious. At times, complete concealment against detection may be gained with no construction. In terrain where natural cover is plentiful, this is a simple task. Even in areas where natural cover is scarce, concealment may be achieved through use of terrain irregularities. Regardless of the activity involved, evaders must always be mentally aware of their positions (Figure 23-13).

Figure 23-33 Location

23.11.1.2. Camouflage Discipline

Camouflage discipline is the avoidance of activity which changes the appearance of an area or reveals military objects to the enemy. A well-camouflaged location is only secure as long as it is well maintained. Concealment is worthless if obvious tracks point like directional arrows to the heart of the location or if signs of occupancy are permitted to appear in the vicinity. Tracks, debris, and terrain disturbances are the most common signs of activity. Therefore, natural lines in the terrain should be used. If practical, exposed tracks should be camouflaged by brushing or beating them out. If leaving tracks is unavoidable, they should be placed where they will be least noticed and partially concealed (along logs, under bushes, in deep grass, in shadows, etc.). If tracks cannot be concealed, brushing them out or slapping vegetation against them will help them disintegrate quickly. Use of a Hudson Bay duffel or brush tied to the feet will disguise boot prints (Figure 23-14).

Figure 23-4 Camouflage Discipline

23.11.1.2.1. Light Discipline

At night, the most important discipline is light discipline. Lights at night not only disclose the evaders' position but also hinder the evaders' ability to detect the enemy. Even on the darkest nights, eyes grow accustomed to the lack of light in approximately 30 minutes. Every time a match is lit or a flashlight is used, the eyes must go through the complete process of getting adjusted to the darkness again. Smoking and lights should be prohibited at night in areas in close proximity to the enemy because the light is impossible to conceal. Additionally, a cigarette light aggravates the situation by creating a reflection which completely illuminates the face. The smell of the evader's foreign tobacco would stand out even if the enemy is smoking.

23.11.1.2.2. Infrared Equipment

Evaders should be especially careful at night due to infrared and low light detection equipment which may be used by the enemy. Keeping close to the ground and using terrain masking for concealment provide the best protection.

23.11.1.2.3. Sound Discipline

IP must maintain good sound discipline while evading. A simple act such as snoring may prove fatal. Calling to one another, talking, and even whispering should be kept to a minimum. Evaders should avoid sound-producing activities. Walking on hard surfaces which produce noises should be avoided. Hand signals or signs should be used when possible during group travel. Individual equipment should be padded and fastened in such a manner as to prevent banging noises.

23.11.1.3. Camouflage Construction

The evader should consider the following points regarding camouflage construction:

- Take advantage of all natural concealment.

- Don't over-camouflage. Too much is as obvious as too little.
- When using natural camouflage, remember that it fades and wilts, so change it regularly.
- If taking advantage of shadows and shade, remember they shift with the Sun.
- Do not expose anything that may shine.
- Break up outlines of manmade objects.
- When observing an area, do so from a prone position, while in cover.
- Match vegetation used as camouflage with that in the immediate locale, and when moving from position to position, change camouflage to blend with each new area's vegetation types.

23.11.2. Individual Camouflage

Evaders must know how to use location, camouflage discipline, and camouflage construction to achieve the best personal camouflage for their situation.

23.11.2.1. Form

Form is basic shape (body outline) and height. Three things which give an evader away in terms of form are to reveal outline of head and shoulders, to present straight lines of sides, and to allow the inverted "V" of the crotch and legs to be distinguishable. If staying in shadows, blending with background, adopting body positions other than standing erect, and other behavioral procedures alone are inadequate. They can be camouflaged by using "add-ons" such as branches or twigs to break up the lines. This addition of vegetation will also help an evader blend in with the background.

23.11.2.2. Background

Effective concealment of evaders depends largely on the choice and proper use of background. Background varies widely in appearance, and evaders may find themselves in a jungle setting, in a barren or desert area, in a farmyard, or in a city street. Each location will require individual treatment because location governs every concealment measure taken by the individual. Clothing which blends with the predominant color of the background is desirable. There will be occasions when the uniform color must be altered to blend with a specific background. The color of the skin must receive individual attention and be toned to blend with the background.

23.11.2.3. Camouflage Techniques

There are certain general aspects of individual body and equipment camouflage techniques which apply almost anywhere. The evaders should take each of the following areas under consideration.

23.11.2.3.1. Exposed Skin

The contrast in tone between the skin of face and hands and that of the surrounding foliage and other background must be reduced. The skin is to be made lighter or darker, as the case may be, to blend with the surrounding natural tones. The shine areas are the forehead, the cheekbones, nose, and chin. These areas should have a dark color. The shadow areas such as around the eyes, under the nose and under the chin should have a light color. The hands, arms, and any other exposed areas of skin must also be toned down to blend with the surroundings. Burnt cork, charcoal, mud, camouflage stick, berry stains, carbon paper, and green vegetation can all be used as toning materials.

23.11.2.3.2. Mosquito Face Net

A mesh mosquito face net, properly toned down to avoid shine, is an effective method of breaking up the outlines of the face and ears.

23.11.2.3.3. Facial Camouflage

Two primary methods of facial camouflage have been found to be successful patterns. They are the "blotch" method for use in deciduous forests/deserts, and the "slash" method for use in coniferous forests (Figure 23-15).

Figure 23-55 Camouflaged Faces

23.11.2.3.4. Slash and Blotch Method

Applications of these two patterns are simply modified appropriately to whatever environment the evader is in. In the jungle, a broader slash method would be used to cover exposed skin. In the desert or barren snow, a wide blotch, and in grass areas, a thin-type slash. To further break up the outline of the facial features, a flop hat or other loosely fitting hat may help. A beard that is not neatly trimmed may also aid in camouflaging exposed skin.

23.11.2.3.5. Toning Facial Camouflage

When toning down the skin, evaders should not neglect to pattern all of the skin; for example, the back of the neck, the insides and backs of the ears, and the eyelids. Covering these areas may help somewhat, especially if there is a lack of other material to tone down the skin. Vegetation hung from the hat, collar buttoned and turned up, a scarf, or even earflaps may help. To cover the hands, evaders may use flight gloves, mittens, or loose cloth if unable to tone down wrists, back of hands, and between fingers sufficiently. Evaders should watch for protruding white undergarments, T-shirt, long underwear sleeves, etc. They should also tone down these areas.

23.11.2.3.6. Camouflaging Hair

Lack of hair or light colored hair requires some type of camouflage. This could include those applied to the skin or an appropriate hat, scarf, or mosquito netting.

23.11.2.3.7. Odors

Odors in a natural environment stand out and may give evaders away. Americans are continually surrounded by artificial odors and are not usually aware of them. Human body odor would have to be very strong to be detected by ground troops (searchers) that have been in the field for long periods. The following odors should be of concern to the evader:

1. In non-permissive areas, personnel should always use unscented toiletries and hygiene products. Shaving cream, after shave lotion, perfume, and other cosmetics are to be avoided.
2. The potential evaders should also remember that insect repellent is scented. They should try to use head nets, but if forced to use a repellent, use the camouflage stick which has repellent in it and is the least scented.
3. Tobacco should not be used. The stain and odor should be removed from the body and clothing.
4. Rations, gum, or candy may have strong or sweet smells. Evaders should take care to rinse out their mouths after use.
5. Smoke odors from campfires may permeate clothing, but if the potential searchers use fires for cooking and heat, they probably can't detect it on evaders.

23.11.2.3.8. Preparation of Clothing

Clothing and personal items require attention both before a mission and again if forced to evade. Prior preparation for the IP may include:

- Ensuring that clothing does not smell of laundry products and is in good repair and not worn to the point where it shines or is faded.
- Checking zippers, snaps, and buttons for shine and function.
- Checking rank insignia and patches for light reflection and color. Remove name tag, branch of service tag, and rank (whether they are stripes or metal insignia, bright unit patches, etc.) from the uniforms and place them either in the pack or in a secure pocket.
- Subdue underwear for camouflage in the event that the outer layer is torn.
- Ensure boots are in good repair but not shiny. Shiny eyelets should be repainted. Squeaky boots should be fixed or replaced.
- Sanitize pocket or wallet contents. Remove items which might aid in enemy exploitation attempts of an individual detainee or group; for example, credit cards, photographs, money, and addresses. Evaders should carry only those necessary pieces of identification which will prove a person is a US military member.

23.11.2.3.9. Additional Clothing

Additional clothing such as a hat, extra socks, a scarf, and gloves may be desirable and located in flight clothing pockets, packs, or other locations. In an evasion environment, clothing and equipment need quick camouflaging attention. Anything to be discarded should be hidden at the initial site.

23.11.2.3.10. Sanitizing Clothing

Clothing should be sanitized by removing anything bright or shiny. Evaders should consider camouflaging their clothing (to include boots) just as they would their skin.

23.11.2.3.11. Camouflaging Clothing

One principle of camouflage is to disrupt or conceal uniform color, straight lines, and squares - things rarely found in natural features. Evaders should reduce the tone of all equipment by smearing it with camouflage stick, mud, or with whatever is available, in a mottled pattern. Camouflage clothing and equipment using whatever materials available. This may include dirt, mud, charcoal, vegetation bark, berries, and leaves. These items can be used to cover clothing, soil it, or stain it depending on the item.

23.11.2.3.12. Care and Use of Clothing

Principles and techniques for care and use of clothing and equipment cannot be forgotten or ignored in an evasion situation, although some modifications may be necessary. A number of variables will influence what changes or omissions will be necessary.

23.11.2.3.13. Readily Available Items

Where metal pieces come into contact with one another, there should be padding between them so they will not inadvertently "clank" together. Evaders should place all items needed for environmental protection in the top of the pack where they will be most readily available. The rest of the gear can be used as padding around metal objects. In this manner, with everything stored inside the bundle or pack, it is secure from loss, damage, and enemy observation, as well as being readily available when needed.

23.11.2.3.14. Minimizing Shine and Sound from Equipment

An evader's pockets should be secured and all equipment, including dog tags, arranged so that no jingle or rattle sounds are made. This can be done with cloth, vegetation, padding, or tape. Evaders should also remove jewelry, watches, exposed pens, and glasses if possible. If glasses are required, hat netting or mask may help reduce shine.

23.11.2.3.15. Minimizing Sound of Clothing

Evaders should minimize the sound of clothing brushing together when the body moves. Moving in a careful manner can decrease this sound. Evaders should remember that camouflaged clothing and equipment alone won't conceal, but it must be used intelligently in accordance with the other principles of camouflage and movement. For example, even if evaders are perfectly camouflaged for the arctic, there could still be problems. Because snow country is not all white, shadows and dark objects appear darker than usual. A snowsuit cannot conceal the small patches of shadow caused by the human figure, but that is not necessary if the background contains numerous dark areas. If the background does not contain numerous dark areas, maximum use is to be made of snowdrifts and folds in the ground to aid in individual concealment (Figure 23-16).

Figure 23-66 Evasion in Snow Environment

23.11.2.3.16. Blending

The concept of blending in with the background is indeed an important one for the evader to understand. One major point in blending with the background is not to show a body silhouette. Losing the body silhouette is done by making use of the shadows in the background. From a concealment point of view, backgrounds consist of terrain, vegetation, artificial objects, sunlight, shadows, and color. The terrain may be flat and smooth, or it may be wrinkled with gullies, mounds, or rock outcroppings. Vegetation may be dense jungle growth or no more than small patches of desert scrub growth. The size of artificial objects may range from a signpost to a whole city block. There may be many colors in a single background, and they may vary from the almost black of deep woods to the sand pink of some desert valleys. Blending simply means matching, with as many of these backgrounds as possible and avoiding contrast. If it is necessary for evaders to be positioned in front of a contrasting or fixed background, they must be aware of their position and take cover in the shortest possible time. As in the daytime, silhouette and background at night are still the vital elements in concealment (Figure 23-17). A silhouette is always black against a night sky, and care must be taken at night to keep off the skyline. On moonlit nights, the same precautions must be taken as in daylight. It should be remembered that the position of the enemy observer, and not the topographic crest, fixes the skyline.

Figure 23-77 Silhouetting

23.12. Concealment in Various Geographic Areas

When not otherwise specified, temperate zone terrain is to be assumed in this section. Desert, snow, and ice areas are mostly barren and concealment may require considerable effort. Jungle and semitropical areas usually afford excellent concealment if the IP employs proper evasion techniques.

23.12.1. Adding Vegetation to Clothing

There are some general observations and rules regarding the addition of vegetation to the uniform and equipment. The cycles of the seasons bring marked changes in vegetation, coloring, and terrain pattern requiring corresponding changes in camouflage. Concealment which is provided in wooded areas during the summer is lost when leaves fall in the autumn. This will create a need for additional camouflage construction. Also, vegetation must be of the variety in the evaders' immediate location. It must be changed if it wilts or the evaders move into a different vegetation zone. Evidence of discarding the old and picking the new should be hidden. The vegetation should not be cut; this will give evidence of human presence.

23.12.2. Natural Materials

Any type of material indigenous to the locality of the evaders may be classified as natural material. Natural materials consist of foliage, grasses, debris, and earth. These materials match local colors and textures and when properly used are an aid against both direct and indirect observation. The use of natural materials provides the best type of concealment. The chief disadvantage of natural foliage is that it cannot be prepared ahead of time, is not always available in usable types and quantities, wilts after gathering, and must be replaced periodically. Foliage of coniferous trees (evergreens) retains its camouflage qualities for considerable periods, but foliage that sheds leaves will wilt in a day or less, depending on the climate and type of vegetation (Figure 23-18).

Figure 23-88 Natural Materials for Personal Camouflage

23.12.2.1. Advantage of Using Live Vegetation

The principal advantage in using live vegetation is its ability to reflect infrared waves and to blend in with surrounding terrain. When vegetation is used as garnishing or screening, it must be replaced with fresh materials before it has wilted sufficiently to change the color or the texture. If vegetation is not maintained, it is ineffective. Thorn bushes, cacti, and other varieties of desert growth retain growing characteristics for long periods after being gathered.

23.12.2.2. Arrangement of Foliage

The arrangement of foliage is important. The upper sides of leaves are dark and waxy, the undersides are lighter. In camouflage, foliage must be placed as it appears in its natural growing state, top sides of leaves up and tips of branches toward the outside of the leaves (Figure 23-19).

Figure 23-19 Dark and Light Leaves

23.12.2.3. Matching Existing Foliage

Foliage gathered by IP must be matched to existing foliage. For example, foliage from trees that shed leaves must not be used in an area where only evergreens are growing. Foliage with leaves that feel leathery and tough should be chosen. Branches grow in irregular bunches and, when used for camouflage, must be placed in the same way. When branches are placed to break up the regular, straight lines of an object, only enough branches to do this should be used. The evader must adopt principles that apply to the area and know that the enemy also applies these principles.

23.12.2.4. Securing Vegetation

When vegetation is applied to the body or equipment of evaders, it must be secured to the clothing or equipment in such a way that:

- Any inadvertent movement of material will not attract attention.

- It appears to be part of the natural growth of the area; that is, when the evaders stop to hide, the light undersides of the vegetation are only visible from beneath. After evaders complete their camouflage, they should inspect it from the enemy's point of view. If it does not look natural, it should be rearranged or replaced.

- It does not fall off at the wrong moment and leave evaders exposed, or it should not show evidence of the evaders' passage through the area.

- Too much vegetation can give evaders away.

23.12.3. Cloth Material used for Vegetation Camouflaging

If cloth material is used like vegetation to break up shape and outline and to help blend in with the environment, there are some points evaders should be aware of. Cloth can be used successfully when wrapped around equipment or designed into loose, irregular shaped clothing or accessories (Figure 23-20).

Figure 23-90 Breakup of Shape and Outline

23.12.3.1. Cloth Materials

Some of the materials evaders may use include:

- The colors (green, brown, white) of parachute materials plus the harness. Consider that parachute material tends to shine and the unraveled edges may leave fine filaments of nylon on the ground as evidence of evaders in an area. Parachute material is also very lightweight, and a sudden breeze might cause it to move when movement is not desirable.
- Excess clothing the IP may have pre-packed (scarf, bandana, etc.).
- Burlap, when found. It is used in battle areas in the form of sandbags.
- Note that most artificial material is versatile, but it can have drawbacks.

23.13. Concealment Factors for Areas Other Than Temperate

23.13.1. Desert

Lack of vegetation, high visibility, and bright tone (smooth texture) all emphasize the need for careful selection of a location for a hole-up site. Deep shadows in the desert, strict observance of camouflage discipline, and the skillful use of deception and camouflage materials aid in concealing evaders in a desert area.

23.13.1.1. Desert Shadows

Deserts are sometimes characterized by strong shadows with heavy broken terrain lines and sometimes by a mottled pattern. Each type of desert terrain presents its own problems. To minimize the effect of shadows, use concealment that is afforded by the shadows of gullies, scrub growth, and rocks when possible. Many objects which cannot be concealed from the air can be effectively concealed from the ground. Lack of reference points in the terrain may make them difficult to locate on a map.

23.13.2. Snow and Ice

From the air, snow-covered terrain is an irregular pattern of white, spotted with dark tones produced by objects projecting above the snow, their shadows, and irregularities in the snow-covered surface such as valleys, hummocks, ruts, and tracks. It is necessary, therefore, to make sure dark objects have dark backgrounds for concealment, to control the making of tracks in the snow, and to maintain the snow cover on camouflaged objects. No practical artificial material has yet been developed which will reproduce the texture of snow sufficiently well to be a protection against recognition by aerial observers. IP evading in snow-covered, frozen areas should wear a complete white camouflage outfit. Concealment from direct ground observation is relatively successful with the use of white snowsuits, white pants, and whitewash. These measures offer some protection against aerial detection. White parachute cloth should also be considered. A white poncho-like cape can be made easily from parachute material. A pair of white pants will normally be sufficient in a heavily wooded area. However, following or during a heavy snowfall when the trees are well covered with snow, the wearing of a complete white camouflage suit is necessary to blend in with the background. Other equipment, such as packs, should also be covered with white material.

23.13.3. Mountain Areas above Timberline and Arctic/Antarctic Areas

Lack of vegetation, high visibility, and bright tone all emphasize the need for careful selection of a location for a hole-up site. Rock outcroppings, layering of snow, ice formations, and terrain features may provide opportunities for concealment (Figure 23-21).

Figure 23-101 Above Timberline

23.13.4. Mountain Areas below the Timberline and Subarctic Areas with Snow

Common characteristics of these areas are forests, rivers, lakes, and features such as trails and buildings. The appearance of the area is irregular in pattern and variable in tone and texture providing ample opportunity for the IP to find concealment and camouflage (Figure 23-22).

Figure 23-112 Below Timberline

23.14. Camouflage and Concealment Techniques for Shelters

Unlike shelters built in permissive environments, an evasion shelter is primarily concerned with concealment, not personal comfort or convenience, while still protecting IP from the environment (Figure 23-23). Anyone who is resting or sleeping will not be totally alert, so added precautions may be necessary to maintain security such as having a person stand watch in group situations.

Figure 23-123 Evasion Shelter

23.14.1. Natural Concealment

Evaders should choose natural concealment areas-such as small concealed caves (when there are many), hollow logs, holes, depressions, clumps of trees, or other thick vegetation (tall grass, bamboo, etc.). The site should have as much natural camouflage as possible. There should be cover on all sides. Natural formations or vegetation can protect evaders from aerial observations. The site should be as concealed as possible with a minimum of work. Sites chosen this way will make concealment easier and require less activity and movement. This is most important if the evaders are close to population centers or if the enemy is present.

23.14.2. Awareness of Climatic Conditions

At no time will evaders be able to safely assume they are free from the threat of either ground or aerial observation. Therefore, not only is the shelter area and type determined by the needs of the moment (enemy presence, etc.), but consideration must also be given to the terrain and climatic conditions of the area. Evaders must constantly be aware of climatic conditions, the potential for rescue, how long they may remain in the area, and what types of enemy observation may be employed.

23.14.3. Using Natural Concealment

The shelters may be naturally present, or they may be those which are assembled and camouflaged by the evaders. Full use must be made of concealment and camouflage no matter what types of shelter areas are selected. The use of the natural concealment afforded by darkness, wooded areas, trees, bushes, and terrain features are recommended; however, any method used for blending or hiding from view will increase the chances for success. There is much for evaders to consider concerning the many facets of evasion shelter site selection if they expect to establish and maintain the security of their area (Figure 23-24).

Figure 23-134 Natural Evasion Shelter

23.14.4. Choosing Shelter Location

Evaders should locate their shelter areas in the least obvious locations. The chosen areas should look typical of the whole environment at a distance and not be near prominent landmarks. Areas that look blended may only get a cursory glance. The areas should also be those least likely to be searched. Rough terrain and thickly vegetated areas are examples of the types of areas least likely to be searched. The shelter sites should also be situated so that in the event of impending discovery, the evaders will be able to depart the area via at least one concealed escape route. Avoid areas which may trap the evaders if the enemy discovers the places of concealment.

- The evaders should select concealment sites near the military crest (three-quarters of the way up) of a hill or mountain if cover is available. Noises from ridge to ridge tend to dissipate. Whispers or other sounds made in a valley tend to magnify as listeners get further up a hill. Shelter areas located on a slope are subject to higher daytime and lower nighttime winds, thereby minimizing the chance of detection through the sense of smell.
- If possible, evaders should be in such a position in the shelter area that shadows will fall over the side of the area throughout the day. This can best be done in heavy brush and timber.
- Evaders should try to locate alternate entrance and exit routes along small ridges, bumps, ditches, and rocks to keep the ground around the shelter area from becoming worn and forming "paths" to the site. They should avoid staying in one area so long that it has the appearance of being lived in. Evaders should try to stay away from and out of sight of any open areas. Examples include roads and meadows.
- Waterways such as lakes, large rivers, and streams, especially at the junctions, are dangerous places because of the increased probability of people at these lines of communication. Power and fence lines or any prominent landmarks may indicate places where people may be. Evaders will want to stay clear of these areas. The enemy may patrol bridges frequently. Evaders should avoid any areas close to population centers. The evaders should be able to observe the enemy and their movements and the surrounding country from the hiding area if at all possible.
- A dry, level sleeping spot is ideal, but the ideal spot to provide non-visibility and comfort may be difficult to find. Evaders must have patience and perseverance to stay hidden until danger has passed or until they are prepared and rested enough to safely move on.
- If camouflage is necessary at the shelter site, evaders should keep in mind that they should always construct to blend. They should match the shelter area with natural cover and foliage, remembering that over-camouflage is as bad as no camouflage. Natural materials should be taken from areas of thick growth. Any place from which materials have been taken should also be camouflaged. BLISS is an easy to remember acronym for evasion shelter principles:
    - **B** - Blend
    - **L** - Low silhouette
    - **I** - Irregular shape (outline)
    - **S** - Small
    - **S** - Secluded location
- Other facilities evaders may use, such as cat-hole (immediate action latrines), caches, etc., must

be located and camouflaged in the same manner as the shelter sites. When using these facilities evaders should dogleg through ground cover to use concealment to its best advantage.

23.15. Firecraft under Evasion Conditions

Fire should only be used when it is absolutely necessary in a life or death situation. Potential evaders must understand that the use of fire can greatly increase the probability of discovery and subsequent capture. If a fire is required, selection of location, time, tinder, kindling, fuel, and construction should be major considerations.

23.15.1. Location

Evaders should keep the fire as inconspicuous as possible. The location of an evasion fire is of primary importance. All evasion-type fires must be small and built in an area where the enemy is least likely to see them. If possible, in hilly terrain with cover, the fire should be built on the side of a ridge (military crest). No matter where the fire is built, it should be as small and smokeless as possible (Figure 23-25).

Figure 23-145 Fire

23.15.2. Time

Fires are easier to disguise when the local population is building their fires, during the times of dawn and dusk, and/or during times of bad weather. At these times, there is a haze or vapor trap that hinges in and around hills and depressions and is prevalent on the horizon. Any smoke from the fire will be masked by this haze in the early morning and at sunset. Another method of disguising the smoke from a fire is to build it under a tree. The smoke will tend to dissipate as it rises up through the branches, especially if there is thick growth or the boughs are low hanging. It can be helpful to camouflage the fire with earthen walls, stone fences, bark, brush, or snow mounds to block the light rays and help disperse the smoke.

23.15.3. Tinder, Kindling and Fuel

The best wood to use in an evasion fire is dry, dead hardwood no larger than a pencil with all the bark removed. This wood will produce more heat and burn cleaner with less smoke. Wood that is wet, heavy with sap/pitch, or green will produce large amounts of smoke. When the wood has been gathered, evaders should select small pieces or make small pieces out of the wood collected. Small pieces of wood will burn more rapidly and cleanly thus reducing the chance of smoldering and creating smoke. The wood selected should be stacked such as in a log cabin layout so the fire gets plenty of air as ventilation which will make the fire burn faster with less smoke.

23.15.4. Constructing a Dakota Hole Fire

One type of evasion fire which has the capability of being inconspicuous is called the Dakota Hole Fire (Figure 23-26). After selecting a site for the fire hole, a "fireplace" must be prepared. This is done by digging two holes in the ground, one for air or ventilation and the other to actually lay the fire in. These holes should be roughly eight to 12 inches deep and about 12 inches apart with a wide tunnel dug to connect the bottoms of the holes. The depth of the holes depends on the intended use of the fire. Place dirt on a piece of cloth so it can be used to rapidly extinguish the fire and conceal the fire site. Evaders who build these fires should strive to keep the flame under the surface of the ground. Initially the fire may appear to be smoking due to the moisture in the ground. At night the area may glow, but there should be no visible flame. A fire built in a hole this way will burn fast as all the heat is concentrated in a small area. This is a good type of fire for a single evader as opposed to many persons taking turns cooking over this one hole. Everything about the evasion situation will have to be examined before deciding which fire configuration would be most useful if a fire must be built.

Figure 23-156 Dakota Hole

23.15.4.1. "H" Fire

An evasion fire can also be built just below the ground cover. Here the emphasis is on quick concealment of the area. Some type of screen should be used to hide the flames. The small fire is built on the bare ground after a layer of sod or earth has been sliced and rolled back to each side, often called an "H" fire due to the shape of the cut (Figure 23-27). After use, evaders simply scatter the fire remnants over the bare area and roll or fold the piece of ground back into place.

- If evaders are in areas where holes can't be dug or sod can't be lifted, they will have to settle for some type of screen around the fire. They should also keep the fire small and finish using it quickly.

- In sand or loose rock such as the desert or coastal regions, scrape a shallow hole, using the material scrapped up to be your barrier or screen around the hole, where the fire will be built. When done with the fire, the sides are pushed back in to camouflage the fire.

- No matter what type of evasion fire used all traces of the fire should be removed. Unburned firewood should be buried. Holes should be totally filled in. Placing the soil on a holder, such as bark, an EVC, or piece of equipment, will aid in replacing it. This way there will be no leftover dirt patches on the ground after the holes were filled in. Once evaders feel all available measures have been taken to obliterate any leftover evidence, they should move out of the area if possible. Since evaders can never be positive the fire wasn't detected, they must assume it was spotted and take all necessary precautions.

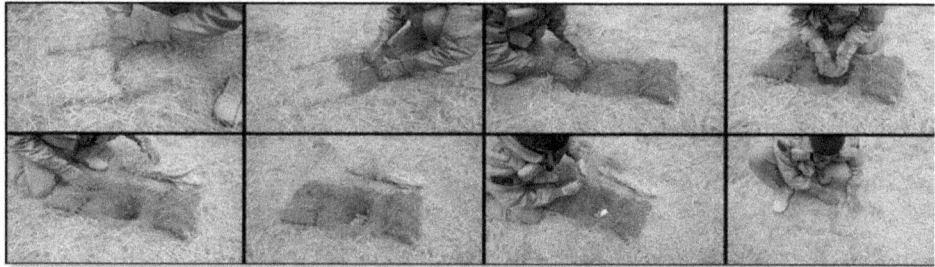

Figure 23-167 "H" Fire

23.16. Sustenance for Evasion

As previously stated, not only do evaders face the problem of remaining undetected by the enemy, they must also have the knowledge which will enable them to live off the land as they evade. They must be prepared to use a wide variety of both "wild" and domestic food sources, obtain water from different sources, and use many methods for preparing and preserving food and water (Figure 23-28).

Figure 23-178 Sustenance for Evasion

23.16.1. Opportunity to Obtain Food

No matter what the circumstances of the evasion situation are, evaders should never miss an opportunity to obtain food. Food will typically be obtained from wild plant and animal sources. If possible, evaders should stay away from domestic plants (crops) and animals. Using wild animals and plant sources for food will reduce the probability of capture. EVCs have graphics and information on edible plants and animals, as well as hazardous life forms.

23.16.2. Animals

Animals can be a source of sustenance for evaders, having more nutritional value for their weight than do plant foods. Evaders may obtain enough animal food in one place to last for several days while they travel. There are several ways which evaders may procure animal foods (trapping, snaring, fishing, hunting, stealing, and poisoning). A few modifications should be considered regarding the use of traps and snares in evasion situation even though the same basic principles apply in both permissive and non-permissive environments.

23.16.2.1. Small Game

Because small game is more abundant than is large game in most areas, evaders should confine traps and snares to the pursuit of small game. There are other advantages to restricting trap and snare size. Evaders will find it is easier to conceal a small trap from the enemy. Small animals make less noise and create less disturbance of the area when caught.

23.16.2.2. Traps

Conversely, there are also a few disadvantages pertaining to the use of snares or traps during evasion episodes. Three disadvantages are: evaders must remain in one place while the snares are working, discovery of the trap or snare, and disturbance of the area.

23.16.3. Fishing

Fishing is another effective means of procuring animal food.

23.16.3.1. Fishing Methods

There are several methods which evaders may use to catch fish. A simple hook and line is one of these methods, another is a "trotline." Evaders may construct a trotline by fixing numerous hooks to a pole and by sliding it into the water from a place of concealment (Figure 23-29).

Figure 23-29 Fishing

1. Nets and traps may also be used. However, they should be set below the water line to avoid detection. Spearing is another option.
2. "Tickling" the fish (Figure 23-30) is also effective if evaders can remain concealed. This method requires no equipment to be successful. (NOTE: Caution should be used when tickling fish in areas with carnivorous fish or reptiles.)
3. Conversely, there are also a few disadvantages pertaining to fishing during evasion episodes. Exposure in open waterways can be very dangerous to the evader. The main disadvantage to fishing is people live by water bodies and travel on them. This greatly increases the chances of being detected. Trotlines, nets, and fish traps may be discovered.

Figure 23-180 Fish Tickling

23.16.4. Using Weapons

Weapons may be used by evaders to procure animals or fish. The best weapons are those which can be operated silently, such as a blowgun, slingshot, bow and arrow, rock, club, or spear. These should be used primarily against small game or fish. One major advantage of using weapons is food can be taken while evaders travel. Using firearms for food procurement should be avoided in an evasion situation due to the risk of detection.

23.16.5. Plant Food Procurement

Plant foods are very abundant in many areas of the world. Evaders may be able to procure plant food types that require no cooking. One advantage of procuring plant foods during evasion is that by collecting natural fruits and nuts, evaders can remain deep in unpopulated areas. In some areas, it is possible to find old garden plots where vegetables may be obtained. When possible, choose foods which can be eaten raw (Refer to chapter 17 - Food.)

23.16.5.1. Disadvantages of Plant Foods Procurement

The disadvantage of plant food procurement is that evaders may not be the only ones looking for food. The natives of the country could also be out looking for food. If natives know of a good area, they may visit that place often. If evaders have been in the area, their presence could be discovered.

23.16.5.2. Considerations and Methods Concerning Plant Food Procurement

- Never take all the plants or fruits from one area, if likely to be detected.
- Pick only a few berries off of any one bush.
- When digging plants, take only one plant, and then move on some distance before digging up another. Try to ensure the area looks undisturbed when departing.
- Camouflage all signs of presence.

23.16.6. Domestic Foods

Most domestic foods must be procured by theft which is very dangerous. However, if proper methods are used and the opportunity presents itself, plants and animals may be stolen. The main reason thieves are captured is the boldness they display after several successful thefts. The basic rules of theft are:include

- If at all possible, the theft should take place at night.
- Evaders should thoroughly observe the area of intended theft from a safe vantage point.
- Evaders should find the vantage points just before dusk and look the place over to make sure everything is the same as it was the last time a check was made.
- Evaders should check for dogs, which could be a big hazard. Barking draws attention; also, some dogs are vicious and can harm evaders. Besides dogs, other animals or fowl can alert the enemy to the evaders' position or can harm evaders.
- Evaders should never return twice to the scene of a theft.
- Every theft should be planned, and after its accomplishment, evaders should leave no evidence of either their presence or the theft itself (Figure 23-31).

- Only small amounts should be taken.

Figure 23-191 Stealing Vegetables

23.16.6.1. Cultivated Plant Foods

When evaders find it necessary to take cultivated plant foods, they should avoid taking the complete plant if it indicates their presence. Taking plants from the inside of the field, not the edge, and leaving the top of plants in place may help conceal the theft, but beware of leaving tracks. When digging plants from garden plots, make sure the plot is old. In many countries the people plant their crops and do not return to the plot until harvest time.

23.16.6.2. Domesticated Animals

The rules of theft also apply to taking domesticated animals. The evader should concentrate on animals that don't make much noise. If a choice has to be made as to which animal to steal, evaders should take the smallest one.

23.16.6.3. Procuring Water

Water is very essential, but it can be difficult to acquire (Figure 23-32). When procuring water, evaders should try to find any water source well away from populated areas. The enemy knows evaders need water and may check all known water sources. No matter where water is procured, evaders should try to remain completely concealed when doing so. The area around the water source should be observed to make sure it isn't patrolled or watched. While obtaining water and when leaving the area, evaders should conceal any evidence of their presence. Good sources of water include trapped rainwater in holes or depressions and plants which contain water.

Figure 23-202 Procuring Water

23.16.7. Preparing Food and Water during Evasion

The preparation and purification of food and water by cooking is a precaution which should be weighed against the possibility of capture. It might be necessary to eat raw plant and animal foods at times. Some plant foods will require cutting off thorns, peeling off the outer layer, or scraping off fuzz before consumption. Raw animal foods may contain parasites and micro-organisms which may not affect evaders for days, weeks, or months, at which time, hopefully, they will be under competent medical care. If the environment in which the parasites live is altered by cooking, cooling, or drying, the organisms may be killed. Meat can be dried and cooled by cutting into thin strips and air-drying. Salting makes the meat more palatable. If the meat must be cooked, small pieces should be cooked over a small hot fire built in an unpopulated area. The best methods of preserving food during evasion are drying or freezing.

23.16.7.1. Purification

In an evasion situation, boiling water for purification should not be used as a method except as a last resort. Iodine tablets are the best method of purification. If evaders do not have purification tablets, and the danger of the enemy detecting their fires is too great, evaders may have to forego purification. The only problem with this is if water is not purified, it may cause vomiting and/or diarrhea. These ailments will slow down the evaders and make them susceptible to dehydration. Aeration and filtration may help to some degree and are better than nothing. If water cannot be purified, evaders should at least try to use water sources which are clean, cold, and clear. Rain, snow, or ice should be used if available.

23.17. Security

When evading for extended periods in enemy-held territories, it becomes essential for evaders to rest. To rest safely, especially if in a group, it is essential to devise and use some sort of early warning system to prevent detection and unexpected enemy infiltration. When establishing an evasion shelter area, there are certain things which should be done (day or night) for security purposes (Figure 23-33).

Figure 23-213 Rest Safely

23.17.1. Scouting the Area

Evaders should scout the area for signs of people. They should pay particular attention to crushed grass, broken branches, footprints, cigarette butts, and other discarded trash. These signs may reveal identity, size, direction of travel, and time of passage of an enemy force. If signs are present, the evader should consider moving to a more secure area.

23.17.2. Alarm System

Once the area has been determined to be fairly secure, some type of alarm system must be devised. A lone evader should use the natural alarm system available. Disturbances in animal life around an evader may indicate enemy activity in the area. Group situations may allow for more security. Two or more evaders may use lookouts or scouts at strategically located observation posts.

23.17.3. Situational Awareness

Situational awareness is another aspect of security. The evader should be aware that, at any time, the area may be overrun, ambushed, or security compromised, making it necessary to vacate the area. If evaders are in a group and future group travel is desired, it is essential that everyone in the group knows and memorizes certain details such as, compass headings or direction of travel, routes of travel, destination descriptions, and rally points (locations where evaders can regroup after separation). Alternate points must be designated in case the original location cannot be reached or if it is compromised by enemy activity.

23.17.4. Establishing a Alternate Plan

Once everyone in the evader's group has reached the final destination, alternate point, or rally point, a new emergency evacuation and rendezvous plan must be established.

- Evaders should always be aware of the next rally point, its location, and direction. These places, which provide concealment and cover, should be designated along the route in case an enemy raid or ambush scatters the group. There should be a rally point for every stage of the journey. Even when approaching the supposed "final destination" of the day, evaders should have an evacuation plan ready.
- Maintaining silence is a very important aspect of security. It is essential to be able to communicate with individual group members and scouts so that everyone is aware of what is going on at all times during evasion. Hand signals are the best method of communication during evasion as they are silent and easily understood (Figure 23-34). Instructions and commands which must be conveyed throughout the entire group include:

    - Freeze.
    - Listen.
    - Take cover.
    - Enemy in sight.
    - Rally.
    - All clear.
    - Right.
    - Left.

Figure 23-224 Hand Signals

23.18. Evasion Movement

Evasion movement is the action of a person who, through training, preparation, and application of natural intelligence, avoids capture and contact with hostiles, both military and civilian. Not only is total avoidance of the enemy desirable for evaders, it is equally important for evaders not to have their presence in any enemy controlled area even suspected. A fleeting shadow, an inopportune movement or sound, and an improper route selection are among a number of things which can compromise security, reveal the presence of evaders, and lead to capture.

23.18.1. Perspective Evader Areas of Interest

One evasion situation will not be identical to another. There are, however, general rules which apply to most circumstances. These rules, carefully observed, will enhance the evaders' chances of returning to friendly control. Evasion begins even before an IP is in or over enemy territory. Two factors which are essential to successful evasion and returning with honor are opportunity and motivation. Pre-mission preparation and knowledge of areas of concern are very important. Pre-mission knowledge gained must be based on the most current information available through directives, area studies, and intelligence briefings. Some areas of interest to the perspective evader include:

23.18.1.1. Topography and Terrain

A potential IP should know the physical features and characteristics, possible barriers, best areas for travel, availability of rescue, and the type of air or ground recovery possible. A potential evader should also know the requirements for long-term unassisted survival in the area of operation.

23.18.1.2. Climate

The typical weather conditions and variations should be known to aid in evasion efforts.

23.18.1.3. People

A very critical consideration is to understand the people in the area. From ethnic and cultural briefs read before the mission, crewmembers should familiarize themselves with the behavior, character, customs, and habits of the people. It may, at times during the evasion episode, be necessary to emulate the natives in these respects (Figure 23-35).

Figure 23-235 Knowing the People

23.18.1.4. Equipment

Personnel should be thoroughly familiar with all of their equipment and where it is located. They should also preplan what equipment should be retained and what should be left behind in an isolating event.

23.18.2. Factors that Influence Decision to Travel

Before addressing evasion movement techniques, it is appropriate to go over some of the factors which influence an evader's decision to travel.

- The first few minutes are the most critical period for the evader. The evader must avoid panic and not take any action without thinking. In these circumstances, the evader must try to recall any previous briefings, standard operating procedures, or training and choose a course of action which will most likely result in return to friendly territory.

- In those first few minutes, there is a great deal for the evader to think about. Quick consideration must be given to landmarks, bearings, and distance to friendly forces and from enemy forces, likely location for recovery, and the initial direction to take for evasion. Knowledge of what to expect is important. When circumstances arise which have been considered in advance, a course of action can be carried out more quickly and easily. Evaders should try to adapt this knowledge and any skills they have to their particular situation. Flexibility is most important, as there are no hard and fast rules governing what may happen in an evasion experience.

- Any movement has the inherent risk of exposure, some specific principles and practices must be observed. Periods of travel are the phases of evasion when evaders are most vulnerable. Many evaders have been captured because they followed the easier or shorter route and failed to employ simple techniques such as watching and listening frequently and seeking concealment sites.
    - Above all, avoid unnecessary movement.
    - When moving, keep off the skyline; use the military crest (three-quarter' way up the hill).

23.19. Observing Terrain

Evaders should visually survey the surrounding terrain from an area of concealment to determine if the route of travel is a safe one. Evaders should first make a quick overall survey for obvious signs of any presence such as unnatural colors, outlines, or movement. This can be done by first looking straight down the center of the area they are observing, starting just in front of their position, and then raising their eyes quickly to the maximum distance they wish to observe. If the area is a wide one, evaders may wish to subdivide it as shown in Figure 23-36.

23.19.1. Steps to Observing the Surrounding Area

All areas may be covered as follows. First, the evader should observe the ground next to them. A strip about six feet deep should be looked at first. They may survey it by looking from right to left parallel to their front. Secondly, observe from left to right over a second strip farther out, but overlapping the first strip. Visually surveying the terrain in this manner should continue until the entire area has been studied. When a suspicious spot has been located, evaders should stop and observe it thoroughly, determine if it is a threat, and then act accordingly.

Figure 23-246 Observing

23.20. Evasion Considerations

- Is the enemy searching for the evader?
- What is the evader's present location?
- Are chances for rescue better in some other place?
- What type of concealment can be afforded in the present location?
- Where is the enemy located relative to the evader's position?

23.20.1. Evaders Steps to Movement

Having considered the necessity and risks of travel and determining that they need to move, evaders must:

- Orient themselves.
- Select a destination, alternates, and the best route.
- Have an alternate plan to cover foreseeable events.

23.20.2. Reasons of Evaders becoming Captured

Cautious execution of plans cannot be overemphasized since capture of evaders has generally been due to one or more of the following reasons:

- Unfamiliarity with emergency equipment.
- Walking on roads or paths.

- Inefficient or insufficient camouflage.
- Lack of patience.
- Noise, movement, or reflection of equipment.
- Failure to have plans if surprised by the enemy.
- Failure to read signs of enemy presence.
- Failure to check and recheck course.
- Failure to stop, look, and listen frequently.
- Neglecting evasion techniques when crossing roads, fences, and streams.
- Leaving signs of travel.
- Underestimating time required to cover distance under varying conditions.

23.20.3. Progress on the Ground

Evaders should understand progress on the ground is measured in stopover points reached. Speed and distance are of secondary importance. Evaders should not let failure to meet a precise schedule inhibit their use of a plan.

23.21. Movement Techniques Which Limit the Potential for Detection of an Evader (Single)

Evaders should constantly be on the lookout for signs of enemy presence. They should look for signs of passage of groups, such as crushed grass, broken branches, footprints, and cigarette butts or other discarded trash. These may reveal identity, size, direction of travel, and time of passage (Figure 23-37).

- Workers in fields and normal activities in villages may indicate absence of the enemy.
- The absence of children in a village is an indication they may have been hidden to protect them from action which may be about to take place.
- The absence of young men in a village is an indication the village may be controlled by the enemy.
- Knowledge of local population and enemy visual signals are very helpful.
- The times evaders choose to travel are as critical as the routes they select. The best time to evade will be dependent on the environment, enemy tactics, and customs of the local population.

Figure 23-257 Signs of Passage

23.21.1. Low Light Conditions

Evaders should try to make use of low light conditions. Low light or darkness provides concealment and in some cases there is also less enemy traffic. If, out of military necessity, the enemy is active during the hours of darkness or low light, evaders may then find it wise to move in the early morning or late afternoon. Night movement is slower, more demanding, and more detailed than daylight movement, but it can be done. If travel is to be done during darkness, the terrain to be traversed should be observed during daylight if possible. While observing the area to be traveled, evaders should give attention to areas offering possible concealment as well as the location of obstacles and hazards they may encounter on their route. If the evader has a map, a detailed study of it should be made. Certain features (ditches, roads, burned-off areas, etc.) may not be on the evaders' map. Such pre-travel reconnoitering of an area will give evaders a head start on knowing how to adapt their travel movements and camouflage from point to point. Low light conditions can degrade the capability of night vision devices (NVD) used by the enemy.

23.21.2. Routes

Evaders should try to memorize the routes they will take and the compass headings to their destinations. This information should not be written down on the map or on other pieces of material.

- If traveling through hilly country that provides cover, the military crest (two thirds of the way up) should be used as it may be the safest route. An evader traveling at this point on a hillside reduces the possibility of silhouetting themselves against the skyline at the top of the hill. Additionally, it enables them to move quickly away from an advancing enemy from above or below (Figure 23-38).

Figure 23-268 Military Crest

- When it is necessary to cross the skyline at a high point in the terrain, an evader should crawl to it and approach the crest slowly using all natural concealment possible. How the skyline is to be crossed depends on whether it is likely the skyline at that point is under hostile observation. Evaders may never be certain any area is not under observation. When a choice of position is possible, the skyline is to be crossed at a point of irregular shapes such as rocks, debris, bushes, fence lines, etc. An evader's route should avoid game trails and human paths on the tops of ridges.

- Evaders should move slowly taking the time to stop, look, and listen every few paces. Additionally, they should avoid making noises that will indicate their presence to the enemy. They should take advantage of all cover to avoid revealing themselves. Background noise can be either a help or hindrance to those who are trying to move quietly - both the evaders and the enemy. Sudden bird and animal cries, or their absence, may alert evaders to the presence of the enemy, but those same signals can also warn the enemy of an approaching or fleeing evader. Sudden movement of birds or animals is also something to look out for.

- If spotted, the evader should leave the area quickly by moving in a zigzag route to their goal if at all possible. If the enemy finds evidence of the evaders' passage, their route may help confuse the pursuer as to direction and goal.

23.21.3. Techniques to Conceal Evidence of Travel

The following are some techniques of limiting or concealing evidence of travel. Evaders should:

- Avoid disturbing any vegetation above knee level. Evaders should not grab at or break off branches, leaves, or tall grass.

- Move gently through tall grass or brush, pushing your feet through the grass rather than stomping it down. Avoid using thrashing movements of your hands and arms which may also disturb vegetation. A walking stick may be used to part the vegetation in front, and then it can be used behind to push the vegetation back to its original position. The best time to move is when the wind is blowing the grass.
- Realize that grabbing small trees or brush as handholds may scuff the bark at eye level. The movement of trees can be spotted very easily from a distance. In snow country, using trees may mark a path of snowless vegetation that can be spotted from a distance when tracks cannot.
- Select firm footing and place the feet lightly but squarely on the surface avoiding the following:
    - Overturning duff, rocks, and sticks.
    - Scuffing bark on logs and sticks.
    - Making noise by breaking sticks.
    - Slipping, this may make noise.
    - Mangling of low grass and bushes that would normally spring back.

23.21.4. Masking Tracks in Snow

Evaders can mask their tracks in snow by:

- Using a zigzag route from one point of concealment to the next and, when possible, placing the unavoidable tracks in the shadows of trees, brush, and along downed logs and snowdrifts.
- Restricting movement before and during snowfall so tracks will fill in. This may be the only safe time to cross roads, frozen rivers, etc.
- Traveling during and after windy periods when wind blows clumps of snow from trees, creating a pockmarked surface which may hide footprints.
- Remembering that manufactured or improvised snowshoes leave bigger tracks, but they are shallower tracks which are more likely to fill in during heavy snowfall. Branches or bough snowshoes make less discernible prints.

23.21.5. Avoiding Tracks in Loose Soil

Evaders' tracks in sand, dust, or loose soil should be avoided or else marked by:

- Using a zigzag route, making use of shadows, rocks, or vegetation to walk on to mask or prevent tracks.
- Wrapping cloth material loosely about the feet as in a Hudson Bay Duffel, in order to make tracks less obvious.

23.21.6. Movement Before or During Wet or Windy Weather

By moving before or during wet or windy weather, evaders may find that their tracks are obliterated or worn down by the elements. Along roadways, evaders should be particularly cautious about leaving their tracks in the soft soil to the side of the road. They should step on sticks, rocks, or vegetation. They shouldn't leave tracks unless there are already tracks of locals on the road and their tracks can be made to look like the existing ones (bare feet, sandals, or enemy footgear). Rolling across the road is a method of avoiding tracks. Walking in wheel ruts with the toes parallel to the road will help conceal tracks. If the road surface is dry, sand, dust, or soil tracks may be brushed or tamped out to make them look old or will help the wind erase them more quickly. This must be done lightly, however, so the tracks do not look as if they have been deliberately swept over. Mud will retain footprints unless the mud is shallow and there is a heavy rain. Evaders should try to go around these areas.

23.22. Movement Techniques Which Limit the Potential for Detection of Evaders (Group)

Many principles and techniques which work for the individual are also appropriate for use in groups. While it is not true that the only safe way to evade is individually, there is a certain danger in moving with a group. It is generally not advisable to evade in a group larger than three. If possible, the senior person should divide the group into pairs. Paired individuals must be compatible. Group evasion can be advantageous because supplementary assistance is available in case of injury, in defense against hostile elements, in travel over rough terrain, and it provides moral support.

- In group evasion, movement becomes critical. But remember, movement attracts attention.
- The intervals and distances between individuals in the group should be made according to the terrain, vegetation, and the light conditions. Intervals will probably need to be greater during the day. The natural but dangerous tendency to "bunch up" is to be avoided when traveling in a group.
- If the undergrowth is light, the interval between evaders must be greater. Added consideration must be taken in deciding whether to travel at night or during the day. The leader will direct and guide the group to and from the best positions. All communications within the group should be made with silent signals only. Security in group evasion is of paramount importance. All members should stay alert. Security posts, lookouts, or guards should be designated for periods of rest or stopping.

23.22.1. Evasion Formations

Various formations are available for use by the evading group during periods of travel. The group must be flexible and able to adapt to changes in the conditions of the situation. The type of formation may also change with the route. In choosing a formation, the following points should be considered:

- Group control and intercommunications
- Security
- Terrain
- Speed in movement
- Visibility
- Weather
- Enemy presence
- The need for dispersion

- Flexibility of change in speed and direction of travel

A formation is merely the arrangement of individuals within a group. This arrangement is designed to give the greatest dispersion consistent with adequate intercommunication, ease and speed of movement, and flexibility to change direction and speed of travel at a moment's notice; that is, close control. Any arrangement which provides the above advantages is satisfactory.

23.22.1.1. Squad File

In a squad file, the evaders are arranged in a single file, or one directly behind the other, at different distances. The distance will vary depending on need for security, terrain, visibility, group control, etc. It is primarily used when moving over terrain which is so restrictive that the squad cannot adopt any other formation. It is also used when visibility is poor and squad control becomes difficult.

23.22.1.1.1. Squad File Advantages

When people are in a squad file, it is easy to control the group and provide maximum observation of the flanks. This is a fast way to travel, especially in the snow.

23.22.1.1.2. Squad File Disadvantages

There are disadvantages to this type of movement. A major one is that this formation is visually eye-catching. All the noise of the group is also concentrated in one place. This type of formation is easily defined and infiltrated. Depending on how many people are in the group, the area they walk through can become very packed down and easily detected by the enemy. If the group in the squad file is small, some members may have to double up on jobs; however, the point or lookout scout should be in the lead and should only perform that job (Figure 23-39).

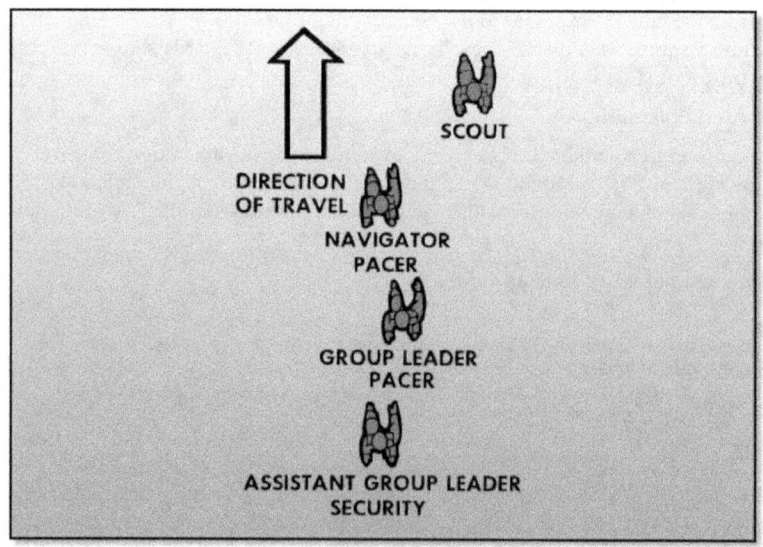

Figure 23-27 Squad File Formation

23.22.1.2. Squad Column

In the squad column, personnel are arranged in two files. The personnel are more closely controlled and yet maneuverable in any direction. There is greater dispersion with reasonable control for all-round security. It is used when terrain and visibility are less restrictive because it provides the best means of moving armed personnel into dispersed all-round security. It is easy to change into either the file or the line.

23.22.1.2.1. Squad Column Advantages

There are many advantages to this type of formation, a major one being a greater dispersal of personnel. Visually, this way of moving is less eye-catching, and the sounds the group makes are less concentrated. With this formation, the rate of travel is reasonable, yet there is no well trampled corridor for enemy trackers to follow.

23.22.1.2.2. Squad Column Disadvantages

There are a few disadvantages to movement in this manner. This formation of travel is hard to control in areas of dense vegetation with poor visibility and makes straying from the group likely. Although the paths may be faint, this mode of travel will also form a large number of trails. The rate of travel will be slower overall when traveling this way. An example of how personnel may be dispersed using the squad column methods is shown in Figure 23-40.

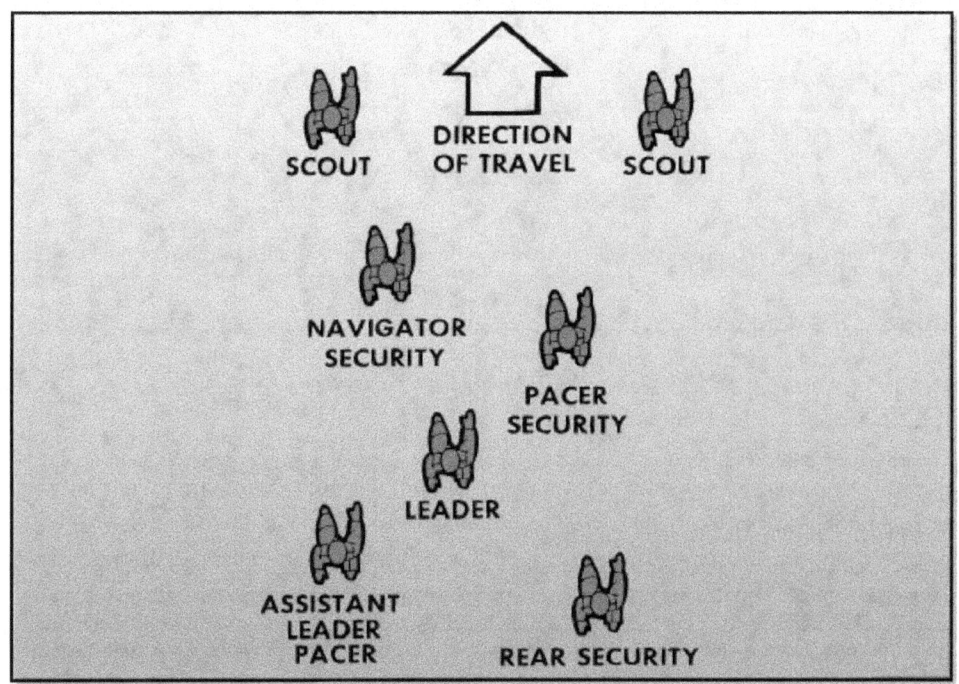

Figure 23-280 Squad Column Formation

23.22.1.3. Squad Line

In the squad line, all personnel are arranged in a line. This formation is used primarily by the Army as an assault formation because it is best for short, fast movements.

23.22.1.4. Squad Line Advantages

The advantage of this formation is that it is the quickest way to cross such obstacles as roads, fences, and small open spaces. It provides for tight control of individual movement while providing security for short-distance moves.

23.22.1.5. Squad Line Disadvantages

The disadvantages are that there are extreme communication and control problems. Some personnel in this formation may be forced to traverse undesirable rough terrain in contrast to the other two formations. Figure 23-41 illustrates the organization of personnel in this type formation. The speed at which these formations will progress will vary with terrain, light, cover, enemy presence, health of personnel, etc.

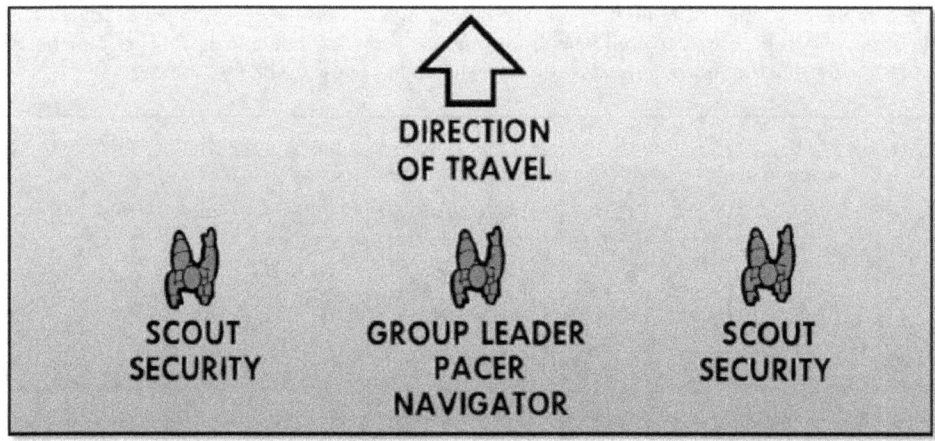

Figure 23-291 Squad Line Formation

23.22.2. Movement Techniques

Regardless of the formation used, the evader should pay particular attention to the technique used to travel. Knowing how to walk or crawl may make the difference between success and failure.

23.22.2.1. Walking Techniques

Correct walking techniques can provide safety and security to the evader. Solid footing can be maintained by keeping the weight totally on one foot when stepping, raising the other foot high to clear brush, grass, or other natural obstacles, and gently letting the foot down, toe first. Feel with the toe to pick a good spot, solid and free of noisy materials. Lower the heel only after finding a solid spot with the toe. Shift the weight and balance in front of the lowered foot and continue. Take short steps to avoid losing balance. When vision is impaired at night, a wand, probe, or staff, should be used. Care should be taken to avoid using the wand, probe, or staff like a walking stick leaving evidence of travel. By moving these aids in a figure-eight motion from near the ground to head height, obstructions may be felt.

23.22.2.2. Crawling Techniques

Another method of movement is by crawling. Crawling is useful when a low silhouette is required. There are times when evaders must move with their bodies close to the ground to avoid being seen and to penetrate some obstacles. There are three ways to do this: the low crawl, the high crawl, and the hands-and-knees position. Evaders should use the method best suited to the condition of visibility, ground cover, concealment available, and speed required.

23.22.2.2.1. The Low Crawl

This can be done either on the stomach or back, depending on the requirement. The body is kept flat and movement is made by moving the arms and legs over the ground (Figure 23-42).

Figure 23-302 Low Crawl

23.22.2.2.2. The High Crawl

This is a position of higher silhouette than the low crawl position, but the high crawl position is lower than the hands and knees position. The body is free of the ground with the weight of the body resting on the forearms and lower legs. Movement is made by alternately advancing the right elbow and left knee, left elbow and right knee, with elbows and knees laid flat on their inside surfaces (Figure 23-43).

Figure 23-313 High Crawl

23.22.2.3. Controlled Movement

The low crawl and high crawl are not always suitable when very near an enemy. They sometimes cause the evader to make a shuffling noise which is easily heard. On the other hand, carefully controlled movement can be made to be silent, and these techniques present the lowest possible silhouettes.

23.22.2.4. The Hands-and-Knees Crawl

This position is used when near an enemy. Noise must be avoided, and a relatively high silhouette is permissible. It should only be used when there is enough ground cover and concealment to hide the higher silhouette involved (Figure 23-44).

Figure 23-324 Hands and Knees Crawl

23.23. Rally Points

If evaders are moving as a group and are forced to disperse, being able to account for everyone after regrouping is important. Likely locations for rallying points are selected during map study or reconnaissance.

23.23.1. Selecting a Rallying Point

Rally points provide a group of evaders with a location to join back up should they become separated. Select rally points that are easily identifiable such as creek/river intersections, GPS coordinates, offsets from road intersections, etc.

- Always select an initial rallying point. If a suitable area for this point is not found during map study or reconnaissance, select it by grid coordinates or in relation to terrain features.
- Select likely locations for rallying points en route.
- Plan for the selection and designation of additional rallying points en route as the evaders reach suitable locations.
- Plan for the selection of rallying points on both near and far sides of danger areas which cannot

be bypassed, such as trails and streams. This may be done by planning that, if good locations are not available, rallying points will be designated in relation to the danger area; for example, "50 yards this side of the trail," or "50 yards beyond the stream."

23.23.2. Use of Rallying Points

If dispersed by enemy activity or through accidental separation, each evader in the group should be prepared to evade, individually, to the regrouping (rallying) point to arrive at a pre-designated time. If this is not possible, the individual will become a "lone evader." The group should not make any effort to locate someone not reaching the rally point. The group should formulate a new plan with new rallying points and clear the area. Rallying points should be changed or updated as they are passed. Points should not be directly on the line or route of travel. By selecting points off line, the job of enemy searchers or trackers is made more difficult and the chance of being "headed off" or "blind stalked" by the enemy is reduced. If the group is dispersed between rallying points en route, the group rallies at the last rallying point or at the next selected rallying point as decided by the group.

23.23.3. Actions at Rallying Points

Actions to be taken at rallying points must be planned in detail. Plans for actions at the initial rallying point and rallying points en route must provide for the continuation of the group as long as there is a reasonable chance to evade as a group. When at the rally point, initial plans should be new plans should be established given the situation. An example of a plan would be for the group to wait for a specified period, after which the senior person present will determine actions to be taken based on personnel and equipment present. Even during movement phases, it is important to check on the presence and status of all group members. A low toned, actual head count starting at the rear of the formation might be one way to do this; hand signals are another.

23.24. Barriers to Evasion Movement

Barriers to evasion can be divided into natural barriers, such as rivers or mountains, and manmade barriers, such as border guards or fences or roads. Some of these obstacles may be helpful while others might be a hindrance.

23.24.1. Rivers and Streams

When crossing rivers and streams, bridges and ferries can seldom be used since the enemy normally establishes checkpoints at these locations. This leaves a choice of fording, swimming, liferaft, or using some improvised method.

23.24.2. Mountains

When traveling in mountainous regions, evaders should not forget to use the military crest if concealment is available. In plains areas, evaders should use depressions, drainages, or other low spots to conceal their movements. Route selection should be planned with the utmost care to avoid unnecessary delays caused by terrain.

23.24.3. Vegetation

Some swamps, drainage areas, and thickets may be too thick for evaders to penetrate, and may require that a detour or alternate route be used. If the vegetation can be moved through, evaders should take care not to leave evidence of passage by disturbing the growth.

23.24.4. Weather

Weather can sometimes be used to screen evaders from the enemy. Certain weather conditions mask the noise made by traveling. Moving during a rainstorm may erase the footprints left by an evader; but after the rain the soft soil will leave definite signs of passage. Thunder may mask the sounds evaders make, but lightning may cause them to be seen. Snowstorms may be used to cover evaders' signs and sounds, but once the storm is over, evaders must use extreme care not to leave a trail.

23.24.5. Manmade Barriers

Evaders may also encounter a wide variety of manmade barriers while traveling within enemy territory or when attempting to leave a controlled area. As a general rule, evaders should not attempt to penetrate these barriers if they can be bypassed. If an analysis of the situation reveals the barriers cannot be bypassed, evaders must be skilled in the methods and techniques for dealing with specific manmade barriers to evasion. If possible, move to a less fortified (controlled) area or find a better damaged area in the barrier.

23.24.5.1. Trip Wires

These wires may be attached to pyrotechnics, booby traps, sensor devices, mines, etc. These wires are normally thin, olive green (or other colors that blend with the environment), strong, and extremely difficult to see. A supple piece of wood with a piece of string attached can be improvised and used as a wand to detect these wires (Figure 23-45).

Figure 23-335 Trip Wires

- A tripwire may be set up to be from one inch to a number of feet above the ground and to extend any number of feet from the device to which it is attached.

- The pressure necessary to activate a sensing device, mine, pyrotechnic device, or other trap-associated device to which the wire may be connected can vary from a few ounces to several pounds. The evader must be very careful when attempting to determine the presence of these devices.

- Once a tripwire has been detected, evaders should move around the wire if possible. If not possible, they should go either over or under it. They should not tamper with or cut the wire. If one device is discovered, be alert for backup devices.

- A number of devices activated by tripwires have a combination pressure-release arming mechanism. Cutting the wire or releasing the tension in the wire may activate the device. Some devices are electrically activated when there is a change in the current flowing through the attached wire - either because the wire is cut, in some way grounded, or otherwise altered. Evaders should take extreme care to avoid touching tripwires, but if contact is made, they must try to avoid sufficient pressure for activation.

23.24.5.2. Illumination Flares

Illumination flares can be activated by evaders themselves by a tripwire, by the enemy in the form of electronically activated ground flares, or by flares dropped from an overflying aircraft. Other overhead flares may be fired by mortars, rifles, artillery, and hand projectors.

23.24.5.2.1.   Overhead Flares

If evaders hear the launching burst of an overhead flare, they should, if possible, get down while it is rising and remain motionless. If evaders are caught in the light of a flare when they blend well with the background, they should freeze in position and not move until the flare goes out. The shadow of a tree will provide some protection. If caught in an open area, evaders may elect to crouch low or lie on the ground and, as a general rule, should not move after the area is illuminated.

23.24.5.2.2.   Ground Flares

If evaders are caught in the light of a ground flare and their position is such that the risk of remaining is greater than that of moving, they should move quickly out of the area. If within a series of obstacles or an obstacle system, evaders must remember running can be extremely dangerous because of the obstacles in the area and the fact that movement, especially fast movement, catches the eyes of an observer. If it is determined they cannot quickly move out of the area because of possible serious injuries due to existing obstacles or because they may be observed by the enemy, evaders should drop to the ground and conceal themselves as much as possible. Evaders should remember the light of a flare (either ground or overhead bursts) is temporarily blinding and the eyes should be covered to conserve night vision.

23.24.5.3. Fences

Various types of chain and wire fences may hinder the progress of evaders who are moving to the safety of friendly areas.

23.24.5.3.1.   Electrical Fences

For indications of electrical fences, security or livestock, evaders should watch for dead animals, insulators, flashes from wires during storms, humming sound, and short circuits causing sparks (Figure 23-46).

Figure 23-346 Electrified Fences

23.24.5.3.2. Barbed Wire

Evaders should use a wand to check for booby traps between strands of multi-strand barbed wire. Generally, they should penetrate the fence under the wire closest to the ground with the body parallel or perpendicular to the wire, depending on circumstances (Figure 23-47). They should lie flat on their backs both to project the lowest possible silhouette and to provide good visibility of the wire against the sky. A stick can sometimes be used to lift the barbed wire. If the lowest barbed wire is close to the ground and is tight, evaders may have to modify their approach to the problem.

Figure 23-357 Penetrating Barbed Wire Fence

23.24.5.3.3. Concertina Wire

Concertina wire is penetrated with the body perpendicular to the wire using a probe to lift the concertina wire if it is not secured to the ground (Figure 23-48). If the wire is secured to the ground, the evader can crawl between the loops. If two loops are not separated enough, they may be tied apart using shoe laces, string, suspension line, or strips of cloth. After passing through, the ties should be removed for future use and to erase evidence of travel.

Figure 23-368 Penetrating Concertina Wire

23.24.5.3.4. Chain Link Fences

Chain link fences are often used around highly sensitive zones, which are typically more highly guarded and patrolled than other areas. There also may be other traps or devices installed. The fence may also be electrified. If the fence must be penetrated, the evader should go under it if possible. If digging is required, the soil should be placed on the opposite side so it can be replaced to remove evidence. Climbing the fence is recommended only as a last resort.

23.24.5.3.5. Rail and Split Rail Fences

Evaders may encounter rail and split-rail fences while evading or escaping. The fences are penetrated by going under or between lower rails if possible. If not, evaders should go over at the lowest point, projecting as low a silhouette as possible (Figure 23-49). They should check between the rails and on the other side of the fence to detect tripwires or booby traps. Firmness of the ground should be checked on both sides of the fence. The body should be parallel to the fence before penetration.

Figure 23-49 Penetrating Log Fence

23.24.5.4. Raked or Plowed Areas

Raked or plowed areas may be found in areas of both low and high density security. If such an area is encountered, it should be avoided if possible. If it can't be avoided evaders should roll across the area, after making sure it is not a mine field, to avoid leaving footprints; or they may side-step, walk backwards, or brush out footprints. Any of the above may be done when it is a requirement not to leave clear-cut evidence of the direction of movement.

23.24.5.5. Roads

Roads are common barriers to evasion and escape. When roads are encountered, evaders should closely observe the road from concealment to determine enemy travel patterns (Figure 23-50). Crossing from points offering best concealment such as bushes, shadows, etc., is best (Figure 23-51). Evaders should cross at straight stretches of road in open country and from the outside to the inside of curves in hilly or wooded areas. This allows the evader to see in both directions so the chance of being spotted or surprised in the open is minimized. Avoid leaving tracks both in the road and on the shoulder of the road.

Figure 23-370 Manmade Barrier to Evasion

Figure 23-381 Crossing Roads

23.24.5.6. Culverts and Drains

Culverts and drains offer excellent means of crossing a road unobserved (Figure 23-52). They should be checked for trip wires and hazardous animals/insects. Beware of getting stuck or leaving evidence of travel.

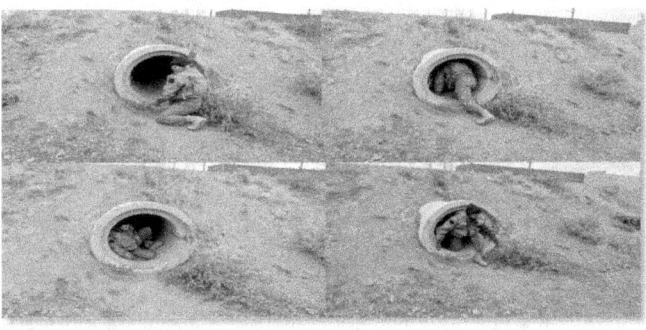

Figure 23-392 Crossing At Culverts

23.24.5.7. Railroad Tracks

Railroad tracks often lie in the path of evaders. If so, evaders should use the same procedures for observation as for roads. If it is determined that tracks are patrolled, a check should be made for booby traps and tripwires between tracks. Aligning the body parallel to the track with face down next to the first track, evaders should carefully move across the first track in a semi-pushup motion, repeating for the second track and subsequent sets of tracks (Figure 23-53). If there is a third rail, they should avoid touching it as it could be electrified. Sound detectors can also be attached and can reveal any crossing if a track is touched. If determined from observation that the tracks are not patrolled, evaders should cross in a normal walking or hands-and-knees manner, attempting to attract as little attention as possible. Evaders should try to keep their hands and feet on the railroad ties to prevent leaving foot or hand prints in the adjacent soil or gravel.

Figure 23-403 Crawling Over Railroad Tracks

23.24.5.8. Ditches

Deep ditches (such as tank traps or natural drainages) may be obstacles with which evaders must deal. Ditches should be entered feet first to avoid injury to the head or upper torso should there be large rocks, barbed wire, or other hazards at the bottom (Figure 23-54). Using a wand to detect tripwires and booby traps in the ditch and on the sides is highly recommended. Maintaining a low silhouette upon exiting the ditch is imperative.

Figure 23-414 Climbing Down into Drainage

23.24.6. Open Terrain

Open terrain complicated by guard towers or walking patrols is a definite hazard to evaders. These areas should be avoided if possible. If it becomes necessary to traverse open terrain or come near guard towers, evaders should stay low to the ground and, when possible, travel at night or during inclement weather. Use terrain masking since night vision devices may be used near border areas (Figure 23-55).

Figure 23-425 Open Terrain

23.24.6.1. Artificial Obstacles

In areas where there is no well-defined terrain feature to indicate the border, artificial obstacles such as electrified or barbed wire fences, augmented with tripwires, anti-personnel mines, or flares may be encountered. Open areas may be patrolled by humans or dogs, or both, particularly during the hours of darkness. The enemy may also use floodlights and plowed strips as aids to detecting evaders. The plan to cross a border must be deliberate and must be designed to take advantage of unusually bad weather (as a major distraction to the enemy force) or areas where security forces are overextended. These areas are usually found where there are natural obstacles.

23.24.6.2. Crossings

Crossings should be made at night, when possible, in battle-damaged areas. If it is necessary to cross, evaders should select a crossing point which offers the best protection and cover. They should then keep the area under close observation to determine:

- The number of guards in the area.
- The manner of their posting.
- Aerial patrols and their frequency.
- The limits of the areas they patrol.
- Location of mines, flares, or tripwires.

23.24.7. FEBA

When evaders are near the FEBA they should also watch out for friendly patrols. If a patrol is spotted, evaders should remain in position and allow the patrol to approach them. When the patrol is close enough to recognize them, evaders should have a white cloth displayed before the patrol gets close enough to see the movement. A patrol may fire at any movement. Shouting or calling out to them jeopardizes both the patrol and the evader. Evaders should stand silently with hands over their head and legs apart so their silhouettes are not threatening. If evaders elect not to make contact with a patrol, they should, if possible, observe their route and approach friendly lines at approximately the same location. This may enable them to avoid mine fields and booby traps. (NOTE: The practice of following any patrol is extremely dangerous. The last persons in line are charged with security, and anyone following them would be considered hostile and eliminated.)

23.24.8. Dogs

Dogs can be a threat to the evader. In this section, the term dog is meant to describe only the animals specifically trained in the areas of patrolling, guarding, and searching. Evaders must be aware they are working against both the dog and its handler. Evaders have a better chance of defeating the dog handler than the dog itself in most cases. If evaders are physically capable, they should attempt to maintain the maximum distance possible from dogs. Moving fast through rugged terrain will slow and probably defeat the handlers of dogs. Here evaders must choose between making mistakes in travel techniques while evading or being caught by dogs if they don't move fast enough.

- Dogs track better when the weather is humid and the air is still - there is less evaporation and dissipation of odor.
- If evaders know they are being followed by dogs, they should try to stay and move in water, if

available, for as long as possible to conceal their tracks and eliminate their scent.

- Evaders should always attempt to move downwind of a dog. This should be attempted when they are traveling in open country or penetrating obstacles such as dog guard posts or border areas. If penetration of obstacles or escape is planned (after careful location and study of guard and dog areas of responsibility and their methods), evaders should select a time for movement when noise will distract the dog to a point away from the planned maneuver.

23.25. Evasion Aids

23.25.1. Personal Survival Kits (PSK)

Personnel may sometimes find it practical to devise and carry compact PSK to complement issued survival equipment. If Personal Recovery Kits (PRK) are provided, potential evaders should be familiar with them, their contents, uses, and limitations.

23.25.2. Maps and Charts

Any maps or charts of the area in an evader's possession should not be marked. A marked map in enemy hands can lead to the compromise of people and locations where assistance was given. Evaders should be wary of even accidentally marking a map; for example, soiled fingers will mark a map just as plainly as a pencil.

23.25.3. Pointee-Talkee

The "pointee-talkee" is a language aid which contains selected phrases in English on one side of the page and foreign language translations on the other side. To use it, evaders determine the question and statement to be used in the English text and then point to its foreign language counterpart. In reply, the natives will point to the applicable phrase in their own language; evaders then read the English translation.

23.25.3.1. "Pointee-Talkee" Limitation

The major limitation of the "pointee-talkee" is in trying to communicate with illiterates. In many countries the illiteracy rate can be astoundingly high, and personnel have to resort to pantomime, sign language, or issued or commercially available Visual Language Survival Guides which have been relatively effective.

23.25.3.2. "Pointee-Talkee" Subheadings

"Pointee-talkee" phrases are presented under the following eight subheadings:

- Finding an interpreter.
- Courtesy phrases.
- Food and drink.
- Comfort and lodging.
- Communications.
- Injury.
- Hostile territory.
- Other military personnel.

23.25.4. Barter Kits

Barter kits are no longer available or an accepted TTP. Barter Kits have resulted in indigenous personnel and enemy military personnel hunting IP for the valuables contained in the kit. This can result in the death or injury of IP.

23.25.5. Other Evasion Aids

Information on other evasion aids and tools is available from unit intelligence personnel and SERE Specialists.

23.25.6. Assisted Evasion

The success of any assisted evasion organization depends almost entirely upon its security. Assisted evasion systems include much planning and work carried out under dangerous conditions. The security of the system often depends upon the evader's cooperation and working knowledge of how it functions, how it may be contacted, and what rules of personal conduct are expected of the evader. History has revealed that in every major conflict there are groups of people in every country who will aid a representative of their government's enemies to evade or escape. The motivating force behind these groups may vary from monetary considerations, hatred of the current leadership, political recognition for their cause, idealistic concepts of government reform, religious or cultural norms, or a combination of some or all of these.

23.25.7. Unplanned Assisted Evasion

These unplanned acts usually fall into one of three categories: Acts of Mercy, Opportunists, or Accidental. These are usually isolated events during which evaders may be provided food, shelter, or medical attention for a brief period of time.

23.25.7.1. Acts of Mercy

Local people may find an exhausted or incapacitated evader and as an act of mercy provide that evader with limited sustenance. This type of assistance is frequently offered with reluctance or under fear of reprisal because an act of providing comfort to the enemy would mean punishment. Unless an evader is in immediate need of medical attention, acts of mercy may consist of only an offer of food followed by an urgent plea that the evader leaves the area immediately. If an evader is physically able to depart with a reasonable chance of evading capture, he or she should do so. An evader should not insist on receiving additional aid other than what is offered by the person who renders assistance.

23.25.7.2. Opportunist

Sometimes local people will know of the presence of an evader in the area and actively seek them out for gain. This gain could be prestige, monetary, political, or some other reason for themselves or organization. The IP can attempt to use their Blood Chit but should always remember this local person could be out for their own gain.

23.25.7.3. Accidental

Accidental contact and assistance could possibly fall into either of the two previous categories but is truly unplanned by the IP. The IP may attempt to use their Blood Chit or determine their best option is to break contact and continue evasion.

23.25.8. Indigenous People

Evaders must understand when dealing with any indigenous personnel while in enemy territory, their own actions will often govern the treatment they will receive at the hands of the local people. How evaders conduct themselves may also have much to do with getting back to their own forces should they fall into the hands of irregulars friendly to the evaders' own cause. The following suggestions may be a useful guide in dealing with these people or groups.

- Evaders should understand that failure to cooperate or obey may result in death.
- Evaders should try to avoid making any promises they cannot personally keep.
- Evaders who support military actions with assisted evasion personnel if captured could face legal repercussions. Under the Geneva Conventions the closest thing to a guarantee of treatment as a military person if captured will be their uniforms. If evaders have the opportunity to influence the group they are with, they should try to encourage them to abide by the four conditions mentioned in the conventions. Evaders must avoid becoming associated in any way with atrocities these groups may commit against civilians, prisoners, or enemy soldiers.
- Evaders should not become involved in political or religious discussions, take sides in arguments, or become involved with the opposite sex.
- Evaders should show consideration for being allowed to share food and supplies. It may also be helpful if evaders understand and show interest in the assister's customs and habits.
- The overall best and safest course for evaders to follow is to exercise self-discipline, display military courtesy, be polite, sincere, and honest. Such qualities are recognized by any group of people throughout the world. The impression left can influence the aid provided to future evaders.

23.25.9. Planned Assisted Evasion

Special Operations Forces may also organize and operate Non-Conventional Assisted Recovery (NAR) mechanisms. They may be limited in nature, such as providing first aid and assistance to reach a national frontier, or they may be linked to larger organizations capable of meeting almost all needs to include returning the evader to friendly control.

23.25.9.1. Making Contact with an Assisted Evasion System

Pre-mission briefings may inform evaders where to go and what actions to take to make contact. After being picked up by an assisted evasion mechanism, evaders will be moved under the control of this mechanism to territory under friendly control or to a recovery area. The organizer of a line in friendly but enemy-occupied territory normally will have arranged a network of spotters who will be especially active when evaders are in the immediate area, but so will the enemy police and counterintelligence organizations. For this reason, certain precautions must be observed when making contact.

23.25.9.2. Approach

Help may be refused by a person simply because he or she thinks someone else has seen the evaders approach to seek assistance. If captured with a local helper, an evader will become a prisoner, but the helper and perhaps an entire family may be more severely punished.

23.25.9.3. Making Contact

Contacts with the local population are discouraged unless observation shows they are dissatisfied with the local governing authority, or previous intelligence has indicated the populace is friendly. The spotter will know evaders are present and will search the immediate area, making frequent visits to designated contact points. Identification signs and countersigns, if used, may be included in the permission briefings. It is seldom advisable to seek first contact in a village, hamlet, or town. Strangers are conspicuous by day, and there may be curfews or other security measures during the hours of darkness. The time of contact should be at the end of the daylight period or shortly thereafter. Darkness will add to the chance of escape, if the contact proves to be unfriendly, and may be advantageous to the contact in providing further assistance.

23.25.9.4. Procedure after Contact

If contact is made, evaders may be told to remain in the vicinity where spotted, or, more likely, they will be taken to a holding area. It must be decided at this time whether or not to trust the contact. If there is any doubt, plans should be made to leave at once. It is also possible that the location may not belong to the organization but rather to someone who will look after evaders until arrangements can be made for the line to identify and accept them.

23.25.9.5. Establishing Identity

Verification of identity will likely be required before an evader is accepted. The constant danger facing the operators of an assisted evasion mechanism is the penetration of the system by enemy agents pretending to be evaders or escapees. Evaders should be prepared to furnish proof of identity or nationality. Since it may lead to later difficulties of identification, evaders should never give a false name - just their name, grade, service number, and date of birth. It is best for them to avoid talking as much as possible.

23.25.9.6. Awaiting Movement

Delays can be expected while proceeding along the mechanism. If the period of waiting is prolonged, frustration and impatience may become unbearable, leading to a desire to leave the holding area. This must not be done, because if seen by other people, the lives of the assisting personnel and the existence of the entire organization may be endangered. Evaders should follow the directions of those assisting them and maintain situational awareness. The assisters should have a plan for rapid evacuation of the area if enemy personnel should raid the holding area; if not, evaders should have a personal plan to include measures for removing all traces of having occupied the area. If the area is being overrun and capture is imminent, evaders must be prepared to fend for themselves. The assisters may attempt to eliminate the security risk to the network.

23.25.9.7. Traveling

It would be a grave breach of faith and security for evaders to discuss any of the earlier stages of the journey. Evaders might be tested to see if they are trustworthy. For security reasons and to protect the compartmentalization of the mechanism, no information should be revealed. It is also useless to ask where the evader is going or how they will eventually reach friendly territory. Evaders should not try to learn or memorize names and addresses and, above all, they shouldn't put these facts or any other information in writing. Evaders should give the impression of having received no assistance from local inhabitants.

23.25.9.8. Fellow Evaders

Caution is required in the case of fellow evaders during assisted evasion unless they are personally known. Even when it has been satisfactorily determined that another person is a genuine evader, no information should be given.

23.25.9.9. Travel with Guides

If under escort, this fact should not be apparent to outsiders. In a public vehicle, for example, evaders should never talk to their guide or appear to be associated with the person in any way unless told to do so. This will lessen the possibility of both the evader and the guide being apprehended if one should arouse suspicion. It should always be possible for the guide to disown an evader if the guide gets into difficulty. When evaders are escorted, they should follow the guide at a safe distance, rather than walk right next to the person, unless instructions indicate the latter action is required.

23.25.9.10. Speaking to Strangers

Evaders should never speak to a stranger if it can be avoided. To discourage conversation in a public conveyance, they can pretend to read or sleep.

23.25.9.11. Personal Articles and Habits

Evaders should not produce articles in public which might show their national origin. This pertains to items such as pipes, cigarettes, tobacco, matches, fountain pens, pencils, and wristwatches that may look out of place or identify them as an evader. Evaders should also ensure their personal habits do not give them away. For example, they should not hum or whistle popular tunes or utter oaths or curse words. Again, in restaurants, imitating local customs in the use of knives and forks and other table manners is advisable. Study of the area before the mission may help evaders avoid making mistakes.

23.25.9.12. Payment

During assisted evasion, evaders should not offer to pay for board, lodging, or other services rendered. These matters will be settled afterwards by those who are directing and financing the line. If in possession of PSKs or PRKs, evaders should keep them as reserves for emergencies. If they have no food reserve, they should try to build up a small stock in case they are forced to abandon the line.

23.25.9.13. Evaders Conduct during Assisted Evasion

- Be polite by local standards.
- Be patient and diplomatic.
- Avoid causing jealousy. Disregard the gender of the assisters.
- Avoid discussions of a religious or political nature.
- Eat and drink if asked, but don't over indulge or become intoxicated.
- Don't take sides in arguments between assisters.
- Don't become inquisitive or question any instruction.
- Help with menial tasks as directed.

- Write or say nothing about the other people or places in the net.
- Don't be a burden; care for self as much as possible.
- Follow all instructions quickly and accurately

23.26. Non-Permissive Recovery

Even though a maximum effort will be made to recover IP, they can jeopardize the whole recovery operation and the lives of rescue personnel by not taking a responsible role in the operation. The responsibilities are many and varied but essential to a successful rescue mission. Evaders should recall that even though they may have little experience participating in actual rescues, they must nonetheless be very proficient in their actions. Evaders should always remember other people are endangering their lives in an attempt to retrieve them.

23.26.1. Proficient of Survival Equipment

IP must be proficient in the operation of all the survival equipment at their disposal. The IP's ability to properly use all available communication and signaling devices could have a significant impact on the overall recovery operation.

23.26.2. Initial Contact

The initial contact with rescue aircraft or other forces in the area must be done as directed by authorities.

23.26.3. Evader Authentication

One important aspect of the recovery process is evader authentication by rescue personnel. IP must be able to authenticate their identity through the use of questions and answers, responding as directed or briefed before the mission. Authentication in a non-permissive environment changes rapidly to reduce the chances of compromising the rescue efforts. IP must keep up with these changes so their rescue can be made without undue danger to rescue personnel. The Isolated Personnel Report (ISOPREP) is a tool designed specifically for recovery.

23.26.4. ISOPREP

The purpose of filling out and using an ISOPREP is positive verification of an evader's identity which is essential before risking search and rescue aircraft or the lives of assisters. The purpose of the photographs, information, and descriptions is to ensure all possible means are used for the proper identification of personnel. The potential IP is responsible for ensuring that the ISOPREP is completely filled out and accurate, and must understand its purpose and use.

23.26.5. Selecting a Evasion Recovery Site

When selecting a site for possible evasion recovery, there is much for evaders to keep in mind. The area they choose could well decide the success or failure of the mission.

- If possible, the potential recovery site should be observed for 24 hours. Evaders should look for signs of enemy or civilian personnel, as well as indicators of local activity such as roads, trails, farming, orchards, etc.
- The recovery site should be observable by aircraft but unobservable from surrounding terrain if possible. There should also be good hiding places around the area. The site should include several escape routes so evaders can avoid being trapped by the enemy if discovered. There

should be a small open area for both signaling and recovery. It would be beneficial if the surrounding terrain provided a masking effect for rescue forces (air or ground) in order to avoid enemy observation and fire.

23.26.6. Ground-to-Air Signals

The size of evasion ground-to-air signals should be as large as possible but must be concealed from people passing by. The contrast these signals make with the surrounding vegetation should be seen from the air only. Any signal displays should be arranged so they can be removed at a moment's notice since enemy aircraft or personnel may transit the area.

23.26.6.1. Noncombat Signals

All of the principles of regular (noncombat) signals should be followed by those building evasion signals. IP may be pre-briefed as to the appearance of specially shaped signals.

23.26.6.2. Evasion Signals

Evasion signals, like all others, must be maintained to be effective. At times, evaders may be instructed to set out their signals according to a prearranged time schedule.

23.26.7. Minimizing Spotting of Signals

Extreme care must be taken to minimize or eliminate chances of enemy elements spotting the signals. For example, the strobe light and mirror can be directed and aimed instead of being used in an indiscriminate manner.

23.26.7.1. Signals in or on Water

If evaders are in or on the water, they should use lights, flares, dye, whistles, etc., with extreme care as they are readily distinguishable over water at great distances.

23.26.7.2. Evasion Radio Procedures

In addition to knowing how to use radios correctly to affect their rescue, evaders should also be familiar with the special points of evasion radio procedures; for example, radios using Direction Finding capability (DF), Personal Location Systems (PLS), and Combat Survivor Evader Locator (CSEL) system capabilities.

23.26.7.3. Signs of Enemy Activity

It is also the evaders' responsibility to communicate all signs of enemy activity such as:

- Locating anti-aircraft emplacements.
- Identifying when they are firing.
- Assisting strikes by spotting hits (high, low, or on target).
- Determining effectiveness of hits.
- Notifying personnel of changes in small arms positions, etc.

23.26.8. Vertical Lift Aircraft Recovery Methods

During a vertical lift aircraft recovery, if at all possible the recovery aircraft will land for extraction. If unable to land, evaders can normally expect to be hoisted by one of the following methods: rescue basket, Stokes litter, horse collar, or forest penetrator.

23.26.9. Fixed-Wing Rescue Capability

There is also fixed-wing capability of rescuing downed crewmembers. Evaders will be prebriefed as to which type rescue vehicle and systems to expect in their areas of operation.

23.27. Ground Rescue

Recoveries by ground forces under austere, field conditions are among the most potentially dangerous situations and require the highest degree of situational awareness from both the IP and the rescue forces. It is important to remember that these forces can be US, foreign, military, or civilian, so specific techniques used in IP recovery can vary greatly. Regardless of the force, IP should be prepared to posture themselves and perform appropriately on land, in water, and with or without communications equipment to insure a successful recovery.

23.27.1. IP with Communication Equipment

IP with communications equipment should follow the procedures outlined in their Evasion Plans of Action (EPA) and the SPINs. Except for dire circumstances, do not attempt 'blind' radio communication outside the windows in your EPA's unless you can see (and positively identify) friendly aircraft or personnel. When contacted by rescue forces, speak in short bursts to reduce the possibility of having your position fixed by enemy forces, even if this means breaking up a single statement into several separate transmissions. It may be wise to cross-authenticate the rescue force via radio prior to making physical contact. Be prepared for the rescue force to ask you to move to a more suitable location for them to positively identify you. If you cannot move or signal from your position, consider vectoring the recovery force to your position. Understand that this is a high-risk technique for the rescue force, and may provoke hostile action if your posture is in any way non-compliant.

23.27.2. IP without Communication Equipment

IP without communications equipment face a more difficult situation. Aircraft can signal via non-electronic means such as message streamers, and can also deliver support bundles by parachute delivery or simply by pushing them out and letting them fall to the ground. You must exercise care to signal rescue forces through other means as well, while staying discreet enough so as to avoid enemy detection. Be sure that your own signals are clear enough so as to be recognized by rescue forces. Stay close enough to your signal that you can see it, but in a location that would prevent someone at the signal from seeing you. Observe potential rescue forces for actions that positively identify them to you. Do not be surprised or afraid if the rescue forces do not "look" like you expected them to. Stay where you are and do not reveal your position until you are called upon by rescue forces. If you cannot move, tell them why and call them to you. Be prepared for a very cautious, aggressive posture from the rescue force.

23.27.3. Contact with Rescue Forces

Upon contact with rescue forces, the way you posture yourself will determine how you are handled. Remember that recovery forces in non-permissive environments will assume that you are a potential threat until proven otherwise. A non-threatening, compliant posture is best. Do not carry weapons in your hands. If you are uncomfortable with placing your weapon on the ground due to suspected imminent enemy action in the area, sling it behind you and place your hands on top of your head as you move toward the recovery force. Have your ID card out and in your hands, along with anything else you believe the recovery force may ask you for to avoid them having to reach into your pockets. Listen carefully and comply with their directions. Do not resist if they attempt to reposition or restrain you, and answer their questions to the best of your ability.

23.28. Water Recovery

Recovery in the water places unique demands on both isolated personnel and recovery forces alike. The lack of cover and the improbability of landing an aircraft require a similarly unique set of procedures and equipment to perform a water recovery. Rescue forces can be delivered by subsurface craft, scuba operations, surface craft, helicopter, or fixed wing assets. Helicopters can employ personnel, hoist cables with various rescue devices, and inflatable water craft. Fixed wing assets can employ freefall bundles, parachute bundles, inflatable watercraft, and personnel parachutists. In all situations, stay where you are and do not attempt to approach the rescue force or recovery platform unless directed to do so. Once rescue forces are in the water be prepared to signal them by any means available. Even waving arms and splashing water can be seen in the day or night as a last resort to be identified. If you are in a survival boat or raft, stay in it unless told to do otherwise. When approached, again take a non-threatening, compliant posture. Humans are naturally more vulnerable in the water than on land, but rescue forces' training and expertise put you at a severe disadvantage. Do not try to grab or hold onto them when they make contact with you. Comply with whatever rescue device they choose to employ to affect your recovery. Understand that this may include being strapped to a backboard if they suspect you may be injured.

Chapter 24

Urban

24. Urban

The United Nations estimates that each day 150,000 people move into or around cities, so by 2025, nearly two-thirds of the world's population will live in or around cities. In regard to future trends, it is estimated 93 percent of urban growth will occur in developing nations, with 80 percent of urban growth occurring in Asia and Africa. This has led military analysts to determine that open battlefields of the past are increasingly shifting to the more complex environments of the city. The IP's determination of a course of action and TTPs during urban conflicts should be guided by the root causes of the conflict such as political, social, cultural, religious, ethnic, or tribal. An urban isolating event could involve any of the non-permissive factors of wartime, peacetime, governmental, terrorists groups, insurgents, and/or criminal organizations.

24.1. Key Components

The urban environment includes the complex and dynamic interaction and relationships between its key components: urban area (man-made constructions), the terrain (natural and man-made), the population density, cultural factors, and the supporting infrastructure (Figure 24-1).

Figure 24-1 Urban Environment

24.1.1. Three Dimensional Urban Battle Spaces

Rescue forces, threat forces, and the IP will conduct operations in a three-dimensional urban battle space: *surface*, *above the surface*, or *below the surface* of the urban area. Additionally, engagements can occur inside and outside of buildings. Multi-story buildings will present the additional possibility of different floors within the same structure being controlled by either friendly or threat forces. The urban IP can break down the three-dimensional urban battle space into urban airspace, super-surface (tops of building), intra-surface (interior of buildings), surface (ground, street, and water level), and sub-surface (underwater and subterranean) (Figure 24-2).

Figure 24-2 Super Surface

24.1.2. Urban Areas

Urban areas usually surround ports and airfields. Troops and aircrew sent in to seize and/or secure these assets are likely to find themselves on the urban battlefield which could potentially lead to an urban isolating event.

24.1.3. Noncombatants

The presence of large numbers of noncombatants and their interaction with friendly or hostile forces plays a critical role in the outcome of the IP.

24.1.4. Captured or Detained Personnel

Captured or detained personnel may end up in an urban area:

- Major prison facilities, that might hold US personnel, are typically located in major cities.
- Even if not held in a major prison facility or urban area, US prisoners or detainees may utilize urban evasion TTPs to facilitate their escape.

24.1.5. Urbanization

In the cities of a developing country, urbanization traditionally forms as a concentration of human activities and settlements around the downtown area. There is a migration (called an in-migration) of people from rural locations and immigrants settle in city centers where industry and jobs are located.

24.1.6. Pre-Mission Planning

Pre-mission planning for a specific urban environment is no different than any pre-mission planning, start with familiarization of the local area. Additional areas of interest for potential IP include:

- Habits, customs, taboos, clothing, and general appearance of indigenous personnel.
- Recognizable features and landmarks such as:
    - Sports fields, cemeteries, statues, prominent buildings (mosques, towers, churches, etc), bridges, and natural features (rivers, mountains, cliffs, etc) (Figure 24-3).
- Areas within the urban area with vegetation that could provide concealment like parks.
    - Deserted battle-damaged areas (Figure 24-4).
    - Subterranean locations like basements, storm drains, sewers tunnels, subways, and, catacombs.

Figure 24-3 Baghdad Soccer Stadium

Figure 24-4 Kabul

- Not only does the IP need to recognize, but they also need to have a fixed orientation of these features/landmarks locations, so they will understand where they are located and where they need to travel in relationship to these features/landmarks. Additionally the potential IP should have a working knowledge of the lines of communication in example public telephones, power lines, pipelines, train and rail tracks, etc (Figure 24-5). Specific landmarks may have other uses besides navigation, an IP may have memorized identifiable exfiltration locations such stadiums, sports fields, parks, etc.

Figure 24-5 Famous Buildings Caracas, Venezuela

- Enemy forces may use the population to provide camouflage, concealment, and deception and look no different than other members of the community. Enemy forces may also use local population as a means of detecting and identifying evaders. Therefore, all indigent personnel should be avoided. Children can especially be hazardous to an urban IP, because unlike an adult, they might comment or act on observed IP actions or peculiarities (Figure 24-6). IP should also have a working knowledge of potential problematic/beneficial areas (depending on the isolating event) and population control measures such as:

    – Patrols or specialized units that look for IP and peculiarities of civil authorities (Figure 24-7).
    – Checkpoints, control zones borders, and false borders.
    – Police, military garrisons, minefields, prisons, potential targets, and air defense sites (Figure 24-8).
    – Religious sites, schools, market places, and food distribution sites (Figure 24-9).
    – Curfews, rationing, and civilian roundups

Figure 24-6 Children

Figure 24-7 Military Patrol

Figure 24-8 Special Police

Figure 24-9 City Market

24.1.6.1. Personal Survival Kit Items

An IP should consider additional personal survival kit items to create a disguise or to help blend into the local population.

24.1.6.2. Extreme Weather Conditions

While possibly be subjected to the normal "extreme" environmental hazards such as extreme weather conditions, during an urban isolating event IP should be aware of the specialized environmental hazards that can be encountered. At all times the IP must weigh capture against possible exposure and injury, keeping in mind these hazards may be an advantage during evasion, since the enemy would most likely avoid them. IP need to be aware of debris, falling rubble, collapsing walls, contaminates, polluted water, edible plants fertilized with human waste, gas leaks, downed power lines, mind-fields, and booby-traps (Figure 24-10).

Figure 24-10 Hazards

24.2. Five Phases of Urban Evasion

Evasion in the urban environment involves a different mindset for the IP. They must recognize and adapt the five phases of evasion to meet specific social-political, cultural, population, climatic, and urban environments they will encounter. For example, hiding in man-made structures, evading by calmly walking in plain-sight and blending into vastly different settings is an example of adapting to an urban setting.

24.2.1. Immediate Action and Initial Movement

Immediate action and initial movement are combined. This is due to the IP's immediate action of moving away from their isolating event location; i.e., leaving behind some type of crashed helicopter, wrecked vehicle, parachute landing area, etc. IP need to assume they have been and are being observed, so having a pre-determined course of action will allow for immediate and quick evasive action.

24.2.1.1. Immediate Departing

Immediately depart the isolating event location since the mechanism of isolation (downed aircraft, disabled vehicle, or parachute) will usually draw in the local population and or enemy forces, becoming the initial and centralized search point.

24.2.1.2. Environmental and Situational Aids

IP should use any environmental and situational aids to concealment, such as darkness, dust, haze, fog, smoke, ground cover, and intensity of any combat action ("fog of war"), to provide evasion windows of opportunity. If ground combat is taking place, IP should asses if staying put in cover or movement will place them in mortal danger (fatal fire from both friend and foe).

24.2.1.3. Historical Searches

Historically searches will begin at the last sighting location with additional forces being placed in possible lanes of travel to intersect the IP. To help IP break contact with initial searches or once spotted, they should change direction radically, possibly "hook" back the direction that they came from, offset left or right.

24.2.1.4. Hole Up/Hide Site

IP considering urban sites to use for a hole up/hide site should weigh general principles behind selecting any evasion hole up site, local customs and culture, population density and controls, as well as the ability to meet their needs. The IP must be able to apply common sense and their imagination when making their selection for a hole up/hide site. Just as in any evasion environments the urban evader should not hide in the only obvious location or where there is a great deal of contrast to call attention to your site (Figure 24-11).

Figure 24-11 Hide Site

24.2.1.4.1. Unoccupied and Undisturbed Areas

While an IP can select any structure for possible hole up/hide site, unoccupied and undisturbed areas might work best, though evaders have successfully hidden in occupied locations using the movements and disturbance of the inhabitants to mask their own. Hole up sites an urban evader could use are old or derelict buildings, sheds, factories, garages, ruins, rubble, subterranean areas, dumps, shops, churches, and alleys.

24.2.1.4.2. Selecting a Specific Hole Up Site

Once a structure or location has been selected a specific hole up site may be found in attics, basements, rooftops, steep slopes, in between walls, under floors, and crawl spaces.

24.2.1.4.3. Multiple Sites

If time and terrain permits the IP should attempt to use multiple sites and rotate between sites (Figure 24-12).

Figure 24-12 Hide Site

24.2.2. Movement

What type(s) of movement an IP uses in an urban area will be dependent on several factors such as the isolating event, the IP's mission, their evasion objectives, density of the local population, and their legal status. Depending on the factors identified above, as well as others, IP could end up using tactical, evasion, disguise, or a combination of all of these movements.

24.2.2.1. Tactical Movement

Tactical movement is not really an IP movement, but may lead to an isolating event. It is when operators have an objective/mission. During the execution of that objective/mission they try and avoid contact with the enemy. If contact is made with the enemy the ground forces usually will attack using "evasion through superior fire power". This technique is very dependent on the mission before the isolating event in example: the mission may have ground forces dressed to blend looking like indigenous homeless at a distance. When ground forces involved in an evasion movement make contact with the enemy they determine a course of action depending on chances of success (Figure 24-13).

Figure 24-13 Tactical Movement

24.2.2.2. Evasion Movement

Evasion movement is when the objective is to avoid any contact with the enemy (locals). It is similar to what has been discussed in the evasion chapter, moving from point of concealment to the next point of concealment avoiding any and all contact with the local population.

24.2.2.3. Disguise Movement

Disguise movement is when the objective is to avoid contact with enemy (locals), but when and if that contact occurs to be perceived as local. The success of the disguise movement is very dependent on the type of contact the IP has prepared for such as blending in to the local population at 20 feet away verses talking to a local. With this an IP has to continual ask themselves if their movement techniques are out of place or suspicious (Figure 24-14).

Figure 24-14 Disguise Movement

24.2.2.4. Mixing and Interspersing Movement Types

The IP should remember that in many cases these movement types can be mixed and interspersed depending on the events, IP and enemy actions, and the rules of engagements.

24.2.2.4.1. Observing for Movement

Prior to any movement the IP should try to observe the area for hours prior to leaving initial hide site. Observe for movement or life threatening obstacles. When moving, try to select the time of day that will be the least noticeable.

- Remember observation can be a two-way street, not using proper concealment and camouflage techniques during observation can lead to the IP being observed. Select the time of day when shadows are cast inside a hole up/hide site and will hide your silhouette.
- Avoid standing or silhouetting yourself in front of any doors, windows, or openings. Stand to the side, utilizing the corners for observation. If possible, utilize already available coverings (doors, curtains, debris, etc) to breakup outline.

24.2.2.4.2. Minimize Signature

When performing evasion movement an IP should stay low to minimize signature. They should stay near battle-damaged area and use clutter in streets and alleys to your advantage, avoiding intact buildings. An IP should utilize buildings, rubble, clutter, foliage, or ditches for terrain masking. And as in all evasion, urban or otherwise, an evader needs to stay alert, ready, and use common sense (Figure 24-15).

Figure 24-15 Evasion Movement

24.2.2.4.3. Moving Inside Stuctures

While moving inside a structure an IP should check structural integrity before entering. They should also avoid making noise by rubbing against wall, weighing the noise created while hiding verse the effectiveness of the hole up/hide site such as in the noise movement in a heating, ventilation and air conditioning (HVAC) ducts creates. An IP can use a small reflective material or their signal mirror to look around corner prior to moving into it. When an IP travels up a stairway they should avoid moving up the center due to noise, hugging the wall may make less sounds, though an IP should consider the construction of stairway and material used by stepping near support frame (nails, bolts, etc.).

24.2.3. Recovery

IP should be aware that that are additional recovery considerations under urban evasion such as rescue forces may have a much more limited time, recovery may tend towards the unconventional or unplanned assistance, and the a movement phase may have to occur to a more rural or isolated area for recovery to occur.

24.2.3.1. Prepare for Types of Recovery Mechanism

As in all isolating events, the IP should be prepared for any and multiple types of recovery mechanism (Figure 24-16).

Figure 24-16 Urban Extrication

24.2.3.2. Signaling and Communication

Additional IP considerations for signaling and communication with recovery forces under urban evasion conditions include:

- Air and light pollution which may adversely affect the effectiveness of signaling devises.

- Structures may limit a signaling devise's angle of visibility similar to being in a deep ravine (Figure 24-17).

- Being near populated areas may also advertently or inadvertently cause of signal interruption or jamming. Of course with the exception of encrypted radios, all other emanations are subject to monitoring.

- As with any environment, signaling in an urban environment must be rapidly discernable by recovery force and compact, reusable, and easy to operate. Visual and ground-to-air signal (GTAS) should be capable of observation from the hide site.

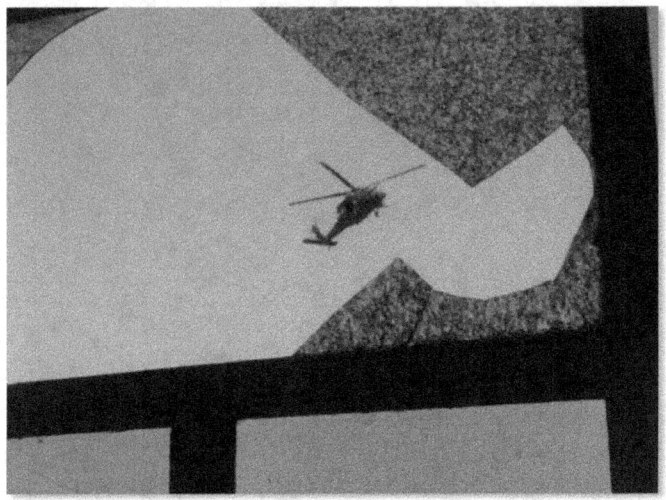

Figure 24-17 Structure Verses Signaling

24.2.3.3. Cell Phones

Depending on the situation leading up to and including the isolating event, an urban IP may find the use of a cell phone to be one of the best forms of communication. In an urban setting a cell phone may be the easiest to blend and to use. An IP must research specific ROEs regarding cell phone use. Additionally if the opportunity presents itself, an IP may find stealing a cell phone for immediate use. In the case of theft, the IP must remember that newer cell phones have built in GPS and anti-theft measures incorporated into the cell service.

24.2.4. BLISS

Urban IP can use a modified "BLISS" principle for urban evasion. This modification can be applied to helping the IP meet all their needs.

- "**B**". The "B" stands for BLEND. The IP needs to blend into the background. Blending can take on many aspects from the more traditional evasion to the IP looking like they belong, so they may not be challenged. This deception works when locals or the enemy look at the IP, but do not see them for what they really are. In most societies sharing public space is an accepted and tolerated fact, so it may be easier for the IP to walk through a street in disguise, rather than sneak around it in the shadows. Avoid drawing attention to yourself; remain in the background (Figure 24-18).

Figure 24-18 Blend

- "L". The "L" is for maintaining a LOW PROFILE. IP need to be patient, polite, and not draw attention. They need to not look or act nervous. A person that moves quickly, frantically, and looks over their shoulder stands out, as opposed to someone who calmly walks like everyone else (Figure 24-19).

    - Avoid children. They are attentive and curious and will point out the peculiar to adults.
    - Avoid cultural taboos; this will avoid situations that demand complicated explanations or actions.

Figure 24-19 Low Profile

- "I". The "I" is for IRREGULAR. If you are performing a repetitive act, you must vary your activity and not become predictable by changing travel routes and patterns. If an IP is maintaining a signal, they should always try to approach and check it at irregular times. In all actions an IP should try and avoid repeat scrutinized by the same people.
- "S". The "S" is for SURVIVAL. The IP must remember their survival training and apply it towards this urban environment to meet their basic needs. Modification to things like personal protection may include using debris and man-made materials to look like the environment. The IP should be aware of preventive measure for localized diseases and hygiene issues. The IP have the additional sustenance problems they will have to overcome in an urban environment, such as competing with the locals for limited resources and ensuring their procurement techniques match local techniques, and purification and preparation practices which do not draw attention.
- "S". The second "S" is for SECLUDED. The IP should try to keep their selves and all that they do out of the way and isolated from others, without being obvious and dramatic about it (Figure 24-20).

Figure 24-20 Secluded

24.2.5. Disguise

IP in urban environments, whether evading or escaping, may need to be able to hide or blend in with the local population as they travel to a recovery site. It is important for potential IP to have an understanding of the culture, threat environment, existing international agreements, and Status of Forces Agreement (SOFA) to affectively blend in. Planning for disguise and understanding the legal ramifications are important parts of pre-mission preparation.

24.2.5.1. Pre-Mission

Having a knowledge of whether a country is a denied area or semi-permissive will help personnel make informed decisions regarding disguise and evasion. In denied countries (Iran and North Korea- currently denied) IP captured by government forces in these countries will face legal actions whether they are in uniform, civilian clothes, or disguise. While IP in semi-permissive environments the actions they engage in will be judged, rather than the fact they were wearing a nontraditional uniform, disguise, or civilian clothing to evade or prevent capture. Potential IP should know what the status is of the country and surrounding countries their mission involves and their legal status in all types of possible isolating events.

24.2.5.2. Wartime

The law of armed conflict makes a distinction between an IP evading initial capture and an IP escaping after capture. Under certain conditions during an escape or an evasion, the adoption of varying degrees of disguise may be logical, appropriate, and legal. For instance, if the population density is such that movement in uniform is not possible, an IP may be required to adopt some sort of disguise to transit the area. Likewise, if contact with an indigenous assistance group has been established, IP may disguise themselves to facilitate movement. In this instance, the judgment of the assistance group regarding adopting a disguise should be respected. However, IP need to understand that in the event of capture they will likely be treated exactly like members of the assistance group, unless they can convince their captors that they are lawful combatants. If the disguise is essentially civilian clothing, the IP should retain at least some of their uniform or personal identification (e.g., ID tags, US Armed Forces/Geneva Conventions ID card, blood chit, etc.) to use as proof of their status in the event of capture.

24.2.5.2.1. Protected Emblems

Red Cross and other protected emblems may not be used as disguises for purposes of evasion or escape in armed conflict. Only bona fide medical personnel, chaplains, and relief agency personnel may wear these emblems during armed conflict.

24.2.5.2.2. Adversary Uniforms

Combatants captured while fighting in the adversary's uniforms have traditionally been subject to criminal prosecution for espionage and war crimes (violation of international law to *"make improper use of"* the uniform of the enemy). IP are permitted to use an adversary's uniform to evade, as long as no other military operations are conducted. Personnel who use the enemy's flag, uniform, insignia, markings, or emblems solely for evasion are not lawfully subject to disciplinary punishment, as long as they do not attack the enemy, gather military information, or engage in similar operations.

24.2.5.2.3. Wearing Civilian Clothing or Nontraditional Uniforms

Wearing civilian clothing or nontraditional uniforms is permissible within limits. There is no requirement in the law of war that combatants must wear a uniform to be accorded POW status; the requirement is that combatants must distinguish themselves from civilians so that civilian populations are not placed at risk of attack. It is permissible for military personnel isolated in hostile territory to wear civilian clothes while evading, though they should avoid combatant or espionage activities while doing so. It is a violation of international law to kill, injure, or capture the enemy by feigning civilian status by wearing civilian clothes. Most importantly, it is the action engaged in while wearing civilian clothing that can be a law of war violation, not the actual wearing of civilian clothing. Evading in a disguise or other nonstandard uniform, for the sake of avoiding capture or during post-escape evasion, are permissible in the law of war.

24.2.5.3. Military Operations Other Than War (MOOTW)

Legal considerations for evasion during MOOTW operations, the national domestic criminal laws of the country in which the operation is taking place and the law of armed conflict applicable to non-international armed conflicts will apply.

24.2.5.3.1. Capture while Wearing Nontraditional Uniform

Capture while wearing a nontraditional uniform or disguise, while not necessarily a crime itself, may complicate the legal issues with the host nation. Again in this situation it is the actions engaged in that will likely have greater consequence with the legal issues. An IP in these types of operations is not entitled to POW status, nor does combatant immunity apply. IP face the possibility of prosecution by the host nation for any warlike or criminal acts committed to include charges of illegal entry, espionage, or theft.

24.2.5.3.2. Operations against Insurgent or Terrorist Groups

Wearing of a nontraditional uniform, civilian clothing, or disguise will likely have little impact on the actions of the insurgent or terrorist groups towards the IP. Knowledge of cultural and social aspects of the local population will aid the IP in using human factors when applying a disguise. The best disguise is designed to make the IP look like everyone else in the area they are evading in. Factors to disguise to remember include:

24.2.5.3.3. Blending into the Crowd

People generally remember a stimulus, not a lack of stimulus. Disguise should help you blend into a crowd.

- The IP could try to look like "untouchables" or homeless people (dregs of society) who are usually avoided or ignored in most cultures (Figure 24-21).

Figure 24-21 Homeless in Colombia

- "Invisible" person. Those that move through an area unnoticed, such as the janitorial service, delivery man, mailman, gasman, utility worker etc. who is basically ignore because they are part of the scenery(Figure 24-22).

Figure 24-22 Street Cleaners Russia

- "Gray" person who looks and acts like everyone else. This is the person that goes unnoticed and no one remembers or notices. The IP should match movements and actions with the flow of society to look like everyone else's. An IP's clothing should match the baseline of society being average and the norm. IP should avoid making eye contact, but remember to mirror actions of those around you (Figure 24-23).

Figure 24-23 Gray Person

24.2.5.4. Varying Conditions

Varying conditions will present unique benefits or problems for the IP's disguise, such factors as time of day, low light, bad weather, and curfews. The IP's decision on the type and time spent on the disguise will depend on what type of likely encounters they planned for. How close these contacts are expected to be will make a determination of how detailed and therefore, how much time and how effective the disguise will be. A disguise made for encounters at 100 feet away is different that a disguise made for encounters at 20 feet away (Figure 24-24).

Figure 24-24 Encounters

- During the Korean Conflict, avoidance of the population was the common rule, so IP tried to stay in remote terrain. Unfortunately, due to population density near the coastal areas (the area needed to travel to return to friendly forces) it became increasingly difficult for IP to avoid any contact with enemy soldiers or the local population. This problem was recognized by several downed flyers, so the IP tried to pose, at least at a distance, as local citizens or Koreans/Chinese soldiers. This was done by use of discarded or stolen clothing items that they found while evading or as preparation for their escape attempt. Captain Ward Millar, an F-80 pilot, was shot down in June 1951, utilized discarded clothing to aid in his escape after being captured. Millar eventually was recaptured, due to injuries slowing his ability to travel, but eventually escaped again and made a successful "homerun" to friendly forces. Captains Clinton Summersill, a USAF T-6 pilot, and Wayne Sawyer, a USA T-6 observer, were shot down in January 1951. They utilized the clothing they had been wearing when they crashed to seem more like two elderly peasants from a distance while evading. The weather was so severe that it was difficult for them to hear and see any enemy patrols as they walked closer to the friendly forces. They successfully used the clothing to try to look like fellow citizens and made a homerun to friendly forces.

- Colonels John Dramesi, a USAF F-105 pilot shot down in April 1967, and Edwin Atterberry, a USAF RF-4C pilot shot down in August 1967, used a disguise as the main focus of their escape attempt from the "Hanoi Hilton". They used a combination of ground iodine pills and redbrick dust to match the average skin color of the North Vietnamese. Sandals modeled after the shoes of the North Vietnamese peasant. They gathered bits of cloth and string and made white "surgical" masks to disguise facial features. Using thread pulled from towels and needles made of copper wire, they fixed their black prison clothes to look like peasant dress. Out of strips of rice-straw pulled from sleeping mats, the IP wove two conical hats. Originally they had camouflage nets made from three blankets with clumps of rice-straw from brooms sew on them, but were forced to turn them over to the rest of their cell mates, so they used mosquito netting with clumps woven into them. The IP also stole a burlap bag, two baskets, and a carrying pole as props to look like traveling peasants. The Colonels moved through the populated area of Cu Luc North Vietnam without raising suspicion, coming within a yard of policemen and others during their disguised evasion movement. They greeted those that walked with simple nods, while any locals who attempted to talk to them were ignored as the pair cong forward at a constant rate. While their disguised allowed them to get out Hanoi, unfortunately they did not travel far enough before going to a hold-up site and were subsequently captured.

- During Desert Storm, Westerners used local apparel to blend in with the population and successfully evade from the Iraqi military.
- Historically, disguise seems to work best for an IP who can blend into the local populace when there is no active search by a trained disciplined force, lines of communication are avoided, when the IP stays far enough away from people that details cannot be discerned and no one tries to engage in conversation, and low light and poor weather are taken advantage of.

24.2.6. Urban Navigation

An urban evader creating the illusion of a "local" walking through an area while trying to stay unnoticed would need to very careful when pulling out their navigation aids like a GPS, map, or compass. The IP casually glancing at their watch with the compass attached might not attract too much attention the first one or two times, but over a short time the risk could build to a dangerous level. The addition of urban navigation aids to the IP's bag of navigation skills may help them to stay *un-lost* when they can't readily get to their normal navigation tools.

24.2.6.1. IP's Situational Awareness

In any immediate action situation, urban ops being one, having an idea of what direction to go towards an extraction point or a safe harbor may be the difference between freedom and captivity. This is part of the IP's situational awareness and their study of the specifics of the urban environment they might find themselves in. Navigation in a city has a great deal of similarities and problems as navigation in any field environment. Prior preparation is always helpful in establishing your orientation with knowledge gained at levels dependent on the depth of the resource. Orientation information can be gained online, with GPS, maps, or from books prior to getting near the city. Prior preparation prevents poor performance.

24.2.6.2. Emotional State of an Evader

The heightened emotional state of an evader would be multiplied in an urban setting and compounded even more if they were lost. Being lost adds to an evader's feelings of fear, anger, depression and sadness about their situation, their emotional state could easily snowball leading to more bad decisions and captivity. Confidence in ones navigation abilities lessens these bad emotions and increases the will to survive.

24.2.6.2.1. History of a City

Newer cities tend to be built systematically, while older cities are built to follow some type of pattern related to the cities original function, so knowing the history of a city and its relative age is useful. What was the city built for? Such as, if a city was part of a major trade route, main infrastructure lines will run this route. If there has been a decline in any specific historic industry this may present locations for blending in or avoiding contact with more "upright" citizens; hiding in deserted structures or industrial sites. This information may tell the IP how the city was developed. What is the city built next too? These facts can tell an IP a great deal on how it was developed by what the city is aligned too i.e., cities built along rivers, large lakes, or the coast will usually develop from these commerce routes.

24.2.6.2.2. Terrain Features in Cities

Cities will also generally follow terrain features, such as hills and mountains, which can make travel confusing, but pre-knowledge of this may help you to avoid this confusion. Obviously an urban evader should take advantage of any natural aids to navigation such as water ways, terrain features, vegetation, prevailing winds, the sun, the moon, and anything else that may help (Figure 24-25).

Figure 24-25 Using Terrain Features

24.2.6.2.3. Urban Landmark

Is the city known for any specific landmarks or identifiable features? This may always give an IP a line of direction no matter where they are. Notice the city's lines of communication and commerce. The IP should try to identify a specific pattern. Is the infrastructure running specific directions? Is there a pattern to the street names or routes? Look for these things to determine any aids to your urban navigation. A simple example is in cities and countries that have not switched to modern analog VHF and UHF TV frequencies television aerials may be directed towards the television station, which historically have been located near the city's center. In the case of TV satellite dishes, they will tend to all be pointed in one general direction within a given city, once cardinal direction has been determined, the IP can continue to use that as a navigational aid in example in South East Asia they are pointed easterly and in South America they are pointed westerly. Another simple tool is using telephone poles and antennas like a giant-sized stick and shadow to get cardinal directions. In many ways, just modifying known field navigation techniques urban operations works for the IP (Figure 24-26).

Figure 24-26 Urban Landmark

24.2.6.3. "Known Point"

Once IP are moving through a city, no matter the situation; establish a known point, use it as a reference to help maintain orientation. This "known point" will depend on the situation the IP finds themselves in, it could be their hotel, the embassy, or some identifiable landmark in the city. The IP should use this "known point" to help maintain a mental map of where they are going and if needed, how to get back. This is considered a primitive means of navigation. The more modernized system is to establish cardinal directions, determining what general cardinal direction the individual is traveling in. The best technique is to combine the two together, this way each system is reinforcing the other. Add the mental map and the cardinal directions, so from the "known point," an IP can figure out where north, south, east and west are. As the IP moves, they need to continue to draw a mental map of where they are in relation to this know point; mentally recording turns and additional recognizable features and updating it with cardinal directions. Modern psychologist determined that doing both of these decreases the chance of getting lost; the more visual stimuli that supports navigation tends to decrease the likelihood of getting lost.

24.2.6.4. City Architecture in Navigation

In addition to overall city architecture in navigation, the IP can also use individual buildings to help them. Historically, architects would orient buildings to take advantage of sunlight. In the temperate and colder climates of the Northern Hemisphere, buildings are generally oriented facing south (direction of greatest sunlight). Living rooms or main rooms will generally face south to take advantage of warmth and light while the kitchens will usually be on the north side. In the temperate and colder climates of the Southern Hemisphere, buildings are generally oriented facing north (direction of greatest sunlight). Living rooms will generally face north to take advantage of warmth and light while the kitchens will usually be on the south side. In areas with strong cold prevailing winds, many older homes have doors and windows on the side facing away from the wind or to the side away from it. In hotter locations the general orientation of homes is focused on avoiding the sun and taking advantage of the prevailing winds. Houses may align to take advantage of the wind and provide shade to the living quarters. So following the constants of the way structures are oriented may aid the IP in maintaining a direction of travel. One potential compass rose in Europe, Central America, and South America are the Christian churches. Christian churches historically were oriented east to west with the rising sun shining on the altar. This may give the IP a quick reference to cardinal directions (Figure 24-27).

Figure 24-27 Direction Indicators

24.2.6.5. Cardinal Direction from Bleaching and Weathering Effects

With modern aspects of heating and cooling newer structures may not attempt to use the sun or follow any type of tradition. The IP then may be able to determine cardinal direction from the bleaching and weathering effects of the sun to paint (especially on darker colors) on buildings, fences, and other structures. Corrosion of iron works and structures may be used, but an IP must keep in mind that the combination of sun, wind, and precipitation may make this harder to judge. Also plants growing in a city will tend to grow towards sunlight; this may give IP an indicator of cardinal directions even during cloudy and overcast days (Figure 24-28). Either way, multiple examples are always best when possible. Studying which sides of a structure shows bleaching or corrosion may lead the IP to determine cardinal directions or aid in maintaining a "heading" of travel without drawing attention to a disguised evader's actions even in situations where the sun is not visible.

Figure 24-28 Urban Plant Indicators

24.2.6.6. Signs of Direction

Urban evasion or avoiding contact can take on many facets, who, what, why, how, and where are too numerous to mention, but like any SERE principle the more knowledge and facts you can bring into a subject the sounder your decisions are. Urban navigation aids are the same, specifically orienting your focus, training yourself to look for those signs of directions can easily make the difference between return to friendly forces or captivity.

24.2.7. Urban Barriers

Specific TTPs for urban barriers along with general principles mentioned in the travel and evasion chapters will assist an urban evader in successfully avoiding capture and returning to friendly forces.

24.2.7.1. AUTO

The AUTO acronym can be used by an urban evader. AUTO stands for Avoid, Under, Through, Over.

- Avoid. Avoid by going around; an IP should not attempt to penetrate barriers if they can be bypassed or avoided because people catch people. An IP may be discovered by inhabitants or their animals.

- Under. Go under a barrier using tunnels, culverts, trestles, or areas where animals may have already tunneled.

- Through. Go through by identifying potential weak entry points, using battle damaged areas, and or using areas where animals may already penetrated through the barrier.

- Over. Go over the barrier. Climbing is used by an IP to access high windows, rooftops, to climb walls, negotiate barriers, etc. Some general rules the IP should use are to select route that is safe, quickly climbed and concealed. Mentally climb route before beginning since once climb is started, it may be impossible to climb back down. Assess the climb based on an estimate of abilities, resources available, and whether solo or in a group. Not everyone can solo climb. IP should also check the integrity of whatever they are going to climb. IP should also have plan for crossing the top and climbing down the other side, the back side of the barrier may be steep, drop-off, or present other hazards. Also the IP should try to get visual of the top of any barrier since it may contain broken glass, razor wire, or expose themselves to enemy.

24.2.7.2. Built in Climbing Aids

Many structures and some barriers already have built in climbing aids. IP can use drainpipes, adjacent trees, available scaffolds, and utility fixtures (pipes, conduit, wires, etc.) as ready-made ladders. In some cases the way the structure is built provides another built-in "ladder" allowing IP to use the structural supports such as cables, girders, beams, etc (Figure 24-29).

Figure 24-29 Ready-Made Ladders

24.2.7.3. Climbing Order

IP may want to use any climbing aids such as wooden pallets, two by fours, pipes, barrels, dumpsters, boxes, improvised rope, grappling hooks (improvised from rucksack, metal pole, rock, wooden plank, etc.), weapons sling, belts, electrical cords, linens, clothing, etc. Even another IP in a group situation can help the other IP to get up onto the barrier. In this case the climbing order should be a consideration. The IP leader may want to send the best climber up first, so they can assist subsequent climbers. Also, having the largest person cross second to last and the smallest/most athletic cross last may be decisions to consider.

24.2.7.4. Climbing Using Hand and Foot Holds

An IP climbing in an urban environment should choose a route offering hand- and foot-holds. When they climb they should maintain three points of contact while carrying their weight over their feet using their hands mostly for balance. The types of holds will vary with the location and type of barrier, the IP will likely find themselves using a mix of push-, pull-, jam-, counter force/opposing pressure, friction (chimney climb), and jumping techniques.

24.2.7.5. Assisted Climbing

In group situations, the type of barrier and the number in the group will determine the type of climbing TTPs used. Climbing can involve one or two people lifting the climber by making a catch for the climber's heel or a supported lift. A supported lift is were the first person to the wall makes a solid frame by putting their feet shoulder width apart (for stability) with their back against the wall and legs bent to create a step for the second person (first climber), the second person steps on the thigh of the first person then stands on their shoulders, and then the first person fully extends upward to give second person a boost. In any type of assisted climbing, the first climber on the wall must also make a solid frame on top before assisting others, they can do this by laying on the wall with their legs straddling both sides, on backside of wall they should extend their leg and arm to wedge and brace for the lift, on the climber side of wall they should bend their leg at knee to hook their foot on top of wall, then they extend their free arm towards climber, and pull while the second climber uses the leg and out-stretched hand of the individual straddling the wall to assist.

24.2.8. Structures

Reasons an IP would enter a structure during an urban evasion event could be to use it as a hide site or hole up area, to regroup, to break contact, as a strong hold point to defend against the enemy, or a combination of these reasons. In most cases using a structure should be considered a last resort since structures are built to house people (people catch people), so what and who are encountered on entering is unknown. Obviously for this reason if an IP determines they must enter a structure they should avoid any that are obviously occupied. IP should also consider that many structures are not occupied 24/7 or even that only part is occupied depending on the time of day, indigenous personnel may show up at any time as in example people may leave their houses empty during the day or may sleep on the roof at night (Figure 24-30).

Figure 24-30 Structures

24.2.8.1. Entering a Structure

How an IP enters should support their type of movement such as if disguise movement is being used the disguised IP would draw less attention by using an unlocked door or possible a large battle damaged hole and just walking in. No matter what type of entry point it is in tactical and evasion movements the IP should always plan a concealed route to the point of entry. Other urban evasion entry points which an IP may use are subsurface holes, windows (subsurface, surface, or upper story), openings from battle damage, HVAC vents, roof tops, or attic entries. While alternate entry points are less likely to be booby trapped, an IP should always be observant for any signs.

24.2.8.2. Locating Hiding Sites and Surveillance Points

Once inside, the IP should locate hide sites and surveillance points. The IP should also determine alternate hide sites, existing points, and escape routes. The IP's movement may also involve going from one structure to another using different routes like sewers, roof top to roof top, shared walls, etc (Figure 24-31).

Figure 24-31 Alternate Exits and Escape Routes

24.2.8.3. Possible Signaling and Recovery within a Structure

Once inside the structure, the IP may want to determine if any other needs can be met within the confines of the structure such as the roof top being used for covert signaling and recovery or finding a water source.

24.2.9. Subsurface

Most heavily populated and older urban environments have some type of extensive subsurface systems of utility tunnels for steam, water, gas, sewage, and storm water. This availability is one of many reasons an IP may decide to use an urban subsurface location, others are it may offer them easy cover, concealed means of travel, or it may seem to be an area with limited contact with people (Figure 24-32).

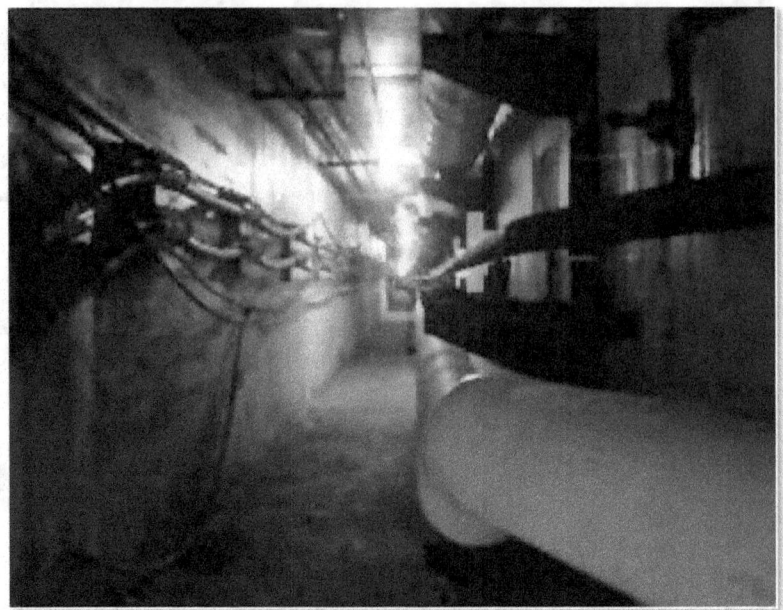

Figure 24-32 City Tunnels

24.2.9.1. Others that use Subsurface Systems

An IP needs to remember that others may use a subsurface system for similar reasons or the very nature of the subsurface system may drive the enemy to protect and guard it as much, if not more, than many surface areas i.e. a city's palatable water system. Other local inhabitants may include homeless, vagrants, and squatter populations which maybe on the increase due to the conflict which caused the isolating event.

24.2.9.2. Hazards and Obstacles

Besides contact with people there are other subsurface hazards and obstacles such as:

- IP may encounter dangerous gas buildup which could lead to asphyxiation or explosion. The IP may have difficult time identifying this since many gases are odorless. The presence of rodents may indicate safe air supply, but many gases will rise and fill the top portion of tunnels leaving breathable air for rodents and several types of rodents can hold their breath for 10-15 minutes.

- Sewers may have times were large quantities of water are released this could be by man-made design or during natural events like rainstorms or melting snow. Sewers in these cases may fill up very rapidly causing a flash flood effect drowning an IP. Even if this does not occur, the amount of normal water activity, flow, and slipperiness may cause difficulties for an IP (Figure 24-33).

- While underground an IP's navigation aids may not work, making it difficult, if not impossible, to maintain direction and knowledge of location.

- Just like an IP's navigational aids, communication equipment may not work. Signals may not broadcast out to the surface, as well as the IP needing to have easy access to the surface to use other signaling devises. Additionally, while subsurface the IP may miss opportunities for recovery.
- Other factors an IP may have to deal with in a subsurface location would be the darkness may require them to use/have a light source., the confined space may provide few exists with little room to maneuver, sounds travel long distances, and problems with animals.

Figure 24-33 Sewer Entrance

24.2.9.3. Entry Points

Subsurface entry points such as subway, underground rail system, sewers, vehicle tunnels, and utility tunnels may have security guards, gates, turnstiles, cameras, as well as be locked, bolted or welded closed. IP may try to access these through manholes, maintenance entry points, subway/rail tunnels (using a disguise), service doors, or culverts. When these access points are secured an IP can attempt to pry open using bolts, metal, sticks, or as a last resort their knife (since it may be easier to break and not replaceable). IP can also try to defeat the lock. The difficulty in this is for the IP to have the time to attempt this since disguising these actions may be impossible (Figure 24-34).

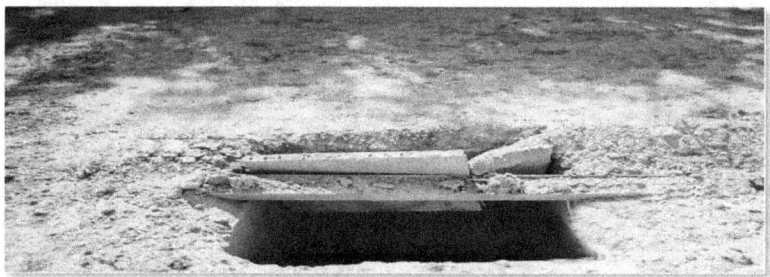

Figure 24-34 Under-Building Entrance

24.2.9.4. Disguising Access to an Entry

Once opened, an IP will usually want to disguise the fact that the entry way has been accessed. In the case of manholes, the IP should pre-position the cover right side up to secure back in place.

Since not every country uses round manhole covers, the IP should be careful not to drop the cover into hole. To make re-opening a manhole cover easier an IP may use a piece of debris to wedge it open or line attached to debris/stick through the "pick holes," which is where an official would insert a hook handle to lift them.

24.2.9.5. Subterranean Locations

Allow gases to dissipate for 15 minutes (time permitting) before entering enclosed subterranean locations. IP should assess depth to reach the floor of the subsurface location to determine a safe entry procedure (ladder, rope, climbing techniques).

24.2.9.6. Transitioning Above and Below the Ground Undetected

If using subsurface location, IP will need to transition between above and below ground undetected. IP need to ensure security and counter observation measures are taken prior to leaving concealment. IP need to stop, look and listen for threats, get a visual of surrounding area, identify the next point of concealment, exit, blend the exit point to look undisturbed, and then move to the new point of concealment.

24.2.10. Ground Vehicle

A vehicle may play the critical role of bringing the potential IP into the urban isolating event or helping to extract the IP from the urban isolating event, it is therefore imperative that judgment, rational thinking and safety permeate all the actions between the individual and a vehicle (Figure 24-35).

Figure 24-35 Operations involving Vehicles

24.2.10.1. Potential IP and a Vehicle

The kind of car, where it is driven, and where it is park all influences the vulnerability to terrorist attempts to kidnap or car bomb the potential IP. If possible, potential IP should use a plain car that doesn't attract attention as a "rich American" this may help lower the terrorist target profile. Consideration should also be given to avoid using government cars that immediately identify an association with the US Government. IP need to understand their options and how to prepare, utilize, and procure a vehicle to meet their needs in case of an isolating event.

24.2.10.1.1. Ground Vehicle Preparation

Ensure all personnel are briefed on emergency procedures for each individual's role with the following possibilities, vehicle under fire operations; vehicle incapacitated operations, PR aspects/procedures, urban evasion operations, and areas identified for recovery (AIR).

- Practice of vehicle to vehicle recovery operations and tactics (with and without hostiles) is advised to help ensure the safety of occupants and prepare potential IP.
- Each potential IP should have their tactical and mandatory emergency equipment. At a minimum, a minimum recommendation for each vehicle should have a medical/trauma kit, hit-n-run bag, GPS, local street maps, and some type of communication devise with communication procedures for multiple contacts. Ensure all vehicle occupants are fully aware of the contents and use of all equipment items. Communication devises should have programmed emergency numbers as well as the numbers of the other vehicle's contact information and a hard copy contact numbers. All items needing it will be fully charged with spare batteries.
- Prior to Traveling. Be familiar with vehicle prior to driving it; found out the vehicle's specific information for operating it. Vehicle will be inspected for all fluid levels to ensure operation of vehicle. All rubbish and trash will be removed from the inside of the vehicle to avoid creating additional shrapnel. Any modifications to interior or exterior of vehicle will be checked (i.e., chicken wire over windows, sandbags in interior, and etc.). Be current with local procedures.
- Identify the number of vehicles and occupants needed to be in compliance with Force Protection standards.
- All occupants are aware of any bona fides, passwords, counter-passwords, and etc needed to ensure safe passage of vehicle through checkpoints and recovery of personnel during a PR event.
- The driver's main job is to drive. Other occupants will have assigned tacks. Front passenger seat "SHOTGUN" will be armed with a 9mm, M-16, or a GAU-5. If SHOTGUN is armed with an M-16, they will sit in the back seat for ease of access to weapon and maneuverability. SHOTGUN will need easy access to weapon as well as the ability to maintain control of weapon. If there are other passengers carried in the vehicle, weapons will be charged and on SAFE. SHOTGUN or if available, other passengers will have access to communication devise to contact other vehicles or follow communication procedures in case of emergency.

24.2.10.1.2. Vehicle Inspection

Always lock the vehicle and place in secure location. If vehicle has to be left in an unsecured location, ensure that it is dusty and/or dirty prior to leaving it so that signs of tampering may be more obvious. Observe the area you have left the vehicle at try to identify memorable features of the surrounding area. Look over any other vehicles parked near the vehicle. Visually review the area within two feet of where the IP parked the vehicle prior to leaving. It is much easier to notice changes and disturbances if known what it was before. Inspecting the vehicle on return should start with a visual inspection and then work its way to a physical contact with the vehicle. The potential IP should break the search down into three areas: the surrounding area around the vehicle, the exterior of the vehicle, and the interior of the vehicle.

- Surrounding area: Look for signs of activity or any additional threats. Be observant for any unusual signs or changes to the area. Look for items that may have fallen out of someone's pockets while they were on the ground around your vehicle.

- Exterior: Inspect the four areas of the exterior front, sides, back, and underside. Each area needs to be looked at and checked for changes. Note any unusual hand prints or signs that your "dirty" vehicle has been disturbed.
- Interior: Inspection of interior breaks down into five areas: under dash, under seats, under baggage, engine compartment, and the trunk.

24.2.10.1.3. Leaving the Departure Point

The lead driver will set the pace through the streets and urban locations. Matching traffic flow is imperative, but *discretion is advised*. The key is to keep moving without becoming a danger to other drivers. Vehicles must stay joined at all costs (within safety guidelines). The lead driver sets the pace, and works the movement to keep both vehicles joined.

24.2.10.1.4. Ground Vehicle Related Isolating Event

One of the most important aspects is recognizing impending hazards BEFORE they become a problem. Early recognition allows the time a potential IP needs to possibly avoid trouble. It is vitally important that drivers as well as passengers recognize and become IMMEDIATELY aware of what they see while driving. Using EYES and MIND to analyze what is seen for potential dangers. It is imperative that all personnel remain alert for anything that appears suspicious. Some examples may include slowing vehicles that contain a large group of men, fast moving vehicles approaching from the rear, crowds on the street that may appear interested in the passage of the vehicles, or the opposite unusual/unexplained absence of local citizens. Decoys have also been used by terrorist against personnel traveling in ground vehicles. Decoys are used to slow the vehicles in order for a passerby to attempt assaulting the vehicle or vehicles. Some examples of decoys that have been used in the past are someone walking an animal across the road, children playing ball in the street, a cyclist falling in front of your car, a flagman or workman stopping your car, an unusual detour, and an auto break down. These decoys are done in attempt to get the vehicle deliberately struck, by other cars or pedestrian traffic that box the vehicle in, allowing for the terrorist to move in closer for a sudden activity such as a kidnapping (Figure 24-36).

Figure 24-36 Decoys

- Reactions must be quick, with full awareness of any governing procedures (i.e., General Orders, specific policies, and or standard ROEs in place) and safety of all the personnel traveling, as well as passersby. Do not attempt to force an issue, but in the same vein, be prepared to be innovative in order to get vehicles and personnel through in a timely and safe manner. If one vehicle becomes incapacitated, the other vehicle acts as cover while the procedures for this event under the current situation are carried out i.e., vehicle to vehicle transfer of personnel under hostile fire.

24.2.10.1.5. Evening Travel

Unless ROEs dictate otherwise, when approaching any guarded gate, driver will dim headlights at a discrete distance so as not to blind the guards. Interior lights will be on to facilitate identification of occupants. Since nighttime operations are considered the most dangerous. The vehicle occupants will be doubly alert for such possibilities as:

- One-light or no-light approaching vehicles crowding the centerline.
- People or stray animals moving across the road.

24.2.10.1.6. Defensive Driving

Terrorist acts against individuals, such as kidnappings, usually occur outside the home and after the victim's habits have been established through surveillance over a period of time. The terrorist will attempt to exploit predictable habits such as routes traveled between home, the work place, and commonly frequented local facilities. The potential IP should vary these routes as much as possible. A potential IP can take certain measures to reduce the chances of being kidnapped from your car or the victim of a car bombing.

- Check occasionally to see if another car is following the potential IP. If the potential IP thinks they are being followed, they should circle the block or change directions several times to confirm the presence of surveillance, make note of a description of the car and its occupants, if possible. While it is okay to let the surveillants know they have been seen, *do not under any circumstances* take any action that might provoke them or that could lead to confrontation. If the surveillants do not stop following, the potential IP should drive directly to the nearest safe haven, such as a US military base or the US Embassy and advise the appropriate security or police authorities. Additionally using a communication devise (cell phone or radio) may also work to notify the US Government of the situation (Figure 24-37).

Figure 24-37 U.S. Embassy

- The potential IP should learn to recognize and be alert to events that could signal the start of a plan to stop the car and take captives, such as a decoy. Such events block and box in the vehicle allowing the terrorist to have a more effective attack. If attacked, the potential IP will have to make instantaneous decisions with very limited time to weigh all the consequences. By mentally and physically rehearsing courses of action to probable scenarios, the potential IP can better react under various similar circumstances. In traumatic and emergency situations, rarely, if ever, do individuals rise to the occasion, what does happen is that individuals tend to psychologically and physiology default to their training.

- Options may be very limited for the potential IP. A potential IP can sound the horn to possibly draw attention, this may, at least, help ensure there will be witnesses to observe and report the individual going from potential IP to IP. The potential IP can attempt to drive away to escape. If the vehicle needs to go over the curb, hit it at a 30-45 degree angle with a maximum speed of 35 mph, increasing speed afterwards. If the path is blocked by a vehicle across the road, the potential IP can, at some risk, ram the blocking vehicle in an effort to spin it out of the way. The potential IP should attempt to hit the other vehicle on an angle, with the impact focused on the wheel you want to move out of the way.

24.2.10.1.7. IP and a Vehicle

If an IP did not have a vehicle prior to an isolating, interaction with a vehicle will occur in two main ways. The first way is as a captive, the enemy has the potential to transport an IP by a vehicle during all or any phase of captivity all types of detention. The second way is the IP may use a vehicle as part of their evasion or escape.

1. Vehicle and Captivity. Once an IP has been forced into a vehicle, depending on the type of detention (wartime, peacetime governmental detention, and terrorist/hostage detention) and the captor, they may be blindfolded, beaten (to cause compliance, docility, or unconsciousness), drugged, or forced to lie face down on the floor of the vehicle. In some instances, hostages have been forced into trunks or specially built compartments for transporting contraband by terrorists. If drugs are administered, do not resist. Their purpose will be to sedate you and make you more manageable. It is probably better to be drugged than to be beaten unconscious. If the IP is conscious, they should follow their captors' instructions. While being confined and transported, the IP should not struggle. The IP should try to remain calm so they can concentrate on surviving. The IP should attempt to visualize the route they are being taken on, making a mental note of turns, street noise, smells, etc. The IP should try to keep track of the amount of time spent between points. This knowledge can assist the IP during escape or be of importance once rescued or returned in an effort to determine where they were held (Figure 24-38).

Figure 24-38 Taken Hostage

2. Vehicle and Evasion/Escape. A useful vehicle in an urban environment may be a bicycle,

motorcycle, car, or truck. An IP may determine to access/use a vehicle while evading in an urban environment for several reasons. The IP may be able to use resources obtained from the vehicle such as water, food, mirrors, flares, blankets, seat covers, foam insulation, clothing, tools, tire iron, jack, etc. If an IP uses a vehicle for resources they should try to make it look undisturbed covering the theft. When the theft of the resources cannot be hidden the IP may want to ensure it looks more like a random act of vandalism or theft verses a theft for basic survival items by an evader (Figure 24-39).

- The IP may determine that a vehicle will make a good hide site or immediate action evasion shelter (Figure 24-40).

- The last thing an IP may use a vehicle for is a tool to help travel facilitating an escape or their evasion. There are as many problems as benefits to using a vehicle for escape and evasion, the IP will need to weigh these to determination a course of action.

Figure 24-39 Vehicle Resources

Figure 24-40 Hiding under Vehicle

- When an IP chooses a vehicle, they should treat it as a potential trap and be ready to encounter people. The IP should select an unoccupied or abandoned vehicle. Observe the vehicle prior to approaching being aware of any occupants (people or animals) and whether the vehicle is guarded, alarmed, under observation, or booby trapped. Another consideration on selecting the vehicle is the ease of entry and based on why the IP is accessing it, the ease of getting it to start to drive away.

- Once the IP has selected the vehicle they should look for an unlocked vehicle or one with an open window. If the vehicle is locked the IP should search for a spare key. If there is no spare key found the IP has two options. Option one is to try to pick or shim the vehicles lock. Option two is to break the vehicle's window or pry open the vehicle's door or trunk. The problem with these two options is that there are tools and resources needed to accomplish these, it takes time, it creates noise, theft draws attention to the IP's presence, and all or any of these may not allow the IP the time to procure from the vehicle or start it to drive away.

24.2.10.2. Public Transportation

Depending on the type of social-political environment, isolating event, and ability of the IP to blend in, an urban evader may find themselves able to use public transportation (Figure 24-41).

- Public transportation may include trains, buses, and ferries. The use of public transportation may be worth the risk to get the IP quickly away from the isolating event while allowing them the opportunity to blend into a crowd. An IP must know if there are any pre-requirements to using these forms of travel such as travel passes, identification checks, paperwork for crossing borders, travel tickets, and even a simple thing like needing exact change may hold up the IP. Of course the IP should have an understanding of the local language and culture.
- An IP must also consider that public transportation hubs may have fixed or incidental security measures such as check points, cameras, biometrics security measures, security details (doing searches and/or checking paperwork), and dog teams (Figure 24-42).
- IP that have used public transportation before have found having a means of avoiding conversations by reading or sleeping to be a practical way to blend in and minimize the potential to make a mistake.

Figure 24-41 Bus Station

Figure 24-42 Security Measures

24.2.11. Urban Foraging

Foraging in an urban environment will involve the IP using skills already discussed in the sustenance and evasion chapters, but there are additional TTPs specific to an urban evader.

24.2.11.1. Urban Food

In locations with a large density population an urban IP may be able to find food and resources discarded by retailers. An urban evader may be able to procure food (and other resources) by pulling it out of the trash (bins), a practice commonly nicknamed "dumpster diving" or "freeganism" in North America and "bin diving" or "skipitarianism" in the UK (so called because the person's diet mostly involves eating out of a skip). Retail suppliers of food such as supermarkets, grocery stores, and restaurants routinely throw away food in perfectly good condition, often because it is approaching its sell-by date (without thereby becoming dangerous), might look unpleasant, be in damaged packaging, or that the food is not entirely sealed (Figure 24-43).

Figure 24-43 Urban Foraging

24.2.11.1.1. Urban Gardens and Vegetation

The IP could steal from urban gardens using TTPs identified in the food and evasion chapters. Many larger urbanized locations will plant vegetation which has edible parts or be used medicinally just for ornamental reasons, while the planter's will not necessarily considered them a resource, using TTPs to avoid being caught an IP can use these to maintain their health (Figure 24-44).

Figure 24-44 Urban Garden

24.2.11.1.2. Urban Game

Many urbanized areas have locations teeming with animal life. From streams, rivers, and ponds that are stocked with fish ready to be caught using concealed fishing techniques to pigeons which can be trapped or hunted. When procuring small game the IP should use TTPs identified in the food and evasion chapters modified to remain concealed and to blend in to the urban environment. Game procurement and preparation will need to take in precautions related to disease and disease carrier possibilities. As mentioned in the evasion food procurement, IP must be on the lookout for guard dogs and other territorial animals, such as geese, which may attack or at the very least sound an alarm (Figure 24-45).

Figure 24-45 Territorial Animals

24.2.11.2. Urban Water

When the situation dictates, it is recommended that an IP modifies evasion water procurement TTPs to met water needs in an urban environment. Depending on the culture, IP procuring water from normal "field" water sources (lakes, springs, streams, rivers, pools, and ponds) in an urban environment may draw attention to themselves. In an urban area, available water sources may be contaminated with chemicals and pollutants. Selecting the safest possible source of water will greatly diminish the danger of ingesting harmful water or getting captured. Recommended water sources include:

24.2.11.2.1. Well Water

Water from wells or other underground sources having natural filtration will be the safer from some chemicals and pollutants produced by the local population or due to the type of conflict occurring (Figure 24-46).

Figure 24-46 Urban Water Source

24.2.11.2.2. Contained Water

Any water in the pipes or containers of abandoned houses or stores should also be free from chemicals and pollutants, and, therefore, relatively safe to drink. However, precautions will have to be taken against bacteria in the water.

24.2.11.2.3. Purify Water

As in any water procurement opportunity, the IP should purify the water whenever possible (see chapter 16 "Water").

MARK C. NOWLAND, Lt Gen, USAF
Deputy Chief of Staff, Operations

Attachment 1
Glossary of References and Supporting Information

References

Air Force Doctrine Document 3-2, Irregular Warfare, 12 July 2016.

Air Force Pamphlet 11-216, *Air Navigation*, 1 Mar 2001.

AFTTP 3-2.26, *Multi-Service Tactics, Techniques, and Procedures for Survival Evasion, and Recovery*, 11 Sep 2012.

JP 1-02, *Department of Defense Dictionary of Military and Associated Terms*, 8 Nov 2010 (As Amended Through 15 Feb 2016).

JP 3-11, *Operations in Chemical, Biological, Radiological, and Nuclear Environments*, 04 October 2013.

JP 3-50, *Personnel Recovery*, 2 Oct 15.

Army Field Manual 3-04.513, *Aircraft Recovery Operations*, Jul 2008

Army Field Manual 3-05.70, *Survival*, 17 May 2002.

Army Field Manual 3-50, *Army Personnel Recovery*, 02 Sep 2014. Training Circular 4-02.3, *Field Hygiene and Sanitation*, 6 May 2015.

Army Field Manual 27-10, *The Law of Land Warfare*, 18 July 1956.

Army Field Manual 90-5, *Jungle Operations*, 16 August 1982. GTA 80-01-003, *Survival, Evasion, and Recovery Tactics, Techniques, and Procedures*, 01 May 2015.

Army Field Manual 21-60, *Visual Signals*, 30 September 1987.

Army Field Manual 90-3, *Desert Operations*, 24 August 1993.

Army Training Circular, *Map Reading and Land Navigation*, 15 November 2013.

Army Training Circular 3-21.75, *The Warrior Ethos and Soldier Combat Skills*, 13 August 2013.

Army Training Circular 3-97.61, *Military Mountaineering*, 26 July 2012.

Army Training Circular 21-3, *Soldier's Handbook for Individual Operations and Survival in Cold Weather Areas*, 17 March 1986.

Army Techniques Publications 3-37.34, *Survivability Options*, 28 June 2013.

Army Techniques Publications 3-90.97, *Mountain Warfare and Cold Weather Operation*, 29 April 2016.

Army Techniques Publications 3-90.4, *Combined Arms Mobility*, 08 March 2016.

Army Techniques Publications 3-50.3, Multi-Service Tactics, Techniques, and Procedures for Survival, Evasion, and Recovery (MCRP 3-02H; NTTP 3-50.3; AFTTP 3-2.26), 11 September 2012.

The Department of Defense Law of War Manual, 12 June 2015 (Updated December 2016).

The Geneva Conventions of 1949 and their Additional Protocols.

Uniform Code of Military Justice (UCMJ).

Commercial Publications:

Abel, Michael, Backpacking Made Easy. Happ Camp CA: Naturegraph Publishers Inc. 1975.

Aleith, R.C., Basic Rock Climbing. New York: Charles Scribner's Sons, 1975.

Baird, P.D., The Polar World. New York: John Wiley and Son's Inc., 1964.

Benton, Allen and William Werner, Field Biology and Ecology. New York: McGraw, Hill Book Company, 1966.

Bergamini, David, The Universe. New York: Life Nature Library, Time Inc., 1966.

Birkby, Robert C., 10th Edition, Bou Scout Handbook. New Jersey: Boy Scouts of America, 1992.

Brower, Kenneth, "A Galaxy of Life Fills the Night," National Geographic, Vol 160 No. 6 (December 1981), 834-847.

Bruemmer, F. Encounter with Arctic Animals. Toronto, Canada: McGraw-Hill Reyerson, Ltd Clark PH.D.

BusinessDictionary.com, http://www.businessdictionary.com/definition/clo.html. (Accessed 21 January 2012).

Carlson, Kurt (1986) *One American Must Die: A Hostage's Personal Account of the Hijacking of Flight 847*. Smithmark Pub. ISBN 0865531617.

Cousteau, Jacques-Yves, "The Ocean," National Geographic, Vol 160 No. 6 (December 1981), 780-791.

Darvill, Fred T., Jr., M.P., Mountaineering Medicine. Skagit Mountain Rescue Unit, Inc., 1969.

Dodge, Natt N., Poisonous Dwellers of the Desert. Arizona: Southwest Parks and Monuments Association, 1972.

Dwight D. Eisenhower: First Inaugural Address; Tuesday, January 20, 1953 http://www.britannica.com/presidents/article-9116870. (Accessed 21 January 2012).

Engel, Leonard, The Sea. New York: Life Nature Library, Time Inc., 1961.

Eugenie. "The Strangest Sea," National Geographic, Vol. 148 No. 3, (September 1975), 388-365.

Faub, P., Ecology. New York: Life Nature Library, Time Inc., 1963.

Fear, Gene, Surviving the Unexpected Wilderness Emergency. Tacoma, Washington: Survival Education Association, 1973.

Fear, Gene, Wilderness Emergency, Tacoma, Washington, Survival Education Association, 1975.

Fear, G. and J. Mitchel, Fundamentals of Outdoor Enjoyment. Tacoma, Washington: Survival Education Association, 1977.

Franklin, Kevin (27 June 1996). "Out There: Hot Summer Reading". Tucson Weekly. Archived from the original on 16 May 2008. http://web.archive.org/web/20080516083042/http://www.tucsonweekly.com/tw/06-27-96/outthere.htm. (Accessed 21 January 2012).

Freeman, Otio W. and H.F. Ranp, Essentials of Geography. New York: McGraw-Hill Book Company, 1959.

Gibson, C.E., Handbook of Knots and Splices. Emerson Books, 1972.

Glasstone, Samuel and Dolan, Philip J., The Effects of Nuclear Weapons. United States Department of Defense and the United States Department of Energy, 1977.

Gore, Rick, A Bad Time to be a Crocodile. National Geographic, Vol 153, No. 1 (January 1978), 90-115.

Halstead, B.W., Dangerous Marine Animals. Cornell Maritime Press, 1959.

Halstead, Bruce W., Poisonous and Venomous Marine Animals of the World. Princeton, New Jersey: Darwin Press, Inc., 1978.

Hanuritz, Bernard and Austin, James, Climatology. New York: McGraw-Hill Book Company, 1944.

Kaplan, M.D. Harold I., Freedman, M.D. Alfred M., Sadock, M.D. Benjamin J., Comprehensive Textbook of Psychiatry/III. Baltimore, Maryland: Williams and Wilkins Company, 1980.

Kasmenn, Bryan, *The needle concept*, July 2003 http://findarticles.com/p/articles/mi_m0IBT/is_6_59/ai_105656746/?tag=content;col1. (Accessed 21 January 2012).

Kearny, Cresson H., Nuclear War Survival. Oregon: NWS Research Bureau Coos Bay.

Kjellstorm, Bjorn, Map and Compass, The Orienteering Handbook. American Orienteering Service, New York, 1955.

Kuhue, Cecil, River Rafting. World Publication Inc., Mt View, California 1979.

Lathrop, Theodore, M.D., Hypothermia: The Killer of the Unexpected. Portland, Oregon, 1972.

Ley, Willy, The Poles. New York: Life Nature Library, Time Inc., 1962.

Leopold, Starker A., The Desert. New York: Life Nature Library, Time Inc., 1962.

Lounsbury, John F. and Lawrence Ogden, Earth Science. New York: Harper and Row, 1969.

Matthews, Samuel W., New World of the Ocean, National Geographic, Vol 160 No. 6 (December 1981), 792-833.

May, W., Mountain Search and Rescue Techniques. Bolder CO: Rocky Mountain Rescue Group, Inc., 1973.

McGinnis, William, White Water Rafting. Time Books New York, NY, 1978.

Mountaineering, The Freedom of the Hills, Fourth Edition. Seattle, Washington: The Mountaineers, 1982.

Nickelsbury, Janet, Ecology: Habitats, Niches, and Food Chain. New York: J.B. Lippincott Company, 1969.

Ormond, C., Complete Book of Outdoor Lore, Outdoor Life. New York: Harper and Row, 1964.

Peterson, Roger Tory., The Birds. New York: Life Nature Library, Time Inc., 1963.

Shanks, Bernard, Wilderness Survival. New York: Universe Books, 1980.

Stephenson, V., Arctic Manual. New York: Greenwood Press, Publishers, Reprinted, 1974.

Stephenson, V., The Friendly Arctic. New York: Greenwood Press, Publishers, Reprint of 1943 Edition.

Strahler, Arthur N., Physical Geography. New York: John Wiley and Sons, Inc., Third Edition 1969.

Strahler, Arthur N., Introduction to Physical Geography. New York: John Wiley and Sons, Inc., 1973.

Stuung, Norman: Curtis, Sl, Perry E., White Water. Collier, McMillam Publishers, New York, NY, 1976.

Tucker, Todd (2006). *The Great Starvation Experiment: Ancel Keys and the Men Who Starved for Science*. New York: Free Press. ISBN 0743270304.

Van Dorn, William G., Oceanography and Seamanship. New York: Dodd, Mead, and Company 1974, 79-94, and 111-128.

Washburn, Bradford, Frostbite. Boston, Museum of Science, 1978.

Watson, Peter, War on the Mind. New York: Basic Books, Inc., Publishers 1978.

Weiner, Michael A., Earth Medicine - Earth Food. New York: MacMillan Publishing Co., Inc., 1980.

Wirth, Eve R., Survival Sense Emergency (May 1982) 38 and 66.

Wolf, A.V., Thirst. Springfield IL: C.C. Thomas, 1958.

American Wilderness, Time Life Books, Time Life Inc., 1972.

Encyclopedia Britannica, Inc., Encyclopedia, William Benton, Publisher, 1972.

National School of Conservation, Conservation

of Natural Resources, Vol 1, Tools and Techniques of Resource Management, Lesson 5. National School of Conservation Inc., Washington, D.C., 1973.

Publication No. 40, Wild, Edible and Poisonous Plants of Alaska. Fairbanks AK: Cooperative Extension Service, University of Alaska.

Special Scientific Reports, Project Mint Julep Part II. Maxwell Air Force Base, Alabama. Research Studies Institute, May 1955.

FAA-H-8083-258, *Private Pilot's Handbook of Aeronautical Knowledge*, 2016.

Other Selected References:

AALTDR 64-23, Project Cold Case, AD 462767, February 1965.

AALTN 57-16, Emergency Food Value of Alaskan Wild Plants. AD 293-31, July 1957.

ADTIC Publication A- 103, Down in the North. Maxwell Air Force Base, Alabama. Research Studies Institute, 1976.

ADTIC Publication A-105, Glossary of Arctic and Subarctic Terms. Maxwell Air Force Base,

Alabama. Air University, 1955.

ADTIC Publication A-107, Man in the Arctic. Maxwell Air Force Base, Alabama. Research Studies Institute, January 1962.

ADTIC Publication D-100, Afoot in the Desert. Maxwell Air Force Base, Alabama. Research Studies Institute, October 1980.

ADTIC Publication D-102, Sun, Sand and Survival. Maxwell Air Force Base, Alabama. Research Studies Institute, 1974.

AGARD Report No. 620, The Physiology of Cold Weather Survival. AD 784-268, April 1973.

Air Force CDC 20450 Intelligence Operations Specialist Vol 2, Maps and Charts. Extension Course Institute, Air Training Command, Gunter AFS, Alabama 36118, March 1979.

AFP 110-31: International Law-The Conduct Of Armed Conflict And Air Operations http://www.cna.org/documents/5500045700.pdf. (Accessed 21 January 2012).

Dayna Curry and Heather Mercer, from the book, *Prisoners of Hope* . Publisher Doubleday. 2002. ISBN 0-385-50783-6

EID Bulletin No. 1, Sharks. Maxwell Air Force- Base, Alabama. Aerospace Studies Institute

EID Bulletin No.2., Poisonous Snakes of North America. Maxwell Air Force Base, Alabama. Aerospace Studies Institute.

EID Bulletin No. 3, Poisonous Snakes of Central and South America. Maxwell Air Force Base, Alabama. Aerospace Studies Institute.

EID Bulletin No. 7, Water Resources. Maxwell Air Force Base, Alabama. Environmental Information Division, July 1969.

EID Bulletin No. 7a, Plant Sources of Water in Southern Asia. Maxwell Air Force Base, Alabama. Aerospace Studies Institute, August 1969.

EID Bulletin, No. 8, Survival Nutrition. Maxwell Air Force Base, Alabama. Environmental Information Division.

EID Bulletin No. 13, Edible And Hazardous Marine Life. Maxwell Air Force Base, Alabama. Aerospace Studies Institute, April 1976.

EID Publication G-104, Airman Against the Sea. Maxwell Air Force Base, Alabama. Aerospace Studies Institute.

EID Publication G- 105, Analysis of Survival Equipment. Maxwell Air Force Base, Alabama. Aerospace Studies Institute, 1957.

EID Publication G-107, Water Survival Field Tests. Maxwell Air Force Base, Alabama. Aerospace Studies Institute, June 1958.

EID Publication T-l00, 999 Survived. AD 727-726, Maxwell Air Force Base, Alabama. Aerospace Studies Institute.

Know Your Knots, Missile Hazard Control Section ATC Sheppard Air Force Base, Texas.

Guerrilla Hostage: 810 Days in Captivity. Publisher Fleming H. Revell. 1999. ISBN 0-8007-5693-2

Laboratory Note CRL-LN-55-21 1, The Will to Survive. Reno, USAF Survival Training School, 1955.

P.O.W.: A Definitive History of the American Prisoner-Of-War Experience in Vietnam, 1964-1973. Publisher: Mcgraw-Hill (Tx); First edition (October 1976). ISBN-10: 0070308314

She Went to War The Rhonda Cornum Story, Publisher Presidio Press, 1992, ISBN 0-9141-463-0

Synopsis of Survival Medicine, Fairchild Air Force Base, Washington, USAF Survival School, 1969.

The Engineering Toolbox.

http://www.engineeringtoolbox.com/clo-clothing-thermal-insulation-d_732.html. (Accessed 21 January 2012).

Attachment 2 - Acronyms

Acronym	Definition
5 A's	Anchor Assistance Air Accessory Bag Assessment
ABCCC	Airborne Battlefield Command and Control Center
ACDE	Aircrew Chemical Defense Ensemble
ADF	Automatic Direction Finding
AGL	Above Ground Level
AIR	Area(s) Identified for Recovery
AO	Area of Operations
AOR	Area of Responsibility
AUTO	A – Around/Avoid U – Under T – Through O – Over
AVPU	Alert, Verbal, Pain, Unconscious
BFT	Blue-Force Tracker
BLISS (Acronym for Urban)	B – Blend into the background L – Low profile I – Irregular S – Survive

Acronym	Definition
	S – Secluded
BMI	Body Mass Index
C2	Command and Control
C2ISR	Command, Control, Intelligence, Surveillance, and Reconnaissance
CAC	Conduct after Capture
CALICS	Communication, Authentication, Location, Intentions, Condition, Situation
CAP	Crisis Action Plan
CBRNE	Chemical, Biological, Radiological, Nuclear, and High Yield Explosives
CD	Chemical Defense
CIAT	International Center for Tropical Agriculture
CII	Critical Information Item
CIK	Crypto Ignition Key
CJCSI	Chairman of the Joint Chiefs of Staff Instruction
COA	Course of Action
CoC	Code of Conduct
COCOM	Combatant Command
CONOPS	Concept of Operations
CONUS	Continental United States
COSPAS	Cosmicheskaya Sistyema Poiska Avariynich Sudov (Russian space system for search of vessels in distress)
CPR	Cardiopulmonary Resuscitation
CRO	Combat Rescue Officer
CSAR	Combat Search and Rescue
CSEL	Combat Survivor Evader Locator
CSI	Contingency SERE Indoctrination
CST	Combat Survival Training
DAR	Directed Area for Recovery

Acronym	Definition
DDES	Distance, Direction, Elevation, and Shape
DF	Direction Finding
DIA	Defense Intelligence Agency
DME	Distance Measuring Equipment
DOD	Department of Defense
DPMO	Defense Prisoner of War/Missing Personnel Office
DRC	Disposition Record Card
E&E	Evasion and Escape
E&R	Evasion and Recovery
EAI	Executive Agent Instruction
EAP	Emergency Action Plan
ECAC	Evasion and Conduct After Capture
ELT	Emergency Locator Transmitter
EMSEC	Emanations Security
EPA	Evasion Plan of Action
ESR	External Supported Recovery
EVC	Evasion Chart
FEBA	Forward Edge of the Battle Area
GNC	Global Navigation Chart
GPRS	Global Personal Recovery System
GPS	Global Positioning System
GPW	Geneva Conventions Relative to the Treatment of Prisoners of War
GTAS	Ground-to-Air Signals
HUMINT	Human Intelligence
IAD	Immediate Action Drill
ICRC	International Committee of the Red Cross
IFAK	Individual First Aid Kit

Acronym	Definition
IFF	Identification, Friend or Foe
INMARSAT	International Maritime Satellite
IP	Isolated Personnel
IPG	Isolated Personnel Guidance
IR	Infrared
ISOPREP	Isolated Personnel Report
ISR	Intelligence, Surveillance, and Reconnaissance
JAOC	Joint Air Operations Center
JNC	Jet Navigation Chart
JOG	Joint Operational Graphic
JPRC	Joint Personnel Recovery Center
JPRCC	Joint Personnel Recovery Coordination Center
JPRSP	Joint Personnel Recovery Support Product
JRCC	Joint Reception Coordination Center
JSRC	Joint Search and Rescue Center
JTTP	Joint Tactics, Techniques, and Procedures
LEU	Low Earth Orbit
LIMFAC	Limiting Factor
LOAC	Law of Armed conflict
LOC	Line of Communications
LOP	Line of Position
LOS	Line of Sight
LPE	Low Probability of Exploitation
LPU	Life Preserver Unit
LUT	Local User Terminal
LZ	Landing Zone
MAJCOM	Major Command
MCC	Mission Coordination Center

Acronym	Definition
MCW	Modulated Continuous Wave
MGRS	Military Grid Reference System
mIRC	Microsoft Internet Relay Chat
MOA	Memorandum of Agreement
MOB	Main Operating Base
MOOTW	Military Operations Other Than War
MOPP	Mission-Oriented Protective Posture
MSL	Mean-Sea-Level
NAR	Nonconventional Assisted Recovery
NGO	Nongovernmental Organization
NSRP	National Search and Rescue Plan
NVD	Night Vision Device
NVG	Night Vision Goggle(s)
OGA	Other Government Agency
ONC	Operational Navigation Chart
OPCON	Operational Control
OPLAN	Operation Plan
OPORD	Operation Order
OPR	Office of Primary Responsibility
OPS	Operations
PAR	Primary Area of Recovery
PET	Polyethylene Terephthalate
PFD	Personal Flotation Device
PLB	Personal Locator Beacon
PLS	Personal Locator System
PR	Personnel Recovery
PRK	Personnel Recovery Kits
PRO	Personnel Recovery Officer

Acronym	Definition
PSK	Personal Survival Kit
PTT	Push-to-Talk
PW	Prisoner of War
RCC	Rescue Coordination Center
RM	Recovery Mechanism
ROE	Rules of Engagement
RSA	Radio Set Adapter
RT	Recovery Team
SA	Situational Awareness
SAFE	Selected Area For Evasion
SAR	Search and Rescue
SERE	Survival, Evasion, Resistance, and Escape
SODIS	Safe Drinking Water in 6 Hours
SOFA	Status of Forces Agreement
SPINS	Special Instructions
TIG	Thermal Insulating Garment
TPC	Tactical Pilotage Chart
TTP	Tactics, Techniques, and Procedures
UAR	Unconventional Assisted Recovery
UARM	Unconventional Assisted Recovery Mechanism
UAV	Unarmed Aerial Vehicle
UCMJ	Uniform Code of Military Justice
UHF	Ultra-high Frequency
USG	United States Government
UTM	Universal Transverse Mercator
VHF	Very High Frequency